PSEUDO-RIEMANNIAN GEOMETRY, δ-INVARIANTS AND APPLICATIONS

PSEUDO-RIEMANNIAN GEOMETRY, δ-INVARIANTS AND APPLICATIONS

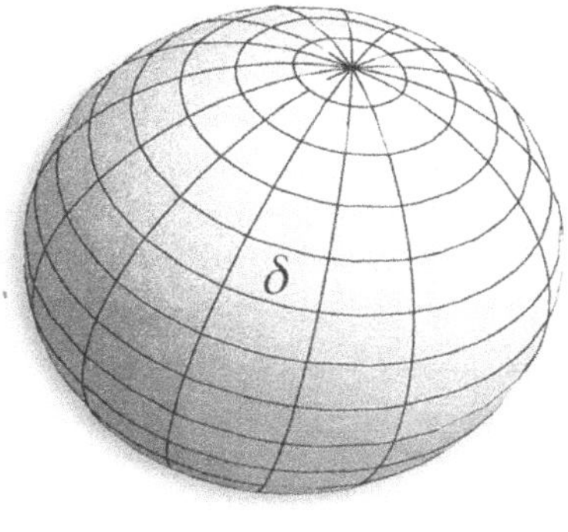

Bang-Yen Chen

Michigan State University, USA

World Scientific

NEW JERSEY • LONDON • SINGAPORE • BEIJING • SHANGHAI • HONG KONG • TAIPEI • CHENNAI

Published by

World Scientific Publishing Co. Pte. Ltd.

5 Toh Tuck Link, Singapore 596224

USA office: 27 Warren Street, Suite 401-402, Hackensack, NJ 07601

UK office: 57 Shelton Street, Covent Garden, London WC2H 9HE

British Library Cataloguing-in-Publication Data
A catalogue record for this book is available from the British Library.

PSEUDO-RIEMANNIAN GEOMETRY, δ-INVARIANTS AND APPLICATIONS

ISBN-13 978-981-4329-63-7
ISBN-10 981-4329-63-0

Printed in Singapore by Mainland Press Pte Ltd.

To Pi-Mei

in appreciation of all her
love and support

Preface

The theory of submanifolds as a field of differential geometry is as old as differential geometry itself, beginning with the theory of curves and surfaces. It was Gauss who proved in 1827 that the intrinsic geometry of a surface S in $\mathbb{E}^3$ can be derived solely from the Euclidean inner product as applied to tangent vectors of S. In 1854, Riemann discussed in his famous inaugural lecture at Göttingen the foundations of geometry, introduced n-dimensional manifolds, formulated the concept of Riemannian manifolds and defined their curvature.

Under the impetus of Einstein's Theory of General Relativity (1915) a further generalization appeared; the positiveness of the inner product was weakened to nondegeneracy. Consequently, one has the notion of pseudo-Riemannian manifolds. Furthermore, inspired by Kaluza–Klein's theory and string theory, mathematicians and physicists study not only submanifolds of Riemannian manifolds but also submanifolds of pseudo-Riemannian manifolds.

The famous Nash embedding theorem published in 1956 was aiming, in the hope that if Riemannian manifolds could be regarded as Riemannian submanifolds, this would then yield the opportunity to use extrinsic help in the study of (intrinsic) Riemannian geometry. However, this hope had not been materialized yet according to [Gromov (1985)]. The main reason for this was the lack of controls of the extrinsic properties of the submanifolds by the known intrinsic invariants. In order to overcome such difficulties as well as to provide answers to an open question on minimal immersions, the author introduced in the early 1990s new types of Riemannian invariants: his so-called δ-curvatures, different in nature from the "classical" Ricci and scalar curvatures. At the same time he was able to establish general optimal relations between his new intrinsic δ-curvatures and the main extrinsic invariants for Riemannian submanifolds. Since then many results concerning the δ-invariants, inequalities, and their applications have been obtained. The main purpose of this book is to provide an extensive and comprehensive survey on pseudo-Riemannian submanifolds and δ-invariants as well as their applications.

It is the author's hope that the reader will find this book both a good introduction to the theories of pseudo-Riemannian submanifolds and of δ-invariants and a useful reference to recent research of both areas.

In concluding the Preface, the author would like to thank World Scientific Publishing for the invitation to undertake this project. He also would like to express his appreciation to Professors D. E. Blair, F. Dillen, I. Dimitric, R. H. Escobales, O. J. Garay, S. Haesen, G. D. Ludden, I. Mihai, M. Petrović-Torgašev, A. Romero, B. Suceava, J. Van der Veken and L. Vrancken for reading parts of the manuscript and offering valuable suggestions. In particular, the author thanks Professor L. Verstraelen for writing an excellent Foreword for this book.

December 2010

B. Y. Chen

Foreword

1. The scientifical community will be delighted with MSU Distinguished Professor Bang-Yen Chen's new book "Pseudo-Riemannian Geometry, δ-invariants and Applications". Its true members will experience this book as an irresistible invitation to deepen their understanding of the concepts and theories under discussion, for their proper value within mathematics (say, for their intrinsic meanings) as well as for their relevance in other sciences and in philosophy (say, for their extrinsic meanings), i.e. as an invitation to continue our kind's great enterprise to ever further study geometry.

This quality of Chen's books naturally comes about as follows. From the combination of his impressive mathematical erudition and his quicksilver talent for research result volcaniclike eruptions of geometrical creativity which on the one hand clarify some regions of long existing mathematical territories that before remained obscure and which on the other hand establish whole new mathematical habitats, some like small islands where it is nice to stroll around and some like vast continents just beginning to be explored. And, then, when, with abundant joy and enthusiasm, he shows the readers such clarifications and gives them guided tours into these new lands, and, since basically what the readers hereby learn is so fundamental and beautiful, during the process of reading they are bound to take in some of the author's scientifical personality itself too. With the present book the author invites all people who really love science to discover the goods of the newly created geometrical continent that could well be named "the world of the curvatures of Chen", and, in addition, to follow how the study of these curvatures so far has shed new light on some old problems and how so far this study has been inspirational for original new studies and reflections in connection with several classical intrinsic and extrinsic geometrical concepts and theories. Essentially, thus could be described the research contents of this book. And, dealing with this subject in a Riemannian setting which includes the indefinite case, the author, like in passing, also filled in various previously existing gaps in the general theory of pseudo-Riemannian submanifolds.

2. It is almost forty years ago now that Bang-Yen Chen's first book on submanifold theory appeared, "Geometry of Submanifolds", which right away became the standard reference for geometry of Riemannian submanifolds of arbitrary dimensions and co-dimensions. As some indication in this respect one could observe that Chen's "Geometry of Submanifolds" figures among the ten "Lehrbücher" mentioned by Dombrowski in his "Differentialgeometrie", (a part of "Ein Jahrhundert Mathematik, 1890/1990"), which text, together with Dombrowski's books "150 years after Gauss' "Disquisitiones generales circa superficies curvas"" and "Wege in euklidischen Ebenen - Kinematik der Speziellen Relativitätstheorie" certainly are highly recommended to read, for their own sake, and, together with other books on geometry and in particular with books on geometry with links to the natural sciences like this new book of Chen, for the sake of improving the appreciation of the contents of these other books by helping to see them in a global mathematical perspective and for a better understanding of their overall importance by putting them in their proper physical and historical contexts.

Starting with and brought about by the "Foundations of Differential Geometry (I, II)" of Kobayashi and Nomizu (1963, 1969) and by the "Geometry of Submanifolds" of Chen (1973), there have been very nice developments indeed in the geometry of submanifolds, that is, in some sense, and at least in my opinion, in the geometry of the human kind. And, there is no doubt at all that the present new book of Chen, which could be positioned between "Lehrbüch" and "Monographie", deals with one of the most important steps in this evolution.

3. "The history of the living world can be summarized as the elaboration of ever more perfect eyes within a cosmos in which there is always more to be seen", (Teilhard de Chardin). The arts and the sciences are amongst the main characteristics of human civilization. The real scientists and the real artists share an immense desire for understanding and beauty. The studies aiming for an ever better understanding, which invariably result in seeing ever more beauty, are amongst the most valuable human activities. The percentage of the members of a human society who really experience great joy when observing the creations of real scientists and of real artists is a scale to measure the degree of civilization of this society.

And here follow the closing sentences of Poincaré's "La Valeur de la Science": "Ce n'est que par la Science et par l'Art que valent les civilisations. On s'est étonné de cette formule: la Science pour la Science; et pourtant cela vaut bien la vie pour la vie, si la vie n'est que misère; et même le bonheur pour le bonheur, si l'on ne croit pas que tous les plaisirs sont de même qualité, si l'on ne veut pas admettre que le but de la civilization soit de fournir de l'alcool aux gens qui aiment à boire. Tout action doit avoir un but. Nous devons souffrir, nous devons travailler, nous devons payer notre place au spectacle, mais c'est pour voir; ou tout au moins pour que d'autres voient un jour. Tout ce qui n'est pas pensée est le pur néant; puisque nous ne pouvons penser que la pensée et que tous les mots dont nous disposons pour parler des choses ne peuvent exprimer que des pensées; dire qu'il y a autre chose que la pensée, c'est donc une affirmation qui ne peut avoir du sens. Et cependant - étrange contradiction pour ceux qui croient au temps–l'histoire géologique nous montre que la vie n'est qu'un court épisode entre deux éternités de mort, et que, dans cet épisode même la pensée consciente n'a duré et ne durera qu'un moment. La pensée n'est qu'un éclair au milieu d'une longue nuit. Mais c'est cet éclair qui est tout". Next along this line follow two more quotes, respectively from Feynman's "The relation of mathematics to physics" and from Weyl's "Philosophy of Mathematics and Natural Science": " ··· (I want) to emphasize the fact that it is impossible to explain honestly the beauties of the laws of nature in a way that people can feel, without their having some deep understanding of mathematics. I am sorry, but this seems to be the case", and, "Exact natural science, if not the most important, is the most distinctive feature of our culture in comparison to other cultures". In the above, at least in my opinion, for specimens of the human kind, "nature" essentially stands for their organized thoughts about their sensations and perceptions of "their worlds outside and inside" and "doing mathematics" basically stands for their thoughtful living in "the universe" of their idealizations and abstractions of these sensations and perceptions. "The history of mathematics is the kernel of the history of human culture, the skeleton which supports and keeps together all the rest of the sciences", (Georges Sarton).

4. To somewhat clearly describe why what could be the essence of geometry, even only in the narrow view through the eyes of a single mind like mine, is an evident mission impossible. But at least just to try to give some indications of what might be alluded to when in this respect having mentioned before as a kind of alternative "the geometry of the human kind", next follow some more quotations and related comments. From Berger's "Peut-on définir la géométrie aujourd'hui?", referring to conversations with Calabi and Bott, respectively, come the following formulations: "Geometry is any branch of mathematics in which you can trace back your primary source of information of intuition in your sensorial experience" and "Geometry is the abstraction of our visual perception of the world we are living in". And, in "The Foundations of Arithmetic", Frege wrote the following: "Empirical propositions hold good of what is physically or psychologically existent, the truths of (Euclidean) geometry govern all that is spatially intuitable, whether existent or product of our fancy. The wildest visions of delirium, the boldest inventions of legend or poetry ··· remain, so long as they remain intuitable, still subject to the axioms of geometry. Conceptual thought can after fashion shake of their yoke, when it postulates, say, a space of four dimensions or of positive curvature. To study such conceptions is not useless by any means; but it is to leave the ground of intuition entirely behind. If we do make use of intuition even here, as an aid, it is still the same old intuition of Euclidean space, the only space of which we can have any picture", what the visual artist Escher compactly expressed when saying that "Our 3D space is the only reality that we know".

Of course, all over in these kinds of thoughts, one is actually concerned with what well might be one of most basic problems in real psychology: scientifically speaking, what are human sensations and perceptions? At this stage, in this respect, here are some quotes from Helmholtz and from Koenderink and van Doorn, respectively, from their opinions on the facts of perception and on the nature of observation: "The impressions of the senses are only signs for the constitution of the external world, the interpretation of which must be learned by experience. ··· we are compelled in every explanation of natural phenomena to leave the sphere of sense, and to pass to things which are not objects of sense, and are defined by abstract conceptions. ··· Perhaps the relation between our senses and the external world may be best enunciated as follows: our sensations are for us only symbols of the objects of the external world, and correspond to them only in some such way as written characters or articulate words to the things they denote. They give us, it is true, information respecting the properties of things without us, but no better information than we give a blind man about color by verbal description", and: "Despite the deep cleft between the realm of smooth functions and observed scalar fields, scientists have not been hesitant to apply the methods of the differential calculus to actual data. That this actually works fine is to be considered a tribute to common sense. However, a more principled approach is desirable. ⋯ Consider a single image ⋯ . In summary, scalar densities can only be defined relative to a fiducial resolution, and the one-parameter family of resolution forms a causal atlas and avoids spurious resolution if and only if the observations take place with a Gaussian weighing function. This structure is well known today and is generally designated as scale space. ⋯ The scale space setting immediately allows us to define differentiation even for scalar fields that are not differentiable at all. ⋯ The point is that we know nothing about the "real" image, except for the observation ⋯ . In particular we have no knowledge whatsoever concerning its smoothness or differentiability. However, we can rest assured that the observed density is smooth by design. For all we know, there might well exist infinitely many "real" images that would yield the identical observation. We shouldn't care, since it is irrelevant. They can't be observed! It is entirely a matter of personal taste whether one wants to believe in the existence of non-observable entities or not".

5. In the chapter on "The Music of the Spheres" of his book "The Ascent of Man", Bronowski wrote the following: "Nature presents us with shapes: a wave, a crystal, the human body, and it is we who have to sense and find the numerical relations in them. Pythagoras was a pioneer in linking geometry with numbers, ⋯ . Pythagoras had proved that the world of sound is governed by exact numbers. He went on to prove that the same thing is true of the world of vision. That is an extraordinary achievement. I look about me; here I am, in this marvelous, colored landscape of Greece, among the wild natural forms, the Orphic dells, the sea. Where under the beautiful chaos can there lie a simple numerical structure? The question forces us back to the most primitive constants in our perception of natural laws. To answer well, it is clear that we must begin from universals of experience. There are two experiences on which our visual world is based: that gravity stands vertical, and that the horizon stands at right angles to it. And, it is that conjunction, those cross-wires in the visual field, which fixes the nature of the right angle; so that if I were to turn the right angle of experience (the direction of "down" and the direction of "sideways") four times, back I come to the cross of gravity and the horizon. The right angle is defined by this fourfold operation, and is distinguished by it from any other arbitrary angle. In the world of vision, then, in the vertical picture plane that our eyes present to us, a right angle is defined by its fourfold rotation back on itself. ⋯ Not only the natural world as we experience it, but the world as we construct it is built on that relation. It has been so since the time that the Babylonians built the Hanging Gardens, and earlier, since the time that the Egyptians built the pyramids. These cultures already knew in a practical sense that there is a builder's set square in which the numerical relations dictate and make the right angle. The Babylonians knew many, perhaps hundreds of formulae for this by 2000 B.C. The Indians and Egyptians knew some. The Egyptians, it seems, almost always used a set square with the sides of the triangle made of three, four, and five units. It was not until 550 B.C. or thereabouts that Pythagoras raised this knowledge out of the world of empirical fact into the world of what we should now call proof. That is, he asked the question, "How do such numbers that make up these builder's triangles flow from the fact that a right angle is what turns four times to point the same way?" His proof, we think, ran something like this. . . . ".

From this perspective, the geometry of 3D Euclidean spaces, i.e. 3D spaces in which distances are determined by the Theorem of Pythagoras, and the geometry hereby induced on their submanifolds, curves and surfaces, seems most natural indeed with reference to our kind's most primitive geometrical sensations and perceptions. And, accordingly, cf. e.g. Frege's above quoted views, for beings of our human kind, one can psychologically well accept the natural generalization and abstraction of and from this geometry of curves and surfaces in 3D Euclidean spaces to the general geometry of submanifolds in Euclidean spaces with any numbers of dimensions and co-dimensions. Geometrical perspective, sections and projections of all kinds, submanifolds of submanifolds, like submanifolds of spheres for instance, etc., thus in a way are all notions and procedures that at least for most of us are pretty common indeed.

6. By the way, amending the above mentioned descriptions by Koenderink and van Doorn of our visual sensations in the sense of smoothing images making use of properly elliptical Gaussians rather than circular ones as done with the psychologist Bart Ons, when then considering the corresponding (Casorati) curvatures as contour indicators in our visual perception, in accordance with Simon Stevin's motto "Whonder en is gheen whonder", the so-called optical visual illusions loose their mystery. This kind of amendment simply takes into account the well documented factual difference in human's appreciations of lengths in horizontal (H) versus vertical (V) directions, the elliptical Gaussians involved therefore having their axes in the HV directions. One might well consider this fact as another manifestation of the gravitational cross in connection moreover with our kind's cultural evolution. To finish this intermezzo, one could reflect further in this context about Fechner's contributions to psycho–aesthetics and to Borisavljević's observation that the human visual field at every instant comes close to an oval inscribed in a golden rectangle.

7. By a formidable, ingenious abstraction and generalization of the inner geometry of 2D surfaces M in 3D Euclidean spaces, which geometry was revealed in Gauss's "disquisitiones" as presented in his talk at Göttingen on October 8, 1827, in his talk at Göttingen on June 10, 1854, "Über die Hypothesen, welche der Geometrie zu Grunde liegen", Riemann initiated Riemannian geometry; also, in this respect, one should not forget Helmholtz's "Über die Tatsachen, welche der Geometrie zu Grunde liegen" which appeared in 1868, that is shortly after Riemann's lecture had been published after all. And in his talk "Raum und Zeit" at Köln on September 21, 1908, Minkowski, (then professor at Göttingen), introduced his Weltaxiom and indefinite metrics.

From the first paragraph of Chern's Introduction to the 2000 Volume 1 of Elsevier's "Handbook of Differential Geometry" (edited by Franki Dillen e.a., and for which Chen wrote Chapter 3, "Riemannian Submanifolds", pp.187–419), comes the following: "While algebra and analysis provide the foundations of mathematics, geometry is at the core. This was already recognized by Euclid, whose book contains a geometrical treatment of the number system". Accordingly, on the real line the additions of real numbers essentially represent translations, and thinking of Simon Stevin's 1586 "law of the parallelogram" for the addition of vectors in the Euclidean plane, which basically once more reflects the gravitational cross, this addition represents translations in the real plane as well, and, by extension, also in dimensions 3 and "n" for that matter. In the school of Pythagoras, it was shown that the side and the diagonal of the regular 5-gon (which was in the logo of this school) are incommensurable line segments, i.e. that their ratio is an irrational number, namely, the golden section, which turned out to be of ever actual distinguished importance in geometry, in the natural sciences and in the arts alike. The theory of Eudoxos properly dealt with the incorporation in the mathematical reality of such pairs of line segments, which earlier had been tacitly assumed not to exist, thereby setting the deductive–postulational trend which aims for "security" of the mathematical reasoning as necessary counterpart for otherwise solely intuitive and eventually too loose ways of imagining and exploring problems in mathematics and in its applications. The crystallization of this mode of thinking in Euclid's "Elements" became a landmark in the history of civilization. And Euclid's complementing "Optics" could be seen as one of the first systematic studies of human vision.

The parallel postulate in planar Euclidean geometry was, as compulsory, pretty intriguing for many a scholar, already since Euclid's time, because it involves a happening taking place at infinity, and our kind is not able to see that far. And this was so good, because, as liberally quoted from Coolidge: "It is to the doubts about Euclid's parallel postulate and the efforts of many thinkers to settle these doubts that we owe the whole modern abstract conception of mathematical science". In 1585 Stevin had published his booklet "De Thiende" which finally definitively settled the concrete performance of all calculations involving real numbers in terms of computations exclusively dealing with natural numbers, by making use of the decimal system. The systematical use of the co-ordinate method of Descartes, which likely may well have originated from the above described human perceptual difficulties with the notion of parallelism, replaced the delicate role played by the axiomatic foundation of Euclidean geometry by the geometrical foundation of the real number system, and, paraphrasing Descartes' opening sentence of his 1637 "Géométrie", by the fact that "all Euclidean geometrical problems can be reduced to the knowledge of the distance between any two points (the Theorem of Pythagoras)". And, then, the mathematical discipline called analysis could be created, focussing respectively on Archimedes' classical problem to determine the tangent lines to arbitrary planar curves at any of their points and Stevin's problem to concretely determine all the spherical loxodromes. The procedures of analysis, for all practical purposes, were reduced to the formula of Taylor-Maclaurin to which Newton was inspired by the consideration of specific values of functions in analogy with the consideration of real numbers in the decimal system. Already before, his teacher Barrow had obtained the Fundamental Theorem of analysis which interrelates integration and differentiation as mutually inverse operations. Next, with vectors by then commonly "being around", the multiplication with the crucial imaginary unit i in the Gaussian plane was found essentially to represent this one's "complex structure", i.e. the rotations over right angles in our natural 2D geometry. In retrospect, the generalizations and abstractions of this Euclidean 2D fact to higher, even and odd dimensional Kählerian and Sasakian geometries cannot be called unnatural, rather to the contrary; and, further in this respect, one could also reflect on Chern's statement "The importance of complex numbers in geometry is a mystery to me" in his "What is Geometry?".

From the before mentioned talk by Minkowski, the opening address will remain actual forever: "M. H. ! Die Anschauugen über Raum und Zeit die ich Ihnen entwickeln möchte, sind auf experimentell–physikalischen Boden erwachsen. Darin liegt ihre Stärke. Ihre Tendenz ist eine Radikale. Von Stund an sollen Raum für sich und Zeit für sich völlig zu Schatten herabsinken, und nur noch eine Art Union der beiden soll Selbständigkeit bewahren". And, our kind's perceptions of the space dimensions, or, say, of the space co-ordinates x, y and z, and of the time dimension, or, say, of the time co-ordinate t, naturally being of different natures, in Minkowski's determination of a Pythagorean measure of the distances in the $(x, y, z; t)$ space-time according to the Weltpostulat, basically the three letter combination which showed up for the very first time, and, as he put it: "Man kann danach das Wesen dieses Postulates mathematisch sehr prägnant in die magische Formel kleiden: 300 000 km $= i$ sek". This work of Minkowski too was a formidable, ingenious extension of our natural measure of distances in space (x, y, z) to a natural measure of distances in space-time $(x, y, z; t)$, supported by the results of physical experiments and by the psychologically distinct appreciations of time and location. And thus began the development of pseudo-Riemannian geometry along the lines of proper, definite Riemannian geometry, and in particular of Lorentzian geometry, from the mathematical side, and, the development of Einstein's general theory of relativity, as its most noteworthy application so far.

Liberally quoting from Chern's already earlier mentioned Introduction: "Riemannian geometry (in its broad sense) is the central topic in geometry in our times"; "A panoramic view of Riemannian geometry" and other survey texts on geometry by Berger are continuous sparkling wells of geometrical delights and insights for the professionals, while Kühnel's "Differentialgeometrie, Kurven–Flächen–Mannigfaltigkeiten" is an excellent textbook (which also got some English translation) on the geometry of curves and surfaces in 3D Euclidean and Minkowski spaces, of hypersurfaces and of nD Riemannian manifolds, of interest to students and professionals alike and whether their interests are in mathematics or in applications of mathematics alike.

In abstract general Riemannian nD spaces M the Riemann–Christoffel curvature tensor R basically is a well defined measure for the deviation of M from being a locally Euclidean space and the Riemannian sectional curvatures $K(p,\pi)$ are well defined as the isometrical invariants determined at a point p of M by the theorema egregium formula for the Gauss curvature at p of the abstract 2D submanifolds of M locally formed by all the geodesics of M passing through p and whose tangent lines there belong to a linear 2D subspace π of the nD tangent space to M at p; and, by a Theorem of Elie Cartan, the information contained in the curvature tensor R and in all Riemannian sectional curvatures $K(p,\pi)$ are the same. Aiming for more concrete geometrical interpretations of these curvatures R and K, independently, Levi-Civita and Schouten in 1917 and 1918 discovered the fundamental notion of the parallel transport of tangent vectors to M along curves in M, i.e. of the Riemannian connection on a Riemannian manifold; alongside, here it could be pointed out that, shortly afterwards, and, as a very far cry from Euclid's parallel postulate indeed, general connections in fibre bundles started to be studied already. Levi-Civita's geometrical interpretation of the curvatures $K(p,\pi)$ concerned his parallelogramoids, while Schouten's interpretation of the curvature tensor R concerned the holonomy of directions, which, in the Koszul formalism standardly became used as the definition of R since the 1950s, based on the work by Nomizu. Schouten's study dealt with abstract Riemannian manifolds as such, whereas Levi-Civita's till some stage made use of the assumption that the Riemannian spaces under investigation could be considered as submanifolds of a Euclidean ambient space. In those days, this was not uncommon and maybe this was even more like a rule than an exception. For instance, in 1871, Suvorov studied the planes for which the sectional curvatures $K(p,\pi)$ at points p of a 3D Riemannian manifold M attain critical values and thus found generally three mutually perpendicular planes π at all points p, one of them yielding the minimal value and one of them yielding the maximal value of the Riemannian curvatures $K(p,\pi)$ at p, and he did so by considering the 3D Riemannian manifold M as a hypersurface in a 4D Euclidean space. It was conjectured almost immediately after the first publications about Riemannian geometry that every Riemannian manifold would be isometrically embeddable in some Euclidean ambient space, and this was settled in the affirmative in 1956, this fundamental matter of fact since then being known as the Theorem of Nash. And, in its light, the "abstract" nD Riemannian geometry can be seen as the intrinsic geometry of nD submanifolds in $(n+m)$D Euclidean spaces, or, still, thus Riemannian geometry can be seen as the inner part of the geometry of submanifolds in Euclidean spaces. Therefore, on top of the other arguments in favour of the importance and relevance of Riemannian geometry, it is such that it is part of the geometry of submanifolds which psychologically after all might well be the most ready abstraction and generalization of the geometry of 2D surfaces in the 3D space of our experience. The Theorem of Nash exclusively concerns positive definite Riemannian manifolds, but later on the Lorentzian and the general indefinite cases were dealt with alike.

8. Above briefly was touched upon a few of the main steps in "the geometry of the humankind", i.e. in "the geometry" done by beings of the species "mens": examples of true science, in the words of Minnaert, "imagination and reason working together, completing and penetrating each other", i.e. examples of doing mathematics in a natural way. The "Neder-Duitse" name for mathematics, which itself comes from the Greek "mathema", standing basically for knowledge and understanding, the combination of which constitutes the foundation of wisdom, is "wiskunde", literally, "the art to achieve wisdom". And the Neder-Duitse name for "geometry" is "meetkunde", which actually expresses much better than the term geometry itself what geometry is really about. "To measure the earth" was and remains a fascinating endeavor, and, for instance, the geometer and cartographer Mercator's conformal planar representations of parts of the earth whereby spherical loxodromes are mapped into straight lines is for always a great example of "meetkunde", both from the pure and from the applied points of view. But the name "meetkunde", literally, stands for "the art to achieve measure, balance, what is just" (meaning here correct as well as precise), or, still, "meetkunde" stands for "the art to see things right or to see things as they really are" (which should be interpreted in view of some above made comments on our sensations and perceptions), and the Dutch "meten" means "to measure", i.e. to take the measure, i.e. to take, in Dutch, "de maat", and "maât" was the name of an Egyptian goddess specialize in justice. Stevin strongly advocated the use of proper terminologies, like wiskunde for mathematics and meetkunde for geometry. Maybe in other languages such better terms are in use as well; it is known that Kepler in vain tried to introduce the term "Mass–Kunst" also in "High Dutch", and, for instance, the Hungarian "mértan" means "meetkunde".

From Freudenthal's initiation to geometry comes the following: "Ornaments from the dawn of mankind seem to support the thesis that symmetry has been the first geometrical idea that arose in the human mind, and which was given expression in human handicraft. Through an infinity of instances from nature and technics, symmetry may be approached by the childish spirit. It is a remarkable fact that planar symmetry with respect to one axis or a number of axes is a strikingly evident notion whereas many people never really become acquainted with central symmetry, though they learn mathematics. Few children discover of their own the congruence of the triangles into which a general parallelogram is divided by a diagonal. The only exceptions are those who have made up parallelograms from triangular tiles and those who have been trained in central symmetry. One may argue that geometry teaching should start with stereometry. Solid bodies are less abstract than plane shapes. The child can grasp them in a literal sense. Stereometry meets the children's creative wishes. Figures are drawn, solids are made. Geometry, as a logical system, is a means - and even the most powerful means - to make children feel the strength of the human spirit, that is: of their own spirit". In his 1952 classic "Symmetry", Weyl wrote the following: "Symmetry is a vast subject, significant in art and nature. Mathematics lies at its root, and it would be hard to find a better one on which to demonstrate the working of the mathematical intellect". Or as it was put by Bronowski: "Symmetry is not merely a descriptive nicety; $\cdots$ it penetrates to the harmony in nature". And a last quote here in this respect is taken from Gromov's 1998 "Possible Trends in Mathematics in the Coming Decades": "The search for symmetries and regularities in the structure of the world will stay at the core of pure mathematics (and physics)".

As Poincaré pointed out: "The object of mathematics is not to reveal the true nature of the things. It rather aims at connecting with each other the laws which, although they are recognizable by experience, we are unable to enunciate without the aid of mathematics", and: "Our ordinary language is too poor, also to vague to express the relations, so suitable, accurate, and full of meaning of the mathematical sciences". In this context, in his "Modern Mathematics" Fuchs wrote that "According to Andreas Speiser, the geometrical figure was historically the first means used in mathematics to untie it from language, even before the formula was applied for the same purpose". Complementing Poincaré's first point in his previous quote, here are further citations in this respect from Hertz's "Principles of Mechanics" and once more from Weyl: "We form for ourselves inner images or symbols of external objects, and the form which we give them is such that the sequences of the images, constructed as a necessity in thought, are equal to the images of the sequences of the external objects, occurring as a necessity in nature", and: "In the natural sciences we touch a sphere which cannot be penetrated by intuitive evidence. Here perception necessarily transforms into symbolic formation. Thus mathematics is brought in by the natural sciences to participate in the process of the theoretical construction of "our" human's universe. This point of view unifies all sciences". Along these lines of thought, the basis of geometry could be described as the intersection in abstracts of the rationales which are developed in physics, psychology, biochemistry, etc., i.e. in the scientific studies by men of the observed natural phenomena. Correspondingly, as Atiyah asserted in his "What is Geometry?" : "Geometry is not so much a branch of mathematics as a way of thinking that permeates all branches".

Most of what was tried to be expressed above about the nature of geometry, is summed up in the following quote from Thom's "Paraboles et Catastrophes" : "Comprendre signifie donc avant tout géométriser", and, therefore, at least as far as "scientifical Poincaré éclairs" are concerned: "Tout ce qui passe la géometrié nous passe", (Blaise Pascal).

9. Chern's "What is Geometry?" lists the following as the major developments in the history of geometry: (1) axioms (Euclid); (2) co-ordinates (Descartes, Fermat); (3) calculus (Newton, Leibniz); (4) groups (Klein, Lie); (5) manifolds (Riemann); (6) fibre bundles (Elie Cartan, Whitney). These developments, in some sense, constitute the solid and beautiful shell in which geometry here and now can safely live and actually may very well enjoy to live and further grow in the world of mathematics and natural sciences. One of the main, like biological driving forces, right from its birth and uptill now, in the life of geometry, is to want to learn ever more about parallelism, which desire may basically be traced back too to our experiences of the gravitational cross, or, in some sense equivalently, to want to learn ever more about curvature. And, above, some attempts were made to try to somewhat illuminate the naturalness for beings of the human kind of the geometry of submanifolds and some of its parts, like, in particular, the Riemannian, affine, projective, Kählerian and Sasakian geometries.

At this stage, it might be appropriate to emphasize that several other classes of geometries very much deserve to be studied as well, their naturalness also being clear and strong, let it be maybe of more sophisticated kinds than the plain psychological ones that were briefly discussed before concerning the geometry of submanifolds, at least as far as things look for the time being. Certainly among these, coming maybe closest to the Riemannian geometries, are the Finsler geometries and Miron's Lagrangian and Hamiltonian geometries. And, other such geometries are for instance also the isotropical geometries, i.e. the geometries of the (pretty anisotropic) spaces (in the sense that in these spaces, right from their point of departure, at every point not all directions have the same properties) whose metric is degenerate; a classic textbook in this respect is Sach's "Isotrope Geometrie des Raumes". By way of example of some natural niceties of such geometries, let us consider now on the standard real 3D space with co-ordinates $(x, y; z)$ the degenerate metric defined by $ds^2 = dx^2 + dy^2$, i.e. consider the Euclidean metric in the "horizontal" (x, y) planes and consider the "null-metric" in the "vertical" z-direction of this 3D space. As discussed by Koenderink and van Doorn in their 2002 article "Image processing done right", this 1-fold isotropical 3D geometry gives the proper setting in which to study the scale space for images given by light intensities $z = I(x, y)$, for some real function I of two real variables. From these intensities, by elliptical Gaussian smoothings as mentioned before, and on a logarithmic scale taking into account the Law of Fechner, result the visual sensation surfaces $z = F(x, y)$ in such 1-fold isotropical 3D space $(x, y; z)$. The perceived contours of planar visual images then do indeed correspond to lines of extremal values of the Casorati curvatures $z = C(x, y)$ of these surfaces $z = F(x, y)$. And, as noticed with Bart Ons, by the trivial rules of how to derive convolutions of functions, this rough geometrical description of human visual sensations and perceptions finally well theoretically explains the neurobiological findings of Barlow, Kuffler e.a. since around 1950 on the workings of ganglion cells in our retina's (related to the Gabor smoothings of the intensities I) and of Hubel–Wiesel e.a. since around 1960 on the workings of neurons in the visual cortex (related to the detection of orientations).

10. Here follows the opening paragraph of Osserman's "Curvature in the Eighties" : "The notion of curvature is one of the central concepts of differential geometry, one could argue that it is the central one, distinguishing the geometrical core of the subject from those aspects that are analytical, algebraic, or topological. In the words of Marcel Berger, curvature is "the number 1 Riemannian invariant and the most natural. Gauss and then Riemann saw it instantly".", (1990). The distinction made herein between the geometrical core of differential geometry from its analytical, algebraic or topological aspects, certainly in view of the fact that a Riemannian nD manifold (M, g) essentially consists of an nD differential manifold M endowed with a Riemannian metric tensor $ds^2 = g = g_{hk}\, dx^h\, dx^k$ as metrical geometrical structure, i.e. with as geometrical structure the quadratic fixation of all infinitesimal distances between pairs of points on M, recalls the very first sentence of Descartes' co-ordinate treatment of his Euclidean "géométrie". And, it seems appropriate to complement this first paragraph with the following lines of Schrödinger's 1940 paper "The General Theory of Relativity and Wave Mechanics", which a.o. also so well illustrate the above quote of Thom: "The most important discoveries are those which in the course of time tend to become tautological. The logical content of Newton's first two laws of motion was to state that a body moves uniformly in a straight line, unless it does something else and that in the latter case we agree upon calling force its acceleration (i.e. basically, the curvature vector field of its trajectory in space, - L.V. -) multiplied by an individual constant. The great achievement was, to concentrate attention on the second derivatives, so that they - not the first or third or fourth, nor any other property of the motion - ought to be accounted for by the environment. The fundamental statements of Einstein's theory of gravitation are of a similar kind. The equations $S_{hk} - \frac{1}{2}\tau g_{hk} = -8\pi T_{hk}$ $(*)$ state, that the contracted curvature tensor (S_{hk}, or, the Ricci tensor, while τ and g_{hk} denote the scalar curvature and the metric, respectively), is either zero or not and that, when and where it is not, we call matter (T_{hk}) the left hand side of equations $(*)$".

In full accordance with our intuition, following the idea's of Kepler and Descartes concerning osculating circles and of Huygens concerning caustics, the curvature of Euclidean planar curves was determined around 1670 by Newton in the sense that he succeeded to derive the formulae for this curvature in terms of the Cartesian equations and in terms of the parameter equations of these curves, which essentially involve the derivatives of order 2 of the functions appearing in these equations. Euler in 1760 then made possible, via his theory of normal sections of 2D surfaces M in Euclidean 3D spaces, to determine all curvatures of such surfaces. Amongst these, the curvature C of Casorati, introduced around 1890, most directly corresponds to our visual intuitive notion of the shape of a surface in space, i.e. of its extrinsic geometry. The best known of these curvatures are the following: (i) the mean curvature H of Germain, introduced in 1831, which measures the tension caused in a surface M as a result of the shape that it assumes in the ambient space, and thus too belongs to the extrinsic geometry, essentially corresponding to the physical impact caused on a surface by its shape, and, (ii) the Gauss curvature K which became properly understood in 1827. And later, likewise, also in other types of geometries, similarly various kinds of curvatures were studied.

In accordance with some previously made comments and observations, now in likely its most familiar formulation: "The book of Nature is written in a mathematical language and the characters are triangles, circles and other geometrical objects", (Galileo), i.e. many natural phenomena can be described in terms of geometrical notions and, in particular, in terms of curvatures. Hence, since moreover many of those phenomena concern realizations of several sorts of extremal values and of certain equilibria or symmetries within given situations, the solutions of many variational principles, beginning with Maupertuis, Fermat, Leibniz, Bernoulli's, etc., yielded spectacular developments in natural philosophy. For instance, several cycloids, catenaries and spirals, with most in particular the spira mirabilis, and, for instance minimal surfaces and surfaces of non-zero constant mean curvature and surfaces of constant Gauss curvature, and their geodesics and eventual loxodromes, showed up in this way as curves and surfaces of very special importance indeed, for geometry and for the applications of geometry in science and in technology alike.

Gauss's "disquisitiones" contained the germ of the notion of differential manifold, i.e. of the mathematical field of "global analysis", and it established the crucial distinction between the intrinsic and extrinsic geometrical properties of 2D surfaces in 3D Euclidean spaces, and so, by extension, in general submanifold theory. By the Theorem of Nash all nD Riemannian manifolds (M, g) can be identified, for their contents as differential manifolds as well as for their contents as geometrical manifolds, i.e. as far as their differential as well as their metrical structures are concerned, with nD submanifolds in $(n+m)$D Euclidean spaces, for appropriate co-dimensions m, and, in general, with many possible different shapes. In this respect and also as a field of interest in submanifold theory for its own sake, the knowledge of close relationships between the various intrinsic and extrinsic geometrical notions that can be defined on submanifolds is of fundamental importance. And, the present book of Professor Bang-Yen Chen is and will remain of great historical value by its presenting together many of the new forms of such knowledge which he discovered and which involve the main intrinsic and extrinsic notions in geometry and by showing the new understandings with which they enrich the geometry of submanifolds and the geometry of Riemann, i.e. by initiating the study of fundamental new regularities and symmetries in geometry. In conclusion, here are just two points on which one could reflect as such, indicatively. As a matter of fact, many results from the book exemplify, involving most basic notions in geometry, a new kind of theory of variations, which could be viewed also in connection with a comment on extrema made before. And, with respect to Berger's contribution to the book "Chern — A Great Geometer of the Twentieth Century" in which he stated that "$\cdots$ knowing K is equivalent to knowing $R \cdots$. But the relations between K and R are subtle and still not completely understood (e.g. what the critical planes of K are, and how they are distributed)", one could think of the general inequalities for the Chen curvatures (e.g. already for $\delta(2)$) and their corresponding ideal submanifolds.

Antwerp, 30 - 11 - 2010

Leopold Verstraelen

Contents

Chapter 1

Pseudo-Riemannian Manifolds

Problems in submanifold theory have been studied since the invention of differential calculus and it started with the differential geometry of plane curves. Fermat (1601–1665) is regarded as a pioneer in this field. Since his time, differential geometry of plane curves, dealing with curvature, circles of curvature, evolutes, envelopes, etc., has been developed as an important part of calculus. Also, the field has been expanded to analogous studies of space curves and surfaces, especially of lines of curvatures, geodesics on surfaces, and ruled and rotational surfaces.

The first major contributor to the subject was Euler (1707–1783). In 1736, Euler introduced arc length and the radius of curvature and so began the study of the intrinsic differential geometry of submanifolds.

It was Gauss (1777–1855) who proved in [Gauss (1827)] that the intrinsic geometry of a surface S in $\mathbb{E}^3$ can be derived solely from the Euclidean inner product as applied to tangent vectors of S [theorema egregium]. Riemann (1826–1866) saw what was needed to generalize the geometry of surfaces in the Euclidean 3-space $\mathbb{E}^3$. In his famous inaugural lecture at Göttingen

"*Über die Hypothesen welche der Geometrie zu Grunde liegen*",

Riemann discussed the foundations of geometry, introduced n-dimensional manifolds, formulated the concept of Riemannian manifolds and defined their curvature [Riemann (1854)].

Under the impetus of Einstein's Theory of General Relativity (1915) a further generalization appeared; the positiveness of the inner product was weakened to nondegeneracy. Consequently, one has the notion of pseudo-Riemannian manifolds.

Pseudo-Riemannian geometry has many important applications for instance, the curvature of spacetimes is essential for the positioning of satellites into orbit around the earth.

1.1 Symmetric bilinear forms and scalar products

A *symmetric bilinear form* on a finite-dimensional real vector space V is a $\mathbf{R}$-bilinear function $B : V \times V \to \mathbf{R}$ such that $B(u, v) = B(v, u)$ for all $u, v \in V$.

A symmetric bilinear form B is called *positive definite* (resp. *positive semi-definite*) if $B(v, v) > 0$ (resp. $B(v, v) \geq 0$) for all $v \neq 0$. Similarly, a symmetric bilinear form B is called *negative definite* (resp. *negative semi-definite*) if $B(v, v) < 0$ (resp. $B(v, v) \leq 0$) for all $v \neq 0$. B is called *nondegenerate* whenever $B(u, v) = 0$ for all $u \in V$ implies $v = 0$.

Definition 1.1. The *index* of a symmetric bilinear form B on V is the dimension of a largest subspace $W \subset V$ on which $B|_W$ is negative definite.

If we choose a basis $v_1, \ldots, v_n$ of V, then the $n \times n$ matrix $(b_{ij}), b_{ij} = B(e_i, e_j)$, is called the *matrix of* B *with respect to* $e_1, \ldots, e_n$. Since B is symmetric, the matrix (b_{ij}) is symmetric. A symmetric bilinear form is nondegenerate if and only if the matrix of B with respect to one basis is invertible.

Definition 1.2. A *scalar product* g on a finite-dimensional real vector space V is a nondegenerate symmetric bilinear form. An *inner product* is a positive definite scalar product.

By a *scalar product space* (V, g) we mean a vector space V equipped with a scalar product g. A subspace U of a scalar product space is called *nondegenerate* if $g|_U$ is nondegenerate.

On a scalar product space V, two vectors $u, v \in V$ are said to be *orthogonal*, denoted by $u \perp v$, if $g(u, v) = 0$. Two subsets $P, Q \subset V$ are called *orthogonal*, denoted by $P \perp Q$, if $g(u, w) = 0$ for all $u \in P$ and $w \in Q$.

For a subspace $U \subset V$, put $U^\perp = \{v \in V : v \perp U\}$. Then $(U^\perp)^\perp = U$.

Lemma 1.1. *A subspace U of a scalar product space V is nondegenerate if and only if V is the direct sum of U and $U^\perp$.*

Proof. Since

$$\dim(U + U^\perp) + \dim(U \cap U^\perp) = \dim U + \dim U^\perp = \dim V,$$

$U + U^\perp = V$ holds if and only if $U \cap U^\perp = \{0\}$ holds. The later condition means that U is nondegenerate. □

On a scalar product space V, the *norm* $||v||$ of a vector v is defined to be $\sqrt{|g(v,v)|}$. A *unit vector* is a vector of norm one. A set of mutually orthogonal unit vectors is called a *orthonormal set.* A set of n orthonormal vectors $e_1, \ldots, e_n$ of V is called an *orthonormal basis* whenever $n = \dim V$.

Lemma 1.2. *A scalar product space V of positive dimension admits an orthonormal basis.*

Proof. Since g is nondegenerate, there is a unit vector $e_1 \in V$. Let U_1 be the subspace spanned by e_1. Then $U_1^\perp$ is a nondegenerate subspace. Thus there is a unit vector $e_2 \in (U_1)^\perp$. The pair $\{e_1, e_2\}$ is an orthonormal basis of Span$\{e_1, e_2\}$. By continuing this process $(n-1)$-times, we obtain an orthonormal basis $e_1, \ldots, e_n$ of V. □

For an orthonormal basis $e_1, \ldots, e_n$ of a scalar product space V, we have

$$g(e_i, e_j) = \epsilon_i \delta_{ij}, \quad \epsilon_i = g(e_i, e_i) = \pm 1,$$

where δ_{ij} is the *Kronecker delta*, which is equal to 1 if $i = j$; and equal to 0 if $i \neq j$. Every vector $v \in V$ can be expressed in a unique way as

$$v = \sum_{i=1}^{n} \epsilon_i g(v, e_i) e_i.$$

For any orthonormal basis $e_1, \ldots, e_n$ of a scalar product space V, the number of negative signs in the signature $(\epsilon_1, \ldots, \epsilon_n)$ is the *index of* V.

A linear transformation $T : V \to W$ between two scalar product spaces is called a *linear isometry* if it preserves the scalar products. Two scalar product spaces are linear isometric if and only if they have the same dimension and the same index.

1.2 Pseudo-Riemannian manifolds

Through out this book, by a *manifold* we mean a connected smooth manifold of dimension ≥ 2 without boundary unless mentioned otherwise. By a *closed manifold* we mean a compact manifold without boundary.

A (*pseudo-Riemannian*) *metric tensor* g on a manifold M is a symmetric nondegenerate $(0, 2)$ tensor field on M of constant index, i.e., g assigns to each point $p \in M$ a scalar product g_p on the tangent space T_pM and the index of g_p is the same for all $p \in M$. Very often, we use $\langle \; , \; \rangle$ as an alternative notation for g. Thus, we have $g(v, w) = \langle u, v \rangle$.

A *pseudo-Riemannian manifold* is by definition a manifold equipped with a metric tensor g. The common value $s\,(0 \leq s \leq \dim M)$ of index on M is called the *index* of M. If $s = 0$, M is called a *Riemannian manifold.* In this case, each g_p is a positive definite inner product on T_pM.

If $s = 1$, M is called a *Lorentz manifold* and the corresponding metric is called *Lorentzian.* A manifold of dimension ≥ 2 admits a Lorentzian metric if and only if it admits a 1-dimensional distribution. A pseudo-Riemannian metric on an even-dimensional manifold M is called a *neutral metric* if its index is equal to $\frac{1}{2}\dim M$.

A pseudo-Riemannian manifold (resp. metric) is also known as a *semi-Riemannian manifold* (resp. metric).

A tangent vector v of a pseudo-Riemannian manifold M is called *space-like* (resp. *timelike*) if $v = 0$ or $\langle v, v\rangle > 0$ (resp. $\langle v, v\rangle < 0$). A vector v is called *lightlike* or *null* if $\langle v, v\rangle = 0$ and $v \neq 0$.

The *light cone* $\mathcal{LC}$ of $\mathbb{E}^n_s$ is defined by

$$\mathcal{LC} = \{v \in \mathbb{E}^n_s : \langle v, v\rangle = 0\}.$$

A curve in a pseudo-Riemannian manifold is called a *null curve* if its velocity vector is a lightlike at each point.

For a timelike vector v in a Lorentzian manifold, the set $\bar{C}(v)$ consists of all causal vectors w with $\langle v, w\rangle < 0$ is called the *time-cone* (or the *causal cone*) containing v. The opposite time-cone is the set $\bar{C}(-v)$ consists of all causal vectors w with $\langle v, w\rangle > 0$.

In each tangent space T_pM of a Lorentz manifold M, there exist two time-cones and there is no intrinsic way to distinguish one from the other, To choose one of them is to time-orient T_pM. Let ζ be a function on M which assigns to each $p \in M$ a time-cone ζ_p in T_pM. ζ is called smooth if for each $p \in M$, there is a smooth vector field V on some neighborhood U of p such that $V_q \in \zeta_q$ for each $q \in U$. Such a smooth function is called a *time-orientation* of M. To choose a specific time-orientation on M is to *time-orient* M.

A *spacetime* is a time-oriented 4-dimensional Lorentz manifold. As with any time-oriented spacetime, the time-orientation is called the *future*, and its negative is the *past.* A tangent vector in a future time-cone is called *future-pointing.* Similarly, a tangent vector in the past time-cone is called *past-pointing.*

A vector in a Lorentzian vector space that is *non-spacelike* (i.e., either lightlike or timelike) is called *causal*. A *causal curve* in a spacetime is a curve whose velocity vectors are all non-spacelike.

Let $\{x_1, \ldots, x_n\}$ be a coordinate system on an open subset $U \subset M$, where $n = \dim M$. Then the components g_{ij} of the metric tensor g on U are given by

$$g_{ij} = \langle \partial_i, \partial_j \rangle, \quad 1 \leq i, j \leq n,$$

where $\partial_i = \partial/\partial x_i$, $i = 1, \ldots, n$. Since g is a symmetric $(0, 2)$ tensor field, we have $g_{ij} = g_{ji}$ for $1 \leq i, j \leq n$. Hence, the metric tensor on U can be written as

$$g = \sum_{i,j=1}^{n} g_{ij} dx_i \otimes dx_j. \tag{1.1}$$

At each point p in the Euclidean n-space $\mathbb{E}^n$, there exists a canonical linear isomorphism from $\mathbb{E}^n$ onto $T_p\mathbb{E}^n$. In terms of natural coordinates on $\mathbb{E}^n$, it sends a vector v to $v_p = \sum v^j \partial_j$. The inner product on $\mathbb{E}^n$ gives rise to a metric tensor on $\mathbb{E}^n$ with

$$\langle v_p, w_p \rangle = \sum_{j=1}^{n} v_j w_j \tag{1.2}$$

with $v = \sum_{j=1}^{n} v_j \partial_j$ and $w = \sum_{j=1}^{n} w_j \partial_j$.

For an integer $s \in [0, n]$, if we change the first s plus signs in (1.2) to minus sign, then it gives rise to a metric tensor

$$\langle v_p, w_p \rangle = -\sum_{j=1}^{s} v_j w_j + \sum_{k=s+1}^{n} v_k w_k \tag{1.3}$$

of index s. The resulting *pseudo-Euclidean space* is denoted by $\mathbb{E}^n_s$.

If $s = 0$, $\mathbb{E}^n_s$ reduces to the Euclidean n-space $\mathbb{E}^n$. If $s = 1$, the pseudo-Euclidean space $\mathbb{E}^n_1$ is called a *Minkowski n-space*. When $n = 4$ and $s = 1$, it is the simplest example of a relativistic spacetime, known as the *Minkowski spacetime*.

1.3 Physical interpretations of pseudo-Riemannian manifolds

1.3.1 *4D spacetimes*

In physics, spacetime is any mathematical model that combines space and time into a single continuum. Spacetime is usually interpreted with space being three-dimensional and time playing the role of a fourth dimension. According to certain Euclidean space perceptions, the universe has three dimensions of space and one dimension of time. By combining space and time into a single manifold, physicists have significantly simplified a large number of physical theories, as well as described in a more uniform way the workings of the universe at both the supergalactic and subatomic levels.

In classical mechanics, the use of Euclidean space instead of spacetime is appropriate, as time is treated as universal and constant, being independent of the state of motion of an observer. Until the beginning of the 20th century, time was believed to be independent of motion, progressing at a fixed rate in all reference frames. However, later experiments revealed that time slowed down at higher speeds. Such slowing is called *time dilation.*

In relativistic contexts, however, time cannot be separated from the three dimensions of space, because the rate at which time passes depends on an object's velocity relative to the speed of light and also on the strength of intense gravitational fields, which can slow the passage of time.

Spacetimes are the arenas in which all physical events take place. An event is a point in spacetime specified by its time and place. In any given spacetime, an event is a unique position at a unique time. A spacetime is independent of any observer. However, in describing physical phenomena (which occur at certain moments of time in a given region of space), each observer chooses a convenient coordinate system. Events are specified by four real numbers in any coordinate system. The *worldline* of a particle is the path that this particle takes in the spacetime and represents the history of the particle.

The geometry of spacetime in *special relativity* is described by the Minkowski spacetime $\mathbb{E}^4_1$. While spacetime can be viewed as a consequence of Albert Einstein's 1905 theory of special relativity, it was first explicitly proposed mathematically by one of his teachers, Hermann Minkowski (1864–1909) in 1908 building on and extending Einstein's work. His concept of Minkowski space is the earliest treatment of space and time as two aspects of a unified whole, the essence of special relativity[1].

A *material particle* in $\mathbb{E}^4_1$ is a timelike future-pointing curve $\alpha : I \to \mathbb{E}^4_1$ such that $||\alpha'(\tau)|| = 1$ for all $\tau \in I$. The parameter τ is called the *proper time* of the particle. Any particle $\beta : I \to \mathbb{E}^4_1$ is a regular curve and its image $\beta(I)$ is the worldline of β. An *observer* is just a material particle.

[1]The Minkowski spacetime is the mathematical setting in which Einstein's theory of special relativity is most conveniently formulated. Originally, Einstein and Laub rejected in [Einstein and Laub (1908)] the four-dimensional electrodynamics of Minkowski as too complicated. Eventually, Einstein agreed in 1912 on the importance of Minkowski's spacetime formalism and used it for his work on the foundations of general relativity.

A choice of an orthonormal basis $\{e_0, e_1, e_2, e_3\}$ for a Minkowski spacetime $\mathbb{E}^4_1$ equipped with a coordinate system $\{x_0, x_1, x_2, x_3\}$ which may be viewed as a preferred coordinate system for a particular observer who perceives himself as at rest. The coordinate x_0 describes time as measured by this observer, and the 3-planes x_0 = constant describe space as viewed by the same observer. This observer's history is described by a curve $\beta(t) = (t, a, b, c)$ at each time t according to his own clock. However, he remains at the same point (a, b, c) in his space. A particle with nonzero rest mass which is in motion relative to our at rest observer will be described in the observer's coordinate system by a parametrized curve $\beta(t) = (t, x_1(t), x_2(t), x_3(t))$. The speed of this particle, as measured by the observer, is the Euclidean speed of $\gamma(t) = (x_1(t), x(t), x_3(t))$.

The spacetime of *general relativity* is a 4-dimensional Lorentzian manifold. In Einstein's general relativity theory, it is assumed that spacetime is curved by the presence of matter-energy, whose curvature is represented by the Riemann curvature tensor [Einstein (1915, 1916)]. Many spacetimes have physical interpretations which most physicists would consider bizarre or unsettling. For example, a compact spacetime contains closed time-like curves, which violate our usual ideas of causality, i.e., future events could affect past ones. For this reason, mathematical physicists usually consider only restricted subsets of all the possible spacetimes. One way to do this is to study "realistic" solutions of the equations of general relativity. Another way is to add some additional "physically reasonable" but still fairly general geometric restrictions and try to prove interesting things about the resulting spacetimes. The latter approach has led to some important results, most notably the Penrose–Hawking singularity theorems.

1.3.2 *Kaluza–Klein theory and pseudo-Riemannian manifolds of higher dimension*

Shortly after the publication of Einstein's theory of General Relativity in [Einstein (1915, 1916)], T. Kaluza (1885–1954) noticed in 1919 that when he solved Einstein's equations using five dimensions, Maxwell's equations for electromagnetism emerged spontaneously. Kaluza wrote to Einstein who encouraged him to publish. This very influential Kaluza's 7 page paper was published in [Kaluza (1921)]. In order to explain *why the extra 5th dimension is unobservable*, O. Klein (1894–1977) suggested in [Klein (1925)] that this extra 5th dimension would be compactified and unobservable on experimentally accessible energy scale. Klein proposed that *the 4th spatial dimension is curled up in a circle of very small radius*, so that a particle moving a short distance along that axis would return to where it began (cf. [Overduina and Wesson (1997)]).

The Kaluza–Klein theory is striking because it has a particularly elegant presentation in terms of differential geometry. In a certain sense, it looks just like ordinary gravity in free space, except that it is phrased in 5D instead of 4D.

To build the Kaluza–Klein theory from differential geometric viewpoint, one picks an invariant metric on the circle S^1 that is the fiber of the $U(1)$-bundle of electromagnetism. An invariant metric g^* is simply one that is invariant under rotations of the circle. Suppose this metric gives the circle S^1 a *total length* Λ, then consider metrics on the bundle P that are consistent with both the fiber metric, and the metric on the underlying 4D spacetime M so that $\pi : P \to M$ is a pseudo-Riemannian submersion.

The original Kaluza–Klein theory identifies Λ with the fiber metric g_{55}, and allows Λ to vary from fiber to fiber. In this case, the coupling between gravity and the electromagnetic field is not constant, but has its own dynamical field, known as the *radion* in physics.

In the formation given above, the extra 5th dimension in Kaluza–Klein theory can be understood to be the circle group $U(1)$, as electromagnetism can essentially be formulated as a gauge theory on a circle bundle with gauge group $U(1)$. Once this geometrical interpretation of Kaluza–Klein theory is understood, it is relatively straightforward to replace $U(1)$ by a general Lie group. Such generalizations are often called *Yang–Mills theories* [Yang and Mills (1954)]. Kaluza and Klein's work was neglected for many years as physicists' attention was directed towards quantum mechanics. This idea that fundamental forces can be explained by additional dimensions did not re-emerge until string theory was developed in the 1960s. This strategy of using higher dimensional pseudo-Riemannian manifolds to unify different forces is now a very active area of research in particle physics. This idea of compactifying the extra dimension has also dominated the search for a unified theory and led to many new developments in string theory.

In recent times, physics and astrophysics have played a central role in shaping the understanding of the universe through scientific observation and experiment. After Kaluza–Klein's theory, the term spacetime has taken on a generalized meaning beyond treating spacetime events with the normal 3+1 dimensions. It becomes the combination of space and time. Other proposed spacetime theories include additional dimensions; normally spatial but there exist some speculative theories that include additional temporal dimensions and even some that include dimensions that are neither temporal nor spatial. How many dimensions are needed to describe the Universe is still an open question.

String theory is a developing theory in particle physics that attempts to reconcile quantum mechanics and general relativity, which has its origins in the *dual resonance model* (i.e., the early investigation of *S-matrix theory* of the strong interaction). String theory hypothesizes that the electrons and quarks within an atom are not 0-dimensional objects, but rather 1-dimensional oscillating line segments (*strings*); most of which are embedded in dimensions that exist only on a scale no larger than the Planck length ($\approx 1.6163 \times 10^{-35}$ m). The earliest string model, Bosonic theory predicts 26 dimensions with 25 dimensions of space and one of time; with M-theory asserting that strings are really 1-dimensional slices of a 2-dimensional membrane vibrating in 11-dimensional space.

1.4 Levi-Civita connection

Let M be an n-manifold. Denote by $\mathcal{F}(M)$ the set of all smooth real-valued functions on M. If f_1, f_2 are smooth functions on M, so is their sum f_1+f_2 and product $f_1 f_2$, The usual algebraic rules hold for these two operations, which make $\mathcal{F}(M)$ a commutative ring. We denote by $\mathcal{X}(M)$ the set of all smooth vector fields. For $V, W \in \mathcal{X}(M)$, the bracket $[V, W]$ is defined by

$$[V, W]_p(f) = V_p(Wf) - W_p(Vf)$$

at each $p \in M$ and $f \in \mathcal{F}(M)$. The bracket operation $[\ ,\ ]$ on $\mathcal{X}(M)$ is $\mathbf{R}$-bilinear and skew-symmetric, and also satisfies the Jacobi identity:

$$[X, [Y, Z]] + [Y, [Z, X]] + [Z, [X, Y]] = 0.$$

These makes $\mathcal{X}(M)$ an infinite-dimensional Lie algebra.

Definition 1.3. An *affine connection* ∇ on a manifold M is a function $\nabla : \mathcal{X}(X) \times \mathcal{X}(M) \to \mathcal{X}(M)$ such that

(1) $\nabla_X Y$ is $\mathcal{F}(M)$-linear in X;
(2) $\nabla_X Y$ is $\mathbf{R}$-linear in Y;
(3) $\nabla_X(fY) = (Xf)Y + f\nabla_X Y$ for $f \in \mathcal{F}(M)$.

$\nabla_X Y$ is called the *covariant derivative* of Y with respect to X.

The *torsion tensor* T of an affine connection ∇ is a tensor of type $(1, 2)$ defined by $T(X, Y) = \nabla_X Y - \nabla_Y X - [X, Y]$.

The following theorem shows that on a pseudo-Riemannian manifold there exists a unique connection sharing two further properties.

Theorem 1.1. *On a pseudo-Riemannian manifold M, there exists a unique affine connection ∇ such that*

(4) ∇ *is torsion-free, i.e.,* $[Y, Z] = \nabla_Y Z - \nabla_Z Y$, *and*
(5) $X\langle Y, Z\rangle = \langle \nabla_X Y, Z\rangle + \langle Y, \nabla_X Z\rangle$

for $X, Y, Z \in \mathcal{X}(M)$.

This unique affine connection ∇ is called the Levi-Civita connection of M and it is characterized by the Koszul formula

$$\begin{aligned} 2\langle \nabla_Y Z, X\rangle &= Y\langle Z, X\rangle + Z\langle X, Y\rangle - X\langle Y, Z\rangle \\ &\quad - \langle Y, [Z, X]\rangle + \langle Z, [X, Y]\rangle + \langle X, [Y, Z]\rangle . \end{aligned} \tag{1.4}$$

Proof. Assume that ∇ is an affine connection on M which satisfies both properties (4) and (5). Then after applying (4) and (5) on the right-hand side of (1.4) we get $2\langle \nabla_Y Z, X\rangle$. Hence, ∇ satisfies the Koszul formula. Therefore, there exists only one affine connection on M which satisfies both properties (4) and (5).

For the existence, let us define $F(Y, Z, X)$ to be the right-hand side of (1.4). A direct computation shows that the function

$$X \mapsto F(Y, Z, X)$$

is $\mathcal{F}(M)$-linear for fixed Y, Z, Hence it is a 1-form. Thus there exists a unique vector field, denoted by $\nabla_Y Z$, such that $2\langle \nabla_Y Z, X\rangle = F(Y, Z, X)$ for all X. Therefore, the Koszul formula holds and from it we can deduce properties (1)-(5). □

On a pseudo-Riemannian manifold M, we shall use the Levi-Civita connection unless mentioned otherwise. Let $\{x_1, \ldots, x_n\}$ be a coordinate system on an open subset U of M. The *Christoffel symbols* for the coordinate system are the real-valued functions Γ_{ij}^k on U such that

$$\nabla_{\partial_i} \partial_j = \sum_{k=1}^{n} \Gamma_{ij}^k \partial_k, \quad 1 \le i, j \le n.$$

Since the connection ∇ is not a tensor, the Christoffel symbols do not obey the usual tensor transformation rule under change of coordinates.

For the Christoffel symbols we have the following.

Proposition 1.1. *Let M be a pseudo-Riemannian n-manifold and let $\{x_1, \ldots, x_n\}$ be a coordinate system on an open subset $U \subset M$. Then*

(1) $\nabla_{\partial_i} Y = \sum_{k=1}^{n} \left\{ \frac{\partial Y_k}{\partial x_i} + \sum_{j=1}^{n} \Gamma_{ij}^k Y_j \right\} \partial_k$ *and*

(2) $\Gamma_{ij}^k = \sum_{t=1}^{n} \frac{g^{kt}}{2} \left\{ \frac{\partial g_{jt}}{\partial x_i} + \frac{\partial g_{it}}{\partial x_j} - \frac{\partial g_{ij}}{\partial x_t} \right\}$,

where $Y = \sum_{j=1}^{n} Y_j \partial_j$ *and* (g^{ij}) *is the inverse matrix of* (g_{ij}).

Proof. Statement (1) is an immediate consequence of property (3) given in Definition 1.3. To prove (2), let us put $X = \partial_t, Y = \partial_i, Z = \partial_j$ in the Koszul formula. Since the brackets are zero, it leaves

$$2 \langle \nabla_{\partial_i} \partial_j, \partial_t \rangle = \frac{\partial g_{jt}}{\partial x_i} + \frac{\partial g_{it}}{\partial x_j} - \frac{\partial g_{ij}}{\partial x_t}.$$

But from the definition of Christoffel symbols we have

$$2 \langle \nabla_{\partial_i} \partial_j, \partial_t \rangle = 2 \sum_{k=1}^{n} \Gamma_{ij}^k g_{kt}.$$

Attacking both equations with $\sum_t g^{tk}$ yields the required formula. □

1.5 Parallel translation

Let $\phi : N \to M$ be a smooth map between two manifolds. The *differential* at a point $p \in N$ is a linear map $\phi_{*_p} : T_pN \to T_{\phi(p)}M$ defined as follows: For each $X \in T_pN$, $\phi_{*_p}X$ is the tangent vector in $T_{\phi(p)}M$ such that

$$(\phi_{*_p}X)f = X(f \circ \phi), \quad \forall f \in \mathcal{F}(N).$$

The *dual* of the differential ϕ_* is denoted by ϕ^*.

For any q-form ω on M, define the q-form $\phi^*\omega$ on N by

$$(\phi^*\omega)(X_1, \ldots, X_q) = \omega(\phi_* X_1, \ldots, \phi_* X_q), \ X_1, \ldots, X_q \in T_pN.$$

A vector field Z on a smooth map $\phi : P \to M$ between two manifolds is a mapping $Z : P \to TM$ such that $\pi \circ Z = \phi$, where π is the projection $TM \to M$. The simplest case of a vector field on a mapping is a vector field Z along a curve $\gamma : I \to M$ defined on an open interval I, where Z smoothly assigns to each $t \in I$ a tangent vector to M at $\gamma(t)$. For instance, the velocity vector field γ' on γ is a vector field on the curve γ. Let $\mathcal{V}(\gamma)$ denote the set consisting of smooth vector field of M along γ.

For a pseudo-Riemannian manifold M, there is a natural way to define the vector rate of change Z' of a vector field $Z \in \mathcal{V}(\gamma)$.

Proposition 1.2. *Let $\gamma : I \to M$ be a curve in a pseudo-Riemannian manifold M. Then there exists a unique function $Z \mapsto Z' = \frac{DZ}{dt}$ from $\mathcal{V}(\gamma) \to \mathcal{V}(\gamma)$ such that*

(1) $(aZ_1 + bZ_2)' = aZ_1' + bZ_2'$,

(2) $(\lambda Z)' = \left(\frac{d\lambda}{dt}\right)Z + \lambda Z'$,

(3) $(V_\gamma)'(t) = \nabla_{\gamma'(t)} V$,

where $a, b \in \mathbf{R}, \lambda \in \mathcal{F}(I), V \in \mathcal{X}(M)$ and $t \in I$. Furthermore, we have

(4) $\frac{d}{dt}\langle Z_1, Z_2\rangle = \langle Z_1', Z_2\rangle + \langle Z_1, Z_2'\rangle$.

Proof. For the uniqueness, let us assume that an induced connection exists which satisfies only the first three properties. We can assume that γ lies in the domain of a single coordinate system $\{x_1, \ldots, x_n\}$. For a vector field $Z \in \mathcal{V}(\gamma)$, we have

$$Z(t) = \sum_{j=1}^{n} (Z(t)x_j)\partial_j = \sum_{j=1}^{n} (Zx_j)(t)\partial_j.$$

Let us denote the component function $Zx_j : I \to \mathbf{R}$ by Z_j. Then, by properties (1), (2) and (3), we find

$$\begin{aligned} Z' &= \sum_{j=1}^{n} \frac{dZ_j}{dt}\partial_j|_\gamma + \sum_j Z_j(\partial_j|_\gamma)' \\ &= \sum_{j=1}^{n} \frac{dZ_j}{dt}\partial_j + \sum_{j=1}^{n} Z_j \nabla_{\gamma'}(\partial_j). \end{aligned}$$

Thus Z' is completely determined by the Levi-Civita connection ∇. This shows uniqueness.

On any subinterval J of I such that $\gamma(J)$ lies in a coordinate neighborhood, let us define Z' by the formula above. Then straightforward computations show that all four properties hold. Now, it follows from the uniqueness that these local definitions of Z' gives rise to a single vector field $Z' \in \mathcal{V}(\gamma)$. □

The $Z' = \frac{DZ}{dt}$ in Proposition 1.2 is called the *induced covariant derivative.* For a vector field Z along a curve γ, we simply write Z' for $\nabla_{\gamma'} Z$ and also γ'' for $\nabla_{\gamma'}\gamma'$. In terms of Christoffel symbols we have

$$Z' = \sum_{k=1}^{n} \left\{ \frac{dZ_k}{dt} + \sum_{i,j=1}^{n} \Gamma_{ij}^k \frac{d(x_i \circ \gamma)}{dt} Z_j \right\} \partial_k.$$

A vector field Z on γ is called *parallel* if $Z' = 0$ holds identically along γ. Hence $Z = \sum_k Z_k \partial_k$ is a parallel vector field if and only if $Z_1, \ldots, Z_n$ satisfy the system of ordinary differential equations:

$$\frac{dZ_k}{dt} + \sum_{i,j=1}^{n} \Gamma_{ij}^k \frac{d(x_i \circ \gamma)}{dt} Z_j = 0, \quad k = 1, \ldots, n.$$

Proposition 1.3. *For $\gamma : I \to M$, $a \in I$ and $z \in T_{\gamma(a)}M$, there exists a unique parallel vector field Z on γ such that $Z(a) = z$.*

Proof. Follows from the fundamental existence and uniqueness theorem of systems of first order linear equations. □

Consider a curve $\gamma : I \to M$. Let $a, b \in I$ and $z \in T_{\gamma(a)}M$. The function

$$P = P_{\gamma(a)}^{\gamma(b)}(\gamma) : T_{\gamma(a)}M \to T_{\gamma(b)}M$$

sending each $z \in T_{\gamma(a)}M$ to $Z(\gamma(b))$ is called *parallel translation along γ* from $\gamma(a)$ to $\gamma(b)$, where Z is the unique parallel vector field along γ such that $Z(a) = z$.

Proposition 1.4. *Parallel translation is a linear isometry.*

Proof. Let $\gamma : I \to M$ be a curve and $p = \gamma(a)$, $q = \gamma(b)$. Let $u, v \in T_pM$ correspond to parallel vector fields U, V. Since $U + V$ is also parallel, we have

$$P(u + v) = (U + V)(b) = U(b) + V(b) = P(u) + P(v).$$

Similarly, we have $P(cu) = cP(u)$. Hence, P is a linear map. For U, V as above, we get

$$\frac{d}{dt} \langle U, V \rangle = \langle U', V \rangle + \langle U, V' \rangle = 0.$$

Thus $\langle U, V \rangle$ is constant. Hence,

$$\langle P(u), P(v) \rangle = \langle U(b), V(b) \rangle = \langle U(a), V(a) \rangle = \langle u, v \rangle,$$

which implies that P is an isometry. □

Definition 1.4. A *geodesic* in a pseudo-Riemannian manifold M is a curve $\gamma : I \to M$ whose velocity vector field γ' is parallel, or equivalently, it satisfies

$$\frac{d^2(x_k \circ \gamma)}{dt^2} + \sum_{i,j=1}^{n} \Gamma_{ij}^k(\gamma) \frac{d(x_i \circ \gamma)}{dt} \frac{d(x_j \circ \gamma)}{dt} = 0 \tag{1.5}$$

for $k = 1, \ldots, n$.

It follows from the existence and uniqueness theorem of linear system of ordinary differential equations that, for any given point $p \in M$ and any given tangent vector $v \in T_pM$, there exists a unit geodesic γ_v such that $\gamma(0) = p$ and $\gamma'(0) = v$. A geodesic with largest possible domain is called a *maximal geodesic.*

A pseudo-Riemannian manifold M for which every maximal geodesic is defined on the entire real line is said to be *geodesic complete* or simply *complete.*

It follows from (1.5) that the geodesic of a pseudo-Euclidean space $\mathbb{E}^m_s$ are straight lines. Thus every pseudo-Euclidean n-plane $\mathbb{E}^m_s$ is geodesically complete.

In general, parallel translation from a point p to another point q depends on the particular curve jointing two points p and q. However, on a pseudo-Euclidean space $\mathbb{E}^m_s$ the natural coordinate vector fields are parallel and hence so their restrictions to any curve. Consequently, parallel translation from a point p to another point q along any curve is just the canonical isomorphism $v_p \to v_q$. This phenomenon is called *distant parallelism.*

For a given $v \in T_pM$, there is a unique geodesic γ_v such that $\gamma_v(0) = p$ with initial tangent vector $\gamma_v'(0) = v$. Let U_p be the set of vectors $v \in T_pM$ such that the geodesic γ_v is defined at least on $[0, 1]$. For a vector $v \in U_p$ the *exponential map* is defined by $\exp_p(v) = \gamma_v(1)$.

Definition 1.5. A subset S of a vector space is called *starshaped about* o if $v \in S$ implies $tv \in S$ for all $t \in [0,1]$. For each point $p \in M$ there exists a neighborhood $\mathcal{U}$ of o in T_pM on which the exponential map $\exp_o$ is a diffeomorphism onto a neighborhood U of p on M. If $\mathcal{U}$ is starshaped about o, then U is called a *normal neighborhood* of p.

Definition 1.6. Let $\{e_1, \ldots, e_n\}$ be an orthonormal basis of T_pM so that $\langle e_i, e_j\rangle = \epsilon_i\delta_{ij}$. The *normal coordinate system* $\{y_1, \ldots, y_n\}$ determined by $e_1, \ldots, e_n$ assigns to each point $q \in U$ the vector coordinates relative to $e_1, \ldots, e_n$ of the corresponding point $\exp_p^{-1}(q) \in \mathcal{U} \subset T_pM$. In other words,

$$\exp_p^{-1}(q) = \sum_{i=1}^{n} y_i(q)e_i, \quad q \in U.$$

The following is well-known.

Proposition 1.5. *Let* $\{y_1, \ldots, y_n\}$ *be a normal coordinate system about a point* $p \in M$. *Then* $g_{ij}(0) = \delta_{ij}$ *and* $\Gamma^i_{jk}(p) = 0$.

Definition 1.7. Let p be a point in a Riemannian manifold M. Let S_r be the hypersphere of T_pM with radius r centered at the origin o. Suppose that r is a sufficiently small positive number such that $\exp_p(S_r)$ lies in a normal coordinate neighborhood of p, then $\exp_p(S_r)$ is called a *geodesic hypersphere.*

1.6 Riemann curvature tensor

Gauss's "theorema egregium" shows that the Gaussian curvature, defined as product of two two principal curvatures, of a surface in a Euclidean 3-space $\mathbb{E}^3$ is an isometric invariant of the surface itself. This lead B. Riemann to his invention of Riemannian geometry, whose most important feature is the generalization of Gaussian curvature to arbitrary Riemannian manifolds. No significant changes are required in extending to pseudo-Riemannian manifolds.

For a pseudo-Riemannian manifold M with Levi-Civita connection ∇, the function $R : \mathcal{X}(M) \times \mathcal{X}(M) \times \mathcal{X}(M) \to \mathcal{X}(M)$ defined by

$$R(X, Y)Z = \nabla_X \nabla_Y Z - \nabla_Y \nabla_X Z - \nabla_{[X,Y]} Z$$

is a $(1, 3)$ tensor field, called the *Riemann curvature tensor.* Sometimes, we put $R(X, Y; Z, W) = \langle R(X, Y)Z, W \rangle$.

Proposition 1.6. *The curvature tensor R satisfies the following properties:*

$$R(u, v)w = -R(v, u)w, \tag{1.6}$$

$$\langle R(u, v)w, z \rangle = - \langle R(u, v)z, w \rangle, \tag{1.7}$$

$$R(u, v)w + R(v, w)u + R(w, u)v = 0, \tag{1.8}$$

$$\langle R(u, v)w, z \rangle = \langle R(w, z)u, v \rangle \tag{1.9}$$

for vectors $u, v, w, z \in T_pM$, $p \in M$.

Proof. Since both ∇ and the bracket operation on vector fields are local operations, it suffices to work on any neighborhood of p. Moreover, because the identities are tensor equations, u, v, w, z can be extended to local vector fields U, V, W, Z on some neighborhood of p in any convenient way. In particular, we may choose the extensions in such way that all of their brackets are zero.

Since $R(U, V)W = [\nabla_U, \nabla_V]W - \nabla_{[U,V]}W$ and the bracket operation is skew-symmetric, (1.6) follows immediately from the definition of the curvature tensor.

For (1.7) we only need to show that $\langle R(u,v)w,z\rangle = 0$ by polarization. By Theorem 1.1(5), we have

$$\begin{aligned}\langle R(U,V)W,W\rangle &= \langle \nabla_U\nabla_V W,W\rangle - \langle \nabla_V\nabla_U W,W\rangle \\ &= \langle \nabla_U W,\nabla_V W\rangle - V\langle \nabla_U W,W\rangle + \langle \nabla_V W,\nabla_U W\rangle - U\langle \nabla_V W,W\rangle \\ &= \frac{1}{2}UV\langle W,W\rangle - \frac{1}{2}VU\langle W,W\rangle = 0.\end{aligned}$$

This proves (1.7), since $[U,V]=0$.

For (1.8) we consider $\mathcal{S}$, the sum of cyclic permutations of U,V,W, to find

$$\begin{aligned}R(U,V)W + R(V,W)U + R(W,U)V &\\ &= \mathcal{S}R(U,V)W \\ &= \mathcal{S}\nabla_U\nabla_V W - \mathcal{S}\nabla_V\nabla_U W \\ &= \mathcal{S}\nabla_U\nabla_V W - \mathcal{S}\nabla_U\nabla_W V \\ &= \mathcal{S}\nabla_U[Y,W] = 0.\end{aligned}$$

If we put

$$S(u,v,w,z) = \langle R(u,v)w,z\rangle + \langle R(v,w)u,z\rangle + \langle R(w,u)v),z\rangle,$$

then a direct computation shows that

$$\begin{aligned}0 &= S(u,v,w,z) - S(v,w,z,u) - S(w,z,u,v) + S(z,u,v,w) \\ &= \langle R(u,v)w,z\rangle - \langle R(v,u)w,z\rangle - \langle R(w,z)u,v\rangle + \langle R(z,w)u,v\rangle.\end{aligned}$$

Thus, by applying (1.6), we obtain (1.9). □

Equation (1.8) is called the *first Bianchi identity.*

Proposition 1.7. *The curvature tensor of a pseudo-Riemannian manifold M satisfies the second Bianchi identity:*

$$(\nabla_W R)(U,V) + (\nabla_U R)(V,W) + (\nabla_V R)(W,U) = 0, \tag{1.10}$$

where $(\nabla_W R)(U,V)$ is defined by

$$\begin{aligned}((\nabla_W R)(U,V))Z = \nabla_W(R(U,V)Z) - R(\nabla_W U,V)Z \\ - R(U,\nabla_W V)Z - R(U,V)(\nabla_W Z).\end{aligned} \tag{1.11}$$

Proof. Clearly, (1.10) is a tensor identity. Let p be a given point in M. we consider a normal coordinate system on a neighborhood of p.

We choose the extensions U,V,W of vectors $u,v,w\in T_pM$ in such way that not only all brackets vanish identically, but also the extensions have constant components with respect to the normal coordinate system. Hence, by Proposition 1.6, we have

$$(\nabla_W R)(U,V)Z = \nabla_W(R(U,V)Z) - R(U,V)(\nabla_W Z)$$

at p, which gives

$$(\nabla_W R)(U,V) = [\nabla_W, R(U,V)] = [\nabla_W, [\nabla_U, \nabla_V]]$$

at p. Thus, summing the above formula over the cyclic permutations of U, V, W yields the required identity at p. □

1.7 Sectional, Ricci and scalar curvatures

Since the Riemann curvature tensor is rather complicated, we consider a simpler real-valued function, the sectional curvature, which completely determines the curvature tensor.

At a point $p \in M$, a 2-dimensional linear subspace π of the tangent space T_pM is called a *plane section*. For a given basis $\{v, w\}$ of the plane section π, we define a real number by

$$Q(v,w) = \langle v,v\rangle \langle w,w\rangle - \langle v,w\rangle^2 .$$

The plane section π is called *nondegenerate* if and only if $Q(u,v) \neq 0$. $Q(u,v)$ is positive when $g|_\pi$ is definite, and is negative when $g|_\pi$ is indefinite. The absolute value $||Q(u,v)||$ is the square of the area of the parallelogram with sides u and v.

For a nondegenerate plane section π at p, the number

$$K(u,v) = \frac{\langle R(u,v)v,u\rangle}{Q(u,v)}$$

is independent of the choice of basis $\{u,v\}$ for π, which is called the *sectional curvature* $K(\pi)$ of π.

A pseudo-Riemannian manifold is said to be *flat* if its sectional curvature K vanishes identically. It is well-known that a pseudo-Riemannian manifold M is flat if and only if the curvature tensor R of M is zero at every point.

For each index s, the pseudo-Euclidean m-space $\mathbb{E}^m_s$ is flat. In fact, the Christoffel symbols all vanish for a natural coordinate system. Hence, the curvature tensor of $\mathbb{E}^m_s$ vanishes identically.

Definition 1.8. A multilinear function

$$F : T_pM \times T_pM \times T_pM \times T_pM \to \mathbf{R}$$

is called *curvature-like* if F satisfies properties (1.6)-(1.9) for the function $(u,v,w,z) \to \langle R(u,v)w,z\rangle$.

For a curvature-like function F we have the following.

Lemma 1.3. *Let M be a pseudo-Riemannian manifold and $p \in M$. If F is a curvature-like function on T_pM such that*

$$K(u,v) = \frac{F(u,v,v,u)}{Q(u,v)}$$

whenever u, v span a nondegenerate plane at p, then

$$\langle R(u,v)w, z\rangle = F(u,v,w,z)$$

for all $u, v, w, z \in T_pM$.

Proof. Put

$$\delta(u,v,w,z) = F(u,v,w,z) - \langle R(u,v)w, z\rangle .$$

Then δ is also curvature-like. Because $\delta(u,v,v,u) = 0$ if u, v span a nondegenerate plane section at p, we obtain $\delta = 0$. □

For sectional curvature K of indefinite Riemannian manifolds, we have the following result [Kulkarni (1979)].

Theorem 1.2. *Let M be a pseudo-Riemannian manifold of dimension ≥ 3 and index $s > 0$. Then, at each point $p \in M$, the following four conditions are equivalent:*

(1) *K is constant;*
(2) *$a \leq K$ or $K \leq b$;*
(3) *$a \leq K \leq b$ on indefinite planes;*
(4) *$a \leq K \leq b$ on definite planes;*

where a and b are real numbers.

It follows from Theorem 1.2 that the sectional curvature of an indefinite Riemannian manifold at each point is unbounded from above and below unless M has constant sectional curvature.

By an *orthonormal frame* on a pseudo-Riemannian n-manifold M, we mean a set consists of n mutually orthogonal unit vector fields $e_1, \ldots, e_n$ on M. We put $\epsilon_j = \langle e_j, e_j\rangle$, $j = 1, \ldots, n$.

For two given vector fields X, Y on M, if we put

$$X = \sum_{j=1}^{n} X_j e_j, \quad Y = \sum_{j=1}^{n} Y_j e_j,$$

then we have

$$\langle X, Y \rangle = \sum_{j=1}^{n} \epsilon_j X_j Y_j.$$

Definition 1.9. The *Ricci tensor* of a pseudo-Riemannian n-manifold M, denoted by Ric, is a symmetric $(0, 2)$ tensor defined by

$$Ric(X, Y) = \text{trace}\{Z \mapsto R(Z, X)Y\},$$

or equivalently,

$$Ric(X, Y) = \sum_{\ell=1}^{n} \epsilon_\ell \langle R(e_\ell, X)Y, e_\ell \rangle, \tag{1.12}$$

where $e_1, \ldots, e_m$ is an orthonormal frame.

It is well-known that $Ric(X, Y)$ is independent of choice of orthonormal frame $e_1, \ldots, e_n$. If the Ricci tensor vanishes, M is called *Ricci flat.*

A flat manifold is certainly Ricci flat. But the converse does not hold. A pseudo-Riemannian manifold M is called an *Einstein manifold* if $Ric = cg$ for some constant c. For a unit vector $u \in TM$, the *Ricci curvature* $Ric(u)$ is defined by $Ric(u) = Ric(u, u)$.

If M is a pseudo-Riemannian manifold of dimension ≥ 3 which satisfies $Ric = fg$ for some function $f \in \mathcal{F}(M)$, then M is always Einsteinian.

Definition 1.10. The *scalar curvature* τ of M is defined by

$$\tau = \sum_{i<j} K(e_i, e_j), \tag{1.13}$$

where $e_1, \ldots, e_n$ is an orthonormal frame of M. The scalar curvature τ is independent of the choice of the orthonormal frame.

1.8 Indefinite real space forms

A pseudo-Riemannian manifold M is said to have *constant curvature* if its sectional curvature is constant.

For a constant c the function F defined by

$$F(u, v, w, z) = c\{\langle u, z \rangle \langle v, w \rangle - \langle u, w \rangle \langle v, z \rangle\}$$

is curvature-like. Thus Lemma 1.3 implies that $F(u, v, v, u) = cQ(u, v)$. Hence, if u, v span a nondegenerate plane section, we have

$$K(u, v) = c = \frac{F(u, v, v, u)}{Q(u, v)}.$$

Consequently, if a pseudo-Riemannian manifold M is of constant curvature c, then its curvature tensor R satisfies

$$R(u,v)w = c\{\langle v,w\rangle\, u - \langle u,w\rangle\, v\}. \tag{1.14}$$

Let $\mathbb{E}^n_t$ be the pseudo-Euclidean n-space equipped with the canonical pseudo-Euclidean metric of index t given by

$$g_0 = -\sum_{i=1}^{t} du_i^2 + \sum_{j=t+1}^{n} du_j^2, \tag{1.15}$$

where $(u_1,\ldots,u_n)$ is a rectangular coordinate system of $\mathbb{E}^n_t$.

Let c be a nonzero real number. We put

$$S^k_s(\mathbf{x}_0,c) = \left\{\mathbf{x}\in\mathbb{E}^{k+1}_s : \langle \mathbf{x}-\mathbf{x}_0, \mathbf{x}-\mathbf{x}_0\rangle = \frac{1}{c} > 0\right\}, \quad s>0, \tag{1.16}$$

$$H^k_s(\mathbf{x}_0,c) = \left\{\mathbf{x}\in\mathbb{E}^{k+1}_{s+1} : \langle \mathbf{x}-\mathbf{x}_0, \mathbf{x}-\mathbf{x}_0\rangle = \frac{1}{c} < 0\right\}, \quad s>0, \tag{1.17}$$

$$H^k(c) = \left\{\mathbf{x}\in\mathbb{E}^{k+1}_1 : \langle \mathbf{x}, \mathbf{x}\rangle = \frac{1}{c} < 0 \text{ and } x_1 > 0\right\}, \tag{1.18}$$

where $\langle\ ,\ \rangle$ is the associated scalar product. $S^k_s(\mathbf{x}_0,c)$ and $H^k_s(\mathbf{x}_0,c)$ are pseudo-Riemannian manifolds of curvature c with index s, known as a *pseudo sphere* and a *pseudo-hyperbolic space*, respectively. The point $\mathbf{x}_0$ is called the center of $S^m_s(\mathbf{x}_0,c)$ and $H^m_s(\mathbf{x}_0,c)$. If $\mathbf{x}_0$ is the origin o of the pseudo-Euclidean spaces, we denote $S^k_s(o,c)$ and $H^k_s(o,c)$ by $S^k_s(c)$ and $H^k_s(c)$, respectively. The pseudo-Riemannian manifolds $\mathbb{E}^k_s$, $S^k_s(c)$, $H^k_s(c)$ are the standard models of the *indefinite real space forms.* In particular, $\mathbb{E}^k_1$, $S^k_1(c)$, $H^k_1(c)$ are the standard models of *Lorentzian space forms.*

Topologically, a de Sitter spacetime S^k_1 is $\mathbf{R}\times S^{k-1}$. Thus, when $k\geq 3$ a de Sitter spacetime is simply-connected. S^4_1 and H^4_1 are known as the *de Sitter spacetime*[2] and the *anti de Sitter spacetime*[3], respectively.

[2] A de Sitter spacetime, named after Willem de Sitter (1872–1934), is a solution to Einstein's field equation. It involves a variation on the spacetime in which spacetime is itself slightly curved even in the absence of matter or energy. In a de Sitter spacetime the dynamics of the universe are dominated by the cosmological constant, which is thought to correspond to dark energy. If the current acceleration of our universe is due to a cosmological constant, then as the universe continues to expand all of the matter and radiation will be diluted. Eventually our universe will have become a de Sitter universe.

[3] An anti de Sitter spacetime is a maximally symmetric vacuum solution of Einstein's field equation with a negative cosmological constant (something not observed in the real-life cosmos). In contrast to de Sitter spacetime, the curvature is hyperbolic instead. An anti de Sitter spacetime can also be thought of as a general relatively like spacetime in which vacuum space itself has negative energy, which causes this spacetime to collapse in on itself at an ever greater rate.

When $s = 0$, the Riemannian manifolds $\mathbb{E}^k$, $S^k(c)$ and $H^k(c)$ are of constant curvature, called *real space forms.* The Euclidean k-space $\mathbb{E}^k$, the k-sphere $S^k(c)$ and the hyperbolic k-space $H^k(c)$ are simply-connected complete Riemannian manifolds of constant curvature $0, c > 0$ and $c < 0$, respectively. A complete simply-connected pseudo-Riemannian k-manifold, $k \geq 3$, of constant curvature c and with index s is isometric to $\mathbb{E}^k_s$, or $S^k_s(c)$ or $H^k_s(c)$ according to $c = 0$, or $c > 0$ or $c < 0$, respectively.

We denote a k-dimensional indefinite space form of curvature curvature c and index s simply by $R^k_s(c)$. We simply denote the indefinite space form $R^k_0(c)$ with index $s = 0$ by $R^k(c)$.

1.9 Lie derivative, gradient, Hessian and Laplacian

Definition 1.11. Let M be a pseudo-Riemannian n-manifold and $f \in \mathcal{F}(M)$. The *gradient* of f, denote by ∇f (or by $\operatorname{grad} f$), is the vector field dual to the differential df. In other word, ∇f is defined by

$$\langle \nabla f, X \rangle = df(X) = Xf \quad \forall X \in \mathcal{X}(M). \tag{1.19}$$

In terms of a coordinate system $\{x_1, \ldots, x_n\}$ of M, we have

$$df = \sum_{j=1}^{n} \frac{\partial f}{\partial x_j} dx_j \ \text{ and } \ \nabla f = \sum_{i,j} g^{ij} \frac{\partial f}{\partial x_i} \partial_j. \tag{1.20}$$

Definition 1.12. If $X \in \mathcal{X}(M)$ and $\{e_1, \ldots, e_n\}$ is an orthonormal frame, the *divergence* of X, denoted by $\operatorname{div} X$, is defined by

$$\operatorname{div} X = \sum_{j=1}^{n} \epsilon_j \left\langle \nabla_{e_i} X, e_i \right\rangle, \tag{1.21}$$

which is independent of the chosen frame.

If we put $X = \sum_{j=1}^n X^j \frac{\partial}{\partial x_j}$, $X_i = \sum_{j=1}^n g_{ij} X^j$, then

$$\operatorname{div} X = \sum_{j=1}^{n} \left\{ \frac{\partial X_i}{\partial x_i} + \sum_{k=1}^{n} \Gamma^j_{jk} X_k \right\}. \tag{1.22}$$

Definition 1.13. The *Hessian* of $f \in \mathcal{F}(M)$, denoted by H^f, is the second covariant differential $\nabla(\nabla f)$, so that

$$H^f(X, Y) = XYf - (\nabla_X Y)f = \langle \nabla_X(\nabla f), Y \rangle \tag{1.23}$$

for vector fields $X, Y \in \mathcal{X}(M)$.

Definition 1.14. The *Laplacian* of $f \in \mathcal{F}(M)$, denoted by Δf, is defined by $\Delta f = -\mathrm{div}(\nabla f)$. In terms of a coordinate system $\{x_1, \ldots, x_n\}$, we have

$$\Delta f = -\sum_{i,j=1}^{n} \left\{ \frac{\partial^2 f}{\partial x_i \partial x_j} - \sum_{k=1}^{n} \Gamma_{ij}^{k} \frac{\partial f}{\partial x_k} \right\}. \tag{1.24}$$

In terms of a natural coordinate system $\{u_1, \ldots, u_n\}$ of $\mathbb{E}_s^n$, we have

$$\nabla f = \sum_{j=1}^{n} \epsilon_j \frac{\partial f}{\partial u_j} \partial_j, \tag{1.25}$$

$$\mathrm{div}\, X = \sum_{j=1}^{n} \frac{\partial X_j}{\partial u_j}, \tag{1.26}$$

$$\Delta f = -\sum_{j=1}^{n} \epsilon_j \frac{\partial^2 f}{\partial u_j^2}. \tag{1.27}$$

Definition 1.15. A *volume element* on a pseudo-Riemannian n-manifold M is a smooth n-form ω such that $\omega(e_1, \ldots, e_n) = \pm 1$ for every orthonormal frame on M.

Volume elements always exist at least locally.

Lemma 1.4. *A pseudo-Riemannian manifold M has a global volume element if and only if M is orientable.*

Proof. If M is a global volume element, then the bases $\{v_1, \ldots, v_n\}$ for each T_pM with $\omega(v_1, \ldots, v_n) > 0$ constitute an orientation of T_pM.

If M is oriented, then for all positively oriented coordinate systems the local volume elements agree on the overlaps, hence give a global volume element. □

On a manifold M the *Lie derivative* $\mathcal{L}$ is a tensor derivation such that for any $V \in \mathcal{X}(M)$ we have

$$\mathcal{L}_V f = Vf \qquad \forall f \in \mathcal{F}(M),$$
$$\mathcal{L}_V X = [V, X] \qquad \forall X \in \mathcal{X}(M).$$

Definition 1.16. A *Killing vector field* on a pseudo-Riemannian manifold is a vector field X for which the Lie derivative of the metric tensor vanishes, i.e., $\mathcal{L}_X g = 0$. A *conformal-Killing vector field* is a vector field X for which $\mathcal{L}_X g = \lambda g$ for some function $\lambda \in \mathcal{F}(M)$.

Under the flow of a Killing vector field X, the metric tensor does not change; thus a Killing vector field is an infinitesimal isometry.

Proposition 1.8. *On a pseudo-Riemannian manifold M, the following three conditions on a vector field X are equivalent:*

(1) *X is a Killing vector field.*
(2) $X\langle V,W\rangle = \langle [X,V],W\rangle + \langle V,[X,W]\rangle, \ \forall V,W \in \mathcal{X}(M)$.
(3) *∇X is a skew-adjoint relative to the metric tensor g, i.e.,*
$$\langle \nabla_V X, W\rangle = -\langle \nabla_W X, V\rangle, \ \forall V,W \in \mathcal{X}(M).$$

Proof. For all $V,W \in \mathcal{X}(M)$, the following are equivalent:
$$\langle \nabla_V X, W\rangle + \langle \nabla_W X, V\rangle = 0;$$
$$\langle \nabla_X V, W\rangle - \langle [X,V],W\rangle = \langle [X,W],V\rangle - \langle \nabla_X W, V\rangle;$$
$$X\langle V,W\rangle = \langle [X,V],W\rangle + \langle V,[X,W]\rangle.$$
In view of the product rule, the last one is equivalent to $\mathcal{L}_X g = 0$. □

Definition 1.17. Let N and M be pseudo-Riemannian manifolds with metrics g_N and g_M. An *isometry* $\psi : N \to M$ is a diffeomorphism that preserves metric tensors, i.e., $\psi^*(g_M) = g_N$.

Definition 1.18. A map $\phi : N \to M$ between two pseudo-Riemannian manifolds is called a *local isometry* at $p \in N$ if there is a neighborhood $U \subset N$ of p such that $\phi : U \to \phi(U)$ is a diffeomorphism satisfying
$$\langle u,v\rangle_p = \left\langle \phi_{*_p}(u), \phi_{*_p}(v)\right\rangle_{\phi(p)}, \tag{1.28}$$
for $u,v \in T_pN$, $p \in N$.

Definition 1.19. A pseudo-Riemannian manifold N is said to be *locally isometric* to a pseudo-Riemannian manifold M if for each point $p \in N$ there exists a neighborhood U of x and a local isometry $\phi : U \to \phi(U) \subset M$.

The following result shows that a local isometry is uniquely determined by its differential map at a point.

Lemma 1.5. *Let $\phi_1, \phi_2 : N \to M$ be two local isometries. If there is a point $p \in N$ such that $\phi_{1*p} = \phi_{2*p}$, then $\phi_1 = \phi_2$.*

Proof. Put $U = \{q \in N : \phi_{1*q} = \phi_{2*q}\}$. Then U is a closed subset of N. Let $q \in U$ and V a normal neighborhood of q. Then for each point $z \in V$ there is a vector $v \in T_qN$ such that $\gamma_v(1) = \exp_q(v) = z$. Hence
$$\phi_1(z) = \phi_1(\gamma_v(1)) = \gamma_{\phi_{1*}v}(1) = \gamma_{\phi_{2*}v}(1) = \phi_2(\gamma_v(1)) = \phi_2(z),$$

which implies that $\phi_1 = \phi_2$ on V. Hence, we have $\phi_{1*z} = \phi_{2*z}$ for all $z \in V$. Thus, U is also an open subset of M. Therefore, we have $U = M$. Consequently, we get $\phi_1 = \phi_2$. □

Definition 1.20. A mapping $\phi : N \to M$ of pseudo-Riemannian manifolds is called *conformal* if $\phi^*(g_M) = fg_N$ for some function $f \in \mathcal{F}(N)$ such that $f > 0$ or $f < 0$. In particular, if the function f is a nonzero real number, then ϕ is called a *homothety*.

Lemma 1.6. *Homothety preserves Levi-Civita connection of pseudo-Riemannian manifolds.*

Proof. Follows immediately from Koszul's formula. □

1.10 Weyl conformal curvature tensor

Let M be a pseudo-Riemannian m-manifold. Associated with the Ricci tensor Ric, define a $(1,1)$-tensor Q by

$$\langle Q(X), Y\rangle = Ric(X,Y).$$

The *Weyl conformal curvature tensor* C is a tensor field of type $(1,3)$ defined by

$$\begin{aligned} C(X,Y)Z = &R(X,Y)Z + \frac{1}{m}\{Ric(X,Z)Y - Ric(Y,Z)X \\ &+ \langle X,Z\rangle QY - \langle Y,Z\rangle QX\} \\ &- \frac{2\tau}{m(m+1)}\{\langle X,Z\rangle Y - \langle Y,Z\rangle X\}. \end{aligned}$$

It is well-known that the Weyl conformal curvature tensor C vanishes identically when $\dim M = 3$ and it is invariant under conformal changes of the metric.

Definition 1.21. A pseudo-Riemannian metric g on a manifold M is called *conformally flat* if it is conformally related with a flat pseudo-Euclidean metric. A manifold with a conformal flat pseudo-Riemannian metric is called a *conformally flat manifold.*

The following result of [Weyl (1918)] is well-known.

Theorem 1.3. *A pseudo-Riemannian manifold M of dimension ≥ 4 is conformally flat if and only if the conformal curvature tensor C vanishes identically.*

Chapter 2

Basics on Pseudo-Riemannian Submanifolds

In the 17th century, Fermat (1601-1665) and Descartes (1596-1650) created the coordinate method in geometry, and Leibniz (1646-1716) and Newton (1642-1727) created the infinitesimal (differential) calculus. Since then differential geometry started as the study of curves in the Euclidean plane and of curves and surfaces in the Euclidean 3-space by means of the techniques of the differential calculus.

With his 1827 fundamental work

"*Disquisitiones generales circa superficies curvas*",

Gauss initiated modern differential geometry [Gauss (1827)]. Of greatest importance in this work was that Gauss demonstrated the existence of an intrinsic geometry of surfaces based only on the measurements of the lengths of arcs in these surfaces. As such, the 2-dimensional Euclidean geometry was generalized to a very large class of surfaces. Furthermore, for the first time Gauss made use of the so called curvilinear coordinates to describe surfaces, which is the origin of a local chart around a point on a manifold. Gauss' work was both original and revolutionary.

The influence of the differential geometry of submanifolds upon branches of mathematics, sciences and engineering has been profound. For instance, the study of geodesics and minimal surfaces is intimately related to dynamics, the theory of functions of a complex variable, calculus of variations, and topology. Furthermore, the application of double helix to the investigation of DNA structure in molecular biology done by biologists Crick and Watson and coding theorist Golomb, among many other scientists, resulting in important advancing in genetic engineering and medicine.

Inspired by the theory of general relativity and string theory, mathematicians and physicists study not only submanifolds of Riemannian manifolds but submanifolds of pseudo-Riemannian manifolds as well.

2.1 Isometric immersions

Let $\phi : N \to M$ be a map between two manifolds. Then ϕ is called an *immersion* if $\phi_{*_p} : T_pN \to T_{\phi(p)}M$ is injective for all $p \in N$. If, in addition, ϕ is a homeomorphism onto $\phi(N)$, where $\phi(N)$ has the subspace topology induced from M, we say that ϕ is an *embedding*.

For most local questions of geometry, it is the same to work with either immersions or embeddings. In fact, if $\phi : N \to M$ is an immersion of a manifold N into another manifold M, then for each point $p \in N$, there exists a neighborhood $U \subset N$ of p such that the restriction $\phi : U \to M$ is an embedding.

For an immersion $\phi : N \to M$, we have $\dim N \le \dim M$. The difference $\dim M - \dim N$ is called the *codimension* of the immersion. Throughout this book we consider only immersions of codimension ≥ 1 unless mentioned otherwise.

Definition 2.1. An immersion $\phi : N \to M$ of a pseudo-Riemannian manifold into another pseudo-Riemannian manifold is called *isometric* if

$$\langle u, v\rangle_p = \langle \phi_{*_p} u, \phi_{*_p} v\rangle_{\phi(p)} \tag{2.1}$$

holds for all $u, v \in T_pN$, $p \in N$.

Let $\phi : N \to M$ be an isometric immersion. Then, for each point $p \in N$, there exists a neighborhood U of p such that $\phi : U \to M$ is an embedding. Thus, each vector $u \in T_pU$ gives rise to a vector $\phi_* u \in T_{\phi(p)}M$. We may identify $u \in T_pN$ with $\phi_* u \in T_{\phi(p)}M$. In this way, each tangent space T_pN is a nondegenerate subspace of $T_{\phi(p)}(M)$. Hence, there is a direct sum decomposition

$$T_{\phi(p)}M = T_pN \oplus T_p^{\perp}N, \tag{2.2}$$

where $T_p^{\perp}N$ is a nondegenerate subspace of $T_{\phi(p)}M$, which is called the *normal subspace* of N at p. Vectors in $T_p^{\perp}N$ are said to be normal to N and those in T_pN are tangent to N. Thus, each vector $v \in T_{\phi(p)}M$ has a unit expression

$$v = \tan v + \operatorname{nor} v, \tag{2.3}$$

where $\tan v \in T_pN$ and $\operatorname{nor} v \in T_p^{\perp}(N)$. The orthogonal projections

$$\tan : T_{\phi(p)}M \to T_pN \quad \text{and} \quad \operatorname{nor} : T_{\phi(x)}M \to T_p^{\perp}N$$

are **R**-linear.

2.2 Cartan–Janet's and Nash's embedding theorems

One of the most fundamental problems in submanifold theory is the problem of isometric immersibility. The earliest publication by L. Schläfli (1814–1895) on isometric embedding appeared in [Schlafli (1873)].

The problem of isometric immersion (or embedding) admits an obvious analytic interpretation; namely, if $g_{ij}(x)$, $x = (x_1, \ldots, x_n)$, are the components of the metric tensor g in local coordinates $x_1, \ldots, x_n$ on a Riemannian n-manifold M, and $u_1, \ldots, u_m$ are the standard Euclidean coordinates of $\mathbb{E}^m$, then the condition for an isometric immersion in $\mathbb{E}^m$ is

$$\sum_{i=1}^{n} \frac{\partial u_j}{\partial x_i}\frac{\partial u_k}{\partial x_i} = g_{jk}(x),$$

i.e., we have a system of $\frac{1}{2}n(n+1)$ nonlinear partial differential equations in m unknown functions. If $m = \frac{1}{2}n(n+1)$, then this system is definite and so we would like to have a solution. Schläfli asserted that any Riemannian n-manifold can be isometrically imbedded in Euclidean space of dimension $\frac{1}{2}n(n+1)$. Apparently, it is appropriate to assume that he had in mind of analytic metrics and local analytic embeddings. This was later called *Schläfli's conjecture.*

M. Janet (1888–1984) published in [Janet (1926)] a proof of Schläfli's conjecture which states that a real analytic Riemannian n-manifold can be locally isometrically embedded into any real analytic Riemannian manifold of dimension $\frac{1}{2}n(n+1)$. É. Cartan (1869–1951) revised Janet's paper with the same title in [Cartan (1927)]; while Janet wrote the problem in the form of a system of partial differential equations which he investigated using rather complicated methods, Cartan applied his own theory of Pfaffian systems in involution. Both Janet's and Cartan's proofs contained obscurities. C. Burstin get rid of them in [Burstin (1931)]. This result of Cartan–Janet implies that every Einstein n-manifold ($n \geq 3$) can be locally isometrically embedded in $\mathbb{E}^{n(n+1)/2}$.

The Cartan–Janet theorem is dimensionwise the best possible, i.e., there exist real analytic Riemannian n-manifolds which do not possess smooth local isometric embeddings into any Euclidean space of dimension strictly less than $\frac{1}{2}n(n+1)$. Not every Riemannian n-manifold can be isometrically immersed in $\mathbb{E}^m$ with $m \leq \frac{1}{2}n(n+1)$. For instance, not every Riemannian 2-manifold can be isometrically immersed in $\mathbb{E}^3$.

A global isometric embedding theorem was proved by J. F. Nash which states as follows [Nash (1956)].

Theorem 2.1. *Every compact Riemannian n-manifold can be isometrically embedded in any small portion of a Euclidean N-space* $\mathbb{E}^N$ *with* $N = \frac{1}{2}n(3n+11)$. *Every non-compact Riemannian n-manifold can be isometrically embedded in any small portion of a Euclidean m-space* $\mathbb{E}^m$ *with* $m = \frac{1}{2}n(n+1)(3n+11)$.

R. E. Greene improved Nash's result in [Greene (1970)] and proved that every non-compact Riemannian n-manifold can be isometrically embedded in the Euclidean N-space $\mathbb{E}^N$ with $N = 2(2n+1)(3n+7)$. Also, it was proved independently in [Greene (1970)] and [Gromov and Rokhlin (1970)] that a local isometric embedding from a Riemannian n-manifold into $\mathbb{E}^{\frac{1}{2}n(n+1)+n}$ always exist.

Concerning the isometric embedding of pseudo-Riemannian manifolds, the following existence theorem was proved independently in [Clarke (1970); Greene (1970)].

Theorem 2.2. *Any pseudo-Riemannian n-manifold* M_t^n *with index t can be isometrically embedded in a pseudo-Euclidean N-space* $\mathbb{E}_s^N$, *for* N, s *large enough. Moreover, this embedding may be taken inside any given open set in* $\mathbb{E}_s^N$.

2.3 Gauss' formula and second fundamental form

Let $\phi : N \to M$ be an isometric immersion. The Levi-Civita connection of the ambient manifold M will be denoted by $\tilde{\nabla}$. Since the discussion is local, we may assume, if we want, that N is embedded in M.

Let X be a vector field tangent to N. A vector field $\tilde{X}$ on M is called an *extension* of X if its restriction to $\phi(N)$ is X. If $\tilde{X}$ and $\tilde{Y}$ are extensions of vector fields X and Y on N, respectively, then $[\tilde{X}, \tilde{Y}]\,|_N$ is independent of the extensions. Moreover, we have

$$[\tilde{X}, \tilde{Y}]\,|_N = [X, Y]. \tag{2.4}$$

If X and Y are local vector fields of N and $\tilde{X}$ and $\tilde{Y}$ are local extensions of X and Y to M. Then the restriction of $\tilde{\nabla}_{\tilde{X}}\tilde{Y}$ on N is independent of the extensions $\tilde{X}, \tilde{Y}$ of X, Y. This can be seen as follows: Let $\tilde{X}_1$ be another extension of X, then we have $\tilde{\nabla}_{\tilde{X}-\tilde{X}_1}\tilde{Y} = 0$ on N since $\tilde{X} - \tilde{X}_1 = 0$ on N. Thus, $\tilde{\nabla}_{\tilde{X}}\tilde{Y} = \tilde{\nabla}_{\tilde{X}_1}\tilde{Y}$. Moreover, it follows from (2.4) and Theorem 1.1(4) that the restriction of $\tilde{\nabla}_{\tilde{X}}\tilde{Y}$ on N is also independent of the extension $\tilde{Y}$.

We define $\nabla'_X Y$ and $h(X,Y)$ by

$$\nabla'_X Y = \tan \nabla_{\tilde{X}} \tilde{Y} \quad \text{and} \quad h(X,Y) = \text{nor}\, \nabla_{\tilde{X}} \tilde{Y}, \tag{2.5}$$

so that we have the following *formula of Gauss*

$$\nabla_{\tilde{X}} \tilde{Y} = \nabla'_X Y + h(X,Y). \tag{2.6}$$

Proposition 2.1. *Let $\phi : N \to M$ be an isometric immersion of a pseudo-Riemannian manifold N into a pseudo-Riemannian manifold M. Then*

(i) ∇' *defined in* (2.5) *is the Levi-Civita connection of N and*
(ii) $h(X,Y)$ *is $\mathcal{F}(M)$-bilinear and symmetric.*

Proof. To prove statement (i), we verify properties (1)-(3) of Definition 1.3 and properties (4) and (5) of Theorem 1.1. Properties (1) and (2) of Definition 1.3 follow from the corresponding properties of ∇ on M and linearity of the projection $\tan : T_{\phi(p)}M \to T_p N$.

To verify property (3), let f be a function in $\mathcal{F}(N)$. Then

$$\nabla_X (fY) = (Xf)Y + f\nabla'_X Y.$$

Thus, after taking the tangential components of both sides, we get property (3) of Definition 1.3.

Next, we prove property (4) of Theorem 1.1. Let us write

$$\nabla_{\tilde{X}} \tilde{Y} = \nabla'_X Y + h(X,Y), \tag{2.7}$$

$$\nabla_{\tilde{Y}} \tilde{X} = \nabla'_Y X + h(Y,X). \tag{2.8}$$

Since ∇ is the Levi-Civita connection of M, it follows form (2.4), (2.7) and (2.8) that

$$[X,Y] = \nabla'_X Y - \nabla'_Y X, \tag{2.9}$$

$$h(X,Y) = h(Y,X), \tag{2.10}$$

which imply property (4) and that $h(X,Y)$ is symmetric.

To prove property (5) we start with

$$X \langle \tilde{Y}, \tilde{Z} \rangle = \langle \nabla_X \tilde{Y}, \tilde{Z} \rangle + \langle \tilde{Y}, \nabla_X \tilde{Z} \rangle. \tag{2.11}$$

From (2.6) we have

$$\langle \nabla_X \tilde{Y}, \tilde{Z} \rangle = \langle \nabla'_X Y, Z \rangle + \langle h(X,Y), Z \rangle = \langle \nabla'_X Y, Z \rangle. \tag{2.12}$$

Similarly, we have

$$\langle \tilde{Y}, \nabla_X \tilde{Z} \rangle = \langle Y, \nabla'_X Z \rangle. \tag{2.13}$$

Hence, we obtain from (2.11)-(2.13) that

$$X\langle Y,Z\rangle = \langle \nabla'_X Y, Z\rangle + \langle Y, \nabla'_X Z\rangle,$$

which is property (5).

Finally, we show that $h(X,Y)$ is $\mathcal{F}(N)$-bilinear. The additivity in X or Y is obvious. Now, for any $f \in \mathcal{F}(N)$, we have

$$\begin{aligned}\nabla'_{fX}Y + h(fX,Y) &= \nabla_{fX}\tilde{Y} = f\nabla_X\tilde{Y}\\ &= f(\nabla'_X Y + h(X,Y)),\end{aligned}$$

which gives $h(fX,Y) = fh(X,Y)$. By symmetry, we also have

$$h(X,fY) = fh(X,Y).$$

□

We define $h : T(N)\times T(N) \to T^{\perp}N$ as the *second fundamental form* of N for the given immersion.

Let $e_1,\ldots,e_n$ and $e_{n+1},\ldots,e_m$ be orthonormal bases of the tangent space T_pN and of the normal space $T_p^{\perp}N$ at a point $p \in N$. If we put $h_{ij}^r = \langle h(e_i,e_j),e_r\rangle$; $i,j = 1,\ldots,n; r = n+1,\ldots,m$, then we have

$$h(e_i,e_j) = \sum_{r=n+1}^{n} \epsilon_r h_{ij}^r e_r, \quad \epsilon_r = \langle e_r,e_r\rangle. \tag{2.14}$$

We call h_{ij}^r the *coefficients of the second fundamental form.*

2.4 Weingarten's formula and normal connection

Let $\phi : N \to M$ be an isometric immersion. If ξ is a normal vector field of N in M and X is a tangent vector field of N, then we may decompose $\nabla_X\xi$ as

$$\nabla_X\xi = -A_\xi(X) + D_X\xi, \tag{2.15}$$

where $-A_\xi(X)$ and $D_X\xi$ are the tangential and normal components of $\nabla_X\xi$. It is easy to verify that $A_\xi(X)$ and $D_X\xi$ are smooth vector fields on N whenever X and ξ are smooth.

Equation (2.13) is known as *formula of Weingarten* and A_ξ is called the *Weingarten map* or *shape operator* at ξ.

Proposition 2.2. *Let $\phi : N \to M$ be an isometric immersion of a pseudo-Riemannian manifold N into a pseudo-Riemannian manifold M. Then*

(a) *$A_\xi(X)$ is $\mathcal{F}(N)$-bilinear in ξ and X; hence, at each point $p \in N$, $A_\xi(X)$ depends only on ξ_p and X_p;*

(b) *For a normal vector field* ξ *and tangent vectors* X, Y *of* N, *we have*

$$\langle h(X,Y), \xi \rangle = \langle A_\xi(X), Y \rangle \, ; \tag{2.16}$$

(c) D *is a metric connection on the normal bundle* $T^\perp N$ *with respect to the induced metric on* $T^\perp N$, *i.e.,* $D_X\langle \xi, \eta \rangle = \langle D_X \xi, \eta \rangle + \langle \xi, D_X \eta \rangle$ *holds for any tangent vector field* X *and normal vector fields* ξ, η.

Proof. (a) Let f, k be two functions in $\mathcal{F}(N)$. We have

$$\begin{aligned} \nabla_{fX}(k\xi) &= f\nabla_X(k\xi) = f\{(Xk)\xi + k\nabla_X \xi\} \\ &= f(Xk)\xi - fkA_\xi(X) + fkD_X\xi, \end{aligned}$$

which implies that

$$A_{k\xi}(fX) = fkA_\xi(X), \tag{2.17}$$

$$D_{fX}(k\xi) = f(Xk)\xi + fkD_X\xi. \tag{2.18}$$

Thus $A_x(X)$ is $\mathcal{F}(N)$-bilinear in ξ and X, since additivity is trivial.

(b) For arbitrary $X, Y \in \mathcal{X}(N)$, we have

$$\begin{aligned} 0 &= \langle \nabla_X Y, \xi \rangle + \langle Y, \nabla_X \xi \rangle \\ &= \langle \nabla'_X Y, \xi \rangle + \langle h(X,Y), \xi \rangle - \langle Y, A_\xi(X) \rangle + \langle Y, D_X \xi \rangle \\ &= \langle h(X,Y), \xi \rangle - \langle Y, A_\xi(X) \rangle \end{aligned}$$

which gives (2.16).

(c) It follows from (2.18) that D defines an affine connection on $T^\perp N$. Moreover, for any normal vector fields ξ and η, we have

$$\nabla_X \xi = -A_\xi(X) + D_X \xi \quad \text{and} \quad \nabla_X \eta = -A_\eta(X) + D_X \eta.$$

Hence

$$\langle D_X \xi, \eta \rangle + \langle \xi, D_X \eta \rangle = \langle \nabla_X \xi, \eta \rangle + \langle \xi, \nabla_X \eta \rangle = X \langle \xi, \eta \rangle.$$

Thus, D is a metric connection on the normal bundle with respect to the induced metric on $T^\perp N$. □

Definition 2.2. An isometric immersion $\phi : N \to M$ is called *totally geodesic* if the second fundamental form vanishes identically, i.e., $h \equiv 0$.

For a normal vector field ξ on N, if $A_\xi = \rho I$ for some $\rho \in \mathcal{F}(N)$, then ξ is called an *umbilical section*, or N is said to be umbilical with respect to ξ. If the submanifold N is umbilical with respect to every local normal vector field, then N is called a *totally umbilical submanifold*.

The *mean curvature vector* H of N in M is defined by

$$H = \left(\frac{1}{n}\right) \operatorname{trace} h, \tag{2.19}$$

where $n = \dim N$. If $\{e_1, \ldots, e_n\}$ is an orthonormal frame of N, then

$$H = \left(\frac{1}{n}\right) \sum_{j=1}^{n} \epsilon_j h(e_j, e_j). \tag{2.20}$$

The second fundamental form of a totally umbilical submanifold satisfies

$$h(X, Y) = \langle X, Y \rangle H, \quad X, Y \in TN. \tag{2.21}$$

Definition 2.3. A pseudo-Riemannian submanifold N is called *minimal* if the mean curvature vector H vanishes identically, i.e., $H \equiv 0$. And N is called *quasi-minimal* (or *pseudo-minimal*) if $H \neq 0$ and $\langle H, H \rangle = 0$ at each point of N [Rosca (1972)].

Definition 2.4. The metric connection D defined by (2.15) is called the *normal connection*. A normal vector field ξ on N is said to be *parallel in the normal bundle*, or simply *parallel* if $D\xi = 0$ holds identically. In particular, N is said to have *parallel mean curvature vector* if $DH = 0$ holds identically.

Let $\phi : N \to M$ be an isometric immersion of a pseudo-Riemannian n-manifold N into a pseudo-Riemannian $(n+k)$-manifold M. For a given orthonormal local frame $\xi_1, \ldots, \xi_k$ of the normal bundle $T^\perp N$, we put

$$A^C = \sum_{r=1}^{k} \epsilon_r A^2_{\xi_r}, \quad \epsilon_r = \langle \xi_r, \xi_r \rangle ,$$

which is independent of the choice of local orthonormal frame field. A^C is self-adjoint $(1, 1)$ tensor field on N, which is called the *Casorati operator* of N in M.

The eigenvectors and eigenvalues of A^C are called the *principal Casorati directions* and *principal Casorati curvatures*. If N is spacelike, the principal Casorati curvatures $c_1, \ldots, c_n$ are real-valued functions satisfying

$$c_1 + \cdots + c_n = C,$$

where $C = \langle h, h \rangle / n$, known as the *Casorati curvature* (cf. [Haesen et al. (2008)]). The original idea of Casorati curvatures for surfaces in $\mathbb{E}^3$ was due to F. Casorati (1835-1890) (cf. [Casorati (1890)]).

2.5 Shape operator of pseudo-Riemannian submanifolds

A shape operator of a Riemannian submanifold is always diagonalizable, but this is not the case for a shape operator of a pseudo-Riemannian submanifolds, in particular, for a Lorentzian submanifold. In order to express shape operator of a pseudo-Riemannian submanifolds, we need the following algebraic lemma from [O'Neill (1983)].

Lemma 2.1. *A linear operator S on $V \approx \mathbb{E}^n_s$ is self-adjoint if and only if V can be expressed as a direct sum of subspaces V_k that are mutually orthogonal (hence nondegenerate) and S-invariant and each $S|_{V_k}$ has matrix of form* either

$$\begin{pmatrix} \lambda & & & & 0 \\ 1 & \lambda & & & \\ & \ddots & \ddots & & \\ & & 1 & \lambda & \\ 0 & & & 1 & \lambda \end{pmatrix}$$

relative to a basis $v_1, \ldots, v_r$ $(r \geq 1)$ with all scalar products zero except $\langle v_i, v_j \rangle = \epsilon = \pm 1$ if $i + j = r + 1$, or

$$\begin{pmatrix} a & b & & & & & & & \\ -b & a & & & & & & & \\ 1 & 0 & a & b & & & & 0 & \\ 0 & 1 & -b & a & & & & & \\ & & 1 & 0 & a & b & & & \\ & & 0 & 1 & -b & a & & & \\ & & & & & & \ddots & & \\ & 0 & & & & & 1 & 0 & a & b \\ & & & & & & 0 & 1 & -b & a \end{pmatrix}, \qquad b \neq 0,$$

relative to a basis $v_1, w_1, \ldots, v_m, w_m$ with all scalar products zero except $\langle v_i, v_j \rangle = 1 = -\langle w_i, w_j \rangle$ if $i + j = m + 1$. (Here r, ϵ and m depend on k.)

In particular, if V is Lorentzian, we have the following.

Lemma 2.2. *A self-adjoint linear operator S of an n-dimensional vector space V with a Lorentzian scalar product $\langle\ ,\ \rangle$ can be put into one of the following four forms :*

$$\text{I.}\ S \sim \begin{pmatrix} a_1 & & & 0 \\ & a_2 & & \\ & & \ddots & \\ 0 & & & a_n \end{pmatrix}, \qquad \text{II.}\ S \sim \begin{pmatrix} a_0 & 0 & & & 0 \\ 1 & a_0 & & & \\ & & a_3 & & \\ & & & \ddots & \\ 0 & & & & a_n \end{pmatrix},$$

$$\text{III. } S \sim \begin{pmatrix} a_0 & 0 & 0 & & & 0 \\ 0 & a_0 & 1 & & & \\ -1 & 0 & a_0 & & & \\ & & & a_4 & & \\ & & & & \ddots & \\ 0 & & & & & a_n \end{pmatrix}, \quad \text{IV. } S \sim \begin{pmatrix} a_0 & b_0 & & & 0 \\ -b_0 & a_0 & & & \\ & & a_3 & & \\ & & & \ddots & \\ 0 & & & & a_n \end{pmatrix},$$

where b_0 is assumed to be nonzero. In cases I *and* IV, *S is represented with respect to an orthonormal basis $\{v_1, \ldots, v_n\}$ satisfying $\langle v_1, v_1 \rangle = -1$, $\langle v_i, v_j \rangle = \delta_{ij}$, $\langle v_1, v_i \rangle = 0$ for $2 \le i, j \le n$, while in cases* II *and* III *the basis $\{v_1, \ldots, v_n\}$ is pseudo-orthonormal satisfying $\langle v_1, v_1 \rangle = 0 = \langle v_2, v_2 \rangle = \langle v_1, v_i \rangle = \langle v_2, v_i \rangle$, for $3 \le i \le n$, $\langle v_1, v_2 \rangle = -1$,* and *$\langle v_i, v_j \rangle = \delta_{ij}$ otherwise.*

2.6 Fundamental equations of Gauss, Codazzi and Ricci

Let $\phi : N \to M$ be an isometric immersion of a pseudo-Riemannian manifold N into a pseudo-Riemannian manifold M and let R be the curvature tensor of M. Then, for $X, Y, Z \in \mathcal{X}(N)$, we have

$$R(X,Y)Z = \nabla_X \nabla_Y Z - \nabla_Y \nabla_X Z - \nabla_{[X,Y]} Z. \tag{2.22}$$

Thus, by Gauss' formula, we find

$$\begin{aligned} R(X,Y)Z &= \nabla_X (\nabla'_X h(Y,Z) + h(Y,Z)) - \nabla_Y (\nabla'_X Z + h(X,Z)) \\ &\quad - \nabla'_{[X,Y]} Z - h([X,Y],Z) \\ &= R(X,Y)Z + h(X, \nabla'_Y Z) - h(Y, \nabla'_X Z) \\ &\quad - h([X,Y],Z) + \nabla_X h(Y,Z) - \nabla_Y h(X,Z). \end{aligned}$$

By using Weingarten's formula we find

$$\begin{aligned} R(X,Y)Z =& R'(X,Y)Z - A_{h(Y,Z)} X + A_{h(X,Z)} Y \\ &+ h(X, \nabla'_Y Z) - h(Y, \nabla'_X Z) - h([X,Y],Z) \\ &+ D_X h(Y,Z) - D_Y h(X,Z). \end{aligned} \tag{2.23}$$

Hence we obtain the following.

Theorem 2.3. *Let $\phi : N \to M$ be an isometric immersion of a pseudo-Riemannian manifold N into a pseudo-Riemannian manifold M. Then for vector fields X, Y, Z, W tangent to N, we have*

$$\begin{aligned} R'(X,Y;Z,W) = R(X,Y;Z,W) &+ \langle h(X,W), h(Y,Z) \rangle \\ &- \langle h(X,Z), h(Y,W) \rangle, \end{aligned} \tag{2.24}$$

$$(R(X,Y)Z)^{\perp} = (\bar{\nabla}_X h)(Y,Z) - (\bar{\nabla}_Y h)(X,Z), \tag{2.25}$$

where $R'(X,Y;Z,W) = \langle R'(X,Y)Z,W\rangle$, R' *the curvature tensor of* N, $(R(X,Y)Z)^{\perp}$ *is the normal component of* $R(X,Y)Z$, *and* $\bar{\nabla}h$ *denotes the covariant derivative of* h *with respect to the van der Waerden-Bortolotti connection* $\bar{\nabla} = \nabla \oplus D$, *i.e.*,

$$(\bar{\nabla}_X h)(Y,Z) = D_X h(Y,Z) - h(\nabla'_X Y, Z) - h(Y, \nabla'_X Z). \tag{2.26}$$

Equations (2.24) and (2.25) are called the *equations of Gauss* and *equation of Codazzi*, respectively.

If ξ and η are normal vector fields of N, we have

$$\begin{aligned} R(X,Y;\xi,\eta) &= \langle \nabla_X\nabla_Y\xi,\eta\rangle - \langle \nabla_Y\nabla_X\xi,\eta\rangle - \langle \nabla_{[X,Y]}\xi,\eta\rangle \\ &= \langle \nabla_Y(A_\xi X),\eta\rangle - \langle \nabla_X(A_\xi Y),\eta\rangle + \langle \nabla_X D_Y\xi,\eta\rangle \\ &\quad - \langle \nabla_Y D_X\xi,\eta\rangle - \langle D_{[X,Y]}\xi,\eta\rangle \\ &= \langle h(Y,A_\xi X),\eta\rangle - \langle h(X,A_\xi Y),\eta\rangle + \langle D_X D_Y\xi,\eta\rangle \\ &\quad - \langle D_Y D_X\xi,\eta\rangle - \langle D_{[X,Y]}\xi,\eta\rangle . \end{aligned}$$

If R^D denotes the curvature tensor of the normal bundle $T^{\perp}N$, i.e.,

$$R^D(X,Y)\xi = D_X D_Y\xi - D_Y D_X\xi - D_{[X,Y]}\xi, \tag{2.27}$$

then

$$R^D(X,Y;\xi,\eta) = R(X,Y;\xi,\eta) + \langle [A_\xi, A_\eta](X), Y\rangle , \tag{2.28}$$

where $[A_\xi, A_\eta] = A_\xi A_\eta - A_\eta A_\xi$.

Equation (2.28) is called the *equation of Ricci.* The three equations of Gauss, Codazzi and Ricci are known as the *fundamental equations.*

Proposition 2.3. *Let* N^n_s *be a pseudo-Riemannian n-manifold with index s isometrically immersed in an indefinite real space form* $R^m_s(c)$ *of constant curvature c. Then the Ricci tensor of* N *satisfies*

$$\begin{aligned} Ric\,(Y,Z) =&(n-1)\,\langle Y,Z\rangle\, c + n\,\langle H, h(Y,Z)\rangle \\ &- \sum_i \epsilon_i\,\langle h(Y,e_i), h(Z,e_i)\rangle , \end{aligned}$$

where $\{e_1,\ldots,e_n\}$ *is an orthonormal frame of* N.

Proof. The equation of Gauss yields

$$\begin{aligned} Ric\,(Y,Z) =& \sum_{i=1}^n \epsilon_i R(e_i,Y;Z,e_i) + n\,\langle H, h(Y,Z)\rangle \\ &- \sum_i \epsilon_i\langle h(Y,e_i), h(Z,e_i)\rangle. \end{aligned}$$

Combining this with (1.14) gives the Proposition. □

An immediate consequence of Proposition 2.3 is the following.

Corollary 2.1. *If N^n_s is a minimal submanifold of $\mathbb{E}^m_s$, then*

$$Ric \leq 0, \tag{2.29}$$

with the equality holding identically if and only if N^n_s is totally geodesic.

Proposition 2.4. *Let N^n_t be an n-dimensional pseudo-Riemannian submanifold of an indefinite real space form $R^m_s(c)$. Then the scalar curvature, the mean curvature vector, and the second fundamental form of N^n_t satisfy*

$$\tau = \frac{n^2}{2}\langle H, H\rangle - \frac{1}{2}S_h + \frac{n(n-1)}{2}c, \tag{2.30}$$

where S_h is defined as

$$S_h = \sum_{i,j=1}^{n} \epsilon_i\epsilon_j \langle h(e_i,e_j), h(e_i,e_j)\rangle, \quad \epsilon_i = \langle e_i, e_i\rangle, \tag{2.31}$$

and $e_1, \ldots, e_n$ is an orthonormal frame of N^n_t.

Proof. Let $e_1, \ldots, e_n$ be an orthonormal frame of N^n_t. Then the equation of Gauss gives

$$\begin{aligned} \sum_{i,j=1}^{n} \epsilon_i\epsilon_j \langle R'(e_i,e_j)e_j, e_i\rangle &= \sum_{i,j=1}^{n} \epsilon_i\epsilon_j \langle R(e_i,e_j)e_j, e_i\rangle \\ &+ \sum_{i,j=1}^{n} \langle \epsilon_i h(e_i,e_i), \epsilon_j h(e_j,e_j)\rangle - \sum_{i,j=1}^{n} \epsilon_i\epsilon_j \langle h(e_i,e_j), h(e_i,e_j)\rangle. \end{aligned} \tag{2.32}$$

Since the sectional curvature K of N satisfies

$$K(e_i \wedge e_j) = \epsilon_i\epsilon_j \langle R'(e_i,e_j)e_j, e_i\rangle, \tag{2.33}$$

we find from (2.32) that

$$2\tau = \sum_{i,j=1}^{n} K(e_i \wedge e_j) = n(n-1)c + n^2\langle H, H\rangle - S_h,$$

which gives (2.30). □

An immediate consequence of Proposition 2.4 is the following.

Corollary 2.2. *If N^n_s is an n-dimensional minimal submanifold of an indefinite real space form $R^{n+r}_s(c)$, then*

$$2\tau \leq n(n-1)c. \tag{2.34}$$

Similarly, if N^n_s is a minimal submanifold in $R^{n+r}_{s+r}(c)$, then

$$2\tau \geq n(n-1)c. \tag{2.35}$$

Either equality holds identically if and only if N^n_s is totally geodesic.

Another application of Proposition 2.4 is the following.

Proposition 2.5. *Let N be a submanifold of a real space form $R^{n+r}(c)$ of constant curvature c. Then the scalar curvature of N satisfies*

$$\tau \leq \frac{n(n-1)}{2}(||H||^2 + c), \quad n = \dim N, \tag{2.36}$$

with the equality holding at a point p if and only if p is a totally umbilical point.

Proof. Choose an orthonormal basis $e_1, \ldots, e_n, e_{n+1}, \ldots, e_{n+r}$ at p such that e_{n+1} is parallel to the mean curvature vector and $e_1, \ldots, e_n$ diagonalize the shape operator $A_{n+1} = A_{e_{n+1}}$. Then we obtain from Proposition 2.4 that

$$n^2||H||^2 = 2\tau + \sum_{i=1}^{n} a_i^2 + \sum_{\alpha=n+2}^{r} \sum_{i,j=1}^{n} (h_{ij}^{\alpha})^2 - n(n-1)c, \tag{2.37}$$

where $a_1, \ldots, a_n$ are eigenvalues of A_{n+1}.

On the other hand, it follows from the Cauchy-Schwarz inequality that

$$\sum_{i=1}^{n} a_i^2 \geq n||H||^2, \tag{2.38}$$

with the equality holding if and only if $a_1 = a_2 = \cdots = a_n$.

Combining (2.37) and (2.38) gives

$$n(n-1)||H||^2 \geq 2\tau - n(n-1)c + \sum_{\alpha=n+2}^{r} \sum_{i,j=1}^{n} (h_{ij}^{\alpha})^2, \tag{2.39}$$

which implies inequality (2.36).

If the equality sign of (2.36) holds at a point $p \in N$, then it follows from (2.38) and (2.39) that $A_{n+2} = \cdots = A_{n+r} = 0$ and $a_1 = \ldots = a_n$. Therefore, p is a totally umbilical point.

The converse is trivial. □

Proposition 2.5 has some nice applications, for instance, it implies the following.

Corollary 2.3. *If the scalar curvature of an n-dimensional Riemannian submanifold N of $\mathbb{E}^m$ satisfies*

$$\tau \geq \frac{n(n-1)}{2}$$

at a point p, then every isometric immersion of N into any Euclidean space satisfies

$$||H||^2 \geq 1 \ \text{ at } p \tag{2.40}$$

regardless of codimension.

In particular, if $\tau = \frac{n(n-1)}{2}$ *on* N*, then* $||H||^2 \geq 1$ *holds identically on* N*, with the equality holding identically if and only if* N *is an open part of a standard unit hypersphere in a totally geodesic* $\mathbb{E}^{n+1} \subset \mathbb{E}^m$.

Proof. Inequality (2.40) follows immediately from Proposition 2.5. If the scalar curvature of N is $\frac{n(n-1)}{2}$ at each point and if $||H|| = 1$ holds identically on N, then N is totally umbilical. Thus, N is an open portion of an ordinary unit n-sphere. □

Similar to Proposition 2.5, we also have the following.

Corollary 2.4. *Let* N *be a spacelike submanifold of an indefinite real space form* $R_r^{n+r}(c)$ *of constant curvature* c*. Then*

$$||H||^2 \leq \frac{2\tau}{n(n-1)} - c, \quad n = \dim N, \tag{2.41}$$

with equality holding at a point $p \in N$ *if and only if* p *is a totally umbilical point.*

For further applications of Proposition 2.5, see [Chen (1996a)].

2.7 Fundamental theorems of submanifolds

Now, we can state the fundamental theorems of submanifolds as follows. For the proofs see, for instance, [Bishop and Crittenden (1964); Eschenburg and Tribuzy (1993); Wettstein (1978)].

Theorem 2.4. (Existence) *Let* (N_t^n, g) *be a simply-connected pseudo-Riemannian* n*-manifold with index* t*. Suppose that there exists an* $(m-n)$*-dimensional pseudo-Riemannian vector bundle* $\nu(N_t^n)$ *with index* $s-t$ *over* N_t^n *and with curvature tensor* R^D *and also exists a* $\nu(N_t^n)$*-valued symmetric* $(0,2)$ *tensor* h *on* N_t^n*. For a cross section* ξ *of* $\nu(N_t^n)$*, define* A_ξ *by* $g(A_\xi X, Y) = \langle h(X,Y), \xi \rangle$*, where* $\langle\ ,\ \rangle$ *is the fiber metric of* $\nu(N_t^n)$*. If they satisfy (2.24), (2.25) and (2.28), then* N_t^n *can be isometrically immersed in an* m*-dimensional indefinite real space form* $R_s^m(c)$ *of constant curvature* c *in such way that* $\nu(N_t^n)$ *is the normal bundle and* h *is the second fundamental form.*

Theorem 2.5. (Uniqueness) *Let $\phi, \phi' : N_t^n \to R_s^m(c)$ be two isometric immersions of a pseudo-Riemannian n-manifold N_t^n into an indefinite space for $R_s^m(c)$ of constant curvature c with normal bundles ν and ν' equipped with their canonical bundle metrics, connections and second fundamental forms, respectively. Suppose there is an isometry $\phi : N_t^n \to N_t^n$ such that ϕ can be covered by a bundle map $\bar{\phi} : \nu \to \nu'$ which preserves the bundle metrics, the connections and the second fundamental forms. Then there is an isometry Φ of $R_s^m(c)$ such that $\Phi \circ \phi = \phi'$.*

Two submanifolds N_1 and N_2 of a pseudo-Riemannian manifold M are said to be *congruent* if there is an isometry of M which carries one to the other. Congruent submanifolds have the same intrinsic and extrinsic geometry.

2.8 A reduction theorem of Erbacher–Magid

Let $\mathbf{R}_{i,j}^n$ denote the affine n-space equipped with the metric whose canonical form is

$$\begin{pmatrix} O_j & & \\ & -I_i & \\ & & I_{n-i-j} \end{pmatrix},$$

where I_k is the $k \times k$ identity matrix and O_j is the $j \times j$ zero matrix.

The metric is non-degenerate if and only if $j = 0$. The j in $\mathbf{R}_{i,j}^n$ measures the degenerate part. The metric of $\mathbf{R}_{i,1}^n = \mathbf{R}_0 \times \mathbb{E}_i^{n-1}$ vanishes on the first factor $\mathbf{R}_0$ and it is the standard pseudo-Euclidean metric with index i on the second factor $\mathbb{E}_i^{n-1}$.

Denote the natural embedding $\iota : \mathbf{R}_{i,1}^n \to \mathbb{E}_{i+1}^{n+1}$ of $\mathbf{R}_{i,1}^n$ into $\mathbb{E}_{i+1}^{n+1}$ given by

$$\iota((x_1, x_2, \ldots, x_n)) = (x_1, x_2, \ldots, x_n, x_1) \in \mathbb{E}_{i+1}^{n+1} \tag{2.42}$$

for $(x_1, \ldots, x_n) \in \mathbf{R}_{i,1}^n$. Then the light-like vector $\zeta_0 = (1, 0, \ldots, 0, 1)$ is a normal vector of $\mathbf{R}_{i,1}^n$ in $\mathbb{E}_{i+1}^{n+1}$.

Let $\phi : N \to M$ be an isometric immersion of a pseudo-Riemannian manifold into another pseudo-Riemannian manifold. At each point $p \in N$, the *first normal space* $N^1(p)$ is defined to be the orthogonal complement of $N^0(p) = \{\xi \in T_p^\perp(N) : A_\xi = 0\}$.

Definition 2.5. Let $\phi : N \to M$ be an isometric immersion of a pseudo-Riemannian manifold into another pseudo-Riemannian manifold. The first normal spaces are called *parallel* if, for any curve σ joining any two points $p, q \in N$, the parallel displacement of normal vectors along σ with respect to the normal connection maps $N^1(p)$ onto $N^1(q)$.

The following result is the reduction theorem of Erbacher-Magid (see [Erbacher (1971); Magid (1984)]).

Theorem 2.6. *Let $\phi : N_i^n \to \mathbb{E}_s^m$ be an isometric immersion of a pseudo-Riemannian n-manifold N_i^n with index i into $\mathbb{E}_s^m$. If the first normal spaces are parallel, then there exists a complete $(n+k)$-dimensional totally geodesic submanifold E^* such that $\psi(N_i^n) \subset E^*$, where k is the dimension of the first normal spaces.*

Proof. Under the hypothesis, the dimension of N^1 is a constant, say k. If ξ is a normal vector field such that $\xi \in N^1(p)$ for each $p \in N_i^n$, then $D_X\xi \in N^1(p)$ for all $X \in T_pN_i^n$. Thus, the first normal spaces $N^1(p)$ form a parallel normal subbundle. Since D is a metric connection, the subspaces $N^0(p)$ are also parallel with respect to the normal connection.

Let p_0 be a point of N_i^n. Consider the $(n+k)$-dimensional subspace E of $\mathbb{E}_s^m$ through $\phi(p_0)$ which is perpendicular to $N^0(p_0)$, i.e.,

$$E = T_{p_0}(N_i^n) \oplus N^1(p_0).$$

Then the degenerate part of E is $N^0(p_0) \cap N^1(p_0)$. Now, we claim that $\phi(N_i^n) \subset E$. This can be proved as follows: Let $\beta(t)$ be any curve in N_i^n starting at p_0. For any $\xi_0 \in N^0(p_0)$, let ξ_t be the parallel displacement of ξ_0 along $\beta(t)$, so that $\xi_t \in N^0(\beta(t))$. For the pseudo-Euclidean connection ∇, we have

$$\nabla_{\beta'(t)}\xi_t = -d\phi(A_{\xi_t}(\beta'(t))) + D_{\beta'(t)}\xi_t = 0,$$

which means that ξ_t is parallel in $\mathbb{E}_s^m$. Thus it is a constant vector. Now, we have

$$\frac{d}{dt}\langle \phi(\beta(t)) - \phi(p_0), \xi_0\rangle = \langle d\phi(\beta'(t)), \xi_0\rangle = \langle d\phi(\beta'(t)), \xi_t\rangle = 0.$$

Thus $\phi(\beta(t))$ lies in E. Since this is true for arbitrary curve $\beta(t)$ in N_i^n, $\phi(N_i^n) \subset E$. □

In Erbacher-Magid's reduction theorem, $E^* = \mathbf{R}_{s,t}^{n+k}$ for some s, t and t need not be zero.

Definition 2.6. A pseudo-Riemannian submanifold N of a pseudo-Riemannian manifold is called *parallel* if $\bar{\nabla} h = 0$ identically.

The following is an easy consequence of the Reduction Theorem.

Corollary 2.5. *Let $\phi : N \to \mathbb{E}^m_s$ be an isometric immersion of a pseudo-Riemannian n-manifold N into pseudo-Euclidean m-space $\mathbb{E}^m_s$. If ϕ is a parallel immersion, then there exists a complete $(n+k)$-dimensional totally geodesic submanifold $E^* \subset \mathbb{E}^m_s$ such that $\phi(N) \subset E^*$, where k is the dimension of the first normal spaces.*

Proof. If ϕ is parallel, then $\bar{\nabla} h = 0$. Thus, it follows from (2.26) that

$$D_X h(Y, Z) = h(\nabla'_X Y, Z) + h(Y, \nabla'_X Z)$$

for $X, Y, Z \in \mathcal{X}(N)$. Hence, the first normal spaces are parallel. Therefore, the corollary follows from the Reduction Theorem. □

2.9 Two basic formulas for submanifolds in $\mathbb{E}^m_s$

The following *formula of Beltrami* for pseudo-Riemannian submanifolds in the pseudo-Euclidean space $\mathbb{E}^m_s$ is fundamental.

Proposition 2.6. *Let $\mathbf{x} : N \to \mathbb{E}^m_s$ be an isometric immersion of a pseudo-Riemannian n-manifold N into a pseudo-Euclidean space. Then*

$$\Delta \mathbf{x} = -nH, \tag{2.43}$$

where H is the mean curvature vector of the immersion.

Proof. Let v be any given vector in $\mathbb{E}^m_s$ and $p \in N$. If $\{e_1, \ldots, e_n\}$ is an orthonormal basis of T_pN, then we may extend $e_1, \ldots, e_n$ to an orthonormal frame $E_1, \ldots, E_n$ such that $\nabla'_{E_i} E_j = 0$ at p for $i, j = 1, \ldots, n$, where ∇' is the Levi-Civita connection of N. Then we have

$$\begin{aligned}(\Delta \langle \mathbf{x}, v\rangle)_p &= -\sum_{i=1}^n \epsilon_i e_i \langle E_i, v\rangle = -\sum_{i=1}^n \epsilon_i \langle \nabla_{e_i} E_i, v\rangle \\ &= -\sum_{i=1}^n \epsilon_i \langle h(e_i, e_i), v\rangle = -n \langle H, v\rangle (p).\end{aligned} \tag{2.44}$$

Since both $\Delta \mathbf{x}$ and H are independent of the choice of the local basis, we have $\langle \Delta \mathbf{x}, v\rangle = -n \langle H, v\rangle$ for any v. Because the scalar product $\langle\ ,\ \rangle$ is nondegenerate, (2.44) implies (2.43). □

An immediate consequence of proposition 2.6 is the following.

Corollary 2.6. *Every spacelike minimal submanifold N in a pseudo-Euclidean space $\mathbb{E}^m_s$ is non-compact.*

Proof. If N is a minimal submanifold, it follows from Proposition 2.6 that each natural coordinate function of $\mathbb{E}^m_s$, restricted to N, is a harmonic function. Thus, when N is spacelike and compact, each such function is constant. Thus, N must be a point, which is a contradiction since each submanifold is assumed to be of dimension at least one. □

Another easy application of proposition 2.6 is the following.

Corollary 2.7. *Every spacelike minimal submanifold N in a pseudo-hyperbolic space $H^m_s(-1)$ is non-compact.*

Proof. Let $\phi : N \to H^m_s(-1)$ be a minimal isometric immersion of a compact Riemannian n-manifold N into $H^m_s(-1)$. Without loss of generality we may regard $H^m_s(-1)$ as a hypersurface of $\mathbb{E}^{m+1}_{s+1}$ via (1.17). Then the mean curvature vector $\tilde{H}$ of N in $\mathbb{E}^{m+1}_{s+1}$ is $\mathbf{x} = \iota \circ \phi$. Thus, Proposition 2.6 yields $\Delta\mathbf{x} = -n\mathbf{x}$. Hence, the Laplacian Δ has an eigenfunction with eigenvalue equal to $-n$. But this is impossible since $n > 0$. □

The following formula was obtained in [Chen (1979b, 1986)].

Proposition 2.7. *Let $\mathbf{x} : N \to \mathbb{E}^m_s$ be an isometric immersion of a pseudo-Riemannian n-manifold N into a pseudo-Euclidean space. Then*

$$\Delta H = \Delta^D H + \sum_{i=1}^{n} \epsilon_i h(A_H e_i, e_i) + (\Delta H)^T, \tag{2.45}$$

where

$$(\Delta H)^T = \frac{n}{2}\nabla\langle H, H\rangle + 2\,\mathrm{trace}\, A_{DH} \tag{2.46}$$

is the tangential component of ΔH, Δ^D is the Laplacian associated with the normal connection D and $e_1, \ldots, e_n$ is an orthonormal tangent frame.

Proof. Let v be a given constant vector field in $\mathbb{E}^m_s$ and let $e_1, \ldots, e_n$ be an orthonormal basis of T_pN, $p \in N$. We may extend $e_1, \ldots, e_n$ to an orthonormal frame $E_1, \ldots, E_n$ such that $\nabla'_{E_i} E_j = 0$ at p for $i, j = 1, \ldots, n$.

For any vector fields $X, Y \in \mathcal{X}(N)$, it follows from Weingarten formula that

$$X\langle H, v\rangle = \langle D_X H, v\rangle - \langle A_H X, v\rangle. \tag{2.47}$$

Thus

$$\begin{aligned} YX\langle H, v\rangle = &\langle D_Y D_X H, v\rangle - \langle \nabla'_Y(A_H X), v\rangle \\ &- \langle A_{D_X H} Y, v\rangle - \langle h(A_H X, Y), v\rangle. \end{aligned} \tag{2.48}$$

Therefore, it follows from the definition of Δ that

$$\Delta H = \Delta^D H + \sum_{i=1}^{n} \epsilon_i \{h(A_H e_i, e_i) + (\nabla'_{e_i} A_H) e_i + A_{D_{e_i} H} e_i\}. \tag{2.49}$$

Put

$$\text{trace}(A_{DH}) = \sum_i \epsilon_i A_{D_{e_i} H} e_i, \quad \text{trace}(\nabla A_H) = \sum_i \epsilon_i (\nabla'_{e_i} A_H) e_i. \tag{2.50}$$

Then

$$\begin{aligned}\langle \text{trace}(\nabla' A_H), X\rangle &= \sum_i \epsilon_i \left\langle (\nabla'_{e_i} A_H) e_i, X \right\rangle \\ &= \sum_i \epsilon_i \{\langle (\nabla'_X A) e_i, e_i\rangle - \langle A_{D_X H} e_i, e_i\rangle + \left\langle A_{D_{e_i} H} e_i, X \right\rangle\} \\ &= \sum_i \{\left\langle (\bar{\nabla}_X h)(e_i, e_i), H\right\rangle + \left\langle A_{D_{e_i} H} e_i, H\right\rangle\} \\ &= \sum_i \{\langle D_X h(e_i, e_i), H\rangle + \left\langle A_{D_{e_i} H} e_i, H\right\rangle\} \\ &= n \langle D_X H, H\rangle + \langle \text{trace}(A_{DH}), X\rangle\} \\ &= \frac{n}{2} X\langle H, H\rangle + \langle \text{trace}(A_{DH}), X\rangle,\end{aligned}$$

which implies that

$$\text{trace}(\nabla' A_H) = \frac{n}{2} \nabla\langle H, H\rangle + \text{trace}(A_{DH}). \tag{2.51}$$

From (2.49), (2.50) and (2.51) we obtain (2.45). □

An immediate consequence of Proposition 2.7 is the following.

Corollary 2.8. *Let N be a pseudo-Riemannian submanifold of a pseudo-Euclidean space. If N has parallel mean curvature vector, then $(\Delta H)^T = 0$ and*

$$\Delta H = \sum_{i=1}^{n} \epsilon_i h(A_H e_i, e_i), \tag{2.52}$$

where $e_1, \ldots, e_n$ is an orthonormal frame of TN.

According to Proposition 2.6 a pseudo-Riemannian submanifold N in $\mathbb{E}^m_s$ is minimal if and only if $\Delta \mathbf{x} = 0$, holds identically.

Definition 2.7. Let $\mathbf{x} : N \to \mathbb{E}^m_s$ be an isometric immersion of a pseudo-Riemannian n-manifold N into a pseudo-Euclidean space. Then N is called a *biharmonic submanifold* if and only if

$$\Delta H = 0 \ \text{ (or equivalently, } \Delta^2 \mathbf{x} = 0) \tag{2.53}$$

holds identically.

The following two corollaries follow easily from Proposition 2.7.

Corollary 2.9. *Let* $\mathbf{x} : N \to \mathbb{E}^m_s$ *be an isometric immersion of a pseudo-Riemannian n-manifold N into a pseudo-Euclidean space* $\mathbb{E}^m_s$*. Then N is biharmonic if and only if it satisfies the following fourth order strongly elliptic semi-linear PDE system:*

$$\Delta^D H + \sum_{i=1}^{n} \epsilon_i h(A_H e_i, e_i) = 0, \tag{2.54}$$

$$n\,\nabla\langle H, H\rangle + 4\,\mathrm{trace}\, A_{DH} = 0. \tag{2.55}$$

Corollary 2.10. *Every biharmonic hypersurface of* $\mathbb{E}^{n+1}$ *with constant mean curvature is minimal.*

Proof. Under the hypothesis, we have $DH = 0$. Thus, (2.54) implies that $\mathrm{trace}(A_H^2) = 0$. Hence, N must be minimal in $\mathbb{E}^{n+1}$. □

Both formulas (2.43) and (2.45) have many other nice applications, in particular, to submanifolds of finite type (see Chapter 7).

2.10 Relationship between squared mean curvature and Ricci curvature

The *relative null subspace* $\mathcal{N}_p$ of a pseudo-Riemannian submanifold N of another pseudo-Riemannian manifold at a point $p \in N$ is defined by

$$\mathcal{N}_p = \{X \in T_pN : h(X, Y) = 0\; \forall Y \in T_pN\},$$

where h is the second fundamental form of N in M. The dimension ν_p of $\mathcal{N}_p$ is called the *relative nullity* at p.

Let T^1N denote the unit tangent bundle of N. The following result provides an optimal relationship between the squared mean curvature and Ricci curvature of N.

Theorem 2.7. *Let N be an n-dimensional spacelike submanifold of an indefinite real space form* $R^{n+r}_r(c)$*. Then*

(1) *for any* $u \in T^1_pN$, $p \in N$, *we have*

$$\langle H, H\rangle \le \frac{4}{n^2}\{Ric(u) - (n-1)c\} \quad \text{at } p; \tag{2.56}$$

(2) *if* $H(p) = 0$*, then a unit tangent vector* $u \in T^1_pN$ *satisfies the equality case of* (2.56) *if and only if u lies in the relative null subspace* $\mathcal{N}_p$*;*

(3) *the equality case of* (2.56) *holds identically for all unit vectors in* T_p^1N *if and only if either* p *is a totally geodesic point or* $n = 2$ *and* p *is a totally umbilical point.*

Proof. Under the hypothesis, let u be a unit vector in T_p^1N. We choose an orthonormal basis $e_1, \ldots, e_n, e_{n+1}, \ldots, e_{n+r}$ at p such that $e_1, \ldots, e_n$ are tangent to N. Then we find from Proposition 2.4 that

$$n^2 \langle H, H\rangle = 2\tau + S_h - n(n-1)c, \tag{2.57}$$

where S_h is defined by (2.31). If we put $h_{ij}^{\alpha} = \langle h(e_i, e_j), e_\alpha\rangle$, then we find from (2.57) that

$$\begin{aligned}
n^2\langle H, H\rangle &= -\sum_{\alpha=n+1}^{r} \Big\{(h_{11}^{\alpha})^2 + (h_{22}^{\alpha} + \cdots + h_{nn}^{\alpha})^2 + 2\sum_{i<j}(h_{ij}^{\alpha})^2\Big\} \\
&\quad + 2\tau + 2\sum_{\alpha=n+1}^{r}\sum_{2\le i<j\le n} h_{ii}^{\alpha}h_{jj}^{\alpha} - n(n-1)c \\
&= 2\tau - \frac{1}{2}\sum_{\alpha=n+1}^{r}\{(h_{11}^{\alpha} + \cdots + h_{nn}^{\alpha})^2 + (h_{11}^{\alpha} - h_{22}^{\alpha} - \cdots - h_{nn}^{\alpha})^2\} \\
&\quad - 2\sum_{\alpha=n+1}^{r}\sum_{i<j}(h_{ij}^{\alpha})^2 + 2\sum_{\alpha=n+1}^{r}\sum_{2\le i<j\le n} h_{ii}^{\alpha}h_{jj}^{\alpha} - n(n-1)c \\
&\le 2Ric(e_1) + \frac{n^2}{2}\langle H, H\rangle - 2\sum_{r=n+1}^{m}\sum_{j=2}^{n}(h_{1j}^{r})^2 - 2(n-1)c \\
&\le 2Ric(e_1) + \frac{n^2}{2}\langle H, H\rangle - 2(n-1)c.
\end{aligned} \tag{2.58}$$

Since e_1 can be chosen to be any arbitrary unit tangent vector at p, (2.58) yields (2.56). This gives (1).

It follows from (2.58) that the equality case of (2.56) holds at p if and only if

$$\begin{aligned}
&h_{12}^{\alpha} = \cdots = h_{1n}^{\alpha} = 0 \\
&h_{11}^{\alpha} = h_{22}^{\alpha} + \cdots + h_{nn}^{\alpha}, \quad \alpha = n+1, \ldots, r.
\end{aligned} \tag{2.59}$$

If $H(p) = 0$, (2.59) implies that e_1 lies in the relative null space N_p. Conversely, if e_1 lies in the relative null space, (2.59) holds automatically due to $H(p) = 0$. This proves (2).

Now, assume that the equality case of (2.56) holds identically for all unit tangent vectors at p. Then, for any $\alpha = n+1, \ldots, r$, we have

$$h_{ij}^{\alpha} = 0, \quad i \neq j, \tag{2.60}$$

$$h_{11}^{\alpha} + \cdots + h_{nn}^{\alpha} = 2h_{ii}^{\alpha}, \quad i = 1, \ldots, n. \tag{2.61}$$

Conditions (2.60) and (2.61) imply that either p is a totally geodesic point or $n = 2$ and p is a totally umbilical point. The converse of this is trivial. Thus we have (3). □

Theorem 2.7 implies immediately the following.

Corollary 2.11. *Let N be a spacelike submanifold of $\mathbb{E}^{n+r}_r$. Then*

$$\langle H, H\rangle \leq \left(\frac{4}{n^2}\right) \min_u Ric(u),$$

where $n = \dim N$ and u runs over all unit tangent vector.

Similarly, we also have the following [Chen (1999a)].

Theorem 2.8. *Let N be an n-dimensional submanifold of a real space form $R^{n+r}(c)$. Then*

(1) *for any $u \in T^1_pN$, $p \in N$, we have*

$$||H||^2 \geq \frac{4}{n^2}\{Ric(X) - (n-1)c\} \quad \text{at } p; \tag{2.62}$$

(2) *if $H(p) = 0$, then a unit tangent vector $u \in T_pN$ satisfies the equality case of (2.62) if and only if $u \in \mathcal{N}_p$;*
(3) *the equality case of (2.62) holds identically for all unit vectors in T_pN if and only if either p is a totally geodesic point or $n = 2$ and p is a totally umbilical point.*

Corollary 2.12. *Let N be a submanifold of $\mathbb{E}^{n+r}$. Then*

$$||H||^2 \geq \left(\frac{4}{n^2}\right) \max_u Ric(u), \tag{2.63}$$

where $n = \dim N$ and u runs over all unit tangent vectors at p.

Remark 2.1. There exist many examples of submanifolds in a Euclidean m-space which satisfy the equality case of (2.63) identically. Two simple such examples are the spherical hypercylinder $S^2(c) \times \mathbb{R}$ and the round hypercone in $\mathbb{E}^4$.

Let N be a pseudo-Riemannian n-manifold and $p \in N$. By an ℓ-plane section we mean an ℓ-dimensional linear subspace of T_pN. An ℓ-plane section L^ℓ is called *nondegenerate* if each 2-dimensional subspace of L^ℓ is nondegenerate. A nondegenerate ℓ-plane section admits a nondegenerate scalar product induced from the pseudo-Riemannian metric of N.

For each unit vector X in a nondegenerate ℓ-plane section L^ℓ, we may choose an orthonormal basis $e_1, \ldots, e_\ell$ of L^ℓ such that $e_1 = X$. Define the *Ricci curvature* Ric_{L^ℓ} of L^ℓ at X by

$$Ric_{L^\ell}(X) = K_{12} + \cdots + K_{1\ell}, \tag{2.64}$$

where $K_{ij} = K(e_i \wedge e_j)$. We call such a curvature an *ℓ-Ricci curvature*. The *scalar curvature* $\tau(L^\ell)$ of the ℓ-plane section is defined by

$$\tau(L^\ell) = \sum_{1 \le i < j \le \ell} K_{ij}. \tag{2.65}$$

Remark 2.2. In general, given an integer k, $2 \le k \le n-1$, there *does not* exist a positive constant, say $C(n,k)$, such that

$$||H||^2(p) \ge C(n,k) \max_{L^k, X} Ric_{L^k}(X), \tag{2.66}$$

where L^k runs over all k-plane sections in T_pM^n and X runs over all unit tangent vectors in L^k. This can be seen from the following example.

Example 2.1. Let $\phi : B \to \mathbb{E}^4$ be a minimal hypersurface whose shape operator is non-singular at a point $p \in B$. Then by the minimality there exist two principal directions at p, say e_1, e_2, such that their corresponding principal curvatures κ_1, κ_2 are of the same sign. Here, by a *principal curvature* we mean an eigenvalue of the shape operator of B. This implies that the sectional curvature $K_{12} := K(e_1, e_2)$ at p is positive. Now, consider the minimal hypersurface in $\mathbb{E}^{n+1}$ which is given by the product of $\phi : B \to \mathbb{E}^4$ and the identity map $\iota : \mathbb{E}^{n-3} \to \mathbb{E}^{n-3}$.

For any integer k, $2 \le k \le n-1$, the maximum value of the k-th Ricci curvatures of $N = B \times \mathbb{E}^{n-3}$ at a point (p, q), $q \in \mathbb{E}^{n-3}$ is given by $K_{12} = \kappa_1\kappa_2$ which is positive. Since $H = 0$, this shows that there does not exist any positive constant $C(n,k)$ which satisfies (2.66).

2.11 Relationship between shape operator and Ricci curvature

We define an invariant θ_ℓ on a pseudo-Riemannian n-manifold N by

$$\theta_\ell(p) = \left(\frac{1}{\ell - 1}\right) \inf_{L^\ell, X} Ric_{L^\ell}(X), \quad p \in N, \tag{2.67}$$

where L^ℓ runs over all nondegenerate ℓ-plane sections at p and X runs over all unit vectors in L^ℓ.

The following theorem provides a sharp relationship between θ_ℓ and the shape operator A_H [Chen (1996a, 1999a)].

Theorem 2.9. *Let N be a submanifold of a real space form $R^{n+r}(c)$ of constant curvature c. For an integer ℓ, $2 \leq \ell \leq n = \dim N$, we have:*

(1) *if $\theta_\ell(p) \neq c$ at a point $p \in N$, then the shape operator satisfies*

$$A_H > \frac{n-1}{n}(\theta_\ell - c)I \ \ at\ p, \tag{2.68}$$

where I denotes the identity map of T_pN;

(2) *if $\theta_\ell(p) = c$, then $A_H \geq 0$ at p, i.e., A_H is nonnegative-definite at p;*

(3) *a unit vector $u \in T_p^1N$ satisfies*

$$A_H u = \frac{n-1}{n}(\theta_\ell(p) - c)u$$

if and only if $\theta_\ell(p) = c$ and u lies in the relative null space at p;

(4) *$A_H \equiv \frac{n-1}{n}(\theta_\ell - c)I$ at p if and only if p is a totally geodesic point, i.e., the second fundamental form vanishes identically at p.*

Proof. Let $\{e_1, \ldots, e_n\}$ be an orthonormal basis of T_pN. Denote by $L_{i_1 \cdots i_\ell}$ the ℓ-plane section spanned by $e_{i_1}, \ldots, e_{i_\ell}$. It follows from (2.64) and (2.65) that

$$\tau(L_{i_1 \cdots i_\ell}) = \frac{1}{2} \sum_{i \in \{i_1, \cdots, i_\ell\}} Ric_{L_{i_1 \cdots i_\ell}}(e_i), \tag{2.69}$$

$$\tau(p) = \frac{(\ell-2)!(n-\ell)!}{(n-2)!} \sum_{1 \leq i_1 < \cdots < i_\ell \leq n} \tau(L_{i_1 \cdots i_\ell}). \tag{2.70}$$

By combining (2.67), (2.69) and (2.70), we find

$$\tau(p) \geq \frac{n(n-1)}{2} \theta_\ell(p). \tag{2.71}$$

On the other hand, from Proposition 2.5, we have

$$||H(p)||^2 \geq \frac{2}{n(n-1)} \tau(p) - c. \tag{2.72}$$

From (2.71) and (2.72), we obtain $||H||^2(p) \geq \theta_\ell - c$. This shows that $H(p) = 0$ may occurs only when $\theta_\ell(p) \leq c$. Consequently, if $H(p) = 0$, (1) and (2) hold automatically. Therefore, without loss of generality, we may assume $H(p) \neq 0$.

Let us choose an orthonormal basis $e_1, \dots, e_n, e_{n+1}, \dots, e_{n+r}$ at p such that e_{n+1} is parallel to the mean curvature vector $H(p)$ and $e_1, \dots, e_n$ diagonalize the shape operator A_H. Then we have

$$A_{n+1} = \begin{pmatrix} a_1 & 0 & \dots & 0 \\ 0 & a_2 & \dots & 0 \\ \vdots & \vdots & \ddots & \vdots \\ 0 & 0 & \dots & a_n \end{pmatrix}, \tag{2.73}$$

$$A_\alpha = (h^\alpha_{ij}), \ \sum_{i=1}^n h^\alpha_{ii} = 0, \ \ \alpha = n+2, \dots, r. \tag{2.74}$$

From the equation of Gauss we get

$$a_i a_j = K_{ij} - c + \sum_{\alpha=n+2}^r (h^\alpha_{ij})^2 - \sum_{\alpha=n+2}^r h^\alpha_{ii} h^\alpha_{jj} \tag{2.75}$$

for $1 \le i \ne j \le n$. From (2.75) we obtain

$$\begin{gathered} a_1(a_{i_2} + \dots + a_{i_\ell}) = Ric_{L_{1i_2\cdots i_\ell}}(e_1) - (\ell - 1)c \\ + \sum_{\alpha=n+2}^r \sum_{j=2}^\ell (h^\alpha_{1i_j})^2 - \sum_{\alpha=n+2}^r \sum_{j=2}^\ell h^\alpha_{11} h^\alpha_{i_j i_j}, \\ 1 < i_2 < \dots < i_\ell, \end{gathered} \tag{2.76}$$

which yields

$$\begin{aligned} a_1(a_2 + \dots + a_n) = & \frac{1}{\binom{n-2}{\ell-2}} \sum_{2 \le i_2 < \dots < i_\ell \le n} Ric_{L_{1i_2\cdots i_\ell}}(e_1) \\ & - (n-1)c + \sum_{\alpha=n+2}^r \sum_{j=1}^n (h^\alpha_{1j})^2. \end{aligned} \tag{2.77}$$

Using (2.67) and (2.77) we find

$$a_1(a_2 + \dots + a_n) \ge (n-1)(\theta_\ell - c), \tag{2.78}$$

with the equality holding if and only if

$$\begin{gathered} Ric_L(e_1) = 0, \ h^\alpha_{1j} = 0, \\ \alpha = n+2, \dots, n+r; \ \ j = 2, \dots, n, \end{gathered} \tag{2.79}$$

for any ℓ-plane section L which contains e_1.

Inequality (2.78) implies that

$$\begin{aligned} a_1(a_1 + \dots + a_n) &\ge (n-1)(\theta_\ell - c) + a_1^2 \\ &\ge (n-1)(\theta_\ell - c). \end{aligned} \tag{2.80}$$

Similar inequalities hold when 1 were replaced by $j \in \{2, \ldots, n\}$. Hence, we have

$$a_j(a_1 + \cdots + a_n) \geq (n-1)(\theta_\ell - c) + a_j^2,$$
$$j = 1, \ldots, n,$$

which yields

$$A_H \geq \frac{n-1}{n}(\theta_\ell - c)I. \tag{2.81}$$

Now, suppose that the equality case of (2.81) is achieved for some unit vector $u \in T_pN$. Then at least one of the n eigenvalues of A_H is equal to $\frac{n-1}{n}(\theta_\ell - c)$. Without loss of generality, we may assume a_1 is such an eigenvalue. Then we have

$$a_1(a_1 + \cdots + a_n) = (n-1)(\theta_\ell - c). \tag{2.82}$$

On the other hand, from (2.80) and (2.82) we obtain $a_1 = 0$ and $\theta_\ell = c$. Moreover, in this case it follows from (2.79) that e_1 must lies in the relative null space N_p. Hence, (1) and (2) follow. Moreover, this also implies one part of (3). The remaining part of (3) is obvious.

Now, if

$$A_H = \frac{n-1}{n}(\theta_\ell - k)I$$

holds identically at p, then every tangent vector of N at p lies in the relative null space N_p, according to (3). Thus, p is a totally geodesic point.

Conversely, if p is a totally geodesic point, then $\theta_\ell = k$ and $A_H = 0$ which imply $A_H = \frac{n-1}{n}(\theta_\ell - c)I$ at p. Hence we obtain (4). □

One may apply Theorem 2.9 to obtain a lower bound of the eigenvalues of the shape operator A_H for an arbitrary isometric immersion of a given Riemannian n-manifold, regardless of codimension. For instance, Theorem 2.9 implies immediately the following.

Corollary 2.13. *If $\phi : N \to \mathbb{E}^m$ is an isometric immersion of an open portion of the unit n-sphere into the Euclidean m-space $\mathbb{E}^m$ with arbitrary codimension. Then every eigenvalue of A_H is greater than $(n-1)/n$.*

For an n-dimensional submanifold N in $\mathbb{E}^m$, let $\mathbb{E}^{n+1}$ be the linear subspace of dimension $n+1$ spanned by the tangent space at a point $p \in N$ and the mean curvature vector $H(p)$ at p.

Geometrically, the shape operator A_{n+1} of N in $\mathbb{E}^m$ at p is the shape operator of the orthogonal projection of N into $\mathbb{E}^{n+1}$. Moreover, it is known that if the shape operator of a hypersurface in $\mathbb{E}^{n+1}$ is definite at a point p, then it is strictly convex at p. For this reason a submanifold N in $\mathbb{E}^m$ is said to be *H-strictly convex* if the shape operator A_H is positive-definite at each point in N.

Theorem 2.9 implies immediately the following.

Corollary 2.14. *Let N be a Riemannian n-manifold whose ℓ-Ricci curvature is positive for some $\ell \in [2, n]$. Then every isometric immersion of N into any Euclidean space is H-strictly convex, regardless of codimension.*

In particular, every isometric immersion of a Riemannian manifold with positive Ricci curvature into any Euclidean space is H-strictly convex.

Remark 2.3. The estimate of A_H given in Theorem 2.9 is sharp. Here we provide an example to illustrate this fact.

Example 2.2. Consider a hyper-ellipsoid in $\mathbb{E}^{n+1}$ defined by

$$ax_1^2 + x_2^2 + \cdots + x_{n+1}^2 = 1, \tag{2.83}$$

where $0 < a < 1$. The principal curvatures $a_1, \ldots, a_n$ of the hyperellipsoid are given by

$$a_1 = \frac{a}{(1 + a(a-1)x_1^2)^{3/2}},$$
$$a_2 = \cdots = a_n = \frac{1}{(1 + a(a-1)x_1^2)^{1/2}}.$$

Thus, for any ℓ, $2 \leq \ell \leq n$, the ℓ-Ricci curvatures at a point p satisfies

$$Ric_{L^\ell}(u) \geq (\ell - 1)\theta_\ell := \frac{(\ell-1)a}{(1 + a(a-1)x_1^2)^2} > 0 \tag{2.84}$$

for each ℓ-plane section L^ℓ and unit vector u in L^ℓ and, moreover, the eigenvalues $\kappa_1, \ldots, \kappa_n$ of the shape operator A_H are given by

$$\begin{aligned} \kappa_1 = \cdots = \kappa_{n-1} &= \frac{a + (n-1)(1 + a(a-1)x_1^2)}{n(1 + a(a-1)x_1^2)^2}, \\ \kappa_n &= \frac{a(a + (n-1)(1 + a(a-1)x_1^2))}{n(1 + a(a-1)x_1^2)^3}. \end{aligned} \tag{2.85}$$

From (2.84) and (2.85) it follows that

$$A_H > \left(\frac{n-1}{n}\right)\theta_\ell I_n$$

and

$$\kappa_1 - \frac{n-1}{n}\theta_\ell = \frac{a^2}{n(1 + a(a-1)x_1^2)^3} \overset{a \to 0}{\longrightarrow} 0.$$

Consequently, the estimate of A_H in Theorem 2.9 is sharp.

2.12 Cartan's structure equations

Let N be a pseudo-Riemannian n-submanifold of a pseudo-Riemannian m-manifold M. Denote by ∇ the Levi-Civita connection of M. Choose a local orthonormal frame $e_1, \ldots, e_n, e_{n+1}, \ldots, e_m$ of N such that $e_1, \ldots, e_n$ are tangent to N and $e_{n+1}, \ldots, e_m$ are normal to N. Let $\omega^1, \ldots, \omega^n$ be the dual frame of $e_1, \ldots, e_n$, i.e., $\omega_i(e_j) = \delta_{ij}$.

We shall make use of the following convention on the ranges of indices unless mentioned otherwise:

$$1 \leq \alpha, \beta, \gamma, \ldots \leq m; \; 1 \leq i, j, k, \ell \leq n; \; n+1 \leq r, s, t, \ldots \leq m.$$

Put $\langle e_\alpha, e_\beta \rangle = \epsilon_\alpha \delta_{\alpha\beta}$; $\alpha, \beta = 1, \ldots, m$, and

$$\begin{aligned} \nabla e_j &= \sum_j \epsilon_k \omega_j^k e_k + \sum_r \epsilon_r \omega_j^r e_r, \\ \nabla e_r &= \sum_k \epsilon_k \omega_r^k e_k + \sum_s \epsilon_s \omega_r^s e_s. \end{aligned} \tag{2.86}$$

The 1-forms ω_α^β, $1 \leq \alpha, \beta \leq m$, are called the *connection forms*. From (2.86) we find

$$\omega_j^k = -\omega_k^j, \quad \omega_j^r = -\omega_r^j, \quad \omega_r^s = -\omega_s^r. \tag{2.87}$$

For the second fundamental form h of N, if we put

$$h(e_i, e_j) = \sum_r \epsilon_r h_{ij}^r e_r, \tag{2.88}$$

then it follows from (2.15) and (2.16) that

$$\omega_j^r(e_i) = -\langle \nabla_{e_i} e_r, e_j \rangle = h_{ij}^r.$$

Combining this with (2.87) gives

$$\omega_j^r = \sum_i h_{ij}^r \omega^i. \tag{2.89}$$

The *Cartan structure equations* are then given by

$$d\omega^i = -\sum_j \epsilon_j \omega_j^i \wedge \omega^j, \tag{2.90}$$

$$d\omega_j^i = \sum_{k,\ell,r} \epsilon_r (h_{ik}^r h_{j\ell}^r - h_{i\ell}^r h_{jk}^r) \omega^k \wedge \omega^\ell - \sum_k \epsilon_k \omega_k^i \wedge \omega_j^k + \Omega_j^i, \tag{2.91}$$

$$d\omega_i^r = -\sum_{j,k} \epsilon_j h_{jk}^r \omega^k \wedge \omega_i^j + \sum_{j,s} \epsilon_s h_{ij}^s \omega^j \wedge \omega_s^r + \Omega_i^r, \tag{2.92}$$

$$d\omega_s^r = \sum_{i,k,\ell} \epsilon_i (h_{ik}^r h_{i\ell}^s - h_{i\ell}^r h_{ik}^s) \omega^k \wedge \omega^\ell - \sum_t \epsilon_t \omega_t^r \wedge \omega_s^t + \Omega_s^r, \tag{2.93}$$

where

$$\Omega_\beta^\alpha = \frac{1}{2} \sum_{j,k} K_{\beta jk}^\alpha \omega^j \wedge \omega^k, \quad R(e_\alpha, e_\beta) e_\gamma = \sum_\delta \epsilon_\delta K_{\beta\gamma\delta}^\alpha e_\delta. \tag{2.94}$$

Those Ω_β^α are known as the *curvature 2-forms* of M, restricted to N.

Chapter 3

Special Pseudo-Riemannian Submanifolds

3.1 Totally geodesic submanifolds

The simplest submanifolds are totally geodesic submanifolds.

Proposition 3.1. *Let $\phi : N \to M$ be an isometric immersion of a pseudo-Riemannian manifold N into a pseudo-Riemannian manifold M. Then the following three statements are equivalent.*

(1) *N is a totally geodesic submanifold of M;*
(2) *Geodesics of N are geodesics of M;*
(3) *For any $p \in N$ and any $v \in T_pN$, the geodesic γ_v with $\gamma_v(0) = p$ and $\gamma_v'(0) = v$ lies locally in N.*

Proof. Since N is totally geodesic in M, the second fundamental form vanishes. Thus, if γ is a curve of N, then $\nabla_{\gamma'}\gamma' = \nabla'_{\gamma'}\gamma'$, where ∇ and ∇' are the Levi-Civita connections of M and N, respectively. Hence, γ is a geodesic of M if and only if it is geodesic of N. This proves the equivalence of (1) and (2).

Now, suppose that $\gamma : I \to N$ is the geodesic of N with initial velocity vector v, then γ is also a geodesic of M. Hence, by the uniqueness of geodesics, we obtain (3).

If (3) holds, Gauss' formula implies that $h(v, v) = 0$. Thus, for vectors $v, w \in T_pN$,

$$\begin{aligned} 0 &= h(v + w, v + w) = h(v, v) + 2h(v, w) + h(w, w) \\ &= 2h(v, w). \end{aligned} \tag{3.1}$$

Since this is true for any $p \in N$ and any $v, w \in T_pN$, we find $h = 0$ by polarization. Thus N is totally geodesic in M. □

Proposition 3.2. *Up to rigid motions, an n-dimensional totally geodesic pseudo-Riemannian submanifold of a pseudo-Euclidean space $\mathbb{E}^m_s$ is an open portion of a pseudo-Euclidean linear subspace $\mathbb{E}^n_t$ of $\mathbb{E}^m_s$.*

Proof. Obviously, every pseudo-Euclidean linear subspace $\mathbb{E}^n_t$ of $\mathbb{E}^m_s$ is a totally geodesic submanifold of $\mathbb{E}^m_s$.

Now, assume that N is an n-dimensional totally geodesic pseudo-Riemannian submanifold of $\mathbb{E}^m_s$ such that the origin o of $\mathbb{E}^m_s$ lies in N. Then $T_o(N)$ is a nondegenerate subspace of $T_o(\mathbb{E}^m_s)$. Because geodesics of $\mathbb{E}^m_s$ are lines, it follows from Proposition 3.1 that N is an open portion of the pseudo-Euclidean subspace whose tangent space at o is $T_o(N)$. □

Recall that the pseudo m-sphere $S^m_s(c)$ is defined by

$$S^m_s(c) = \left\{ \mathbf{x} = (x_1, \ldots, x_{m+1}) \in \mathbb{E}^{m+1}_s : \langle \mathbf{x}, \mathbf{x} \rangle = \frac{1}{c} > 0 \right\}.$$

For $n < m$ and $0 \leq t \leq s$,

$$\left\{ (x_1, \ldots, x_t, 0, \ldots, 0, x_{s+1}, \ldots, x_{n+1}, 0, \ldots, 0) \in S^m_s(c) \right\}.$$

defines a pseudo-Riemannian manifold of constant curvature c with index t, which is totally geodesic in $S^m_s(c)$. We call this totally geodesic submanifold of $S^m_s(c)$ a *pseudo n-subsphere* of $S^m_s(c)$.

Analogously, we call

$$\left\{ (x_1, \ldots, x_{t+1}, 0, \ldots, 0, x_{s+1}, \ldots, x_{n+1}, 0, \ldots, 0) \in H^m_s(c) \right\}.$$

a *pseudo-hyperbolic n-subspace* of $H^m_s(c)$.

Similar to Proposition 3.2, we have the following.

Proposition 3.3. *Up to rigid motions, an n-dimensional totally geodesic pseudo-Riemannian submanifold of a pseudo m-sphere $S^m_s(c)$ is an open portion of a pseudo n-subsphere of $S^m_s(c)$.*

Proposition 3.4. *Up to rigid motions, an n-dimensional totally geodesic pseudo-Riemannian submanifold of a pseudo-hyperbolic m-space $H^m_s(c)$ is an open portion of a pseudo-hyperbolic n-subspace of $H^m_s(c)$.*

Remark 3.1. The classification of totally geodesic submanifolds of symmetric spaces are rather complicated (cf. [Chen and Nagano (1977, 1978)]). Thus, the (M_+, M_-)-method for the classification of such submanifolds was introduced in [Chen and Nagano (1978)] (see also, [Chen (1987c)]).

3.2 Parallel submanifolds of (indefinite) real space forms

Clearly, totally geodesic submanifolds are parallel submanifolds. All parallel submanifolds of Euclidean spaces are classified in [Ferus (1974)]. Affine subspaces of $\mathbb{E}^m$ and symmetric R-spaces minimally embedded in a hypersphere of $\mathbb{E}^m$ as described in [Takeuchi and Kobayahi (1968)] are parallel submanifolds of $\mathbb{E}^m$. The class of *symmetric R-spaces* includes:

(a) all Hermitian symmetric spaces of compact type;
(b) Grassmann manifolds $O(p+q)/O(p)\times O(q), Sp(p+q)/Sp(p)\times Sp(q)$;
(d) the classical groups $SO(m)$, $U(m)$, $Sp(m)$;
(d) $U(2m)/Sp(m)$, $U(m)/O(m)$;
(e) $(SO(p+1)\times SO(q+1))/S(O(p)\times O(q))$, where $S(O(p)\times O(q))$ is the subgroup of $SO(p+1)\times SO(q+1)$ consisting of matrices of the form

$$\begin{pmatrix} \epsilon & 0 & & \\ 0 & A & & \\ & & \epsilon & 0 \\ & & 0 & B \end{pmatrix}, \quad \epsilon=\pm 1, \quad A\in O(p), \quad B\in O(q);$$

(f) the Cayley projective plane $\mathcal{O}P^2$;
(g) the three exceptional spaces $E_6/Spin(10)\times T, E_7/E_6\times T$, and E_6/F_4.

D. Ferus proved that essentially these submanifolds exhaust all parallel submanifolds of $\mathbb{E}^m$ in the following.

Theorem 3.1. *A parallel submanifold N of the Euclidean m-space $\mathbb{E}^m$ is congruent to*

(1) $N=\mathbb{E}^{m_0}\times N_1\times\cdots\times N_k\subset\mathbb{E}^{m_0}\times\mathbb{E}^{m_1}\times\cdots\times\mathbb{E}^{m_k}=\mathbb{E}^m$, $k\geq 0$, *or to*
(2) $N=N_1\times\cdots\times N_k\subset\mathbb{E}^{m_1}\times\cdots\times\mathbb{E}^{m_k}=E^m$, $k\geq 1$,

where each $N_i\subset\mathbb{E}^{m_i}$ is an irreducible symmetric R-space.

Notice that in case (1) N is not contained in any hypersphere of $\mathbb{E}^m$, but in case (2) N is contained in a hypersphere of $\mathbb{E}^m$.

If the unit $(m-1)$-sphere $S^{m-1}(1)$ is regarded as an ordinary hypersphere of $\mathbb{E}^m$, then a submanifold $N\subset S^{m-1}(1)$ is parallel if and only if $N\subset S^{m-1}(1)\subset\mathbb{E}^m$ is a parallel submanifold of $\mathbb{E}^m$. Hence, Ferus' result implies the following.

Theorem 3.2. *A submanifold N of $S^{m-1}(1)$ is a parallel submanifold if and only if N is obtained by a submanifold of type (2) in Theorem 3.1.*

A submanifold N of a real space form $R^m(c)$ is called *full* if it is not contained in any totally geodesic hypersurface of $R^m(c)$. Parallel submanifolds in hyperbolic spaces have been determined in [Takeuchi (1981)].

Theorem 3.3. *Let $H^m(c)$, $c < 0$, be the hyperbolic m-space defined by*

$$H^m(c) = \left\{(x_0, \ldots, x_m) \in \mathbb{E}_1^{m+1} : x_1^2 + \cdots + x_m^2 - x_0^2 = \frac{1}{c},\ x_0 > 0\right\}.$$

If N is a complete, full parallel submanifold of $H^m(c)$, then

(i) *if N is not contained in any complete totally umbilical hypersurface of $H^m(c)$, then N is congruent to the product*

$$H^{m_0}(c_0) \times N_1 \times \cdots \times N_s \subset H^{m_0}(c_0) \times S^{m-m_0-1}(c') \subset H^{m_0}(c)$$

with $c_0 < 0$, $c' > 0$, $c_0{}^{-1} + c'^{-1} = c^{-1}$, $s \geq 0$, where $N_1 \times \cdots \times N_s \subset S^{m-m_0-1}(c')$ is a parallel submanifold as described in Ferus' result; and

(ii) *if N lies in a complete totally umbilical hypersurface M of $H^m(c)$, then M is isometric to a complete simply-connected real space form $R^{m-1}(\bar{c})$, where $\bar{c} = c + ||H||^2$, H is the mean curvature vector, and $N \subset R^{m-1}(\bar{c})$ is of the type described in* (i) *when $\bar{c} < 0$, or described by Theorem 3.1 when $\bar{c} = 0$, or by Theorem 3.2 when $\bar{c} > 0$.*

The following lemma is useful.

Lemma 3.1. *Let $\mathbf{x} : N \to \mathbb{E}_s^m$ be an isometric immersion of a pseudo-Riemannian manifold N into $\mathbb{E}_s^m$ and let $\{x_1, \ldots, x_n\}$ be a coordinate system of N such that g_{ij}, $1 \leq i, j \leq n$, are constants. Then N is a parallel submanifold in $\mathbb{E}_s^m$ if and only if $\mathbf{x}_{ijk}$, $1 \leq i, j, k \leq n$, are tangent to N, where $\mathbf{x}_{ijk} = \frac{\partial^2 \mathbf{x}}{\partial x_i \partial x_j \partial x_k}$.*

Proof. Under the hypothesis, we have $\nabla_{\partial_i} \partial_j = 0$. It follows from (2.26) that $\bar{\nabla} h = 0$ if and only if $D_{\partial_k} h(e_i, e_j) = 0$ for $i, j, k = 1, \ldots, n$. Thus, we find from from (2.6) that $\mathbf{x}_{ij} = h(e_i, e_j)$. Hence, by Weingarten's formula, we have $\mathbf{x}_{ijk} = -A_{h(\partial_i, \partial_j)} \partial_k$, which is tangent to N for each i, j, k.

Conversely, if $\mathbf{x}_{ijk}$ are tangent to N, then by $\mathbf{x}_{ij} = h(e_i, e_j)$, we have $D_{\partial_k} h(\partial_i, \partial_j) = 0$. Thus $\bar{\nabla} h = 0$. □

Remark 3.2. The explicit classifications of parallel pseudo-Riemannian submanifolds in indefinite real space forms are much more complicated than parallel submanifolds in real space forms. Nevertheless, the classification of parallel pseudo-Riemannian surfaces in 4-dimensional indefinite space forms were done in a series of papers [Chen and Van der Veken (2009);

Chen, Dillen and Van der Veken (2010); Chen (2010d,e,g)] (see also [Graves (1979a,b); Magid (1984)]). The complete explicit classification of spacelike and Lorentz parallel surfaces in indefinite real space forms with arbitrary codimension and arbitrary index were achieved in [Chen (2010a,b)].

3.3 Totally umbilical submanifolds

The following result is analogous to Proposition 3.1 [Ahn et al. (1996)].

Proposition 3.5. *Let $\phi : N_t^n \to M_s^m$ be an isometric immersion of a pseudo-Riemannian n-manifold N_t^n with index $t \in [1, n-1]$ into another pseudo-Riemannian manifold M_s^m. Then N_t^n is totally umbilical if and only if null geodesics of N_t^n are geodesics of M_s^m.*

Proof. Under the hypothesis, assume that N_t^n is totally umbilical in M_s^m. If $\gamma : I \to N_t^n$ is a null geodesic of N_t^n, then $\gamma'(t)$ is a null vector for each $t \in I$. Then it follows from (2.21) that $h(\gamma'(t), \gamma'(t)) = 0$. Thus, $\nabla_{\gamma'}\gamma' = 0$, which shows that γ is also a geodesic of M_s^m.

Conversely, if null geodesics of N_t^n are geodesics of M_s^m, then $h(v, v) = 0$ for null vectors v of N_t^n. At a point $p \in N_t^n$, let us choose an orthonormal basis $\{e_1, \ldots, e_n\}$ of $T_pN_t^n$ such that $\langle e_i, e_i\rangle = -1$ for $i = 1, \ldots, t$ and $\langle e_j, e_j\rangle = 1$ for $j = t+1, \ldots, n$. Then $e_i \pm e_j$ are null vectors for $i \in \{1, \ldots, t\}$ and $j \in \{t+1, \ldots, n\}$. Thus, $h(e_i \pm e_j, e_i \pm e_j) = 0$, which implies that

$$h(e_i, e_j) = 0, \quad h(e_i, e_i) + h(e_j, e_j) = 0. \tag{3.2}$$

If $t \geq 2$, then $e_{i_1} + e_{i_2} + \sqrt{2}e_n$ is a null vector for $1 \leq i_1 \neq i_2 \leq s$. Thus, by applying (3.2), we find $h(e_{i_1}, e_{i_2}) = 0$. Similarly, if $n - t \geq 2$, we have $h(e_{j_1}, e_{j_2}) = 0$ for $t + 1 \leq j_1 \neq j_2 \leq n$. Consequently, we obtain (2.21). Thus, N_t^n is totally umbilical in M_s^m. □

Lemma 3.2. *Let $\phi : N \to R_s^m(c)$ be an isometric immersion of a pseudo-Riemannian n-manifold N into an indefinite real space form $R_s^m(c)$. If N is totally umbilical, then*

(1) *H is a parallel normal vector field, i.e., $DH = 0$;*
(2) *$\langle H, H\rangle$ is constant;*
(3) *ϕ is a parallel immersion, i.e., $\bar{\nabla}h = 0$ identically on N;*
(4) *N is of constant curvature $c + \langle H, H\rangle$;*
(5) *$A_H = \langle H, H\rangle I$;*
(6) *N is a parallel submanifold.*

Proof. Under the hypothesis, we have (2.21). Thus

$$(\bar{\nabla}_Z h)(X,Y) = \langle X,Y\rangle\, D_Z H. \tag{3.3}$$

Hence, by the equation of Codazzi, we find $\langle X,Y\rangle D_Z H = \langle Z,Y\rangle D_X H$ for $X,Y,Z \in TN$. Since $\dim N > 1$, this shows that $DH = 0$. Therefore we get (1). Statement (2) follows immediately from (1) and the fact that D is a metric connection. (3) follows from (1) and (3.3). And (4) is an easy consequence of (2) and the equation of Gauss. Statement (5) follows immediately from (2.16) and (2.21). Finally, statement (6) follows from (2.26), (2.21) and statement (1). □

Lemma 3.3. *If $\phi : N \to \mathbb{E}^m_s$ is a totally umbilical immersion of a pseudo-Riemannian n-manifold N into the pseudo-Euclidean space $\mathbb{E}^m_s$, then*

(1) *the mean curvature vector H of N satisfies $\Delta H = bH$, where b is the constant equals to $n\langle H,H\rangle$;*
(2) *N lies in an $(n+1)$-dimensional totally geodesic submanifolds of $\mathbb{E}^m_s$ as a hypersurface.*

Proof. (1) follows easily from formula (2.45) and Lemma 3.2. (2) follows Reduction Theorem of Erbacher-Magid. □

Proposition 3.6. *Let $\mathbf{x} : N \to \mathbb{E}^m_s$ be an isometric immersion of a pseudo-Riemannian n-manifold N with index t into a pseudo-Euclidean m-space $\mathbb{E}^m_s$. If $n > 1$ and N is totally umbilical in $\mathbb{E}^m_s$, then N is congruent to an open portion of one of the following submanifolds:*

(1) *a totally geodesic pseudo-Euclidean subspace $\mathbb{E}^n_t \subset \mathbb{E}^m_s$;*
(2) *a pseudo n-sphere $S^n_t(c)$ lying in a totally geodesic pseudo-Euclidean $(n+1)$-subspace $\mathbb{E}^{n+1}_t \subset \mathbb{E}^m_s$;*
(3) *a pseudo-hyperbolic n-space $H^n_t(c)$ lying in a totally geodesic pseudo-Euclidean $(n+1)$-subspace $\mathbb{E}^{n+1}_{t+1} \subset \mathbb{E}^m_s$;*
(4) *a flat quasi-minimal submanifold defined by*

$$\left(\sum_{i=1}^{t} x_i^2 - \sum_{j=t+1}^{n} x_j^2, x_1, \ldots, x_t, 0, \ldots, 0, x_{t+1}, \ldots, x_n, \sum_{i=1}^{t} x_i^2 - \sum_{j=t+1}^{n} x_j^2\right).$$

The last case occurs only when $s > t$.

Proof. Let N be a totally umbilical pseudo-Riemannian submanifold of $\mathbb{E}^m_s$. Assume that the index of N is t. Then Lemma 3.2 implies that $\langle H,H\rangle$ is a constant, $DH = 0$ and $\mathbf{x}$ is a parallel immersion.

Case (a): $H = 0$. In this case N is totally geodesic. Thus, we obtain (1) by Proposition 3.1.

Case (b): $H \neq 0$. Corollary 2.1 implies that $\mathbf{x}(N)$ lies in a $(n+1)$-dimensional totally geodesic submanifold $E^* \subset \mathbb{E}^m_s$. From Lemma 3.2 we have $\nabla_X H = -\langle H, H\rangle X$ for $X \in T(N)$.

Case (b.1): $\langle H, H\rangle = \epsilon r^2$, $r > 0, \epsilon = \pm 1$. In this case, $\mathbf{y} = \mathbf{x} + (\epsilon/r^2)H$ is a constant vector, say $\mathbf{x}_0$. By applying a suitable translation we have $\mathbf{x}_0 = 0$. Thus $\langle \mathbf{x}, \mathbf{x}\rangle = \epsilon/r^2$, which gives case (2) or case (3) according to H is spacelike or timelike, respectively.

Case (b.2): *H is lightlike.* In this case, E^* is a totally geodesic $\mathbf{R}^{n+1}_{1,t}$. Since H is a constant lightlike vector and N is totally umbilical, (2.21) and Gauss' equation imply that N is flat. Thus, locally there is a natural coordinate system $\{x_1, \ldots, x_n\}$ such that the metric tensor g_0 of N is given by

$$g_0 = -\sum_{i=1}^{t} dx_i^2 + \sum_{j=t+1}^{n} dx_j^2. \tag{3.4}$$

Hence, it follows from Proposition 1.1, (2.6), (2.21) and (3.4) that

$$\begin{aligned} &\mathbf{x}_{x_i x_i} = H, \ \ i = 1, \ldots, t; \ \ \mathbf{x}_{x_k, x_\ell} = 0, \ \ 1 \leq k \neq \ell \leq n, \\ &\mathbf{x}_{x_j x_j} = -H, \ \ j = t+1, \ldots, n. \end{aligned} \tag{3.5}$$

After solving system (3.5) we find

$$\mathbf{x} = \mathbf{c}_0 + \sum_{k=1}^{n} \mathbf{c}_k x_k + \frac{H}{2}\sum_{i=1}^{t} x_i^2 - \frac{H}{2}\sum_{j=t+1}^{n} x_j^2 \tag{3.6}$$

for some vectors $\mathbf{c}_0, \mathbf{c}_1, \ldots, \mathbf{c}_n \in \mathbb{E}^m_s$. Since H is a constant lightlike vector, without loss of generality we may put $H = (2, 0, \ldots, 0, 2) \in \mathbb{E}^m_s$. Hence, after choosing suitable initial conditions, we obtain (4). □

Remark 3.3. A different statement is given in [Ahn et al. (1996)].

Lemma 3.4. *Let $\mathbf{x} : N \to \mathbb{E}^m_s$ be an isometric immersion of a pseudo-Riemannian n-manifold N into $\mathbb{E}^m_s$ and $f_1, \ldots, f_\ell \in \mathcal{F}(N)$. Then*

$$\mathbf{y} = (f_1, \ldots, f_\ell, \mathbf{x}, f_\ell, \ldots, f_1) : N \to \mathbb{E}^{m+2\ell}_{s+\ell} \tag{3.7}$$

is a parallel isometric immersion if and only if the Hessian $\mathcal{H}_{f_1}, \ldots, \mathcal{H}_{f_\ell}$ of $f_1, \ldots, f_\ell$ are covariant constant and $\mathbf{x}$ is a parallel immersion.

Proof. Since the metric of $\mathbb{E}^{m+2\ell}_{s+\ell}$ is

$$g_0 = -\sum_{i=s+\ell}^{s+\ell} dx_i^2 + \sum_{j=s+\ell+1}^{m+2\ell} dx_j^2 \tag{3.8}$$

and $\mathbf{x}$ is an isometric immersion, (3.7) is an isometric immersion. It follows from (2.6) and (3.7) that the second fundamental form $h_{\mathbf{y}}$ of $\mathbf{y}$ is

$$h_{\mathbf{y}} = (\mathcal{H}_{f_1}, \ldots, \mathcal{H}_{f_\ell}, h_{\mathbf{x}}, \mathcal{H}_{f_\ell}, \ldots, \mathcal{H}_{f_1}), \tag{3.9}$$

where $h_{\mathbf{x}}$ is the second fundamental form of $\mathbf{x}$. Consequently, $\mathbf{y}$ is a parallel immersion if and only if the Hessians $\mathcal{H}_{f_1}, \ldots, \mathcal{H}_{f_\ell}$ are covariant constant and $\mathbf{x}$ is a parallel immersion. □

3.4 Totally umbilical submanifolds of $S^m_s(1)$ and $H^m_s(-1)$

Let $\phi : N \to S^m_s(\frac{1}{r^2})$ (resp. $\phi : N \to H^m_s(-\frac{1}{r^2})$) be an isometric immersion of a pseudo-Riemannian n-manifold N into $S^m_s(\frac{1}{r^2})$ (resp. $H^m_s(-\frac{1}{r^2})$). Denote by

$$\mathbf{x} = \iota \circ \phi : N \to \mathbb{E}^{m+1}_s, \qquad (\text{resp. } \mathbf{x} = \iota \circ \phi : N \to \mathbb{E}^{m+1}_{s+1})$$

the composition of ϕ with the inclusion map $\iota : S^m_s(\frac{1}{r^2}) \subset \mathbb{E}^{m+1}_s$ via (1.16) (resp. $\iota : H^m_s(-\frac{1}{r^2}) \subset \mathbb{E}^{m+1}_{s+1}$ via (1.17)). Let $\nabla', \tilde{\nabla}$ and ∇ be the Levi-Civita connections of N, $\mathbb{E}^{m+t}_i$ and $S^m_s(\frac{1}{r^2})$ (resp. of $H^m_s(-\frac{1}{r^2})$). Denote by D' the normal connection of N in $S^m_s(\frac{1}{r^2})$ (resp. in $H^m_s(-\frac{1}{r^2})$) and by D the corresponding quantities of N in $\mathbb{E}^{m+1}_s$ (resp. in $\mathbb{E}^{m+1}_{s+1}$) via $\mathbf{x}$.

Lemma 3.5. *Let $\phi : N \to S^m_s(\frac{1}{r^2})$ (resp. $\phi : N \to H^m_s(-\frac{1}{r^2})$) be an isometric immersion of a pseudo-Riemannian n-manifold N into $S^m_s(\frac{1}{r^2})$ (resp. Ninto $H^m_s(-\frac{1}{r^2})$). Then the second fundamental form h and the mean curvature vector H of N in $\mathbb{E}^{m+1}_s$ (resp. N in $\mathbb{E}^{m+1}_{s+1}$) via $\mathbf{x}$ are related with the second fundamental form h' and the mean curvature vector H' of N in $S^m(\frac{1}{r^2})$ (resp. N in $H^m_s(-\frac{1}{r^2})$) by*

$$h(X,Y) = h'(X,Y) - \frac{\epsilon}{r^2}\langle X,Y\rangle \mathbf{x}, \tag{3.10}$$

$$H = H' - \frac{\epsilon}{r^2}\mathbf{x}, \tag{3.11}$$

where $\epsilon = 1$ or -1, depending on $\phi : N \to S^m_s(\frac{1}{r^2})$ or $\phi : N \to H^m_s(-\frac{1}{r^2})$. Moreover, we have $DH = D'H'$.

Proof. Under the hypothesis, the positive vector field is a normal vector field of N which is normal to $S^m_s(\frac{1}{r^2})$ or $H^m_s(-\frac{1}{r^2})$. Since $\tilde{\nabla}_X \mathbf{x} = X$ for $X \in T(N)$, the Weingarten formula yields

$$A_{\mathbf{x}} = -I, \quad DH = D'H', \tag{3.12}$$

where A is the Weingarten map in $\mathbb{E}^{m+1}_s$ or in $\mathbb{E}^{m+1}_{s+1}$. Hence, it follows from the formula of Gauss and (3.12) that

$$\tilde{\nabla}_X Y = \nabla_X Y - \epsilon r^2 \mathbf{x} = \nabla'_X Y + h(X,Y) - \frac{\epsilon}{r^2}\mathbf{x}, \tag{3.13}$$

which gives (3.10). By taking the trace of (3.10) we get (3.11). □

Corollary 3.1. *If $\phi : N \to S^m_s(\frac{1}{r^2})$ (resp. $\phi : N \to H^m_s(-\frac{1}{r^2})$) is an isometric immersion of a pseudo-Riemannian manifold N into $S^m_s(\frac{1}{r^2})$ (resp. N into $H^m_s(-\frac{1}{r^2})$) and $\iota : S^m_s(\frac{1}{r^2}) \subset \mathbb{E}^{m+1}_s$ (resp. $\iota : H^m_s(-\frac{1}{r^2}) \subset \mathbb{E}^{m+1}_{s+1}$) is the inclusion map defined in Section 1.6, then*

(1) *ϕ has parallel mean curvature vector if and only if $\mathbf{x} = \iota \circ \phi$ has parallel mean curvature vector;*
(2) *ϕ is a parallel immersion if and only if $\mathbf{x} = \iota \circ \phi$ is a parallel immersion;*
(3) *ϕ is totally umbilical if and only if $\mathbf{x} = \iota \circ \phi$ is totally umbilical.*

Proof. Follows immediately from Lemma 3.5 and (3.13). □

Proposition 3.7. *Let $\phi : N^n_t \to S^m_s(1) \subset \mathbb{E}^{m+1}_s$ be an isometric immersion of a pseudo-Riemannian n-manifold N into a pseudo-Riemannian m-space $S^m_s(1)$ with $n > 1$. If N^n_t is totally umbilical in $S^m_s(1)$, then it is congruent to an open portion of one of the following submanifolds:*

(1) *a totally geodesic pseudo subsphere $S^n_t(1) \subset S^m_s(1)$;*
(2) *a pseudo sphere $S^n_t(\frac{1}{r^2})$, $r \in (0,1)$, lying in a totally geodesic pseudo-Euclidean $(n+2)$-subspace $\mathbb{E}^{n+2}_t \subset \mathbb{E}^{m+1}_s$ as*
$$\left\{\left(\mathbf{y}, \sqrt{1-r^2}\right) \in \mathbb{E}^{n+2}_t : \langle \mathbf{y}, \mathbf{y}\rangle = r^2,\ \mathbf{y} \in \mathbb{E}^{n+1}_t\right\};$$
(3) *a pseudo sphere $S^n_t(\frac{1}{r^2})$, $r > 1$, lying in a totally geodesic pseudo-Euclidean $(n+2)$-subspace $\mathbb{E}^{n+2}_{t+1} \subset \mathbb{E}^{m+1}_s$ as*
$$\left\{\left(\sqrt{r^2-1}, \mathbf{y}\right) \in \mathbb{E}^{n+2}_{t+1} : \langle \mathbf{y}, \mathbf{y}\rangle = r^2,\ \mathbf{y} \in \mathbb{E}^{n+1}_t\right\};$$
(4) *a pseudo-hyperbolic space $H^n_t(-\frac{1}{r^2})$, $r > 0$, lying in a totally geodesic pseudo-Euclidean $(n+2)$-subspace $\mathbb{E}^{n+2}_{t+1} \subset \mathbb{E}^{m+1}_s$ as*
$$\left\{\left(\mathbf{y}, \sqrt{1+r^2}\right) \in \mathbb{E}^{n+2}_{t+1} : \langle \mathbf{y}, \mathbf{y}\rangle = -r^2,\ \mathbf{y} \in \mathbb{E}^{n+1}_{t+1}\right\};$$
(5) *a flat totally umbilical submanifold lying in a totally geodesic pseudo-Euclidean $(n+3)$-subspace $\mathbb{E}^{n+3}_{t+1} \subset \mathbb{E}^{m+1}_s$ as*
$$\left\{\left(r\langle \mathbf{y}, \mathbf{y}\rangle - rb - \frac{r}{4}, r\mathbf{y}, \sqrt{1+br^2}, r\langle \mathbf{y}, \mathbf{y}\rangle - rb + \frac{r}{4}\right) \in \mathbb{E}^{n+2}_{t+1} : \mathbf{y} \in \mathbb{E}^n_t\right\},$$
where b, r are real numbers satisfying $br^2 \geq -1$ and $r > 0$;

(6) *a flat totally umbilical submanifold lying in a totally geodesic pseudo-Euclidean* $(n+3)$*-subspace* $\mathbb{E}^{n+3}_{t+2} \subset \mathbb{E}^{m+1}_s$ *as*

$$\Big\{\Big(r\langle \mathbf{y},\mathbf{y}\rangle + rb - \frac{r}{4}, \sqrt{br^2-1}, r\mathbf{y}, r\langle \mathbf{y},\mathbf{y}\rangle + rb + \frac{r}{4}\Big) \in \mathbb{E}^{n+2}_{t+2} : \mathbf{y} \in \mathbb{E}^n_t\Big\},$$

where b, r *are real numbers satisfying* $br^2 \geq 1$ *and* $r > 0$.

Proof. In view of Corollary 3.1, it suffices to look for totally umbilical submanifolds of $\mathbb{E}^{m+1}_s$ from Proposition 3.6 such that, up to dilations and rigid motions, the totally umbilical submanifolds lie in $S^m_s(1) \subset \mathbb{E}^{m+1}_s$.

Obviously, a totally geodesic $\mathbb{E}^n_t$ in $\mathbb{E}^{m+1}_s$ does not lie in any pseudo hypersphere of $\mathbb{E}^{m+1}_s$. From Proposition 3.6(2) we get (2) and (3) of the Proposition. Similarly, we obtain (4) form Proposition 3.6(3). Finally, (5) and (6) are obtained from Proposition 3.6(4) after applying a suitable dilation and a rigid motion. □

Proposition 3.8. *Let* $\phi : N^n_t \to H^m_s(-1)$ *be an isometric immersion of a pseudo-Riemannian* n*-manifold* N^n_t *into a pseudo-hyperbolic* m*-space* $H^m_s(-1)$*. If* $n > 1$ *and* N^n_t *is totally umbilical in* $H^m_s(-1) \subset \mathbb{E}^{m+1}_{s+1}$*, then it is congruent to an open portion of one of the following submanifolds:*

(1) *A totally geodesic pseudo-hyperbolic* n*-space* $H^n_t(-1) \subset H^m_s(-1)$;

(2) *A pseudo* n*-sphere* $S^n_t(\frac{1}{r^2})$, $r > 0$*, lying in a totally geodesic pseudo-Euclidean* $(n+2)$*-subspace* $\mathbb{E}^{n+2}_{t+1} \subset \mathbb{E}^{m+1}_{s+1}$ *as*

$$\Big\{\Big(\sqrt{1+r^2}, \mathbf{y}\Big) \in \mathbb{E}^{n+2}_{t+1} : \langle \mathbf{y},\mathbf{y}\rangle = r^2, \mathbf{y} \in \mathbb{E}^{n+1}_{t+1}\Big\};$$

(3) *A pseudo-hyperbolic* n*-space* $H^n_t(-\frac{1}{r^2})$, $r > 1$*, lying in a totally geodesic pseudo-Euclidean* $(n+2)$*-subspace* $\mathbb{E}^{n+2}_{t+1} \subset \mathbb{E}^{m+1}_{s+1}$ *as*

$$\Big\{\Big(\mathbf{y}, \sqrt{r^2-1}\Big) \in \mathbb{E}^{n+2}_{t+1} : \langle \mathbf{y},\mathbf{y}\rangle = -r^2, \mathbf{y} \in \mathbb{E}^{n+1}_{t+1}\Big\};$$

(4) *A pseudo-hyperbolic* n*-space* $H^n_t(-\frac{1}{r^2})$, $r > 0$*, lying in a totally geodesic pseudo-Euclidean* $(n+2)$*-subspace* $\mathbb{E}^{n+2}_{t+2} \subset \mathbb{E}^{m+1}_{s+1}$ *as*

$$\Big\{\Big(\sqrt{1-r^2}, \mathbf{y}\Big) \in \mathbb{E}^{n+2}_{t+2} : \langle \mathbf{y},\mathbf{y}\rangle = -r^2, \mathbf{y} \in \mathbb{E}^{n+1}_{t+1}\Big\};$$

(5) *A flat totally umbilical submanifold lying in a totally geodesic pseudo-Euclidean* $(n+3)$*-subspace* $\mathbb{E}^{n+3}_{t+1} \subset \mathbb{E}^{m+1}_{s+1}$ *as*

$$\Big\{\Big(r\langle \mathbf{y},\mathbf{y}\rangle - rb - \frac{r}{4}, r\mathbf{y}, \sqrt{br^2-1}, r\langle \mathbf{y},\mathbf{y}\rangle - rb + \frac{r}{4}\Big) \in \mathbb{E}^{n+3}_{t+2} : \mathbf{y} \in \mathbb{E}^n_t\Big\},$$

where b, r *are real numbers satisfying* $br^2 \geq 1$ *and* $r > 0$;

(6) *A flat totally umbilical submanifold lying in a totally geodesic pseudo-Euclidean $(n+3)$-subspace $\mathbb{E}^{n+3}_{t+2} \subset \mathbb{E}^{m+1}_{s+1}$ as*

$$\Big\{\Big(r\langle \mathbf{y},\mathbf{y}\rangle - rb - \frac{r}{4}, \sqrt{1-br^2}, r\mathbf{y}, r\langle \mathbf{y},\mathbf{y}\rangle - rb + \frac{r}{4}\Big) \in \mathbb{E}^{n+3}_{t+3} : \mathbf{y} \in \mathbb{E}^n_t\Big\},$$

where b, r are real numbers satisfying $br^2 \leq 1$ and $r > 0$.

Proof. Analogous to the proof of Proposition 3.7 it suffices to look for totally umbilical submanifolds of $\mathbb{E}^{m+1}_{s+1}$ from the list of Proposition 3.6 such that, up to dilations and rigid motions, the totally umbilical submanifolds lie in $H^m_s(-1) \subset \mathbb{E}^{m+1}_{s+1}$. Clearly, totally geodesic $\mathbb{E}^n_t$ in $\mathbb{E}^{m+1}_s$ does not lie in any pseudo-hyperbolic hypersurface of $\mathbb{E}^{m+1}_{s+1}$. From Proposition 3.6(2) we obtain (2) of the Proposition. Similarly, we obtain (3) and (4) form Proposition 3.6(3). (5) and (6) are obtained from Proposition 3.6(4) after applying suitable dilation and rigid motion. □

3.5 Pseudo-umbilical submanifolds of $\mathbb{E}^m_s$

Definition 3.1. A non-minimal pseudo-Riemannian submanifold N of a pseudo-Riemannian manifold M is called a *pseudo-umbilical submanifold* if there exists a function λ in $\mathcal{F}(N)$ such that

$$\langle h(X,Y), H\rangle = \lambda \langle X, Y\rangle, \quad X, Y \in \mathcal{X}(N). \tag{3.14}$$

The following three results classify pseudo-umbilical submanifolds with parallel mean curvature vector in indefinite real space forms.

Proposition 3.9. *Let $\mathbf{x} : N \to \mathbb{E}^m_s$ be an isometric immersion of a pseudo-Riemannian submanifold N into a pseudo-Euclidean m-space $\mathbb{E}^m_s$. Then $\mathbf{x}$ is a pseudo-umbilical immersion with parallel mean curvature vector if and only if one of the following three cases occurs:*

(1) *N is a minimal submanifold of a pseudo hypersphere $S^{m-1}_s(\mathbf{x}_0, \frac{1}{r^2})$ for some $\mathbf{x}_0 \in \mathbb{E}^m_s$ and $r > 0$;*
(2) *N is a minimal submanifold of a pseudo-hyperbolic hyperspace $H^{m-1}_{s-1}(\mathbf{x}_0, -\frac{1}{r^2})$ for some $\mathbf{x}_0 \in \mathbb{E}^m_s$ and $r > 0$;*
(3) *$\mathbf{x}$ is congruent to $(f, \mathbf{z}, f)$, where $f \in \mathcal{F}(N)$, Δf is a nonzero real number and $\mathbf{z} : N \to \mathbb{E}^{m-2}_{s-1}$ is a minimal isometric immersion.*

Case (2) occurs only when $s \geq 1$ and case (3) occurs only when $s \geq 1$ and $m \geq \dim N + 2$.

Proof. Assume that N is a pseudo-umbilical submanifold of $\mathbb{E}^m_s$ with parallel mean curvature vector. Then $X\langle H, H\rangle = 2\langle H, D_X H\rangle = 0$ for any $X \in T(N)$. Thus, $\langle H, H\rangle$ is constant.

Case (a): $\langle H, H\rangle \neq 0$. We put

$$\langle H, H\rangle = \frac{\epsilon}{r^2}, \tag{3.15}$$

where $\epsilon = 1$ or -1 depending on H is spacelike or timelike. On the other hand, it follows from (3.14) that $A_H = \lambda I$ for some function $\lambda \in \mathcal{F}(N)$. Thus, we find from (3.15) that $\epsilon = \lambda r^2$. Let us put

$$\hat{\mathbf{x}} = \mathbf{x} + \epsilon r^2 H. \tag{3.16}$$

Then $\tilde{\nabla}_X \hat{\mathbf{x}} = X - \epsilon r^2 A_H X = 0$. Thus, $\hat{\mathbf{x}}$ is a constant vector, say $\mathbf{x}_0 \in \mathbb{E}^m_s$. Hence

$$\langle \mathbf{x} - \mathbf{x}_0, \mathbf{x} - \mathbf{x}_0\rangle = r^4 \langle H, H\rangle = \epsilon r^2,$$

which implies that N lies either in $S^{m-1}_s(\mathbf{x}_0, \frac{1}{r^2})$ or in $H^{m-1}_s(\mathbf{x}_0, -\frac{1}{r^2})$, according to H is spacelike or timelike, respectively. Since H is either normal to $S^{m-1}_r(\mathbf{x}_0, \frac{1}{r^2})$ or to $H^{m-1}_s(\mathbf{x}_0, -\frac{1}{r^2})$, it follows from Lemma 3.5 that N is minimal in $S^{m-1}_r(\mathbf{x}_0, \frac{1}{r^2})$ or in $H^{m-1}_s(\mathbf{x}_0, -\frac{1}{r^2})$.

Case (b): *H is lightlike.* It follows from (3.14) that $A_H = \langle H, H\rangle I = 0$. Combining this with $DH = 0$ implies that H is a lightlike constant vector, say $\zeta_0 \in \mathbb{E}^{m+1}_s$. Thus, $X\langle \mathbf{x}, \zeta_0\rangle = 0$ for any $X \in T(N)$. So, $\langle \mathbf{x}, \zeta_0\rangle = c$ for some real number c.

If we put $\zeta_0 = (1, 0, \ldots, 0, 1) \in \mathbb{E}^m_s$ and $\mathbf{x} = (x_1, \ldots, x_m)$, then we obtain $x_m = x_1 + c$. By applying a suitable translation, we find $c = 0$. Thus, the immersion $\mathbf{x} : N \to \mathbb{E}^m_s$ takes the form

$$\mathbf{x} = (f, \mathbf{z}, f), \tag{3.17}$$

where f is function on N and $\mathbf{z} : N \to \mathbb{E}^{m-2}_{s-1}$ is an isometric immersion. Now, by applying the Laplace operator Δ to (3.17) we find from Beltrami's formula that $nH = (-\Delta f, nH_{\mathbf{z}}, -\Delta f)$, where $H_{\mathbf{z}}$ is the mean curvature vector of $\mathbf{z}$. Since $H = (1, 0, \ldots, 0, 1)$, we find $H_{\mathbf{z}} = 0$ and $\Delta f = -nr$. Thus, $\mathbf{z}$ is a minimal immersion and Δf is a nonzero constant.

The converse can be verified easily. □

3.6 Pseudo-umbilical submanifolds of $S^m_s(1)$ and $H^m_s(-1)$

Proposition 3.10. *Let $\phi : N \to S^m_s(1)$ be an isometric immersion of a pseudo-Riemannian submanifold N into the pseudo m-sphere $S^m_s(1)$. Then*

ϕ is a pseudo-umbilical immersion with parallel mean curvature vector if and only if N lies in a non-totally geodesic, totally umbilical hypersurface of $S_s^m(1)$ as a minimal submanifold.

Proof. Let $\phi : N \to S_s^m(1)$ be an isometric immersion of a pseudo-Riemannian submanifold N into $S_s^m(1)$. Then it follows from Lemma 3.5 that N is a pseudo-umbilical submanifold with parallel mean curvature vector in $S_s^m(1)$ if and only if N is a pseudo-umbilical submanifold of $\mathbb{E}_s^{m+1}$ via the composition $\mathbf{x} = \iota \circ \phi$, where ι is the inclusion $S_s^m(1) \subset \mathbb{E}_s^{m+1}$.

Let N be a pseudo-umbilical submanifold with parallel mean curvature vector in $S_s^m(1)$. Then the mean curvature vector H of N in $\mathbb{E}_s^{m+1}$ satisfies

$$A_H = \lambda I, \quad DH = 0, \quad \lambda \in \mathbf{R}. \tag{3.18}$$

Case (1): *H is spacelike.* If we put $\langle H, H\rangle = r^{-2}$, then as in the proof of Proposition 3.9, we have

$$\mathbf{x} - \mathbf{x}_0 = -r^2 H \tag{3.19}$$

for some vector $\mathbf{x}_0$. Hence, N is a minimal submanifold of the pseudo hypersphere $S_s^m(\mathbf{x}_0, \frac{1}{r^2})$. Since N lies in $S_s^m(1) \cap S_s^m(\mathbf{x}_0, \frac{1}{r^2})$, we find

$$\langle \mathbf{x}, \mathbf{x}\rangle = 1 \quad \text{and} \quad \langle \mathbf{x} - \mathbf{x}_0, \mathbf{x} - \mathbf{x}_0\rangle = \frac{1}{r^2}, \tag{3.20}$$

which imply that

$$\langle \mathbf{x}, \mathbf{x}_0\rangle = c, \quad c = \frac{1 + \langle \mathbf{x}_0, \mathbf{x}_0\rangle}{2} - \frac{1}{2r^2}. \tag{3.21}$$

Since N is non-minimal in $S_s^m(1)$, $\mathbf{x}_0 \neq 0$.

Case (1.1): $\langle \mathbf{x}_0, \mathbf{x}_0\rangle = k^2 > 0$. If we put $\mathbf{x}_0 = (0, \ldots, 0, k^{-1}) \in \mathbb{E}_s^{m+1}$, then the first equation in (3.21) implies that the last canonical coordinate x_{m+1} of N satisfies $x_{m+1} = ck$. Hence, N is contained in $\mathcal{H} = S_s^m(1) \cap E$, where E is the hyperplane defined by $x_{m+1} = ck$. An easy computation shows that $\xi = c\mathbf{x} - \mathbf{x}_0$ is a normal vector field of $\mathcal{H}$ in $S_s^m(1)$. Moreover, it follows from Gauss' formula that $A_\xi = -cI$. Hence $\mathcal{H}$ is a totally umbilical hypersurface of $S_s^m(1)$. Since N is non-minimal in $S_s^m(1)$, $\mathcal{H}$ must be a non-totally geodesic, totally umbilical hypersurface. Also, since H lies in Span $\{\mathbf{x}, \xi\}$, N is minimal in $\mathcal{H}$. Therefore, N lies in a non-totally geodesic, totally umbilical hypersurface of $S_s^m(1)$ as a minimal submanifold.

Case (1.2): $\langle \mathbf{x}_0, \mathbf{x}_0\rangle = -k^2 < 0$. By putting $\mathbf{x}_0 = (k^{-1}, 0, \ldots, 0)$ and applying the same arguments as case (1.1), we get the same conclusion as case (1.1).

Case (1.3): $\mathbf{x}_0$ *is lightlike.* We find from (3.21) that $\langle \mathbf{x}, \mathbf{x}_0 \rangle = c \in (-\infty, \frac{1}{2})$, which defines a hyperplane $\mathcal{E}$ of $\mathbb{E}^{m+1}_s$. Hence, N lies in $\mathcal{H}_1 = S^m_s(1) \cap \mathcal{E}$. It is easy to verify that $\eta = \mathbf{x}_0 - c\mathbf{x}$ is a normal vector field of $\mathcal{H}_1$ in $S^m_s(1)$ such that $\langle \eta, \eta \rangle = -c^2$. Since $\tilde{\nabla}_X \eta = -cX$ for $X \in T(N)$, we get $A_\eta = cI$. This shows that $\mathcal{H}_1$ is totally umbilical in $S^m_s(1)$. Because N is non-minimal in $S^m_s(1)$, we have $c \neq 0$. Moreover, it follows from (3.19) and Lemma 3.5 that the mean curvature vector H' of N in $S^m_s(1)$ is η. Hence, we have the same conclusion as case (1.1).

Case (2): *H is timelike.* A similar arguments as case (1) gives the same conclusion as case (1).

Case (3): *H is lightlike.* Just like in case (b) of the proof of Proposition 3.9, H is a constant lightlike vector in $\mathbb{E}^{m+1}_s$, say ζ_0, and that we have $\langle \mathbf{x}, \zeta_0 \rangle = c$ for some real number c. Let $\mathcal{E}$ denote the hyperplane defined by $\langle \mathbf{x}, \zeta_0 \rangle = c$. Then, N is contained in $\mathcal{H}_1 = S^m_s(1) \cap \mathcal{E}$. It is easy to verify that $\eta = \zeta_0 - c\mathbf{x}$ is a normal vector field of $\mathcal{H}_1$ in $S^m_s(1)$ with $\langle \eta, \eta \rangle = -c^2$. Since $\tilde{\nabla}_X \eta = -cX$, we get $A_\eta = cI$. Thus, $\mathcal{H}_1$ is totally umbilical in $S^m_s(1)$. Because N is non-minimal in $S^m_s(1)$, we find $c \neq 0$. Hence, $\mathcal{H}_1$ is a non-totally geodesic, totally umbilical hypersurface. Moreover, it follows from (3.19) and Lemma 3.5 that the mean curvature vector of N in $S^m_s(1)$ is exactly η. Therefore, N is minimal in $\mathcal{H}_1$.

The converse is easy to verify. □

Proposition 3.11. *Let $\phi : N \to H^m_s(-1)$ be an isometric immersion of a pseudo-Riemannian manifold N into the pseudo-hyperbolic m-space $H^m_s(-1)$. Then ϕ is a pseudo-umbilical immersion with parallel mean curvature vector if and only if N is contained in a non-totally geodesic, totally umbilical hypersurface of $H^m_s(-1)$ as a minimal submanifold.*

Proof. This can be done in the same way as Proposition 3.10 □

By applying Proposition 3.10 we have the following.

Corollary 3.2. *Let $\phi : N \to S^m_s(1)$ be an isometric immersion of a pseudo-Riemannian manifold N into $S^m_s(1)$. Then ϕ is a pseudo-umbilical immersion with unit timelike parallel mean curvature vector if and only if N lies in a flat totally umbilical hypersurface of $S^m_s(1)$ as a minimal submanifold.*

Proof. We follow the same notations as in the proof of Proposition 3.10. Assume $\phi : N \to S^m_s(1)$ is a pseudo-umbilical immersion with unit timelike parallel mean curvature vector H'. Then the mean curvature vector H of N in $\mathbb{E}^{m+1}_s$ is $H' - \mathbf{x}$. Since H' is a unit timelike vector field, as in the proof of Proposition 3.9, we see that H is a lightlike constant vector, say $\zeta_0 \in \mathbb{E}_s^{m+1}$. Clearly, we have $\langle \mathbf{x}, \zeta_0 \rangle = -1$. So, N lies in $\mathcal{H}_1 = S_s^m(1) \cap \mathcal{E}$, where $\mathcal{E}$ is defined by $\langle \mathbf{x}, \zeta_0 \rangle = -1$.

Since $\eta = \zeta_0 + \mathbf{x}$ is a unit timelike normal vector field of $\mathcal{H}_1$ in $S_s^m(1)$ and $A_\eta = -I$, the second fundamental form $\hat{h}$ of $\mathcal{H}_1$ in $S_s^m(1)$ satisfies $\hat{h}(X,Y) = -\langle X,Y \rangle \eta$, $X,Y \in T\mathcal{H}_1$. Therefore, the equation of Gauss shows that $\mathcal{H}_1$ is a flat totally umbilical hypersurface. The same argument as the proof of Proposition 3.10 implies that N is minimal in $\mathcal{H}_1$.

The converse is easy to verify. □

Corollary 3.3. *Let $\phi : N \to H_s^m(-1)$ be an isometric immersion of a pseudo-Riemannian submanifold N into $H_s^m(-1)$. Then ϕ is a pseudo-umbilical immersion with unit spacelike parallel mean curvature vector if and only if N lies in a flat totally umbilical hypersurface of $H_s^m(-1)$ as a minimal submanifold.*

Proof. This can be done in the same way as Corollary 3.2. □

3.7 Minimal Lorentz surfaces in indefinite real space forms

The following lemma is an easy consequence of a result of [Larsen (1996)].

Lemma 3.6. *Locally there exists a coordinate system $\{x, y\}$ on a Lorentz surface N_1^2 such that the metric tensor is given by*

$$g = -E^2(x,y)(dx \otimes dy + dy \otimes dx) \tag{3.22}$$

for some positive function $E(x,y)$.

Proof. It is known that locally there exist isothermal coordinates (u,v) on a Lorentz surface N_1^2 such that the metric tensor takes the form:

$$g = F(u,v)^2(-du \otimes du + dv \otimes dv) \tag{3.23}$$

for some positive function F [Larsen (1996)]. Thus, by putting $x = u + v, y = u - v$, we obtain (3.22) from (3.23) with $E(x,y) = F(x,y)/\sqrt{2}$. □

The Levi-Civita connection of metric tensor (3.22) satisfies

$$\nabla_{\frac{\partial}{\partial x}} \frac{\partial}{\partial x} = \frac{2E_x}{E} \frac{\partial}{\partial x}, \ \nabla_{\frac{\partial}{\partial x}} \frac{\partial}{\partial y} = 0, \ \nabla_{\frac{\partial}{\partial y}} \frac{\partial}{\partial y} = \frac{2E_y}{E} \frac{\partial}{\partial y} \tag{3.24}$$

and the Gaussian curvature K is given by

$$K = \frac{2EE_{xy} - 2E_xE_y}{E^4}. \tag{3.25}$$

If we put

$$e_1 = \frac{1}{E}\frac{\partial}{\partial x}, \quad e_2 = \frac{1}{E}\frac{\partial}{\partial y}, \tag{3.26}$$

then $\{e_1, e_2\}$ forms a pseudo-orthonormal frame such that

$$\langle e_1, e_1\rangle = \langle e_2, e_2\rangle = 0, \ \langle e_1, e_2\rangle = -1. \tag{3.27}$$

We define the *connection 1-form* ω by the following equations:

$$\nabla_X e_1 = \omega(X)e_1, \quad \nabla_X e_2 = -\omega(X)e_2. \tag{3.28}$$

From (3.24) and (3.26) we find

$$\begin{aligned} \nabla_{e_1} e_1 &= \frac{E_x}{E^2} e_1, \ \nabla_{e_2} e_1 = -\frac{E_y}{E^2} e_1, \\ \nabla_{e_1} e_2 &= -\frac{E_x}{E^2} e_2, \ \nabla_{e_2} e_2 = \frac{E_y}{E^2} e_2. \end{aligned} \tag{3.29}$$

By comparing (3.28) and (3.29), we get

$$\omega(e_1) = \frac{E_x}{E^2}, \quad \omega(e_2) = -\frac{E_y}{E^2}. \tag{3.30}$$

Let $\psi : N_1^2 \to M_s^m$ be an isometric immersion of N_1^2 into a pseudo-Riemannian m-manifold M_s^m with index s. Then it follows from (2.19) and (3.27) that the mean curvature vector is given by

$$H = -h(e_1, e_2). \tag{3.31}$$

Thus, N_1^2 is a minimal surface of M_s^m if and only if $h(e_1, e_2) = 0$ identically.

The following result completely classifies minimal Lorentz surfaces in an arbitrary indefinite pseudo-Euclidean m-space $\mathbb{E}_s^m$ [Chen (2011a)].

Theorem 3.4. *A Lorentz surface in a pseudo-Euclidean m-space $\mathbb{E}_s^m$ is minimal if and only if, locally, the immersion takes the form*

$$L(x, y) = z(x) + w(y),$$

where z and w are null curves satisfying $\langle z'(x), w'(y)\rangle \neq 0$.

Proof. Let $L : N_1^2 \to \mathbb{E}_s^m$ be an isometric immersion of a Lorentz surface N_1^2 into a pseudo-Euclidean m-space $\mathbb{E}_s^m$. Then, according to Lemma 3.6, we may choose a local coordinate system $\{x, y\}$ on N_1^2 such that

$$g = -E^2(x, y)(dx \otimes dy + dy \otimes dx) \tag{3.32}$$

Then we have (3.24)-(3.31). If N_1^2 is a minimal surface, then it follows from (3.31) that $h(e_1, e_2) = 0$ holds. Hence, we may put

$$h(e_1, e_1) = \xi, \ \ h(e_1, e_2) = 0, \ \ h(e_2, e_2) = \eta \tag{3.33}$$

for some normal vector fields ξ, η. After applying Gauss' formula, (3.24), (3.26) and (3.33), we obtain

$$L_{xx} = \frac{2E_x}{E} L_x + E^2 \xi, \quad L_{xy} = 0, \quad L_{yy} = \frac{2E_y}{E} L_y + E^2 \eta. \tag{3.34}$$

After solving the second equation in (3.34), we find

$$L(x, y) = z(x) + w(y) \tag{3.35}$$

for some vector-valued functions $z(x), w(y)$. Thus, by applying (3.23) and (3.35), we obtain $\langle z', z' \rangle = \langle w', w' \rangle = 0$, $\langle z', w' \rangle = -E^2$. Therefore, z and w are null curves satisfying $\langle z', w' \rangle \neq 0$.

Conversely, if $L : N_1^2 \to \mathbb{E}_s^m$ is an immersion of a Lorentz surface N_1^2 into $\mathbb{E}_s^m$ such that $L = z(x) + w(y)$ for some null curves z, w satisfying $\langle z', w' \rangle \neq 0$, then $\langle L_x, L_x \rangle = \langle L_y, L_y \rangle = 0, \langle L_x, L_y \rangle \neq 0$, $L_{xy} = 0$. Thus, N_1^2 is surface with induced metric given by

$$g = F(x, y)(dx \otimes dy + dy \otimes dx)$$

for some nonzero function F. Hence, after applying (3.31) and $L_{xy} = 0$, we conclude that L is a minimal immersion of a Lorentz surface. □

Remark 3.4. If $m = 3$, this theorem is due to [McNertney (1980)].

For flat minimal Lorentz surfaces in $\mathbb{E}_s^m$, we have [Chen (2011a)].

Corollary 3.4. *A flat Lorentz surface in a pseudo-Euclidean m-space $\mathbb{E}_s^m$ is minimal if and only if, locally, the immersion takes the form*

$$L(x, y) = z(x) + w(y), \tag{3.36}$$

where z and w are null curves satisfying $\langle z', w' \rangle = const. \neq 0$.

Proof. Let $L : N_1^2 \to \mathbb{E}_s^m$ be an isometric immersion of a flat Lorentz surface N_1^2 into a pseudo-Euclidean m-space $\mathbb{E}_s^m$. Then we may choose a local coordinate system $\{x, y\}$ on N_1^2 satisfying

$$g = -(dx \otimes dy + dy \otimes dx) \tag{3.37}$$

Then we find from (3.34) that the immersion L satisfies

$$L_{xx} = \xi, \quad L_{xy} = 0, \quad L_{yy} = \eta \tag{3.38}$$

for some normal vector fields ξ, η. After solving the second equation in (3.38) we find

$$L(x, y) = z(x) + w(y) \tag{3.39}$$

for some vector functions z, w. Thus, by applying (3.37), we find $\langle z', z'\rangle = \langle w', w'\rangle = 0$ and $\langle z', w'\rangle = -1$. Consequently, z and w are null curves satisfying $\langle z'(x), w'(y)\rangle = -1$.

Conversely, consider a map L defined by (3.36) such that z and w are null curves satisfying $\langle z'(x), w'(y)\rangle = const. \neq 0$. Then we have

$$\langle L_x, L_x\rangle = \langle L_y, L_y\rangle = 0, \quad \langle L_x, L_y\rangle = const. \neq 0.$$

Thus, with respect to the induced metric, (3.36) defines an isometric immersion of a flat Lorentz surface N_1^2 into $\mathbb{E}_s^m$. The remaining part follows from Theorem 3.4. □

Since every totally geodesic Lorentz surface in an indefinite space form $R_s^m(c)$ is of constant curvature c, a natural question is the following.

Question 3.1. *Besides totally geodesic ones how many minimal Lorentz surfaces of constant curvature c in $R_s^m(c)$ are there ?*

Corollary 3.4 provides the answer to this basic question for $c = 0$. For $c \neq 0$, we have the following two results from [Chen (2011a)].

Theorem 3.5. *Let N be a Lorentz surface of constant curvature one. Then an isometric immersion $\psi : N \to S_s^m(1)$ is minimal if and only if one of the following three cases occurs:*

(a) *N is an open portion of a totally geodesic $S_1^2(1) \subset S_s^m(1)$;*
(b) *the immersion $L = \iota \circ \psi : N \to S_s^m(1) \subset \mathbb{E}_s^{m+1}$ is locally given by*

$$L(x, y) = \frac{z(x)}{x + y} - \frac{z'(x)}{2},$$

where $z(x)$ is a spacelike curve with constant speed 2 lying in the light cone $\mathcal{LC}$ satisfying $\langle z'', z''\rangle = 0$ and $z''' \neq 0$;
(c) *the immersion $L = \iota \circ \psi : N \to S_s^m(1) \subset \mathbb{E}_s^{m+1}$ is locally given by*

$$L(x, y) = \frac{z(x) + w(y)}{x + y} - \frac{z'(x) + w'(y)}{2},$$

where z and w are curves in $\mathbb{E}_s^{m+1}$ satisfying

$$\begin{aligned} &\left\langle \frac{z(x) + w(y)}{x + y} - \frac{z'(x) + w'(y)}{2}, \frac{z(x) + w(y)}{x + y} - \frac{z'(x) + w'(y)}{2} \right\rangle = 1, \\ &2\langle z + w, z'''\rangle = (x + y)\langle z' + w', z'''\rangle, \\ &2\langle z + w, w'''\rangle = (x + y)\langle z' + w', w'''\rangle. \end{aligned} \tag{3.40}$$

Theorem 3.6. *Let N be a Lorentz surface of constant Gauss curvature -1. Then an isometric immersion $\psi : N \to H^m_s(-1)$ is a minimal immersion if and only if one of the following three cases occurs:*

(a) *N is an open portion of a totally geodesic $H^2_1(-1) \subset H^m_s(-1)$;*
(b) *the immersion $L = \iota \circ \psi : N \to H^m_s(-1) \subset \mathbb{E}^{m+1}_{s+1}$ is locally given by*

$$L(x, y) = z(x) \tanh\Big(\frac{x+y}{\sqrt{2}}\Big) - \frac{z'(x)}{\sqrt{2}},$$

where $z(x)$ is a timelike curve with constant speed $\sqrt{2}$ lying in the light cone $\mathcal{LC} \subset \mathbb{E}^{m+1}_{s+1}$ satisfying $\langle z'', z''\rangle = 4$ and $z''' \neq 2z'$;
(c) *the immersion $L = \iota \circ \psi : N \to H^m_s(-1) \subset \mathbb{E}^{m+1}_{s+1}$ is locally given by*

$$L(x, y) = (z(x) + w(y)) \tanh\left(\frac{x+y}{\sqrt{2}}\right) - \frac{z'(x)+w'(y)}{\sqrt{2}},$$

where z and w are curves satisfying

$$\begin{aligned}
&\left\langle (z+w)\tanh\left(\frac{x+y}{\sqrt{2}}\right) - \frac{z'+w'}{\sqrt{2}}, (z+w)\tanh\left(\frac{x+y}{\sqrt{2}}\right) - \frac{z'+w'}{\sqrt{2}} \right\rangle = -1, \\
&\sqrt{2}\,\langle z+w, 2z' - z'''\rangle \tanh\Big(\frac{x+y}{\sqrt{2}}\Big) = \langle z'+w', 2z' - z'''\rangle\,, \\
&\sqrt{2}\,\langle z+w, 2w' - w'''\rangle \tanh\Big(\frac{x+y}{\sqrt{2}}\Big) = \langle z'+w', 2w' - w'''\rangle\,. \qquad (3.41)
\end{aligned}$$

Remark 3.5. It was shown in [Chen (2011a)] that there exist many minimal Lorentz surfaces of classes (a), (b) and (c) for Theorems 3.5 and 3.6.

Remark 3.6. A natural extension of minimal surfaces are surfaces with parallel mean curvature vector. The classification of such surfaces is much more complicated than minimal ones. Nevertheless, the complete classifications of spacelike and Lorentz surfaces with parallel mean curvature vector in indefinite real space forms were done in [Chen (2009b,c, 2010c)] (see also [Fu and Hou (2010)]). For a recent survey on these, see [Chen (2010j)].

3.8 Marginally trapped surfaces and black holes

The idea of an object with gravity strong enough to prevent light from escaping was proposed in 1783 by a British astronomer Michell (1724-1793), and independently by a French mathematician Laplace (1749-1823) in 1795. Black holes, as currently understood, are described by Einstein's general theory of relativity; it points towards the existence of black holes[1]. The theory predicts that when a large enough amount of mass is present in a sufficiently small region, all paths through space are warped inwards towards the center of the volume, preventing all matter and radiation within it from escaping.

According to the American Astronomical Society, every large galaxy has a super massive black hole ($\sim 10^5$ - $10^9\, M_{sun}$) at its center. The black hole's mass is proportional to the mass of the host galaxy, suggesting that the two are linked very closely. Black holes can't be seen, because everything that falls into them, including light, is trapped. But the swift motions of gas and stars near an otherwise invisible object allows astronomers to calculate that it's a black hole and even to estimate its mass.

The concept of *trapped surfaces*, introduced in [Penrose (1965)] plays very important role in the theory of cosmic black holes. If there is a massive source inside the surface, then close enough to a massive enough source, the outgoing light rays may also be converging; a trapped surface. Everything inside is trapped. Nothing can escape, not even light. It is believed that there will be a marginally trapped surface, separating the trapped surfaces from the untrapped ones, where the outgoing light rays are instantaneously parallel. The surface of a black hole is the *marginally trapped surface*. As times develops, the marginally trapped surface generates a hypersurface in spacetime, a *trapping horizon*.

From mathematical point of view, the light converging condition means that the mean curvature vector, which measures the tension of the surface coming from the surrounding space, is timelike. If the mean curvature vector is future-pointing or past-pointing all over the surface, the trapped surface is accordingly called *future trapped* or *past trapped*. Thus, *"a surface is marginally trapped" means that the mean curvature vector field is lightlike at each point on the surface* (cf. e.g. [Senovilla (2002)]).

According to Beltrami's formula, an isometric immersion $\phi : N \to \mathbb{E}^4_s$ of a pseudo-Riemannian surface N is minimal if and only if ϕ is harmonic. The following result relates marginally trapped surfaces with biharmonic surfaces [Chen and Ishikawa (1991)].

[1]Although Einstein was aware that his field equation implies that a large star would collapse in on itself and form a black hole, he denied several times that black holes could form. Somehow, he convinced that something like an explosion would always occur to throw off mass and prevent the formation of a black hole. In 1939, he published an article [Einstein (1939)] which argues that a star collapsing would spin faster and faster, spinning close to the speed of light with almost infinite energy well before the point where it is about to collapse into a black hole. His conclusion is understood to be wrong.

Theorem 3.7. *Let $\phi : N \to \mathbb{E}_1^4$ be a biharmonic surface in Minkowski spacetime $\mathbb{E}_1^4$ with flat normal connection. Then ϕ is marginally trapped if and only if, up to rigid motions of $\mathbb{E}_1^4$, the surface is given by*

$$\phi(u,v) = (f(u,v), u, v, f(u,v)),$$

where f is proper biharmonic function on N, i.e., $\Delta f \neq 0$ and $\Delta^2 f = 0$.

The following three results on marginally trapped surfaces are obtained in [Chen (2009e); Chen and Van der Veken (2010)].

Proposition 3.12. *Let $\phi : N \to \mathcal{LC}$ be an isometric immersion of a spacelike surface N into the light cone $\mathcal{LC} = \{y \in \mathbb{E}_1^4 : \langle y, y \rangle = 0\}$. Then N is marginally trapped in $\mathbb{E}_1^4$ if and only if N is flat.*

Proposition 3.13. *Let N be a spatial surface in the de Sitter space-time $S_1^4(1)$. If N lies in $\{(y,1) \in S_1^4(1) \subset \mathbb{E}_1^5 : y \in \mathbb{E}_1^4, \langle y, y \rangle = 0\}$, then N is marginally trapped in $S_1^4(1)$ if and only if it is of curvature one.*

Proposition 3.14. *Let N be a spatial surface in the anti-de Sitter space-time $H_1^4(-1)$. If N lies in $\{(1,y) \in H_1^4(-1) \subset \mathbb{E}_2^5 : y \in \mathbb{E}_2^4, \langle y, y \rangle = 0\}$, then N is marginally trapped if and only if it is of curvature -1.*

For marginally trapped surfaces with positive nullity in the Minkowski spacetime, we have the following [Chen and Van der Veken (2007a)].

Theorem 3.8. *Up to Minkowskian motions, there exist two families of marginally trapped surfaces with positive relative nullity in $\mathbb{E}_1^4$:*

(1) *a surface defined by $\big(f(x), x, y, f(x)\big)$, where $f(x)$ is a differentiable function with $f''(x)$ being nowhere zero;*
(2) *a surface defined by*

$$\left(\int_0^x r(x)q'(x)dx + q(x)y, y\cos x - \int_0^x r(x)\sin x dx, \right.$$
$$\left. y \sin x + \int_0^x r(x)\cos x dx, \int_0^x r(x)q'(x)dx + q(x)y \right),$$

where $q(x)$ and $r(x)$ are functions defined on an open interval $I \ni 0$ such that $q''(x) + q(x) \neq 0$ for each $x \in I$;

Conversely, every marginally trapped surface with positive relative nullity in the Minkowski spacetime $\mathbb{E}_1^4$ is congruent to an open portion of a surface obtained from the two families.

Marginally trapped surfaces with parallel mean curvature vector in $\mathbb{E}^4_1$ were classified in [Chen and Van der Veken (2010)].

Theorem 3.9. *Let N be a marginally trapped surface with parallel mean curvature vector in the Minkowski spacetime $\mathbb{E}^4_1$. Then, with respect to suitable Minkowskian coordinates (t, x_2, x_3, x_4) on $\mathbb{E}^4_1$, N is an open part of one of the following six types of surfaces:*

(1) *a flat parallel biharmonic surface given by*

$$\frac{1}{2}\Big((1-b)u^2+(1+b)v^2, (1-b)u^2+(1+b)v^2, 2u, 2v\Big), \ b \in \mathbf{R};$$

(2) *a flat parallel surface given by* $a\big(\cosh u, \sinh u, \cos v, \sin v\big)$, $a > 0$;
(3) *a flat surface given by $(f(u,v), u, v, f(u,v))$, where f is a function on N such that $\Delta f = c$ for some nonzero real number c;*
(4) *a flat surface lying in the light cone $\mathcal{LC}$;*
(5) *a non-parallel surface lying in the de Sitter spacetime $S^3_1(c)$ for some $c > 0$ such that the mean curvature vector H' of N in $S^3_1(c)$ satisfies $\langle H', H'\rangle = -c$;*
(6) *a non-parallel surface lying in the hyperbolic space $H^3(-c)$ for some $c > 0$ such that the mean curvature vector H' of N in $H^3(-c)$ satisfies $\langle H', H'\rangle = c$.*

Conversely, all surfaces of types (1)–(6) above give rise to marginally trapped surfaces with parallel mean curvature vector in $\mathbb{E}^4_1$.

One may define marginally trapped surfaces in a pseudo-Riemannian m-manifold as a spacelike surface with lightlike mean curvature vector. For marginally trapped surfaces in $\mathbb{E}^m_s$ with $m \geq 5$ we have [Chen (2009e)].

Theorem 3.10. *Let $\phi : N \to \mathbb{E}^m_s$ be a marginally trapped surface in $\mathbb{E}^m_s$. If N has parallel mean curvature vector, then either N is congruent to*

(1) *a surface given by $\phi = (f, \psi, f)$, where f is a function satisfying $\Delta f = b$ for some real number $b \neq 0$ and $\psi : N \to \mathbb{E}^{m-2}_{s-1}$ is an isometric minimal immersion, or*
(2) *N is a marginally trapped surface lying in a totally geodesic $\mathbb{E}^4_1 \subset \mathbb{E}^m_s$.*

A spacetime is called *strictly stationary* if it contains a Killing vector field which is time-like everywhere.

The following non-existence result for strictly stationary space-time was proved in [Mars and Senovilla (2003)].

Theorem 3.11. *There do not exist closed marginally trapped surfaces in strictly stationary space-times.*

Remark 3.7. The *boost group* in Minkowski space-time $\mathbb{E}^4_1$ is defined by

$$G = \left\{ \begin{pmatrix} \cosh\theta & \sinh\theta & 0 & 0 \\ \sinh\theta & \cosh\theta & 0 & 0 \\ 0 & 0 & 1 & 0 \\ 0 & 0 & 0 & 1 \end{pmatrix} : \theta \in \mathbf{R} \right\}.$$

It is known in general relativity theory that the inertial vacuum state of Minkowski space-time is a thermal state when analyzed regarding the notion of time translations defined by a one-parameter family of Lorentz boosts. The orbits of these boost isometries correspond to a family of uniformly accelerating observers. This result, known as the *Unruh effect.*

Marginally trapped surfaces in $\mathbb{E}^4_1$ invariant under boost group were studied in [Haesen and Ortega (2007)]. In particular, it was proved in [Haesen and Ortega (2007)] that there exist no G-invariant extremal surfaces in Minkowski space-time with constant Gaussian curvature.

3.9 Quasi-minimal surfaces in indefinite space forms

There is an important subject closely related with marginally trapped surfaces in spacetimes; namely, quasi-minimal surfaces. Recall that a *Lorentz surface* in a pseudo-Riemannian manifold is called *quasi-minimal* if its mean curvature vector is lightlike at each point.

Quasi-minimal surfaces with parallel mean curvature vector in $\mathbb{E}^4_2$ were classified in [Chen and Garay (2009)].

Theorem 3.12. *Let $\phi : N \to \mathbb{E}^4_2$ be a flat quasi-minimal surface in $\mathbb{E}^4_2$. If N has parallel mean curvature vector, then ϕ is congruent to an open portion of one of surfaces of the following seven types:*

(1) *a surface defined by $\frac{1}{\sqrt{2}}\big(xy+f(x)+k(y), x+y, x-y, xy+f(x)+k(y)\big)$, for some functions $f(x), k(y)$;*
(2) *a surface defined by*

$$\frac{1}{2ab}\Big(2ab\cos ax\cos by - \sin ax\sin by, 2ab\cos ax\sin by + \sin ax\cos by, 2ab\cos ax\cos by + \sin ax\sin by, 2ab\cos ax\sin by - \sin ax\cos by\Big),\ \ a,b>0;$$

(3) *A surface defined by*

$$\frac{1}{2ab}\Big(2ab\cos ax\cosh by-\sin ax\sinh by,2ab\cos ax\sinh by \\ +\sin ax\cosh by,2ab\cos ax\cosh by+\sin ax\sinh by, \\ 2ab\cos ax\sinh by-\sin ax\cosh by\Big),\ \ a,b>0;$$

(4) *A surface defined by*

$$\frac{1}{2ab}\Big(2ab\cosh ax\cosh by-\sinh ax\sinh by,2ab\cosh ax\sinh by \\ +\sinh ax\cosh by,2ab\cosh ax\cosh by+\sinh ax\sinh by, \\ 2ab\cosh ax\sinh by-\sinh ax\cosh by\Big),\ \ a,b>0;$$

(5) *A surface defined by* $\phi(x,y)=z(x)y+w(x)$, *where* $z(x)$ *is a null curve in the light-cone* $\mathcal{LC}$ *and* $w(x)$ *is a null curve such that* $\langle z',w'\rangle=0$, $\langle z,w'\rangle=-1$ *and* $z''(x)+\beta(x)z(x)=0$ *for some function* $\beta(x)$*;*
(6) *A surface defined by* $\phi=z(y)\cos ax+w(y)\sin ax$, *where* z,w *are null curves lying in* $\mathcal{LC}$ *such that* $\langle z,w\rangle=z''+\delta z=w''+\delta w=0$ *and* $\langle z,w'\rangle=\frac{1}{a}$ *for a non-constant function* $\delta(y)$ *and a positive number* a*;*
(7) *A surface defined by* $\phi=z(y)\cosh ax+w(y)\sinh ax$, *where* z,w *are null curves lying in* $\mathcal{LC}$ *such that* $\langle z,w\rangle=z''+\delta z=w''+\delta w=0$ *and* $\langle z,w'\rangle=\frac{1}{a}$ *for a non-constant function* $\delta(y)$ *and a positive number* a.

Biharmonic marginally quasi-minimal surfaces in $\mathbb{E}^4_2$ were classified in [Chen (2009d)].

Theorem 3.13. *A quasi-minimal surface* N *in* $\mathbb{E}^4_2$ *is biharmonic if and only if* N *congruent to one of the following surfaces:*

(1) *a surface given by* $\frac{1}{\sqrt{2}}\big(\varphi(x,y),x+y,x-y,\varphi(x,y)\big)$, *where* $\varphi(x,y)$ *is a function satisfying* $\varphi_{xy}\neq 0$ *and* $\varphi_{xxyy}=0$ *on an open domain* $U\subset\mathbb{E}^2_1$*;*
(2) *a surface given by* $\phi(x,y)=z(x)y+w(x)$, *where* z *is a null curve in* $\mathcal{LC}$ *and* w *is a null curve such that* $\langle z',w'\rangle=0$ *and* $\langle z,w'\rangle=-1$.

Several other important families of quasi-minimal surfaces have also been classified. For examples, quasi-minimal surfaces with constant Gauss curvature in $\mathbb{E}^4_2$ were classified in [Chen (2008b, 2009f); Chen and Yang (2010)]. Moreover, quasi-minimal Lagrangian surfaces and quasi-minimal slant surfaces in complex space forms were classified respectively in [Chen and Dillen (2007); Chen and Mihai (2009)].

Chapter 4

Warped Products and Twisted Products

One of the most fruitful generalizations of the direct product of two pseudo-Riemannian manifolds is the warped product defined in [Bishop and O'Neill (1964)]. The concept of warped products appeared in the mathematical and physical literature long before [Bishop and O'Neill (1964)], e.g. warped products were called *semi-reducible (pseudo) Riemannian spaces* in [Kruchkovich (1957)]. The notion of warped products plays very important roles in differential geometry as well as in mathematical physics, especially in general relativity.

A warped product is a pseudo-Riemannian manifold whose metric tensor can be written in the form:

$$g = \sum_{i,j} g_{ij}(y)dy^i \otimes dy^j + f(y) \sum_{s,t} g_{st}(x)dx^s \otimes dx^t,$$

where the warped geometry decomposes into a product of the "y" geometry and the "x" geometry, except that the second part is warped, i.e., it is rescaled by a scalar function of the other coordinates "y". If we substitute the variable y for time variable t and x for a 3-dimensional spatial space, then the first part becomes the effect of time in Einstein's curved space. How it curves space will define one or the other solution to a spacetime model. For that reason different models of spacetime in general relativity are often expressed in terms of warped geometry.

Many basic solutions of the Einstein field equations are warped products. For instance, both Schwarzschild's and Robertson-Walker's models in general relativity are warped products. Schwarzschild's spacetime[1] is

[1]Schwarzschild metric is the first non-trivial exact solution to Einstein's field equations discovered in [Schwarzschild (1916)] by the astro-physicist Karl Schwarzschild. This solution laid the groundwork for the description of the final stages of gravitational collapse, and the objects known today as black holes (cf. [Hawking and Ellis (1973)]).

the best relativistic model that describes the outer space around a massive star or a black hole and the Robertson-Walker model describes a simply-connected homogeneous isotropic expanding or contracting universe.

Basic properties on warped products can be found in [Bishop and O'Neill (1964); O'Neill (1983)].

Twisted products provide another natural extensions of direct products. The notion of twisted products extends the notion of warped products in a very natural way.

4.1 Basics of warped products

Let B and F be two pseudo-Riemannian manifolds of positive dimensions equipped with pseudo-Riemannian metrics g_B and g_F, respectively, and let f be a positive smooth function on B. Consider the product manifold $B \times F$ with its natural projection $\pi : B \times F \to B$ and $\eta : B \times F \to F$.

The *warped product* $M = B \times_f F$ is the manifold $B \times F$ equipped with the pseudo-Riemannian structure such that

$$\langle X, X\rangle = \langle \pi_*(X), \pi_*(X)\rangle + f^2(\pi(x))\,\langle \eta_*(X), \eta_*(X)\rangle$$

for any tangent vector $X \in TM$. Thus we have

$$g = g_B + f^2 g_F. \tag{4.1}$$

The function f is called the *warping function* of the warped product. When $f = 1$, $B \times_f F$ is a direct product.

For a warped product $B \times_f F$, B is called the *base* and F the *fiber*. The leaves $B \times \{q\} = \eta^{-1}(q)$ and the fibers $\{p\} \times F = \pi^{-1}(p)$ are pseudo-Riemannian submanifolds of M. Vectors tangent to leaves are called *horizontal* and those tangent to fibers are called *vertical*. We denote by $\mathcal{H}$ the orthogonal projection of $T_{(p,q)}M$ onto its horizontal subspace $T_{(p,q)}(B\times\{q\})$ and by $\mathcal{V}$ the projection onto the vertical subspace $T_{(p,q)}(\{p\} \times F)$.

If $u \in T_pB$, $p \in B$ and $q \in F$, then the *lift* $\bar{u}$ of u to (p,q) is the unique vector in $T_{(p,q)}M$ such that $\pi_*(\bar{u}) = u$. For a vector field $X \in \mathcal{X}(B)$, the lift of X to M is the vector field $\bar{X}$ whose value at each (p,q) is the lift of X_p to (p,q). The set of all horizontal lifts is denoted by $\mathfrak{L}(B)$. Similarly, we denote by $\mathfrak{L}(F)$ the set of all vertical lifts.

For $\bar{X}, \bar{Y} \in \mathfrak{L}(B)$ and $\bar{V}, \bar{W} \in \mathfrak{L}(F)$, we have

$$[\bar{X}, \bar{Y}] = [X, Y]^- \in \mathfrak{L}(B), \tag{4.2}$$

$$[\bar{V}, \bar{W}] = [V, W]^- \in \mathfrak{L}(F), \tag{4.3}$$

$$[\bar{X}, \bar{V}] = 0, \tag{4.4}$$

where $[X, Y]^-$ denotes the lift of $[X, Y]$.

Lemma 4.1. *If $\lambda \in \mathcal{F}(B)$, then the gradient of the lift $\lambda \circ \pi$ of λ to $M = B \times_f F$ is the lift to M of the gradient of λ on B.*

Proof. If v is a vertical vector tangent to M, it follows from $\pi_*(v) = 0$ that

$$\langle \nabla(\lambda \circ \pi), v \rangle = v(\lambda \circ \pi) = \pi_*(v)\lambda = 0.$$

Thus, $\nabla(\lambda \circ \pi)$ is a horizontal vector. Hence, if z is horizontal, we have

$$\begin{aligned} \langle \pi_*(\nabla(\lambda \circ \pi)), \pi_*(z) \rangle &= \langle \nabla(\lambda \circ \pi), z \rangle \\ &= z(\lambda \circ \pi) = (\pi_* z)\lambda \\ &= \langle \nabla\lambda, \pi_* z \rangle , \end{aligned}$$

which implies that at each point, $\pi_*(\nabla(\lambda \circ \pi)) = \nabla\lambda$. □

Based on Lemma 4.1, we may simply write λ for $\lambda \circ \pi$ and $\operatorname{grad}\lambda$ for $\operatorname{grad}(\lambda \circ \pi)$. The Levi-Civita connection ∇ of $M = B \times_f F$ is related with the Levi-Civita connections of B and F as follows.

Proposition 4.1. *For $X, Y \in \mathfrak{L}(B)$ and $V, W \in \mathfrak{L}(F)$, we have on $B \times_f F$ that*

(1) $\nabla_X Y \in \mathfrak{L}(B)$ *is the lift of* $\nabla_X Y$ *on* B*;*
(2) $\nabla_X V = \nabla_V X = \left(\frac{Xf}{f}\right) V$*;*
(3) $\operatorname{nor}(\nabla_V W) = h(V, W) = -\frac{\langle V, W \rangle}{f} \nabla f$*;*
(4) $\tan(\nabla_V W) \in \mathfrak{L}(F)$ *is the lift of* $\nabla'_V W$ *on* F*, where* ∇' *is the Levi-Civita connection of* F*.*

Proof. Property (1) can be proved as follows. From Koszul's formula we find

$$2 \langle \nabla_X Y, V \rangle = \langle V, [X, Y] \rangle - V \langle X, Y \rangle$$

due to $[X, V] = [Y, V] = 0$. Since X, Y are lifts from B, $\langle X, Y \rangle$ is constant on fibers. Because V is vertical, $V\langle X, Y \rangle = 0$. But $[X, Y]$ is tangent to leaves, $\langle V, [X, Y] \rangle = 0$. Hence, $\langle \nabla_X Y, V \rangle = 0$ for all $V \in \mathfrak{L}(F)$. This shows that $\nabla_X Y$ is horizonal. Since each $\pi|_{B \times q}$ is an isometry, we obtain properties (1).

From $[X,V]=0$, we find $\nabla_X V=\nabla_V X$. Since these vector fields are vertical, we obtain $\langle \nabla_X V, Y\rangle = -\langle V, \nabla_X, Y\rangle = 0$. Thus, by the Koszul formula, we get

$$2\langle \nabla_X V, W\rangle = X\langle V, W\rangle . \tag{4.5}$$

On the other hand, by the definition of the warped product metric tensor, we find

$$\langle V, W\rangle_{(p,q)} = f^2(p)\langle V_q, W_q\rangle .$$

So, after writing f for $f\circ\pi$, we have $\langle V,W\rangle = f^2(\langle V,W\rangle\circ\eta)$. Hence

$$\begin{aligned} X\langle V,W\rangle &= X(f^2(\langle V,W\rangle\circ\eta) = 2fXf(\langle V,W\rangle\circ\eta) \\ &= 2\frac{Xf}{f}\langle V,W\rangle . \end{aligned}$$

Combining this with (4.5) give property (2).

By property (2) we find

$$\langle \nabla_V W, X\rangle = -\langle W, \nabla_V X\rangle = -\frac{Xf}{f}\langle V,W\rangle .$$

Thus, after applying Lemma 4.1, we find $Xf=\langle \nabla f, X\rangle$ on M as on B. Hence, for any X, we obtain

$$\langle \nabla_V W, X\rangle = -\frac{\langle V,W\rangle}{f}\langle \nabla f, X\rangle$$

which implies property (3).

Since V and W are tangent to all fibers, $\tan(\nabla_V W)$ is the fiber covariant derivative applied to the restrictions of V and W to that fiber. Therefore, we have property (4). □

4.2 Curvature of warped products

Consider a warped product $M = B\times_f F$. The *lift* $\tilde{T}$ of a covariant tensor T on B to M is its pullback $\pi^*(T)$ under the projection $\pi: M\to B$. Let ${}^B R$ and ${}^F R$ denote the lifts to M of the curvature tensors of B and F, respectively.

The next result provides the curvature of a warped product $M=B\times_f F$ in terms of its warping function f and the curvature tensors ${}^B R$ and ${}^F R$ of B and F.

Proposition 4.2. *Let $M = B\times_f F$ be a warped product of two pseudo-Riemannian manifolds. If $X,Y,Z\in\mathfrak{L}(B)$ and $U,V,W\in\mathfrak{L}(F)$, then*

(1) *$R(X,Y)Z \in \mathfrak{L}(B)$ is the lift of ${}^{B}R(X,Y)Z$ on B;*

(2) $R(X,V)Y = \dfrac{H^f(X,Y)}{f}V$;

(3) $R(X,Y)V = R(V,W)X = 0$;

(4) $R(X,V)W = -\dfrac{\langle V,W\rangle}{f}\nabla_X(\nabla f)$;

(5) $R(V,W)U = {}^{F}R(V,W)U + \dfrac{\langle \nabla f, \nabla f\rangle}{f^2}\{\langle V,U\rangle W - \langle W,U\rangle V\}$,

where R is the curvature tensor of M and H^f is the Hessian of f.

Proof. Since the projection $\pi : M \to B$ is isometric on each leaf, ${}^{B}R$ gives the the Riemannian curvature tensor of each leaf. Because leaves are totally geodesic in M, ${}^{B}R$ agrees with the curvature tensor R of M on horizontal vectors. Thus we have (1).

For $X, Y \in \mathfrak{L}(B)$ and $V \in \mathfrak{L}(F)$, we have $[V,X] = 0$ and so

$$R(X,V)Y = \nabla_X\nabla_V Y - \nabla_V\nabla_X Y.$$

Hence, by Proposition 4.1, we find

$$\begin{aligned}\nabla_X\nabla_V Y &= \nabla_X((Y(\ln f)V) \\ &= (XY(\ln f))V + (Y(\ln f))\nabla_X V \\ &= \left\{XY(\ln f) + (Yf)Xf^{-1}\right\}V + (X(\ln f))(Y(\ln f))V \\ &= (XY(\ln f))V.\end{aligned}$$

Thus we find

$$R(X,V)Y = (XY(\ln f))V - \nabla_V\nabla_X Y. \tag{4.6}$$

On the other, since $\nabla_X Y \in \mathfrak{L}(B)$, we have $\nabla_V\nabla_X Y = (\nabla_X Y(\ln f))V$. Combining this with (4.6) gives (2).

To prove (3) we assume that $[V,W] = 0$. Then we have

$$R(V,W)X = \nabla_V\nabla_W X - \nabla_W\nabla_V X.$$

Since

$$\nabla_V\nabla_W X = (VX(\ln f))W + (X(\ln f))\nabla_V W = (X(\ln f))\nabla_V W,$$

$X(\ln f)$ is constant on fibers. Therefore, by Proposition 1.7, we obtain

$$\begin{aligned}R(V,W)X &= (X(\ln f))\nabla_V W - \nabla_W((X(\ln f))V) \\ &= (X(\ln f))(\nabla_V W - \nabla_W V) \\ &= (X(\ln f))[V,W] = 0.\end{aligned}$$

To show $R(X,Y)V=0$, first we find

$$R(X,Y;V,W)=R(V,W;X,Y)=0$$

by applying Proposition 4.1. Then, from (1) we find

$$R(X,Y;V,Z)=-R(X,Y;Z,V)=0.$$

Consequently, we obtain $R(X,Y)V=0$. This gives (3).

To prove (4), first we notice that $R(X,V)W$ is horizontal, since $R(X,V;W,U)=R(W,U;X,V)=0$ according to (3). Thus, by the first Bianchi identity, we find $R(X,V)W=R(X,W)V$. Thus, after using (2), we obtain

$$\begin{aligned} R(X,V;W,Y) &= R(V,X;Y,W) \\ &= -\frac{H^f(X,Y)}{f}\langle V,W\rangle \\ &= -\frac{\langle V,W\rangle}{f}\langle \nabla_X(\nabla f),Y\rangle . \end{aligned}$$

Since $R(X,V)W$ is horizontal and equation holds for every Y, we have (4).

For (5), we observe that $R(V,W)U$ is a vertical vector field since $R(V,W;U,X)=-R(V,W;X,U)=0$ by (3). Now, because the projection $\eta: M\to F$ is a homothety on fibers, ${}^FR(V,W)U\in\mathfrak{L}(F)$ is the application to V,W,U of the curvature tensor of each fiber. Consequently, ${}^FR(V,W)U$ and $R(V,W)U$ are related by the equation of Gauss. Finally, by combining this with the fact the second fundamental form of the fibers satisfies $h(V,W)=-(\langle V,W\rangle/f)\nabla f$, we obtain (5). □

From Proposition 4.2 we have the following.

Corollary 4.1. *On a warped product $M=B\times_f F$ with $k=\dim F>1$, let X,Y be horizontal vectors and V,W vertical vectors. Then the Ricci tensor Ric of M satisfies*

(1) $Ric(X,Y)={}^BRic(X,Y)-\dfrac{k}{f}H^f(X,Y)$;

(2) $Ric(X,V)=0$;

(3) $Ric(V,W)={}^FRic(V,W)-\left\{\dfrac{\Delta f}{f}-(k-1)\dfrac{\langle\nabla f,\nabla f\rangle}{f^2}\right\}\langle V,W\rangle$,

where ${}^BRic$ and ${}^FRic$ are the lifts of the Ricci curvatures of B and of F, respectively.

For an isometric immersion $\phi: N\to M$ and $\varphi\in\mathcal{F}(M)$, we denote by $D\varphi$ the $T^\perp N$-component of the gradient $\nabla\varphi$.

4.3 Warped product immersions

The notion of warped products can be naturally extended to multiply warped products as follows:

Definition 4.1. Let $N_1, \cdots, N_\ell$ be ℓ pseudo-Riemannian manifolds and let $N = N_1 \times \cdots \times N_\ell$ be the Cartesian product of $N_1, \ldots, N_\ell$. Denote by $\pi : N \to N_1$ the canonical projection of N onto N_1 and by $\eta_i : N \to N_i$ the canonical projection of N onto N_i for $i = 2, \ldots, \ell$. If $f_1, \cdots, f_\ell : N_1 \to \mathbf{R}_+$ are positive-valued functions in $\mathcal{F}(N_1)$, then

$$\langle X, Y \rangle = \langle \pi_* X, \pi_* Y \rangle + \sum_{i=2}^{\ell} (f_i \circ \pi)^2 \langle \eta_{i_*} X, \eta_{i_*} Y \rangle \tag{4.7}$$

defines a pseudo-Riemannian metric g on N, called a *multiply warped product metric*. The product manifold N endowed with this metric g, denoted by $N_1 \times_{f_2} N_2 \times \cdots \times_{f_\ell} N_\ell$, which is called a *multiply warped product*.

For a multiply warped product $N_1 \times_{f_2} N_2 \times \cdots \times_{f_\ell} N_\ell$, let $\mathcal{D}_i$ denote the distribution obtained from the vectors tangent to N_i (or more precisely, vectors tangent to the horizontal lifts of N_i).

Definition 4.2. Assume that $\phi : N_1 \times_{f_2} N_2 \times \cdots \times_{f_\ell} N_\ell \to M$ is an isometric immersion of $N_1 \times_{f_2} N_2 \times \cdots \times_{f_\ell} N_\ell$ into a pseudo-Riemannian manifold M. Denote by h the second fundamental form of ϕ. Then ϕ is called *mixed totally geodesic* if $h(\mathcal{D}_i, \mathcal{D}_j) = \{0\}$ holds for distinct $i, j \in \{1, \ldots, \ell\}$.

Let $\nabla^1, R^1, \langle\ ,\ \rangle_1$, etc., be the Levi-Civita connection, the Riemann curvature tensor, the scalar product, etc., of the Riemannian product

$$N_1 \times N_2 \times \cdots \times N_\ell \text{ with } f_2 = \ldots = f_\ell = 1$$

and denote by $\nabla^f, R^f, \langle\ ,\ \rangle_f$, etc., the corresponding quantities of the warped product $N_1 \times_{f_2} N_2 \times \cdots \times_{f_\ell} N_\ell$.

The next lemma provides relations between the Levi-Civita connections ∇^1 and ∇^f and the curvature tensors R^1 and $\tilde{R}^f$ [Ponge and Reckziegel (1993); Nölker (1996); Dobarro and Ünal (2005)].

Lemma 4.2. *Let $N = N_1 \times_{f_2} N_2 \times \cdots \times_{f_\ell} N_\ell$ be a multiply warped product of ℓ pseudo-Riemannian manifolds $N_1, \ldots, N_\ell$. If we put*

$$H_i = -\nabla((\ln f_i) \circ \pi_1), \tag{4.8}$$

then for $X, Y \in \mathcal{X}(N)$ we have

$$\nabla^f_X Y - \nabla^1_X Y = \sum_{i=2}^{\ell} \left(\langle X^i, Y^i \rangle H_i - \langle H_i, X \rangle Y^i - \langle H_i, Y \rangle X^i \right), \tag{4.9}$$

$$\begin{aligned} R^f(X,Y) - R^1(X,Y) &= \sum_{i=2}^{\ell} (\nabla^f_{X^1} H_i - \langle H_i, X \rangle H_i) \wedge Y^i \\ &\quad + \sum_{i=2}^{k} X^i \wedge (\nabla^f_{Y^1} H_i - \langle H_i, Y \rangle H_i) \\ &\quad - \sum_{i,j=2}^{\ell} \langle H_i, H_j \rangle X^i \wedge Y^j, \end{aligned} \tag{4.10}$$

where $\langle\ ,\ \rangle = \langle\ ,\ \rangle_f$, X^i is the N_i-component of X and $X \wedge Y$ is defined by

$$(X \wedge Y)Z = \langle Z, Y \rangle X - \langle Z, X \rangle Y.$$

Proof. Since $\mathcal{D}_i$ is parallel with to $\langle\ ,\ \rangle_1$, we have

$$(\nabla_X Y)^i = \nabla_X Y^i \quad \text{for all } X, Y \in \mathcal{X}(N). \tag{4.11}$$

The tensor field S on N given by the right side of (4.9) is symmetric, hence $\hat{\nabla} = \nabla^1 + S$ is a torsion-free affine connection on N. From (4.7), (4.8) and (4.11) we immediately derive that $\hat{\nabla}$ is metric with respect to $\langle\ ,\ \rangle_f$, thus $\hat{\nabla} = \nabla^f$ by the uniqueness of Levi-Civita connection. This gives (4.9).

Equation (4.10) can be obtained by a lengthy direct calculation. □

The following is the indefinite version of *Moore's lemma* [Moore (1971)] stated in [Magid (1984)].

Lemma 4.3. *Let $\phi : N_1 \times \cdots \times N_\ell \to \mathbb{E}^m$ be an isometric immersion of the direct product of ℓ pseudo-Riemannian manifolds $N_1, \ldots, N_\ell$ into $\mathbb{E}^m_j$. Then ϕ is a mixed totally geodesic immersion if and only if ϕ is a product immersion, i.e., there exist an isometry Ψ of $\mathbb{E}^m$ and isometric immersions $\phi_i : N_i \to \mathbb{E}^{m_i}_{j_i}$, $1 \le i \le \ell$, such that*

$$\Psi \circ \phi(p_1, \ldots, p_\ell) = (\phi_1(p_1), \ldots, \phi_\ell(p_\ell)),$$

where $p_i \in N_i$, $1 \le i \le \ell$.

Definition 4.3. Let $N_1 \times_{\rho_2} N_2 \times \cdots \times_{\rho_\ell} N_\ell$ be a multiply warped product and let $\phi_i : N_i \to M_i$, $i = 1, \cdots, \ell$, be isometric immersions. Define $\sigma_i = \rho_i \circ \phi_1 : N_1 \to \mathbf{R}_+$ for $i = 2, \cdots, \ell$. Then the map

$$\phi : N_1 \times_{\sigma_2} N_2 \times \cdots \times_{\sigma_\ell} N_\ell \to M_1 \times_{\rho_2} M_2 \times \cdots \times_{\rho_\ell} M_\ell$$

defined by

$$\phi(p_1, \cdots, p_\ell) = (\phi_1(p_1), \cdots, \phi_\ell(p_\ell))$$

is an isometric immersion, called a *warped product immersion.*

Definition 4.4. A multiply warped product $M_1 \times_{\rho_2} M_2 \times \cdots \times_{\rho_\ell} M_\ell$ is called a *warped product representation* of a real space form $R^m(k)$ if the warped product $M_1 \times_{\rho_2} M_2 \times \cdots \times_{\rho_\ell} M_\ell$ is an open dense subset of $R^m(k)$.

S. Nölker extended Moore's result to the following [Nölker (1996)].

Theorem 4.1. *Let $\phi : N_1 \times_{\sigma_2} N_2 \times \cdots \times_{\sigma_\ell} N_\ell \to R^m(k)$ be an isometric immersion of a multiply warped product into a complete simply-connected real space form $R^m(k)$ of constant curvature k. If ϕ is mixed totally geodesic, then there is an explicitly constructible isometry ψ of $M_1 \times_{\rho_2} M_2 \times \cdots \times_{\rho_\ell} M_\ell$ of a multiply warped product onto an open dense subset of $R^m(k)$, where M_1 is an open subset of a standard space and $M_2, \ldots, M_\ell$ are standard spaces, and there exist isometric immersions $\phi_i : N_i \to M_i, i = 1, \ldots, \ell$, such that $\sigma_i = \rho_i \circ \phi_1$ for $i = 2, \ldots, \ell$ and $\phi = \psi \circ (\phi_1 \times \cdots \times \phi_\ell)$.*

Here, by a *standard space* we mean a sphere, a Euclidean space or a hyperbolic space of constant curvature.

Let $\phi : N_1 \times_{f_2} N_2 \times \cdots \times_{f_\ell} N_\ell \to M$ be an isometric immersion of a multiply warped product $N_1 \times_{f_2} N_2 \times \cdots \times_{f_\ell} N_\ell$ into a pseudo-Riemannian manifold M. Let h_i denote the restriction of the second fundamental form to $\mathcal{D}_i$, $i = 1, \ldots, \ell$. Denote by $\operatorname{trace} h_i$ the trace of h_i restricted to N_i, i.e.,

$$\operatorname{trace} h_i = \sum_{\alpha=1}^{n_i} h(e_\alpha, e_\alpha)$$

for an orthonormal frame fields $e_1, \ldots, e_{n_i}$ of $\mathcal{D}_i$.

The *partial mean curvature vector* H_i is defined by

$$H_i = \frac{\operatorname{trace} h_i}{\dim N_i}, \quad i = 1, \ldots, \ell. \tag{4.12}$$

Definition 4.5. An immersion $\phi : N_1 \times_{f_2} N_2 \times \cdots \times_{f_\ell} N_\ell \to M$ is called *N_i-totally geodesic* (resp. *N_i-minimal*) if h_i (resp. H_i) vanishes identically.

Notation 4.1. Let h^ϕ denote the second fundamental form of a warped product immersion

$$\phi = (\phi_1 \ldots, \phi_\ell) : N_1 \times_{\sigma_2} N_2 \times \cdots \times_{\sigma_\ell} N_\ell \to M_1 \times_{\rho_2} M_2 \times \cdots \times_{\rho_\ell} M_\ell,$$

and let h^1 denote the second fundamental form of the corresponding direct product immersion

$$(\phi_1 \ldots, \phi_\ell) : N_1 \times \cdots \times N_\ell \to M_1 \times \cdots \times M_\ell.$$

Denote by ∇^1 and ∇^f the Levi-Civita connections of $N_1 \times N_2 \times \cdots \times N_\ell$ and of $N_1 \times_f N_2 \times \cdots \times_{f_\ell} N_\ell$, respectively; and by $\tilde{\nabla}^1$ and $\tilde{\nabla}^\rho$ the Levi-Civita connections of $M_1 \times M_2 \times \cdots \times M_\ell$ and $M_1 \times_{\rho_2} M_2 \times \cdots \times_{\rho_\ell} M_\ell$, respectively.

The second fundamental forms h^ϕ and h^1 are related by the following.

Lemma 4.4. *Let $\phi : N_1 \times_f N_2 \times \cdots \times_{f_\ell} N_\ell \to M_1 \times_{\rho_2} M_2 \times \cdots \times_{\rho_\ell} M_\ell$ be a warped product immersion between pseudo-Riemannian multiply warped products. Then h^ϕ and h^1 are related by*

$$h^\phi(X,Y) = h^1(X,Y), \quad X,Y \in \mathcal{D}_1, \tag{4.13}$$
$$h^\phi(Z,W) = h^1(Z,W) - \langle Z,W\rangle\, D(\ln \rho_i), \quad Z,W \in \mathcal{D}_i, \tag{4.14}$$
$$h^\phi(\mathcal{D}_i,\mathcal{D}_j) = \{0\}, \quad 1 \le i \ne j \le \ell. \tag{4.15}$$

Proof. Follows from Gauss' formula and (4.9). □

Corollary 4.2. *Let $\phi : N_1 \times_f N_2 \times \cdots \times_{f_\ell} N_\ell \to M_1 \times_{\rho_2} M_2 \times \cdots \times_{\rho_\ell} M_\ell$ be a warped product immersion. Then ϕ is totally geodesic if and only if the following two conditions are satisfied:*

(1) *$\phi_1 : N_1 \to M_1$ is totally geodesic and*
(2) *$\phi_i : N_i \to M_i$ is a totally umbilical immersion such that the second fundamental form is given by*

$$h^{\phi_i}(Z,W) = \langle Z,W\rangle\, D(\ln \rho_i), \quad Z,W \in T(N_i),$$

for each $i \in \{2,\ldots,\ell\}$.

Proof. Follows from Lemma 4.4 and the definition of totally umbilical immersions. □

4.4 Twisted products

Twisted products are natural generalizations of warped products, namely the warping function were replaced by twisting function which depend on both factors (cf. [Chen (1981a)]).

Definition 4.6. Let B and F be two pseudo-Riemannian manifolds equipped with pseudo-Riemannian metrics g_B and g_F, respectively, and let f be a positive smooth function on M. The *twisted product* $M = B \times_f F$ is the manifold $B \times F$ equipped with the pseudo-Riemannian metric g given by

$$g(X,Y) = g_B(\pi_*(X), \pi_*(Y)) + f^2 \cdot g_F(\eta_*(X), \eta_*(Y)) \tag{4.16}$$

for tangent vectors $X,Y \in TM$.

When f depends only on B, the twisted product $B \times_f F$ becomes a warped product. If B is a point, the twisted product is nothing but a conformal change of metric on F.

Just like warped products, B is called the *base* and F the *fiber* of the twisted product $B \times_f F$. Both the leaves $B \times q = \eta^{-1}(q)$ and the fibers $p \times F = \pi^{-1}(p)$ are pseudo-Riemannian submanifolds of $B \times_f F$. Vectors tangent to leaves are called *horizontal* and those tangent to fibers are called *vertical*.

We denote by $\mathcal{H}$ the orthogonal projection of $T_{(p,q)}M$ onto its horizontal subspace $T_{(p,q)}(B \times q)$ and by $\mathcal{V}$ the projection onto the vertical subspace $T_{(p,q)}(p \times F)$.

If v is vector tangent to B at $p \in B$ and $q \in F$, then the *lift* $\bar{v}$ of v to (p, q) is the unique vector in $T_{(p,q)}M$ such that $\pi_* \bar{v} = v$. For a vector field $X \in \mathcal{X}(B)$, the lift of X to M is the vector field $\bar{X}$ whose value at each (p, q) is the lift of X_p to (p, q).

The following result can be found in [Chen (1981a)].

Proposition 4.3. *Let $M = B \times_f F$ be a twisted product of two pseudo-Riemannian manifolds. Then*

(1) *leaves are totally geodesic in M;*
(2) *for each $p \in B$, the fiber $\{p\} \times F$ is a totally umbilical submanifold of M with $-(\nabla(\ln f))^H$ as its mean curvature vector, where $(\nabla(\ln f))^H$ is the horizontal component of $\nabla(\ln f)$;*
(3) *fibers have parallel mean curvature vector if and only if f is the product of two positive functions $\lambda \in \mathcal{F}(B)$ and $\mu \in \mathcal{F}(F)$.*

Proof. Since the projection $\pi : M \to B$ is isometric on each leaf, every geodesic of a leaf is a geodesic of M. Thus leaves are totally geodesics.

Let $X, Y, Z \in \mathfrak{L}(B)$ and $V, W \in \mathfrak{L}(F)$. Then $[X, V] = 0$, so

$$\nabla_X V = \nabla_V X. \tag{4.17}$$

From (4.16) we find

$$X\langle V, W\rangle = X(f^2 \langle V, W\rangle_F) = 2(X(\ln f)) \langle V, W\rangle , \tag{4.18}$$

where $\langle \ , \ \rangle_F$ is the scalar product associated with the metric g_F on F.

On the other hand, it follows from (4.17) that

$$\begin{aligned} X\langle V, W\rangle &= \langle \nabla_V X, W\rangle + \langle V, \nabla_W X\rangle \\ &= -\langle \nabla_V W, X\rangle - \langle \nabla_W V, X\rangle \\ &= -2\langle h(V, W), X\rangle , \end{aligned}$$

where h is the second fundamental of fibers.

By comparing this with (4.18) we find

$$h(V,W) = -(\nabla(\ln f))^H \langle V, W\rangle, \tag{4.19}$$

which implies (2).

Now, let us choose $\{e_1,\dots,e_r\}$ to be an orthonormal frame of horizontal spaces. Then the mean curvature vector of fibers can be expressed as

$$H = -\sum_{i=1}^{r} \epsilon_i(e_i(\ln f))e_i. \tag{4.20}$$

Thus, for any vertical vector V, we find

$$\nabla_V H = -\sum_{i=1}^{r} \epsilon_i(Ve_i(\ln f))e_i - \sum_{i=1}^{r} \epsilon_i(e_i(\ln f))\nabla_V e_i. \tag{4.21}$$

Because the leaves are totally geodesic, the last term in (4.21) is vertical. Hence, the mean curvature vector of fibers is a parallel normal vector field if and only if

$$VX(\ln g) = 0 \tag{4.22}$$

for every horizontal vector field X and every vertical vector field V. Consequently, the function f is the product of two positive functions $\lambda \in \mathcal{F}(B)$ and $\mu \in \mathcal{F}(F)$.

The converse is easy to verify. □

Corollary 4.3. *Let $M = B\times_f F$ be a warped product of two pseudo-Riemannian manifolds. Then*

(1) *leaves are totally geodesic in M;*
(2) *fibers are totally umbilical submanifolds with parallel mean curvature vector.*

Proof. Follows immediately from Proposition 4.3. □

Definition 4.7. A pseudo-Riemannian submanifold of a pseudo-Riemannian manifold is called an *extrinsic sphere* if it is a totally umbilical submanifold with parallel mean curvature vector.

Another immediate consequence of Proposition 4.3 is the following.

Corollary 4.4. *Every pseudo-Riemannian n-manifold N can be isometrically embedded in some $(n+p)$-dimensional pseudo-Riemannian manifold M with arbitrary $p \geq 1$ as an extrinsic sphere.*

Remark 4.1. Corollary 4.4 shows that there do not exist Riemannian or topological obstructions for a manifold to be an extrinsic sphere.

4.5 Double-twisted products and their characterization

In [Ponge and Reckziegel (1993)], the notion of twisted products was extended to the notion of double-twisted products.

Definition 4.8. Let M_1 and M_2 be two pseudo-Riemannian manifolds endowed with pseudo-Riemannian metrics g_1 and g_2, respectively. If f_1, f_2 are two positive functions in $\mathcal{F}(M_1 \times M_2)$ and $\pi_i : M \to M_i$ the canonical projection for $i = 1, 2$. Then the *double-twisted product* $M_1 \times_{(f_1,f_2)} M_2$ of (M_1, g_1) and (M_2, g_2) is the manifold $M_1 \times M_2$ equipped with the pseudo-Riemannian metric g defined by

$$g(X,Y) = f_1^2 \cdot g_1(\pi_{1*}X, \pi_{1*}Y) + f_2^2 \cdot g_2(\pi_{2*}(X), \pi_{2*}Y) \qquad (4.23)$$

for tangent vectors $X, Y \in T(M_1 \times M_2)$.

Definition 4.9. A *foliation* on a manifold M is an integrable subbundle $\mathcal{F}$ of the tangent bundle of M, i.e. for any two vector fields X and Y taking values in $\mathcal{F}$, then the Lie bracket $[X, Y]$ takes values in $\mathcal{F}$ as well.

Definition 4.10. A foliation L on a pseudo-Riemannian manifold M is called *totally umbilical*, if every leaf of L is a totally umbilical pseudo-Riemannian submanifold of M. If, in addition, the mean curvature vector of every leaf is parallel in the normal bundle, then L is called a *spheric foliation*, because in this case each leaf of L is an *extrinsic sphere* of M.

If every leaf of L is a totally geodesic submanifold of M, then L is called a *totally geodesic foliation*.

The following two results were proved in [Ponge and Reckziegel (1993)].

Theorem 4.2. *Let g be a pseudo-Riemannian metric on $M_1 \times M_2$. If the canonical foliations L_1 and L_2 intersect perpendicularly everywhere, then g is the metric of*

(a) *a double-twisted product $M_1 \times_{(f_1,f_2)} M_2$ if and only if L_1 and L_2 are totally umbilical foliations;*
(b) *a twisted product $M_1 \times_f M_2$ if and only if L_1 is a totally geodesic and L_2 a totally umbilical foliation;*
(c) *a warped product $M_1 \times_f M_2$ if and only if L_1 is a totally geodesic and L_2 a spheric foliation;*
(d) *a direct product of pseudo-Riemannian manifolds if and only if L_1 and L_2 are totally geodesic foliations.*

Theorem 4.3. *Let (M, g) be a simply-connected pseudo-Riemannian manifold which admits two complementary foliations L and K whose leaves intersect perpendicularly. If L is totally geodesic and K is totally umbilical, then (M, g) is isometric to a twisted product $B \times_f F$ such that L and K correspond to the canonical foliations of the product $B \times F$.*

An immediate consequence of Theorems 4.2 and 4.3 is following result of [Hiepko (1979)].

Theorem 4.4. *Let (M, g) be a simply-connected pseudo-Riemannian manifold which admits two complementary foliations L and K whose leaves intersect perpendicularly. If L is totally geodesic and K is spherical, then (M, g) is isometric to a warped product $B \times_f F$ such that L and K correspond to the canonical foliations of the product $B \times F$.*

The connection and the curvature tensor of a double-twisted product $M_1 \times_{(f_1, f_2)} M_2$ can be expressed in terms of the functions f_1 and f_2 and the connections and the curvature tensors of M_1, M_2 [Ponge and Reckziegel (1993)].

Proposition 4.4. *Let (M_1, g_1) and (M_2, g_2) be two pseudo-Riemannian manifolds and g the metric of double-twisted product $M_1 \times_{(f_1, f_2)} M_2$. Put $U_i = -\nabla(\ln f_i^2)$, where the gradient of $\ln f_i^2$ is calculated with respect to g. Then the Levi-Civita connection ∇ and curvature tensor R of the double-twisted product $M_1 \times_{(f_1, f_2)} M_2$ is related to the Levi-Civita connection $\tilde{\nabla}$ and curvature tensor R of the direct product of (M_1, g_1) and (M_2, g_2) by*

$$\begin{aligned} \nabla_X Y = & \tilde{\nabla}_X Y + \sum_i \{g(P_i X, P_i Y) U_i - g(X, U_i) P_i Y \\ & - g(Y, U_i) P_i X\} \end{aligned}$$

and

$$\begin{aligned} R(X, Y) = & \tilde{R}(X, Y) + \sum \{(\nabla_X U_i - g(X, U_i) U_i) \wedge P_i Y \\ & - (\nabla_Y U_i - g(Y, U_i) U_i) \wedge P_i X\} \\ & + \sum_{i,j} g(U_i, U_j) P_i X \wedge P_j Y, \end{aligned} \tag{4.24}$$

where $u \wedge v$ is the linear map $w \mapsto g(v, w)u - g(u, w)v$ for all $u, v \in T_pM$ and $P_i : TM \to \zeta_i$ is the vector bundle projection related to the splitting $TM = \zeta_1 \oplus \zeta_2$ with $\zeta_1 = \ker(\pi_{2_*})$ and $\zeta_2 = \ker(\pi_{1_*})$.

The $\ker(\pi_{i_*})$ in Proposition 4.4 denotes the *kernel* of π_{i_*}.

Chapter 5

Robertson–Walker Spacetimes

5.1 Cosmology, Robertson–Walker spacetimes and Einstein's field equations

Cosmology is the study of the structure and changes in the universe. Four thousand years ago, the Babylonians were the first to recognize that astronomical phenomena are periodic and they were able to apply mathematics to predict the apparent motions of the moon and the stars and the planets and the sun upon the sky, and could even predict eclipses. In Babylonian cosmology the earth and heavens are a spatial whole, even of round shape, revolving around the "cult-place of the deity" rather than the Earth. But it was the Ancient Greeks who were the first to build a cosmological model within which to interpret these motions. Greek philosopher–scientists set themselves the task of envisioning the universe as a set of physical objects. Aristotle (384BC - 322BC) taught that rotating spheres carried the moon, sun, planets, and stars around a stationary Earth. The Earth was unique because of its central position and its material composition.

In 1543, an Italian astronomer–mathematician Copernicus (1473-1543) published an epochal book before his death

"*De revolutionibus orbium coelestium*",

in which he proposed to switch the places of the earth and the sun. He put the Sun in the center of the Universe and placed the earth in revolution around the Sun. To account for the daily motion of the heavens, he set the earth rotating about its own axis. His work became a landmark in the history of science that is often referred to as the *Copernican Revolution.*

It was in the early 17th century that Galileo Galilei (1564-1642), with the aid of telescope, could deal a fatal blow to the notion that the earth was at the center of the universe. He discovered the four largest moons orbiting the planet Jupiter. And if moons could orbit another planet, why could not the planets orbit the sun? After 1610, he began publicly supporting the heliocentric view, which placed the Sun at the center of the Universe.

In the 17th century, Newton (1643-1727) extended the range of physics from the earth to the solar system. In the static Newtonian world, every particle in the universe attracts every other particle. Matter on the large scale is uniformly distributed, gravitationally balanced but unstable.

Until the beginning of the 20th century, time was believed to be independent of motion. In relativistic contexts, time cannot be separated from the three dimensions of space. Einstein's special relativity [Einstein (1905)] is a theory of the structure of spacetime, which radically reinterpreted Lorentzian electrodynamics by changing the concepts of space and time and abolishing the aether.

Special relativity is based on two postulates which are contradictory in classical mechanics:

1. The laws of physics are the same for all observers in uniform motion relative to one another (*principle of relativity*),

2. The speed of light in a vacuum is the same for all observers, regardless of their relative motion or of the motion of the source of the light.

The resultant theory agrees with experiment better than classical mechanics. In particular, it yielded the 20th century's best-known equation: $E = mc^2$; implying that energy and mass are equivalent and transmutable.

Einstein's General Relativity [Einstein (1916)] is the geometrical theory of nature, especially of gravitation, which unifies special relativity and Newton's law of universal gravitation with the insight that gravitation is not due to a force but rather is a manifestation of curved space and time, with the curvature of spacetime being influenced by the stress-energy density of matter and electromagnetism.

Einstein's General Relativity is distinguished from other theories of gravitation, e.g. Galilei's and Newton' theories of gravitation, by its use of the *Einstein field equation*

$$Ric - \frac{1}{2}\tau g = kT \tag{5.1}$$

to relate the mass-energy tensor T with the Ricci curvature tensor Ric and the metric tensor g of spacetime, where k is called the *Einstein constant of gravitation*. In a vacuum, a region of spacetime with no matter, $T = 0$; so the spacetime is an Einstein space.

The predictions of general relativity differ significantly from those of classical physics, especially concerning the passage of time, the geometry of space, the motion of bodies in free fall, and the propagation of light.

At the beginning of the 20th century, astronomers were unsure of the size of our galaxy. Generally, they believed it was not much greater than a few tens of thousands of light years across. However, astronomers had noted fuzzy patches of light on the night sky, which they called *nebulae.* Some astronomers thought these could be distant galaxies. It was only in the 1920s that E. Hubble (1889-1953) established that some of these nebulae were indeed distant galaxies comparable in size to our own Milky Way. Hubble also made the remarkable discovery that our Universe is expanding! Hubble's undeniable observations that the light from nebulae showed a red shift increasing with distance ruled out the possibility that Einstein's static model represented the real universe. De Sitter's static model, without matter, was also ruled out by new observations. A new estimate, made in 1927, of the mass of our galaxy caused de Sitter to re-examine his assumption. At a meeting in London of the Royal Astronomical Society early in 1930, de Sitter admitted that neither his nor Einstein's solution to the field equations could represent the observed universe.

In fact, a few astronomers had been looking for other solutions to Einstein's field equations. Back in 1922 the Russian meteorologist Friedmann (1888-1925) had published in Zeitschrift für Physik a set of possible mathematical solutions that gave a non-static universe. Einstein noted that this model was indeed a mathematically possible solution to his field equations. In 1927, Lemaître (1894-1966) at the University of Leuven arrived independently at similar results as Friedmann and published them in Annals of the Scientific Society of Brussels. During the 1930s, Robertson (1903-1961) and Walker (1909-2001) explored the problem further. In 1935 they rigorously proved that the Robertson–Walker metric is the only one on a spacetime that is spatially homogeneous and isotropic.

Robertson–Walker spacetimes have another important property. In physics, a *perfect fluid* is a fluid that can be completely characterized by its rest frame energy density ρ and isotropic pressure p. Real fluids are "stick" and contain and conduct heat. Perfect fluids are idealized models in which these possibilities are neglected. Specifically, perfect fluids have no shear stresses, viscosity, or heat conduction. In tensor notation the energy-momentum tensor T of a perfect fluid can be written in the form

$$T = (\rho + p)U^* \otimes U^* + pg,$$

where U^* is the dual 1-form of the velocity vector field of the fluid and g is the metric tensor. For any Robertson–Walker spacetime, the flow given by the vector field ∂_t is that of a perfect fluid.

After the observation of the Hubble redshift indicated that the universe might not be stationary, Einstein abandoned the concept as he had based his theory that the universe is unchanging. The *cosmological constant* Λ was proposed by Einstein as a modification of his original theory to achieve a stationary universe. Einstein's *modified field equation* is

$$Ric - \frac{1}{2}\tau g + \Lambda g = kT. \tag{5.2}$$

When Λ is zero, this reduces to the original field equation. When T is zero, the field equation describes empty space (the vacuum).

5.2 Basic properties of Robertson–Walker spacetimes

In general relativity, a *Robertson–Walker spacetime* is a warped product

$$L_1^4(k,f) := (I \times R^3(k), g), \quad g = -dt^2 + f^2(t)g_k,$$

of an open interval I and a Riemannian 3-manifold $(R^3(k), g_k)$ of constant curvature k, while the warping function f describes expanding or contracting of our Universe. In this chapter we shall consider a *Robertson–Walker spacetime* as a warped product

$$L_1^m(k,f) := (I \times R^{m-1}(k), g), \quad g = -dt^2 + f^2(t)g_k, \tag{5.3}$$

of an open interval I and a real space form $(R^{m-1}(k), g_k)$, where m is any integer ≥ 2. Let ∂_t denote the first coordinate vector field on $L_1^m(k,f)$, known as the *comoving observer field* in general relativity.

A Robertson–Walker spacetime possesses two relevant geometrical features. On one hand, its fibers have constant curvature; thus the spacetime is *spatially homogeneous*. On the other hand, it has a timelike vector field $K = f(t)\partial_t$ which satisfies $\nabla_X K = f'(t)X$ for any X. In particular $\mathcal{L}_K g = 2f'g$, where $\mathcal{L}_K$ is the Lie derivative along K. Hence, K is a conformal-Killing vector field. These properties of K show a certain symmetry of the spacetime metric on $L_1^m(k,f)$.

By a *spacelike slice* we mean a hypersurface of $L_1^{m-1}(k,f)$ given by a fiber $S(t_0) := \{t_0\} \times_{f(t_0)} R^{m-1}(k)$, $t_0 \in I$. The spacelike slice $S(t_0)$ is a manifold of constant curvature whose metric tensor is $f^2(t_0)g_k$.

Definition 5.1. A pseudo-Riemannian submanifold N of $L_1^m(k,f)$ is called *transverse* if it is contained in a *slice* $S(t_0)$ for some $t_0 \in I$. A pseudo-Riemannian submanifold N of $L_1^m(k,f)$ is called a *$\mathcal{H}$-submanifold* if the comoving observer field ∂_t is tangent to N at each point on N.

For a tangent vector X of $L_1^m(k,f)$, we decompose

$$X = \varphi_X \partial_t + X^-, \tag{5.4}$$

where $\varphi_X = -\langle X, \partial_t\rangle$ and X^- is the vertical component of X.

The following two lemmas follow from Propositions 4.1 and 4.2.

Lemma 5.1. *For $V, W \in \mathfrak{L}(R^{m-1}(k))$, we have*

(1) $\nabla_{\partial_t}\partial_t = 0$;
(2) $\nabla_{\partial_t} V = \nabla_V \partial_t = (\ln f)' V$;
(3) $\langle \nabla_V W, \partial_t \rangle = -\langle V, W\rangle (\ln f)'$;
(4) $(\nabla_V W)^-$ *is the lift of* $\nabla'_V W$ *on* $R^{m-1}(k)$,

where ∇' is the Levi-Civita connection on $R^{m-1}(k)$.

Proof. Follows immediately from Proposition 4.1. □

Lemma 5.2. *The curvature tensor R of $L_1^m(k,f)$ satisfies*

(1) $R(\partial_t, V)\partial_t = \dfrac{f''}{f} V$;
(2) $R(V, \partial_t)W = -\langle V, W\rangle \dfrac{f''}{f}\partial_t$;
(3) $R(V, W)\partial_t = 0$;
(4) $R(U, V)W = \dfrac{k + f'^2}{f^2}\{\langle V, W\rangle U - \langle U, W\rangle V\}$

for $U, V, W \in \mathfrak{L}(R^{m-1}(k))$.

Proof. Follows immediately from Proposition 4.2. □

Corollary 5.1. *A Robertson–Walker spacetime $L_1^m(k,f)$ is of constant curvature if and only if the warping function f satisfies $ff'' = f'^2 + k$.*

Proof. Assume $L_1^m(k,f)$ is of constant curvature. Then it follows from Lemma 5.2(1) that the curvature of $L_1^m(k,f)$ is f''/f. On the other hand, it follows from Lemma 5.2(4) that the curvature is also equal to $(k+f'^2)/f^2$. Thus, the warping function satisfies $ff'' = f'^2 + k$.

The converse can be verified by direct computation. □

Remark 5.1. It follows from Corollary 5.1 that

(a) $L_1^m(k,f)$ is flat if and only if $f(t) = at + b$, with $k = -a^2$;
(b) $L_1^m(k,f)$ has constant curvature $c^2 > 0$ if and only if $f(t) = a\cosh(ct) + b\sinh(ct)$ with $k = c^2(a^2 - b^2)$;

(c) $L_1^m(k,f)$ has constant curvature $-c^2<0$ if and only if $f(t)=a\sin(ct)+b\cos(ct)$ with $k=-c^2(a^2+b^2)$.

Lemma 5.3. *Let N be a pseudo-Riemannian submanifold of $L_1^m(k,f)$. Then for $u,v,w\in T_pN$, $p\in N$ and $\xi,\eta\in T_p^{\perp}(N)$ we have:*

(1) $(R(u,v)w)^{\perp}=\Big(\frac{k}{f^2}-(\ln f)''\Big)(\langle u,w\rangle\,\varphi_v-\langle v,w\rangle\,\varphi_u)\partial_t^{\perp}$, *where $(\,\cdot\,)^{\perp}$ is the normal component of $(\,\cdot\,)$;*
(2) $(R(u,v)\xi)^{\perp}=0$;
(3) *the Ricci equation in $L_1^m(k,f)$ is given by*

$$\langle R^D(u,v)\xi,\eta\rangle=\langle[A_\xi,A_\eta]u,v\rangle\,;$$

(4) *the sectional curvature $K(u\wedge v)$ of $L_1^m(k,f)$ with respect to the plane section spanned by two orthonormal vectors $u,v\in T_p(N)$ is given by*

$$K(u\wedge v)=\frac{k+f'^2}{f^2}+(\epsilon_u\varphi_u^2+\epsilon_v\varphi_v^2)\Big(\frac{k}{f^2}-(\ln f)''\Big),$$

where $\epsilon_u=\langle u,u\rangle$ and $\epsilon_v=\langle v,v\rangle$.

Proof. Let $e_1,\ldots,e_n$ be a local orthonormal frame field on N. From (5.4) we find

$$e_j=\varphi_j\partial_t+\hat e_j,\ \ j=1,\ldots,n. \tag{5.5}$$

where $\varphi_j=-\langle e_j,\partial_t\rangle$, $1\le j\le n$. Then we have

$$\langle\hat e_i,\hat e_j\rangle=\epsilon_j\delta_{ij}-\varphi_i\varphi_j,\ \ \epsilon_j=\langle e_j,e_j\rangle\,,\ \ i,j=1,\ldots,n. \tag{5.6}$$

It follows from Lemma 5.2, (5.5) and (5.6) that

$$\begin{aligned}R(e_i,e_j)e_k&=\Big\{\varphi_i\varphi_k\frac{f''}{f}+\frac{f'^2+k}{f^2}(\varphi_i\varphi_k-\epsilon_k\delta_{ik})\Big\}\,\hat e_j\\ &-\Big\{\varphi_j\varphi_k\frac{f''}{f}+\frac{f'^2+k}{f^2}(\varphi_j\varphi_k-\epsilon_k\delta_{jk})\Big\}\hat e_i-(\delta_{ik}\varphi_j-\delta_{jk}\varphi_i)\epsilon_k\frac{f''}{f}\partial_t.\end{aligned} \tag{5.7}$$

Thus, by applying (5.5) and (5.7), we find

$$(R(e_i,e_j)e_k)^{\perp}=(\delta_{ik}\varphi_j-\delta_{jk}\varphi_i)\epsilon_k\Big(\frac{k}{f^2}-(\ln f)''\Big)\partial_t^{\perp}, \tag{5.8}$$

which implies (1) by linearity. By applying Lemma 5.2(4) we find

$$R(u,v)\xi=\frac{ff''+f'^2+k}{f^2}(\varphi_u\hat v-\varphi_v\hat u)\varphi_\xi. \tag{5.9}$$

Combining this with $\hat u=u-\varphi_u\partial_t$ and $\hat v=v-\varphi_v\partial_t$ gives (2). (3) follows from (2) and Ricci's equation. Finally, (4) follows from (5.7). □

For an $\mathcal{H}$-submanifold N, we put $\mathfrak{D}^\perp = \{X \in \mathcal{X}(N) : X \perp \partial_t\}$.

Lemma 5.4. *Let N be an $\mathcal{H}$-submanifold of $R^m(k,f)$. We have:*

(1) *the Levi-Civita connection ∇' of N satisfies*

$$\nabla'_{\partial_t}\partial_t = 0, \quad \nabla'_X Y = (\nabla_X Y)^- + \langle X, Y\rangle (\ln f)'\partial_t \tag{5.10}$$

for vector fields X, Y in $\mathfrak{D}^\perp$, where $(\nabla_X Y)^-$ is the vertical component of $\nabla_X Y$;

(2) *the distribution $\mathfrak{D}^\perp$ is integrable;*

(3) *locally, N is a warped product $I' \times_{\hat f} P^{n-1}$, where I' is an open subinterval of I, P^{n-1} is a submanifold of $R^{m-1}(k)$ and $\hat f = f|_{I'}$;*

(4) *the second fundamental form h of N in $L_1^m(k,f)$ satisfies*

(4.a) *$h(\partial_t,\partial_t) = h(\partial_t, X) = 0$, and*

(4.b) *$h(\bar X,\bar Y)$ is the lift of $h^S(X,Y)$ for $X,Y \in T(P^{n-1})$, where h^S is the second fundamental form of P^{n-1} in $R^{m-1}(k)$ and $\bar X, \bar Y$ are the lift of X, Y to $I' \times_{\hat f} N^{n-1}$.*

Proof. (1) follows immediately from Lemma 5.1. From (4.3) we get $[X,Y] = [X,Y]^- \in \mathfrak{D}^\perp$. Thus, we obtain (2) by Frobenius' Theorem.

For (3) let us observe that Lemma 5.1(1) implies that the rank-one distribution spanned by ∂_t on N is a totally geodesic distribution of N. Moreover, it follows from (1) that the integral manifolds of $\mathfrak{D}^\perp$ are totally umbilical hypersurfaces of N with constant mean curvature. Thus, Hiepko's theorem implies that N is locally the warped product of an open subinterval $I' \subset I$ and an integral manifold P^{n-1} of $\mathfrak{D}^\perp$ with respect to the warping function $\hat f = f|_{I'}$. Since P^{n-1} is perpendicular to $I' \subset I$, P^{n-1} lies in some slice $S(t_0), t_0 \in I'$. Without loss of generality, we may assume that P^{n-1} is a submanifold of $R^{m-1}(k)$. This proves statement (3).

Statement (4) follows from statement (3) and Lemma 5.1. □

Lemma 5.5. *If N is a transverse submanifold of $L_1^m(k,f)$, then*

(1) *the second fundamental form h of N in $L_1^m(k,f)$ and $h^{S(t_0)}$ of N in $S(t_0)$ are related by*

$$h(X,Y) = h^{S(t_0)}(X,Y) + \langle X,Y\rangle (\ln f)'\partial_t, \quad X, Y \in \mathcal{X}(N);$$

(2) *the normal connection D of N in $R^m(k,f)$ and $D^{S(t_0)}$ of N in $S(t_0)$ satisfy $D_X\partial_t = 0$ and $D_X\xi = D_X^{S(t_0)}\xi$ for every $X \in \mathcal{X}(N)$ and ξ orthogonal to ∂_t;*

(3) *the normal curvature tensor* R^D *of* N *in* $R^m(k,f)$ *and* $R^{D^{S(t_0)}}$ *of* N *in* $S(t_0)$ *satisfy* $R^D(X,Y)\xi = R^{D^{S(t_0)}}(X,Y)\xi$ *for* $X, Y \in \mathcal{X}(N)$ *and* ξ *orthogonal to* ∂_t.

Proof. Statements (1) and (2) follow easily from Lemma 5.1; and statement (3) is an easy consequence of statement (2). □

The next two corollaries follow immediately from Lemma 5.5(1).

Corollary 5.2. *A transverse submanifold* N *of* $L_1^m(k,f)$ *is non-totally geodesic unless* N *lies in a slice* $S(t_0)$ *with* $f'(t_0) = 0$ *as a totally geodesic submanifold.*

Corollary 5.3. *If a transverse submanifold of* $L_1^m(k,f)$ *is totally umbilical, then it lies in a slice* $S(t_0)$ *as a totally umbilical submanifold.*

5.3 Totally geodesic submanifolds of RW spacetimes

Definition 5.2. A submanifold N of a pseudo-Riemannian manifold M is called *curvature-invariant*if $R(u,v)(T_pN) \subset T_pN$ for any $u, v \in T_pN$ at each point $p \in N$, where R is the curvature tensor of M.

Lemma 5.6. *Assume that* $L_1^m(k,f) = I \times_f R^{m-1}(k)$ *contains no open subsets of constant curvature. Then a curvature-invariant pseudo-Riemannian submanifold* N *with* $\dim N \geq 2$ *in* $L_1^m(k,f)$ *is either*

(a) *a transverse submanifold of* $L_1^m(k,f)$ *or*
(b) *an* $\mathcal{H}$*-submanifold of* $L_1^m(k,f)$.

Proof. Assume that $L_1^m(k,f)$ contains no open subsets of constant curvature. Then Corollary 5.1 implies that the warping function f satisfies $(\ln f)'' \neq k/f^2$ on an dense subset of the open interval I.

Let N be a pseudo-Riemannian submanifold of $L_1^m(k,f)$ and let ∂_t^T and $\partial_t^\perp$ be the tangential and the normal component of ∂_t. If N is a curvature-invariant submanifold, then Lemma 5.3 yields

$$(\langle u,w\rangle \langle v,\partial_t\rangle - \langle v,w\rangle \langle u,\partial_t\rangle)\partial_t^\perp = 0, \quad u,v,w \in T_pN, \ p \in N.$$

Hence, at each point $p \in N$, ∂_t is either normal to N (i.e. $\partial_t^T = 0$) or ∂_t is tangent to N (i.e. $\partial_t^\perp = 0$); otherwise choosing $u = w \perp v = \partial_t^T \neq 0$ in the above expression would lead to $\partial_t^\perp = 0$, a contradiction. Thus, by continuity, ∂_t is either normal to N at each point on N or tangent to N at each point on N. Hence, we have either (a) or (b). □

The following result classifies totally geodesic spacelike submanifolds of a Robertson–Walker spacetime.

Proposition 5.1. *If $L_1^m(k,f)$ contains no open subsets of constant curvature, then $L_1^m(k,f)$ admits a spacelike totally geodesic submanifold of dimension ≥ 2 if and only if f has a critical point.*

Further, the only spacelike totally geodesic submanifolds of dimension $\geq$ 2 in $L_1^m(k,f)$ are spacelike slices $S(t_0)$ with $f'(t_0)=0$ and totally geodesic submanifolds which lie in some spacelike slice.

Proof. Let $L_1^m(k,f)$ be a Robertson–Walker spacetime which contains no open subsets of constant curvature. If N is a totally geodesic spacelike submanifold, then it is curvature-invariant by the equation of Codazzi. Hence, it follows from Lemma 5.6 that N is a transverse submanifold. Thus, N lies in a spacelike slice $S(t_0)$, for some t_0. Since N is totally geodesic in $L_1^m(k,f)$, Corollary 5.2 implies $f'(t_0)=0$ and N is totally geodesic in $S(t_0)$.

The converse follows immediately from Lemma 5.5(1). □

Proposition 5.2. *Suppose that $L_1^m(f,c)=I\times_f R^{m-1}(k)$ is a Robertson–Walker spacetime which contains no open subsets of constant curvature. If N is an n-dimensional totally geodesic Lorentzian submanifold of $L_1^m(k,f)$, then N is an open portion of a warped product $I'\times_{\hat f} P^{n-1}\subset I\times_f R^{m-1}(k)$, where I' is an open subinterval of I, P^{n-1} is a totally geodesic submanifold of $R^{m-1}(k)$ and $\hat f=f|_{I'}$.*

Proof. Under the hypothesis, N is curvature-invariant. Thus, by Lemma 5.6, N is an $\mathcal{H}$-submanifold. Hence, ∂_t is tangent to N at each point. Thus, according to Lemma 5.4(3), N is an open portion of $I'\times_{\hat f} P^{n-1}$ for some submanifold P^{n-1} of $R^{m-1}(k)$. Because N is totally geodesic in $L_1^m(k,f)$, P^{n-1} is totally geodesic in $R^{m-1}(k)$. □

5.4 Parallel submanifolds of RW spacetimes

Proposition 5.3. *If a Robertson–Walker spacetime $L_1^m(k,f)$ contains no open subsets of constant curvature, then a pseudo-Riemannian submanifold of $L_1^m(k,f)$ is a parallel submanifold if and only if it is one of the following:*

(a) *A transverse submanifold lying in a slice $S(t_0)$ of $L_1^m(k,f)$ as a parallel submanifold.*

(b) *An $\mathcal{H}$-submanifold which is locally a warped product $I \times_f P^{n-1}$, where I is an open interval and P^{n-1} is a submanifold of $R^{m-1}(k)$. Further,*

(b.1) *if $f' \neq 0$ on I, then $I \times_f P^{n-1}$ is totally geodesic in $L_1^m(k,f)$;*

(b.2) *if $f' = 0$ on I, then P^{n-1} is a parallel submanifold of $R^{m-1}(k)$.*

Proof. Assume that $L_1^m(k,f)$ contains no open subsets of constant curvature. Then, by Corollary 5.1, f satisfies $(\ln f)'' \neq k/f^2$ on I. Suppose that N is a parallel submanifold, then Codazzi's equation shows that N is curvature-invariant. Hence, N is either a transverse submanifold or an $\mathcal{H}$-submanifold according to Lemma 5.6.

Case (1): *N is a transverse submanifold.* In this case, N lies in a slice $S(t_0)$ of $L_1^m(k,f)$, $t_0 \in I$. Let X, Y, Z be tangent to N. Then, Lemma 5.1 implies that the second fundamental form h of N in $L_1^m(k,f)$ is given by

$$h(X,Y) = \hat{h}(X,Y) + \langle X,Y\rangle\,(\ln f)'\partial_t, \tag{5.11}$$

where $\hat{h}$ is the second fundamental form of N in $S(t_0)$. So, after applying Lemma 5.1, we find

$$\begin{aligned}\nabla_X(h(Y,Z)) = \nabla_X(\hat{h}(Y,Z)) &+ \langle \nabla_X Y, Z\rangle\,(\ln f)'\partial_t \\ &+ \langle Y, \nabla_X Z\rangle\,(\ln f)'\partial_t + \langle Y,Z\rangle\,(\ln f)'^2 X.\end{aligned}$$

Thus $D_X(h(Y,Z)) = D_X\hat{h}(Y,Z) + (\langle \nabla'_X Y, Z\rangle + \langle Y, \nabla'_X Z\rangle)(\ln f)'\partial_t$, which implies that

$$\begin{aligned}(\bar{\nabla}_X h)(Y,Z) &= D_X(\hat{h}(Y,Z) + \langle Y,Z\rangle\,(\ln f)'\partial_t) - h(\nabla'_X Y, Z) - h(Y, \nabla'_X Z) \\ &= D_X(\hat{h}(Y,Z)) + \langle \nabla'_X Y, Z\rangle\,(\ln f)'\partial_t - h(\nabla'_X Y, Z) \\ &\quad + \langle Y, \nabla'_X Z\rangle\,(\ln f)'\partial_t - h(Y, \nabla'_X Z) \\ &= D_X(\hat{h}(Y,Z)) - \hat{h}(\nabla'_X Y, Z) - \hat{h}(Y, \nabla'_X Z) \\ &= D_X(h^{S(t)}(Y,Z)) - h^{S(t)}(\nabla'_X Y, Z) - h^{S(t)}(Y, \nabla'_X Z) \\ &= (\bar{\nabla}_X h^{S(t)})(Y,Z).\end{aligned}$$

Hence, N is a parallel submanifold in $L_1^m(k,f)$ if and only if N is a parallel submanifold in the slice $S(t)$. This gives (a).

Case (2): *N an $\mathcal{H}$-submanifold.* We follow the same notation as Lemma 5.4. Since N an $\mathcal{H}$-submanifold, the second fundamental form h of N in $L_1^m(k,f)$ lies in the space of lifts of $\mathcal{X}(R^{m-1}(k))$. Moreover, according to Lemma 5.4(3), locally N is a warped product $I' \times_f P^{n-1}$, where I' is an open subinterval of I and P^{n-1} is a submanifold of $R^{m-1}(k)$. Because N is parallel, $(\bar{\nabla}_X h)(Y,Z)=0$ for $X,Y,Z\in TN$. Thus

$$0=(\bar{\nabla}_{\partial_t}h)(Y,Z)=D_{\partial_t}h(Y,Z)-h(\nabla'_{\partial_t}Y,Z)-h(Y,\nabla'_{\partial_t}Z) \tag{5.12}$$

for $Y,Z\in\mathfrak{D}^{\perp}$. Now, by applying Lemma 5.1(2), we obtain

$$D_{\partial_t}h(Y,Z)=2(\ln f)'h(Y,Z). \tag{5.13}$$

On the other hand, Weingarten's formula and Lemma 5.1(2) give

$$-A_{\xi}(\partial_t)+D_{\partial_t}\xi=\nabla_{\partial_t}\xi=(\ln f)'\xi \tag{5.14}$$

for ξ normal to N, which yields

$$A_{\xi}(\partial_t)=0,\quad D_{\partial_t}\xi=(\ln f)'\xi. \tag{5.15}$$

Case (2.a): *$f'\neq 0$ on I.* Combining (5.13) and the second equation of (5.15) gives $h(Y,Z)=0$, $Y,Z\in\mathfrak{D}^{\perp}$. Hence, by using the first equation of (5.15), we find $h=0$. Thus, N is totally geodesic, which gives (b.1).

Case (2.b): *$f'=0$ on I.* It follows from (5.15), Lemmas 5.1 and 5.4 that

$$\begin{aligned}&\nabla'_{\partial_t}\partial_t=\nabla'_{\partial_t}X=\nabla'_X\partial_t=D_{\partial_t}\xi=0,\\&\nabla'_XY=(\nabla_XY)^{-},\quad h(\partial_t,\partial_t)=h(\partial_t,X)=0\end{aligned} \tag{5.16}$$

for $X,Y\in\mathfrak{D}^{\perp}$. Now, it follows from (5.12) and (5.16) that

$$(\bar{\nabla}_{\partial_t}h)(\,\cdot\,,\cdot\,)=(\bar{\nabla}_Xh)(\partial_t,Y)=(\bar{\nabla}_Xh)(\partial_t,\partial_t)=0, \tag{5.17}$$

$$(\bar{\nabla}_Xh)(Y,Z)=(\bar{\nabla}_Xh^S)(Y,Z). \tag{5.18}$$

Therefore, after applying the assumption $\bar{\nabla}h=0$, we conclude that P^{n-1} is a parallel submanifold of $R^{m-1}(k)$. This gives (b.2).

The converse can be verified easily. □

Remark 5.2. Most results given in sections 5.2-5.4 are based on [Chen and Van der Veken (2007b); Chen and Wei (2009)].

5.5 Totally umbilical submanifolds of RW spacetimes

Proposition 5.4. *Let $L_1^m(k,f)$ be a Robertson–Walker spacetime which contains no open subsets of constant curvature. Then a pseudo-Riemannian submanifold of $L_1^m(k,f)$ is totally umbilical with parallel mean curvature if and only if it is one of the following two types of submanifolds:*

(i) *a totally geodesic $\mathcal{H}$-submanifold of $L_1^m(k,f)$, or*

(ii) *a submanifold lying in a slice $S(t_0)$ as a totally umbilical submanifold with parallel mean curvature vector.*

Proof. Under the hypothesis, assume that N is a totally umbilical submanifold of $L_1^m(k,f)$ with parallel mean curvature vector. Then the second fundamental form h of N satisfies

$$h(X,Y) = \langle X,Y\rangle H, \quad X,Y \in \mathcal{X}(N). \tag{5.19}$$

Thus $(\bar\nabla_X h)(Y,Z) = \langle Y,Z\rangle D_X H = 0$ for $X,Y,Z \in \mathcal{X}(N)$, which implies that N is a parallel submanifold. Hence, by Proposition 5.3, N is either a transverse submanifold or an $\mathcal{H}$-submanifold.

If N is an $\mathcal{H}$-submanifold, then it follows from Lemma 5.5(1) that $h(\partial_t,\partial_t) = 0$. Combining this with the total umbilicity of N implies that M is totally geodesic. This gives (i). If N is a transverse submanifold of $L_1^m(k,f)$, it lies in a slice $S(t_0)$. Thus, by Lemma 5.5, we have

$$h(X,Y) = h^{S(t_0)}(X,Y) + \langle X,Y\rangle (\ln f)'\partial_t.$$

Thus N is totally umbilical in $S(t_0)$. We also have $H = H^{S(t_0)} + (\ln f)'\partial_t$, where $H^{S(t_0)}$ is the mean curvature vector of N in $S(t_0)$. Therefore we find $h^{S(t_0)}(X,Y) = \langle X,Y\rangle H^{S(t_0)}$. Now, by applying Lemma 5.1, we get $D_X H = D_X H^{S(t_0)}$. Consequently, N is totally umbilical with parallel mean curvature vector in $S(t_0)$. This gives (ii).

The converse can be easily verified. □

Proposition 5.5. *If $L_1^m(k,f)$ contains no open subsets of constant curvature, then a pseudo-Riemannian submanifold N of $L_1^m(k,f)$ is totally umbilical with constant mean curvature if and only if either*

(i) *N lies in a slice $S(t_0)$ as a totally umbilical submanifold with parallel mean curvature vector, or*
(ii) *N is a totally umbilical submanifold of $L_1^m(k,f)$ with vertical mean curvature vector field, i.e., $H \perp \partial_t$.*

Proof. Under the hypothesis on $L_1^m(k,f)$, if N is totally umbilical in $L_1^m(k,f)$, then we have $(\bar\nabla_X h)(Y,Z) = \langle Y,Z\rangle D_X H$. Thus, for orthonormal vectors X,Y, we obtain $(\bar\nabla_X h)(Y,Y) - (\bar\nabla_Y h)(X,Y) = D_X H$. Hence, it follows from the equation of Codazzi and Lemma 5.3(1) that

$$D_X H = \Big(\frac{k}{f^2} - (\ln f)''\Big)\langle Y,Y\rangle \langle X,\partial_t\rangle\, \partial_t^{\perp} \tag{5.20}$$

for any $X \in \mathcal{X}(N)$. Hence, we find

$$X\langle H,H\rangle = 2\Big(\frac{k}{f^2} - (\ln f)''\Big)\langle Y,Y\rangle \langle X,\partial_t\rangle \langle \partial_t, H\rangle .$$

Thus if $\langle H,H\rangle$ is constant, then $\langle X,\partial_t\rangle\langle\partial_t,H\rangle=0$ holds at each point on N. Thus if we put $U=\{p\in M:\langle\partial_t,H\rangle\neq 0\,\text{at}\,p\}$, then U is an open subset of N; moreover, ∂_t is perpendicular to U.

If $U=\emptyset$, we get (ii). If $U=N$, then N is transverse. So, it lies in a slice, say $S(t_o)$. Since N is totally umbilical in a real space form $S(t_o)$, the mean curvature vector is a parallel normal vector field. This gives (i).

If U is neither empty nor N, then each connected component U^o of U is a transverse submanifold. Thus U^o lies in some slice $S(t^o)$. Hence, by Lemma 5.5(1), the mean curvature vector $H^{S(t^o)}$ of U^o in $S(t^o)$ satisfies

$$H=H^{S(t^o)}+(\ln f)'\partial_t. \tag{5.21}$$

Obviously, the value of $(\ln f)'$ on U^o is the constant $(\ln f)'(t^o)$.

On the other hand, since the mean curvature vector H on $M-U$ is perpendicular to ∂_t, (5.21) implies that $(\ln f)'=0$ on the boundary of $M-U$. Thus, by the continuity of H, $(\ln f)'(t^o)$ is zero. This leads to a contradiction; namely, $\langle\partial_t,H\rangle=\left\langle\partial_t,H^{S(t^o)}\right\rangle=0$ on U^o.

The converse is easy to verify. □

Contrast to totally umbilical submanifolds in $R^m(c)$, there exist totally umbilical submanifolds in $L_1^m(k,f)$ with non-constant mean curvature.

Example 5.1. Let $f(t)$ be a positive function with $f''>0$ on an open interval $I\ni 0$ and $(t,x_2,\ldots,x_m)$ a natural coordinate system of $I\times\mathbb{E}^{m-1}$. The metric tensor of $L_1^m(0,f)=I\times_f\mathbb{E}^{m-1}$ is

$$\tilde g=-dt^2+f^2(t)\sum\nolimits_{j=2}^{n}dx_j^2. \tag{5.22}$$

Consider the immersion $\phi:I\times\mathbf{R}^{n-1}\to L_1^m(0,f)$ defined by

$$\phi(s,u_2,\ldots,u_n)=\left(s,b\int_0^s\frac{dt}{f(t)},u_2,\ldots,u_n,0,\ldots,0\right),\ \mathbf{R}\ni b>1. \tag{5.23}$$

Then

$$\phi_s=(1,bf(s)^{-1},0,\ldots,0),\ \ \phi_{u_j}=(0,\ldots,0,\overbrace{1}^{(j+1)-th},0,\ldots,0), \tag{5.24}$$

for $j=2,\ldots,n$. The metric tensor induced from (5.28) via ϕ is

$$g=(b^2-1)ds^2+f^2(s)\sum\nolimits_{j=2}^{n}du_j^2, \tag{5.25}$$

and

$$\xi_1=\frac{1}{\sqrt{b^2-1}}(b,f^{-1},0,\ldots,0),\ \ \xi_r=f^{-1}(0,\ldots,0,\overbrace{1}^{r-th},0,\ldots,0),$$
$$r=n+2,\ldots,m,$$

are orthonormal normal vector fields to N in $L_1^m(0,f)$. A straight-forward computation shows that the second fundamental form of ϕ satisfies

$$\begin{aligned} &h(\partial_s,\partial_s)=b\sqrt{b^2-1}\frac{f'}{f}\xi_1,\ h(\partial_s,\partial_{u_i})=0,\ i=2,\ldots,n,\\ &h(\partial_{u_2},\partial_{u_2})=\cdots=h(\partial_{u_n},\partial_{u_n})=\frac{bff'}{\sqrt{b^2-1}}\xi_1,\\ &h(\partial_{u_i},\partial_{u_j})=0,\ \ 2\le i\ne j\le n. \end{aligned} \tag{5.26}$$

Hence ϕ is a totally umbilical immersion such that

$$h(X,Y)=\frac{bf'}{\sqrt{b^2-1}f}\langle X,Y\rangle\,\xi_1. \tag{5.27}$$

Since f'' is nowhere zero, (5.27) implies that ϕ is totally umbilical with non-constant mean curvature. Therefore, we have $DH\neq 0$.

Proposition 5.6. *Assume that $L_1^m(k,f)$ contains no open subsets of constant curvature and N is a totally umbilical pseudo-Riemannian submanifold of $L_1^m(k,f)$ with $\dim N\ge 3$. Then N is an Einstein manifold if and only if N is a transverse submanifold.*

Proof. If $L_1^m(k,f)$ contains no open subsets of constant curvature and N is a totally umbilical in $L_1^m(k,f)$ with $\dim N\ge 3$, then it follows from (5.19) and Lemma 5.3(4) that the sectional curvature $K^N(u\wedge v)$ of N with respect to orthonormal vectors u,v is given by

$$K^N(u\wedge v)=\langle H,H\rangle+\frac{k+f'^2}{f^2}+(\epsilon_u\varphi_u^2+\epsilon_v\varphi_u^2)\Big(\frac{k}{f^2}-(\ln f)''\Big). \tag{5.28}$$

If we choose an orthonormal basis $e_1,\ldots,e_n$ of T_pN such that $e_2,\ldots,e_n$ are perpendicular to ∂_t, then the Ricci curvature of N satisfies

$$\begin{aligned} &Ric(e_1)=(n-1)\left\{\langle H,H\rangle+\frac{k+f'^2}{f^2}+\epsilon_1\varphi_{e_1}^2\Big(\frac{k}{f^2}-(\ln f)''\Big)\right\},\\ &Ric(e_j)=(n-1)\left\{\langle H,H\rangle+\frac{k+f'^2}{f^2}\right\}+\epsilon_1\varphi_{e_1}^2\Big(\frac{k}{f^2}-(\ln f)''\Big), \end{aligned} \tag{5.29}$$

for $j=2,\ldots,n$. Since $n\ge 3$, it follows from (5.29) and the Einstein condition that $\varphi_{e_1}=0$. Hence N is a transverse submanifold.

Conversely, if N is a totally umbilical transverse submanifold of $L_1^m(k,f)$, then N lies in a slice $S(t_0)$ for some $t_0\in I$. Since $S(t_0)$ is a real space form, Gauss' equation implies that N is of constant curvature. Hence, N is an Einstein manifold. □

5.6 Hypersurfaces of constant curvature in RW spacetimes

Definition 5.3. An n-dimensional pseudo-Riemannian submanifold in a pseudo-Riemannian $(n+1)$-manifold is called *quasi-umbilical* if the shape operator of the submanifold has an eigenvalue with multiplicity $\geq n-1$.

For hypersurfaces of constant curvature in a Robertson–Walker spacetime, we have the following result of [Chen and Wei (2009)].

Proposition 5.7. *Assume $L_1^{n+1}(k,f), n \geq 4$, contains no open subsets of constant curvature. Let N be a spacelike hypersurface of constant curvature in $L_1^{n+1}(k,f)$. Then*

(a) *N is a quasi-umbilical hypersurface; thus N has a principal curvature λ with multiplicity $\geq n-1$;*
(b) *At each point $p \in N$ the eigenspace $E_p(\lambda)$ associated with λ is a vertical subspace.*

Proof. Under the hypothesis, let $p \in N$. Choose orthonormal eigenvectors $e_1, \ldots, e_n$ of the shape operator of N with $\lambda_1, \ldots, \lambda_n$ as their corresponding eigenvalues. Then it follows from Gauss' equation and Lemma 5.3(4) that the sectional curvature of N satisfies

$$K^N(e_i \wedge e_j) = \lambda_i \lambda_j + (\varphi_i^2 + \varphi_j^2)\left(\frac{k}{f^2} - (\ln f)''\right) + \frac{k + f'^2}{f^2}, \tag{5.30}$$

for $1 \leq i \neq j \leq n$, where $\varphi_i = \varphi_{e_i}$. Because N is of constant curvature, expressing $K(e_i \wedge e_j) = K(e_i \wedge e_k)$ in terms of (5.30) gives

$$\lambda_i(\lambda_j - \lambda_k) = (\varphi_k^2 - \varphi_j^2)\left(\frac{k}{f^2} - (\ln f)''\right) \tag{5.31}$$

for $i \neq j, i \neq k$, $i,j,k \in \{1, \ldots, n\}$. Hence, by first shifting the index j to k and k to ℓ in (5.31), then shifting the index i to j, j to k and k to ℓ in (5.31), and taking the difference of these two resultant equations, we find

$$(\lambda_i - \lambda_j)(\lambda_k - \lambda_\ell) = 0 \text{ for } i,j \neq k \text{ and } i,j \neq \ell. \tag{5.32}$$

If $\lambda_1 = \cdots = \lambda_n$, then we obtain the first assertion. Otherwise, without loss of generality, we may assume $\lambda_1 \neq \lambda_2$, then (5.32) implies $\lambda_3 = \ldots = \lambda_n$, and $(\lambda_1 - \lambda_3)(\lambda_2 - \lambda_4) = 0$. Hence, we have either $\lambda_1 = \lambda_3 = \lambda_4 = \cdots = \lambda_n$, or $\lambda_2 = \lambda_3 = \lambda_4 = \cdots = \lambda_n$. In both cases, N has a principal curvature λ of multiplicity $\geq n-1$. This gives (a).

It follows from (5.31) that φ_X^2 is independent of the unit vector X in $E_p(\lambda)$. Since $\dim E_p(\lambda) \geq n-1 \geq 3$, there exists a unit vector $X \in E_p(\lambda)$ with $\varphi_X = 0$. Hence, we get $\varphi_X = -\langle X, \partial_t \rangle = 0$ for $X \in E_p(\lambda)$. Thus ∂_t is orthogonal to $E_p(\lambda)$. This gives (b). □

5.7 Realization of RW spacetimes in pseudo-Euclidean spaces

Now, we provide explicit realizations of Robertson–Walker spacetimes $L_1^n(k,f)$ as pseudo-Riemannian submanifolds in pseudo-Euclidean spaces.

Case (A): $k = r^{-2}, r > 0$. Let I be an open interval containing 0 and $\iota_1 : S^{n-1}(k) \subset \mathbb{E}^n$ the inclusion map given by

$$S^{n-1}(k) = \{\mathbf{x} \in \mathbb{E}^n : \langle \mathbf{x}, \mathbf{x} \rangle = k > 0\}.$$

Consider the map $\phi_+ : L_1^n(k,f) = I \times_f S^{n-1}(k) \to \mathbb{E}_1^{n+1}$ defined by

$$\phi_+(t,p) = \left(\int_0^t \sqrt{1+f'(u)^2}du, rf(t)\iota_1(p) \right). \tag{5.33}$$

Then ϕ is an embedding. Moreover, we have

$$\tilde{\nabla}_{\partial/\partial t}\phi_+ = (\sqrt{1+f'(t)^2}, rf'(t)\iota_1), \quad \tilde{\nabla}_X \phi_+ = (0, rf(t)d\iota_1(X)) \tag{5.34}$$

for X tangent to $S^{n-1}(k)$. It follows from (5.33)-(5.34) that the induced metric tensor via ϕ_+ is exactly the metric tensor of $L_1^n(k,f)$. It is direct to show that ϕ_+ is a quasi-umbilical isometric embedding.

Case (B): $k = -r^{-2}, r > 0$. Let I be an open interval containing 0. Denote by ι_2 the inclusion map $H^{n-1}(k) \subset \mathbb{E}_1^n$ defined by

$$H^{n-1}(k) = \{\mathbf{x} \in \mathbb{E}_1^n : \langle \mathbf{x}, \mathbf{x} \rangle = -r^{-2} < 0\}.$$

Consider the map $\phi_- : I \times_f H^{n-1}(k) \to \mathbb{E}_2^{n+1}$ defined by

$$\phi_-(t,p) = \left(\int_0^t \sqrt{1-f'(u)^2}du, rf(t)\iota_2(p) \right).$$

Just like ϕ_+, the map ϕ_- is a quasi-umbilical isometric embedding.

Case (C): $k = 0$. The Robertson–Walker spacetime is the warped product $L_1^n(0,f) = I \times_f \mathbb{E}^{n-1}$. We may assume that I containing 0. Consider the map $\phi : L_1^n(0,f) = I \times_f \mathbb{E}^{n-1} \to \mathbb{E}_1^{2n-1}$ defined by

$$\phi(t, u_2, \ldots, u_n) = \left(\int_0^t \sqrt{1+(n-1)f'(u)^2}du, f(t)\cos u_2, \right.$$
$$\left. f(t)\sin u_2, \ldots, f(t)\cos u_n, f(t)\sin u_n \right).$$

Then ϕ is an isometric immersion.

Chapter 6

Hodge Theory, Elliptic Differential Operators and Jacobi's Elliptic Functions

Hodge theory is a branch of algebraic geometry, algebraic topology and complex manifold theory that deals with the decomposition of the cohomology groups of an complex projective algebraic variety. The theory was developed in the 1930s by Scottish mathematician William V. D. Hodge (1903–1975) as an extension of de Rham cohomology. Hodge theory has major applications on Riemannian manifolds, Kähler manifolds and algebraic geometry. The theory was very influential on subsequent beautiful and important work done by Kunihiko Kodaira (1915-1997).

Elliptic differential operators are differential operators that resemble the Laplacian. They can be defined by a positivity condition on the coefficients of the highest-order derivatives, which implies the key property that the principal symbol is invertible, or equivalently that, there are no real characteristic directions. Elliptic regularity implies that their solutions tend to be smooth functions if the coefficients in the operator are smooth. The eigenspaces and eigenvalues of elliptic differential operators have many nice properties. Elliptic differential operators are typical of potential theory, and they appear frequently in electrostatics and continuum mechanics.

One of the greatest accomplishments of Carl Jacobi (1804-1851) was his original theory of elliptic functions and their relation to theta functions. Elliptic functions were studied by many of the great mathematicians of the 19th century, including Abel, Cauchy and Weierstrass. Such functions are very useful in mathematics, physics and engineering, e.g. the equations of motion are integrable in terms of elliptic functions in the well known cases of the pendulum, Kepler's problem, the Euler top and the symmetric Lagrange top in a gravitational field. Theta functions are of great importance in mathematical physics, in particular in fluid mechanics, because of their role in the inverse problem for periodic and quasi-periodic flows.

6.1 Operators d, $*$ and δ

In differential geometry, the *exterior derivative* d extends the concept of the differential of a function on a smooth manifold M, which is a form of degree zero, to differential forms of higher degree.

The exterior derivative d has the property that $d^2 = 0$ and is the differential (coboundary) used to define de Rham cohomology on forms. Integration of forms gives a natural homomorphism from the de Rham cohomology to the singular cohomology of a smooth manifold. The exterior derivative of a differential form of degree k is a differential form of degree $k+1$.

If f is in $\mathcal{F}(M)$, then the exterior derivative of f is its differential of f. That is, df is the unique 1-form such that, for every $X \in \mathcal{X}(M)$,

$$df(X) = Xf,$$

where Xf is the directional derivative of f in the direction of X. Thus the exterior derivative of a 0-form is a 1-form.

The *exterior derivative* is defined to be the unique $\mathbf{R}$-linear mapping from k-forms to $(k+1)$-forms satisfying the following properties:

(1) df is the differential of f for $f \in \mathcal{F}(M)$;
(2) $d(df) = 0$ for any $f \in \mathcal{F}(M)$;
(3) $d(\alpha \wedge \beta) = d\alpha \wedge \beta + (-1)^p(\alpha \wedge d\beta)$, where α is a p-form, i.e., d is a derivation of degree one on the exterior algebra of differential forms.

We denote the space of all k-forms on M by $\Omega^k(M)$. To deal with differential forms, one can work entirely in a local coordinate system $x_1, \ldots, x_n$ on a manifold M. First, the coordinate differentials $dx_1, \ldots, dx_n$ form a basic set of 1-forms within the coordinate chart. Given a multi-index $(i_1, \ldots, i_k)$ with $1 \le i_j \le n$ for $1 \le j \le k$, the exterior derivative of a k-form $\omega = \sum_{1 \le i_1 < \cdots < i_k \le n} f_{i_1 \cdots i_k} dx_{i_1} \wedge \cdots dx_{i_k}$ is

$$d\omega = \sum_{i=1}^{n} \sum_{1 \le i_1 < \cdots < i_k \le n} \frac{\partial f_{i_{i_1} \cdots i_1}}{\partial x_i} dx_i \wedge dx_{i_1} \wedge \cdots dx_{i_k}.$$

Alternatively, an explicit formula can be given for the exterior derivative of a k-form ω, when paired with $k+1$ arbitrary smooth vector fields $V_0, \ldots, V_k \in \mathcal{X}(M)$. In this case, we have

$$\begin{aligned} d\omega(V_0, V_1, \ldots, V_k) &= \sum_{i=0}^{k} (-1)^i V_i(\omega(V_0, \ldots, \hat{V}_i, \ldots, V_k)) \\ &\quad + \sum_{i<j} (-1)^{i+j} \omega([V_i, V_j], V_0, \ldots, \hat{V}_i, \ldots, \hat{V}_j, \ldots, V_k), \end{aligned} \tag{6.1}$$

where [,] is Lie bracket and $\hat{}$ denotes the omission of that element. In particular, for 1-forms we have

$$d\omega(X,Y) = X\omega(Y) - Y\omega(X) - \omega([X,Y]).$$

Definition 6.1. A differential form ω is called a *closed form* if $d\omega = 0$. The image of d are called *exact forms.*

Closed and exact forms are related, because of the identity $d^2\omega = 0$ for any k-form ω. This implies that every exact form is closed. The converse is true in contractible regions according to the *Poincaré lemma.*

The *de Rham complex* is the cochain complex of exterior differential forms on M, with the exterior derivative as the differential

$$0 \longrightarrow \Omega^0(M) \xrightarrow{d} \Omega^1(M) \xrightarrow{d} \Omega^2(M) \xrightarrow{d} \Omega^3(M) \longrightarrow \cdots \tag{6.2}$$

where $\Omega^0(M) = \mathcal{F}(M)$ is the space of smooth functions on M.

The idea of de Rham cohomology is to classify the different types of closed forms on a manifold M. One performs this classification by saying that two closed forms $\alpha, \beta \in \Omega^k(M)$ are *cohomologous* if they differ by an exact form, i.e., if $\alpha - \beta$ is exact. This classification induces an equivalence relation on the space of closed forms in $\Omega^k(M)$. One then defines the k-th *de Rham cohomology group* to be the set of equivalence classes, i.e., the set of closed forms in $\Omega^k(M)$ modulo the exact forms.

On an oriented Riemannian n-manifold M, if we choose an orthonormal local frame $e_1, \ldots, e_n$ of TM whose orientation is compatible with that of M, then $\omega^1 \wedge \cdots \wedge \omega^n$ is the volume element of M, where $\{\omega^1, \ldots, \omega^n\}$ is the dual frame of $\{e_1, \ldots, e_n\}$. The *Hodge star operator* $*$ is an isomorphism $* : \Omega^k(M) \to \Omega^{n-k}(M)$ defined as follows: For a k-form

$$\omega = \sum\nolimits_{i_1 < \cdots < i_k} f_{i_1 \cdots i_k} \omega^{i_1} \wedge \cdots \wedge \omega^{i_k},$$

we define

$$*\,\omega = \sum\nolimits_{j_1 < \cdots < j_{n-k}} \epsilon_{i_1 \cdots i_k j_1 \cdots j_{n-k}} f_{i_1 \cdots i_k} \omega^{j_1} \wedge \cdots \wedge \omega^{j_{n-k}},$$

where $\epsilon_{i_1 \cdots i_k j_1 \cdots j_{n-k}}$ is 0 if $i_1 \cdots i_k j_1 \cdots j_{n-k}$ does not form a permutation of $1, \ldots, n$; and 1 or -1, according to the permutation is even or odd.

The form $*\,\omega$ is called the *adjoint* of ω. The adjoint of 1 is just the volume form $\omega^1 \wedge \cdots \wedge \omega^n$.

Lemma 6.1. *The star operator $*$ satisfies*

(1) $*(\alpha + \beta) = *\alpha + *\beta$ *and* $*(f\alpha) = f(*\alpha)$*;*

(2) $*(*\alpha) = (-1)^{nk+k}\alpha$;
(3) $\alpha \wedge *\beta = \beta \wedge *\alpha$;
(4) $\alpha \wedge *\alpha = 0$ *if and only if* $\alpha = 0$,

where α *and* β *are* k*-forms and* f *is a* 0*-form.*

Proof. Follows from direct computation. □

Let α and β be k-forms given by

$$\alpha = \sum_{i_1<\cdots<i_k} a_{i_1\cdots i_k}\omega^{i_1} \wedge \cdots \wedge \omega^{i_k}, \quad \beta = \sum_{j_1<\cdots<j_k} b_{j_1\cdots j_k}\omega^{j_1} \wedge \cdots \wedge \omega^{j_k}.$$

Then

$$\alpha \wedge *\beta = \sum_{i_1<\cdots<i_k} a_{i_1\cdots i_k} b_{i_1\cdots i_k} * 1.$$

For α and β we define a global inner product of α and β by

$$(\alpha, \beta) = \int_M \alpha \wedge *\beta, \tag{6.3}$$

whenever the integral converges.

Two k-forms α, β are called *orthogonal* if $(\alpha, \beta) = 0$.

Lemma 6.2. *The star operator* $*$ *satisfies*

(1) $(\alpha, \alpha) \geq 0$ *and is equal to zero if and only if* $\alpha = 0$;
(2) $(\alpha, \beta) = (\beta, \alpha)$;
(3) $(\alpha, \beta_1 + \beta_2) = (\alpha, \beta_1) + (\alpha, \beta_2)$;
(4) $(*\alpha, *\beta) = (\alpha, \beta)$,

where $\alpha, \beta, \beta_1, \beta_2$ *are* k*-forms.*

Proof. Follows from Lemma 6.1. □

The *codifferential* δ on k-forms are defined as

$$\delta\alpha = (-1)^{nk+n+1} * d * \alpha. \tag{6.4}$$

In contrast with the differential operator d, the codifferential operator δ involves the metric structure of M.

Lemma 6.3. *The codifferential operator* $\delta : \Omega^k(M) \to \Omega^{k-1}(M)$ *satisfies*

(1) $\delta(\alpha + \beta) = \delta\alpha + \delta\beta$;
(2) $\delta^2\alpha = 0$;
(3) $*\delta\alpha = (-1)^k d * \alpha$ *and* $*d\alpha = (-1)^{k+1}\delta * \alpha$

where α, β *are* k*-forms.*

Proof. Follows from (6.4), Lemma 6.1 and the property $d^2 = 0$. □

Definition 6.2. A differential form α is called *coclosed* if $\delta\alpha = 0$. If $\alpha = \delta\beta$ for some form β, then α is called *coexact*.

6.2 Hodge–Laplace operator

The Hodge–Laplace operator Δ is defined by $\Delta = d\delta + \delta d$.

Definition 6.3. A form α on M is called *harmonic* if $\Delta\alpha = 0$.

Proposition 6.1. *If M is a compact oriented Riemannian manifold and α and β are forms of degree k and $k+1$, respectively, then*

$$(d\alpha, \beta) = (\alpha, \delta\beta), \tag{6.5}$$

i.e., δ is the adjoint of d, Consequently, the Hodge-Laplace operator Δ is self-adjoint. In particular, the Laplacian acting on $\mathcal{F}(M)$ is self-adjoint.

Proof. Since M is compact, the Stokes theorem implies that

$$\int_M d(\alpha \wedge *\beta) = 0.$$

Thus, by using the properties of d, we find

$$\int_M d\alpha \wedge *\beta = (-1)^{k-1}\alpha \wedge d * \beta.$$

Hence, we obtain (6.5) from (6.4). □

Proposition 6.1 implies immediately the following.

Corollary 6.1. *On a compact oriented Riemannian manifold M, we have*

(1) *A k-form is closed if and only if it is orthogonal to all coexact k-forms;*
(2) *A k-form is coclosed if and only if it is orthogonal to all exact k-forms.*

Another application of Proposition 6.1 is the following.

Proposition 6.2. *On a compact oriented Riemannian manifold, a form is harmonic if and only if $d\alpha = 0$ and $\delta\alpha = 0$.*

Proof. If α is a k-form on M, then Proposition 6.1 implies that

$$(\Delta\alpha, \alpha) = (d\delta\alpha, \alpha) + (\delta d\alpha, \alpha) = (d\alpha, d\alpha) + (\delta\alpha, \delta\alpha).$$

From this we conclude that α is harmonic if and only if α is closed and coclosed. □

An important application of Proposition 6.2 is the following.

Corollary 6.2. *Every harmonic function on a compact Riemannian manifold is a constant.*

The following result is known as the *Divergence Theorem.*

Proposition 6.3. *If X is a vector field on a compact oriented Riemannian manifold M, then*

$$\int_M (\operatorname{div} X) * 1 = 0. \tag{6.6}$$

Proof. By Proposition 6.2 we have

$$\int_M (\operatorname{div} X) * 1 = -\int_M (\delta X^{\#}) * 1 = -(\alpha^{\#}, d1) = 0,$$

where $X^{\#}$ is dual 1-form of X given by $X^{\#}(Y) = g(X, Y)$, $Y \in \mathcal{X}(M)$. □

Corollary 6.3. *If f is differentiable function on a compact Riemannian manifold, then*

$$\int_M (\Delta f) * 1 = 0. \tag{6.7}$$

Proof. Since a differentiable function f on M is a 0-form, $\Delta f = \delta df$, the Divergence Theorem implies (6.7). □

The following result is well-known as Hopf's lemma.

Corollary 6.4. *Let M be a compact Riemannian n-manifold. If f is a differentiable function on M such that $\Delta f \geq 0$ everywhere on M (or $\Delta f \leq 0$ everywhere on M), then f is a constant function.*

Proof. We may assume that M is orientable by taking the 2-fold covering of M if necessary. If $\Delta f \geq 0$ (or $\Delta f \leq 0$), then Corollary 6.3 implies $\Delta f = 0$. Hence, f is a constant function according to Corollary 6.2. □

6.3 Elliptic differential operator

Let U be an open set in $\mathbb{E}^n$ with natural coordinates $x_1, \ldots, x_n$. For each n-tuple $t = (t_1, \ldots, t_n)$ of non-negative integer we put

$$|t| = t_1 + \cdots + t_n, \quad D^t = \frac{\partial^{|t|}}{\partial_1^{t_1} \cdots \partial_n^{t_n}}. \tag{6.8}$$

A *linear differential operator* D of degree r over U takes the form:

$$D = \sum_{|t| \leq r} a_t(p) D^t, \tag{6.9}$$

where $a_t(p) : L \to V$ is a homomorphism of vector spaces and depends differentiably on $p \in U$. For each $\mathbf{y} = (y_1, \ldots, y_n) \in \mathbb{E}^n$, we set

$$\sigma(D, \mathbf{y}) = \sum_{|t| = r} a_t(\mathbf{y}) \mathbf{y}^t, \tag{6.10}$$

where we use the multi-indices: $\mathbf{y}^t = y_1^{t_1} \cdots y_n^{t_n}$. $\sigma(D, \mathbf{y})$ is called the *characteristic polynomial* of D and

$$\sigma(D, \cdot\,) : \mathbb{E}^n \to \mathrm{Hom}(L, V)$$

is called the *symbol* of D.

A differential operator D is called *elliptic* if the characteristic polynomial $\sigma(D, \mathbf{y})$ of D has no real zeros except $\mathbf{y} = 0$ at $p \in U$. The notion of elliptic differential operators has a natural generation to manifolds defined as follows: Let E, F be two (complex) vector bundles over a manifold M. A differential operator is a linear map $D : \Gamma(E) \to \Gamma(F)$ between the spaces of cross sections, which when restricted to each coordinate neighborhood U of M (over which E and F are trivial) is expressible in the form (6.9). The differential operator D is again called *elliptic* if the characteristic polynomial $\sigma(D, \mathbf{y})$ of D has no real zeros except $\mathbf{y} = 0$ at each point $p \in M$.

Perhaps the most important elliptic differential operator is the Laplacian Δ. Take $M = \mathbb{E}^n$ and $E = F$ the trivial line bundle over M. Then

$$\Delta = -\sum_{i=1}^{n} \frac{\partial^2}{\partial x_i^2}.$$

Since $\sigma(\Delta, \mathbf{y}) = -|y|^2 = -\sum y_i^2$, Δ is an elliptic differential operator.

Let E, F be two complex vector bundles over a compact Riemannian manifold M. Let us equipped E and F with *Hermitian metrics*, i.e., metrics g which are compatible with the complex structure J; namely they satisfy $g(JX, JY) = g(X, Y)$. This allows us to define an inner product (,) of on $\Gamma(E)$, e.g. if $s, s' \in \Gamma(E)$,

$$(s, s') = \int_M \langle s, s' \rangle * 1.$$

Let $D : \Gamma(E) \to \Gamma(F)$ be an elliptic differential operator. It is possible to define the *adjoint* of D, $D^* : \Gamma(F) \to \Gamma(E)$, as a differential operator characterized by the property:

$$(Ds, u) = (s, D^* u), \quad s \in \Gamma(E), \; u \in \Gamma(F). \tag{6.11}$$

It can be verified that D^* is also elliptic.

It follows from (6.11) that the kernel, $\ker(D)$, of D and the image, $\mathrm{im}(D^*)$, of D^* are orthogonal with respect to $(\ ,\ )$. Moreover, $\ker(D)$ is precisely the complement of $\mathrm{im}(D^*)$. In fact, if s is orthogonal to $\mathrm{im}(D^*)$, then $0 = (s, D^*Ds) = (Ds, Ds)$, thus $Ds = 0$. Hence, we have

$$\Gamma(E) = \ker(D) \oplus \mathrm{im}(D^*). \tag{6.12}$$

An elliptic differential operator $D : \Gamma(E) \to \Gamma(F)$ has many other important properties. For instance, an elliptic operator D is a *Fredholm operator*, i.e., D is a differential operator which has finite-dimensional kernel and cokernel and it has closed image. From (6.12) we may conclude that for any η in the orthogonal complement of $\ker(D)$, there exists a solution of the equation

$$D^*u = \eta. \tag{6.13}$$

Consider the special case: $E = F$ and $D = D^*$. For each $\lambda \in \mathbf{R}$, we put

$$\Gamma_\lambda = \{s \in \Gamma(E) : Ds = \lambda s\}. \tag{6.14}$$

Let $L^2(E)$ denote the completion of $\Gamma(E)$ with respect to $(\ ,\ )$. Then it is known that there are only countably many λ with $\Gamma_\lambda \neq 0$ and each $\Gamma_\lambda \neq 0$ is a finite-dimensional vector space. Moreover,

$$L^2(E) = \hat{\oplus}_\lambda \Gamma_\lambda, \tag{6.15}$$

where $\hat{\oplus}$ is the completion of orthogonal sum.

Because the Hodge-Laplace operator $\Delta : \Omega^k(M) \to \Omega^{k+1}(M)$ is elliptic and self-adjoint, we have

$$\Omega^k(M) = \hat{\oplus}_i V_{k,i}, \tag{6.16}$$

where $V_{k,i}$ is the i-th eigenspace of Δ acting on k-forms. In fact, the eigenvalues of $\Delta : \Omega^k(M) \to \Omega^k(M)$ satisfies

$$0 \leq \lambda_{k,1} < \lambda_{k,2} < \cdots < \lambda_{k,i} < \cdots \nearrow \infty.$$

Since the kernel of $\Delta : \Omega^k(M) \to \Omega^k(M)$ is finite-dimensional, we have the following well-known results.

Theorem 6.1. *If M is a compact Riemannian manifold, then for each $k \in \{0, 1, \ldots, n\}$ the space of harmonic k-forms is finite-dimensional.*

We simplify denote $V_{0,i}$ by V_i and $\lambda_{0,i}$ by λ_i.

Theorem 6.2. *If M is a compact Riemannian manifold, then*

$$\mathcal{F}(M) = \hat{\oplus}_i V_i. \tag{6.17}$$

Denote by $Spec^k(M)$ the set of all eigenvalues of $\Delta : \Omega^k(M) \to \Omega^k(M)$ enumerated with multiplicity. $Spec^k(M)$ is called the *spectrum of k-forms.* We simply denote $Spec^0(M)$ by $Spec(M)$.

Remark 6.1. *Spectral geometry* is an area concerning relationships between geometric structures of manifolds and spectra of canonically defined differential operators. In particular, between geometric structures of manifolds and the spectra of the Hodge-Laplace operator. This subject concerns with two kinds of questions: direct problems and inverse problems.

Inverse problems seek to identify features of the geometry from information about the eigenvalues of the Laplacian. One of the earliest results was due to Hermann Weyl (1885-1995) who used David Hilbert's theory of integral equation in 1911 to show that the volume of a bounded domain in Euclidean space can be determined from the asymptotic behavior of the eigenvalues for the Dirichlet boundary value problem of the Laplace operator. A refinement of Weyl's asymptotic formula obtained in 1949 by Minakshisundaram and Pleijel produces a series of local spectral invariants involving covariant differentiations of the curvature tensor, which can be used to establish spectral rigidity for a special class of manifolds. However, as the example given in [Milnor (1964)] tells us, the information of eigenvalues is not enough to determine the isometry class of a manifold.

6.4 Hodge–de Rham decomposition and its applications

Let M be a compact oriented Riemannian n-manifold. For each integer $k \in \{0, 1, 2, \ldots, n\}$, denote by $\Omega^k_d(M)$, $\Omega^k_\delta(M)$ and $\Omega^k_H(M)$ the subspaces of $\Omega^k(M)$ consisting of k-forms which are exact, coexact and harmonic, respectively.

Lemma 6.4. *The three subspaces $\Omega^k_d(M), \Omega^k_\delta(M)$ and $\Omega^k_H(M)$ are mutually orthogonal.*

Proof. If $\omega \in \Omega^k_d(M)$, ω is closed. Thus, Corollary 6.1 implies that ω is orthogonal to $\Omega^k_\delta(M)$.

If $\omega = d\alpha$ and $\beta \in \Omega^k_H(M)$, Propositions 6.1 and 6.2 imply that

$$(\omega, \beta) = (d\alpha, \beta) = (\alpha, \delta\beta) = 0.$$

This shows that $\Omega^k_d(M)$ is orthogonal to $\Omega^k_H(M)$. Similar argument applies to the remaining case. □

The following is the *Hodge-de Rham Decomposition Theorem.*

Theorem 6.3. *A k-form α on a compact oriented Riemannian manifold M is uniquely decomposed into the orthogonal sum*

$$\alpha = \alpha_d + \alpha_\delta + \alpha_H, \tag{6.18}$$

where $\alpha_d \in \Omega_d^k(M)$, $\alpha_\delta \in \Omega_\delta^k(M)$ and $\alpha_H \in \Omega_H^k(M)$.

Proof. Theorem 6.1 implies that $\Omega_H^k(M)$ is finite-dimensional. Thus, we may choose an orthonormal basis $\omega^1, \ldots, \omega^h$ of $\Omega_H^k(M)$, $h = \dim \Omega_H^k(M)$. Let α be any k-form on M, we put $\alpha_H = \sum_{i=1}^h (\alpha, \omega^i)\omega^i$. Then $\Delta\alpha_H = 0$. Moreover, $\alpha - \alpha_H$ is orthogonal to $\Omega_H^k(M)$.

Since $\Delta^* = \Delta$, there exists a solution of the equation $\Delta u = \alpha - \alpha_H$. Hence, there is a k-form β such that $\Delta\beta = \alpha - \alpha_H$. If we put

$$\alpha_d = d\delta\beta, \quad \alpha_\delta = \delta d\beta,$$

we obtain (6.18).

Suppose that $\alpha = \alpha'_d + \alpha'_\delta + \alpha'_H$ is another decomposition of α, then

$$(\alpha_d - \alpha'_d) + (\alpha_\delta + \alpha'_\delta) + (\alpha_H - \alpha'_H) = 0.$$

Thus Lemma 6.4 implies that $\alpha_d = \alpha'_d, \alpha_\delta = \alpha'_\delta$ and $\alpha_H = \alpha'_H$. This proves the uniqueness. □

The following result of Hodge-de Rham is an important application of Theorem 6.3.

Theorem 6.4. *Every cohomology class in the k-th cohomology group $H^k(M)$ is represented uniquely by a harmonic k-form.*

Proof. If $\Phi \in H^k(M)$, then $\Phi = \{\alpha\}$ for some closed k-form α. Thus, by Corollary 6.1, we have $\alpha = \alpha_d + \alpha_H$ with $\alpha_d \in \Omega_d^k(M)$ and $\alpha_H \in \Omega_H^k(M)$. Since α_d is exact, Φ is thus represented by α_H.

If β is another k-form which represents Φ, then $\alpha - \beta$ is exact. Then, by Theorem 6.3, we have $\alpha_H = \beta_H$. □

The following result is known as *Poincaré's Duality Theorem.*

Theorem 6.5. *If M is a compact oriented Riemannian n-manifold, then there is a natural isomorphism:*

$$\mu : H^k(M) \cong H^{n-k}(M).$$

Proof. If $\Phi \in H^k(M)$, there exists a hormonic k-form α_H representing Φ. Since $\Delta(*\alpha_H) = *(\Delta\alpha) = 0$, $*\alpha_H$ is also harmonic. Hence, if we put $\mu(\Phi) = [*\alpha_H] \in H^{n-k}(M)$, then μ defines an isomorphism from $H^k(M)$ onto $H^{n-k}(M)$. □

A fundamental result in algebraic topology asserts that there exists a natural isomorphism between the k-th cohomology group $H^k(M)$ and the k-th homology group $H_k(M)$. Thus, the k-th *Betti number* $b_k(M) := \dim H_k(M)$ is equal to $\dim H^k(M)$. Consequently, Theorem 6.5 implies

$$b_k(M) = b_{n-k}(M) \tag{6.19}$$

for any compact oriented n-manifold M.

Another consequence of Hodge-de Rham's decomposition theorem is the following known result.

Corollary 6.5. *If $\hat{M}$ is a compact covering manifold of a compact oriented n-manifold M, then*

$$b_k(M) \leq b_k(\hat{M}) \quad k = 1, \ldots, n-1. \tag{6.20}$$

Proof. We may equip M with a Riemannian metric. For each nonzero harmonic k-form α on M, there exists a periodic extension $\hat{\alpha}$ on $\hat{M}$ given by $\hat{\alpha} = \pi^*\alpha$, where π is the projection of the covering map and π^* is the induced map of π. Clearly, $\hat{\alpha}$ is also a nonzero harmonic k-form. Since linearly independent harmonic forms on M lift to linearly independent harmonic forms on $\hat{M}$, we obtain (6.20) from Theorem 6.4. □

6.5 The fundamental solution of heat equation

Let M be a compact Riemannian manifold. A *heat operator* on M is the operator

$$L = \Delta + \frac{\partial}{\partial t}$$

acting on functions defined on $M \times \mathbf{R}_+$ which is of class C^2 on the first variable and of class C^1 on the second.

The *heat equation* on M works on functions $F : M \times \mathbf{R}_+ \to \mathbf{R}$ which satisfy

$$L(F) = 0, \quad F(p, 0) = f(0), \quad p \in M, \tag{6.21}$$

where $f : M \to \mathbf{R}$ is a given initial condition.

Definition 6.4. A *fundamental solution of the heat equation* on M is a function $h : M \times M \times \mathbf{R}_+ \to \mathbf{R}$ which satisfies the following conditions:

(h_1) h is continuous on the three variables of class C^2 on the first two variables and of class C^1 on the third;

(h_2) $L_2h = 0$, where $L_2 = \Delta_2 + \frac{\partial}{\partial t}$ and Δ_2 is the Laplacian on the second variable;

(h_3) for each $p \in M$, $\lim_{t\to 0^+} h(p,\cdot,t) = \delta_p$, where δ_p is the Dirac distribution at p, i.e., for each function f on M with $p \in \operatorname{supp}(f)$, we have

$$\lim_{t\to 0^+} \int_M h(p,x,t)f(x)dx = f(p),$$

where dx denotes the volume element of the second M.

The following result is well-known.

Theorem 6.6. *A fundamental solution of heat equation on M exists and is unique.*

For an eigenspace V_i of $\Delta : \mathcal{F}(M) \to \mathcal{F}(M)$, we choose an orthonormal basis $\varphi_1^i, \ldots, \varphi_{m_i}^i$ of V_i, $m_i = \dim V_i$. The set of $\{\varphi_\alpha^i\}_{i,\alpha}$ is called an *orthonormal set of eigenfunctions* of Δ. According to Theorem 6.2, associated with a function $f \in \mathcal{F}(M)$ we have

$$f = \sum_{\alpha,i} (\varphi_\alpha^i, f)\varphi_\alpha^i \quad \text{(in the } L^2\text{-sense)}. \tag{6.22}$$

Proposition 6.4. *If $\{\varphi_\alpha^i\}$ is an orthonormal set of eigenfunctions, then for each $(p,x,t) \in M \times M \times \mathbf{R}_+$, the series*

$$\sum_{i,\alpha} e^{-\lambda_i t}\varphi_\alpha^i(p)\varphi_\alpha^i(x) \tag{6.23}$$

converges and the fundamental solution of heat equation is given by

$$h(p,x,t) = \sum_{i,\alpha} e^{-\lambda_i t}\varphi_\alpha^i(p)\varphi_\alpha^i(x). \tag{6.24}$$

The proof of Proposition 6.4 bases on the uniqueness of the fundamental solution of the heat equation. From the definition of L_2 we have

$$\begin{aligned}
L_2&\Bigg(\sum_{i,\alpha} e^{-\lambda_i t}\varphi_\alpha^i(p)\varphi_\alpha^i(x)\Bigg) \\
&= \sum_{i,\alpha} e^{-\lambda_i t}\varphi_\alpha^i(p)(\Delta\varphi_\alpha^i)(x) + \sum_{i,\alpha} \frac{\partial}{\partial t}\left(e^{-\lambda_i t}\right)\varphi_\alpha^i(p)\varphi_\alpha^i(x) \\
&= \sum_{i,\alpha} \lambda_i e^{-\lambda_i t}\varphi_\alpha^i(p)\varphi_\alpha^i(x) - \sum_{i,\alpha} \lambda_i\left(e^{-\lambda_i t}\right)\varphi_\alpha^i(p)\varphi_\alpha^i(x) \\
&= 0.
\end{aligned}$$

Moreover, for each $f \in \mathcal{F}(M)$, we also have

$$\lim_{t\to 0^+} \int \sum e^{-\lambda_i t} \varphi_\alpha^i(p)\varphi_\alpha^i(x) f(x)dx = \lim_{t\to 0^+} \sum e^{-\lambda_i t}(\varphi_\alpha^i, f)\varphi_\alpha^i$$
$$= \sum (\varphi_\alpha^i, f)\varphi_\alpha^i = f.$$

These show that (6.24) satisfies conditions (h_2) and (h_3). In fact (6.24) also satisfies (h_1). Thus, (6.24) is a fundamental solution of the heat equation. Consequently, by the uniqueness, it is $h(p, x, t)$.

By integrating $h(x, x, t)$ over M and using (6.24) we obtain

Proposition 6.5. *For each $t > 0$, the series $\sum_i m(\lambda_i)e^{-\lambda_i t}$ converges and*

$$\int_N h(x, x, t)dt = \sum_i m(\lambda_i)e^{-\lambda_i t}, \tag{6.25}$$

where $m(\lambda_i)$ is the multiplicity of λ_i.

To construct the fundamental solution $h(p, s, t)$ of heat equation, we use a successive approximation method. Although $\mathbb{E}^n$ is not compact, there exists a unique fundamental solution of heat equation on $\mathbb{E}^n$ under the condition to be decreasing at infinity. The solution is given by

$$h_0(p, x, t) = (4\pi t)^{-\frac{n}{2}} e^{-\frac{d(p,x)^2}{4t}},$$

where $d(p, x)$ denotes the Euclidean distance between p and x.

Using the idea that on a Riemannian manifold M, the fundamental solution of heat equation differs little from the pull-back of h_0 by $\exp^{-1}$, one arrives at the following *Asymptotic Expansion of Minakshisundaram-Pleijel* after long computation.

Theorem 6.7. *For each compact Riemannian n-manifold M, there exist constants $a_i's$ $(i = 0, 1, 2, \ldots)$ with*

$$\sum_j m(\lambda_j)e^{-\lambda_j t} \underset{t\to 0}{\sim} (4\pi t)^{-\frac{n}{2}} \sum_{i=0}^{\infty} a_i t^i. \tag{6.26}$$

The first three coefficients a_0, a_1, a_2 are given by

$$a_0 = \int_M *1 = \mathrm{vol}(M); \tag{6.27}$$

$$a_1 = \frac{1}{6}\int_M \tau * 1 \text{ (already folklore in 1965);} \tag{6.28}$$

$$a_2 = \frac{1}{360}\int_M (2||R||^2 - 2||Ric||^2 + 5n^2(n-1)^2\tau^2) * 1, \tag{6.29}$$

where a_2 was determined in [McKean and Singer (1967)]. The fourth coefficient a_3 was determined in [Sakai (1971)].

In particular, formulas (6.27) and (6.28) imply the following.

Corollary 6.6. *If M is a compact Riemannian manifold, then the volume and the total scalar curvature $\int_M \tau * 1$ of M are spectral invariants.*

By a *spectral invariant* we mean a Riemannian invariant which depends only on Spec(M).

6.6 Spectra of some important Riemannian manifolds

If we consider the Laplacian Δ acting on $\mathcal{F}(M)$ of a Riemannian manifold M, then $\Delta = \delta d$. For a function $f \in \mathcal{F}(M)$, df is a 1-form. Denote by $(df)_\#$ the associated vector field of df, i.e., $df(v) = \langle (df)_\#, v\rangle$ for $v \in TM$. Then $\Delta f = -\text{div}(df)_\#$.

Various expressions of Δ.

(a) Since df is a 1-form, the covariant derivative $\nabla(df)$ is a 2-form which is the Hessian of f. The trace of $\nabla(df)$ is $-\Delta f$.

(b) Let $p \in M$ and $u_1, \ldots, u_n$ a normal coordinate system about p. Then

$$(\Delta f)(p) = -\sum_{i=1}^{n} \frac{\partial^2 f}{\partial u_i^2}(p). \tag{6.30}$$

This is equivalent to say that, for each $p \in M$, pick an orthonormal set of geodesics $\{\gamma_i\}$ parametrized by arc length and passing through p at $s = 0$, then

$$(\Delta f)(p) = -\sum_{i=1}^{n} \frac{d^2(f \circ \gamma_i)}{ds^2}(0). \tag{6.31}$$

(c) In terms of a local coordinate system $\{x_1, \ldots, x_n\}$ of M, we have

$$\Delta f = -\frac{1}{\sqrt{\mathfrak{g}}} \sum_{i,j=1}^{n} \frac{\partial\left(\sqrt{\mathfrak{g}}\, g^{ij}(\partial f/\partial x_j)\right)}{\partial x_i}, \tag{6.32}$$

where $\mathfrak{g} = \det(g_{ij})$, g_{ij} the components of metric tensor g with respect to $x_1, \ldots, x_n$ and (g^{ij}) is the inverse matrix of (g_{ij}).

Definition 6.5. Let M and B be two pseudo-Riemannian manifolds. A *pseudo-Riemannian submersion* is a smooth map $\pi : M \to B$ which is onto and satisfies the following three axioms:

(S1) $\pi_*|_p$ is onto for all $p \in M$;
(S2) the fibers $\pi^{-1}(b)$, $b \in B$, are pseudo-Riemannian submanifolds of M;
(S3) π_* preserves scalar products of vectors normal to fibers.

Condition (S2) holds automatically when M is Riemannian. Just like warped and twisted products, vectors tangent to fibers are called *vertical* and those normal to fibers are called *horizontal.*

Proposition 6.6. *Let $\pi : (M, g_M) \to (B, g_B)$ be a Riemannian submersion with totally geodesic fibers. Then*

$$\Delta_M(f \circ \pi) = (\Delta_B f) \circ \pi \tag{6.33}$$

for $f \in \mathcal{F}(B)$.

Proof. Let $p \in M$ and let $e_1, \ldots, e_b, e_{b+1}, \ldots, e_m$ be an orthonormal basis of T_pM such that $e_1, \ldots, e_b$ are horizontal and $e_{b+1}, \ldots, e_m$ are vertical with $b = \dim B$ and $m = \dim M$. Let $\{\gamma_i\}_{i=1,\ldots,m}$ be the corresponding orthonormal set of geodesics through p. Then we find from (6.31) that

$$\Delta_M(f \circ \pi) = -\sum_{i=1}^{b} \frac{d^2}{ds^2}(f \circ \pi \circ \gamma_i) - \sum_{j=b+1}^{m} \frac{d^2}{ds^2}(f \circ \pi \circ \gamma_j).$$

Since $\pi : (M, g_M) \to (B, g_B)$ is a Riemannian submersion with totally geodesic fibers, $\{\pi \circ \gamma_i\}$ form an orthonormal set of geodesic in B through p and $\gamma_{b+1}, \ldots, \gamma_m$ are geodesics of the fiber $\pi^{-1}(\pi(p))$. Thus, we obtain $\Delta_M(f \circ \pi)(p) = (\Delta_B f)(\pi(p))$. Because the Laplacian is well-defined, it is independent of the choice of local coordinates. Thus we have (6.33). □

An eigenfunction of $\Delta : \mathcal{F}(M) \to \mathcal{F}(M)$ is called a *proper function.*

Proposition 6.7. *Let $\pi : (M, g_M) \to (B, g_B)$ be a Riemannian submersion with totally geodesic fibers. Then proper functions of (B, g_B) are those functions $f \in \mathcal{F}(B)$ such that $\pi^*(f) = f \circ \pi$ are proper functions of (M, g_M).*

Proof. If f is a proper function of (B, g_B) with eigenvalue λ, then (6.33) yields $\Delta_M(f \circ \pi) = \lambda f \circ \pi$, Thus $f \circ \pi$ is a proper function of (M, g_M) with eigenvalue λ.

Conversely, if $f \circ \pi$ is a proper function of (M, g_M) with eigenvalue λ, then $\Delta_M(f \circ \pi) = \lambda(f \circ \pi) = (\Delta_B f) \circ \pi$. Since $f \circ \pi$ is constant along fibers, we have $\Delta_B f = \lambda f$. □

A point $p \in S^n$ determines a unit vector $e_{n+1} \in \mathbb{E}^{n+1}$. Let $e_1, \ldots, e_n$ be an orthonormal basis of T_pS^n. Then $e_1, \ldots, e_n, e_{n+1}$ form an orthonormal basis of $T_p\mathbb{E}^{n+1}$. Let $\gamma_1, \ldots, \gamma_n$ be the associated orthonormal set of geodesics of S^n through p. If we regard $e_1, \ldots, e_{n+1}$ as points in $\mathbb{E}^{n+1}$, then the geodesic γ_i through p with velocity vector e_i at p is given by

$$\gamma_i(s) = (\cos s)e_{n+1} + (\sin s)e_i, \quad i = 1, \ldots, n.$$

Let $f \in \mathcal{F}(\mathbb{E}^{n+1})$ and $x_1, \ldots, x_{n+1}$ the Euclidean coordinates associated with $e_1, \ldots, e_{n+1}$. Consider the functions $(f \circ \gamma_i)(s) = f(\gamma_i(s))$. By using the chain rule, we find

$$\frac{d(f \circ \gamma_i)}{ds} = -(\sin s)\frac{\partial f}{\partial x_{n+1}} + (\cos s)\frac{\partial f}{\partial x_i},$$
$$\frac{d^2(f \circ \gamma_i)}{ds^2}(0) = \frac{\partial^2 f}{\partial x_i^2}(p) - \frac{\partial f}{\partial x_{n+1}}(p).$$

Thus, by applying (6.31), we get

$$\Delta(f|_{S^n})(p) = -\sum_{i=1}^{n} \frac{\partial^2 f}{\partial x_i^2}(p) + n\frac{\partial f}{\partial x_{n+1}}(p). \tag{6.34}$$

On the other hand, the Laplacian $\tilde{\Delta}$ of $\mathbb{E}^{n+1}$ satisfies

$$(\tilde{\Delta} f)(p) = -\sum_{j=1}^{n+1} \frac{\partial^2 f}{\partial x_j^2}(p). \tag{6.35}$$

By combining (6.34) and (6.35) we find

$$\Delta(f|_{S^n})(p) = (\tilde{\Delta} f)|_{S^n}(p) + \frac{\partial^2 f}{\partial x_{n+1}^2}(p) + n\frac{\partial f}{\partial x_{n+1}}(p).$$

Consequently, if we denote by r the distance function from a point in $\mathbb{E}^{n+1}$ to the origin, then

$$(\tilde{\Delta} f)|_{S^n} = \Delta(f|_{S^n}) - \frac{\partial^2 f}{\partial r^2}|_{S^n} - n\frac{\partial f}{\partial r}|_{S^n}. \tag{6.36}$$

Consider a homogeneous polynomial $\tilde{P}$ of degree $k \geq 0$ on $\mathbb{E}^{n+1}$. Let $P = \tilde{P}|_{S^n}$. Then $\tilde{P} = r^k P$. Thus we have

$$\frac{\partial \tilde{P}}{\partial r} = kr^{k-1}P, \quad \frac{\partial^2 \tilde{P}}{\partial r^2} = k(k-1)r^{k-2}P. \tag{6.37}$$

By substituting (6.37) into (6.36), we get

$$\tilde{\Delta}\tilde{P}|_{S^n} = \Delta P - k(n+k-1)P. \tag{6.38}$$

In particular, if $\tilde{P}$ is harmonic, we obtain

$$\Delta P = k(n+k-1)P. \tag{6.39}$$

Let $\tilde{\mathcal{H}}_k$ denote the vector space of harmonic homogeneous polynomials of degree k on $\mathbb{E}^{n+1}$ and $\mathcal{H}_k$ the restriction of $\tilde{\mathcal{H}}_k$ on S^n. Then

$$\dim \tilde{\mathcal{H}}_k = \binom{n+k}{k} - \binom{n+k-2}{k-2},$$

where the last term is assumed to be zero for $k = 0, 1$. Because $\dim \tilde{\mathcal{H}}_k > 0$ for $k \geq 0$, we obtain from (6.39) the following well-known result [Berger et al. (1971)].

Proposition 6.8. *The spectrum* $Spec(S^n)$ *of the unit* n*-sphere* S^n *is given by*

$$\lambda_k = k(n+k-1), \quad k \geq 0.$$

The multiplicity $m(\lambda_k)$ *of* λ_k *are given by*

$$m(\lambda_0) = 1, \quad m(\lambda_1) = n+1,$$

$$m(\lambda_k) = \frac{(n+k-2)(n+k-3)\cdots(n+1)n}{k!}(n+2k-1), \quad k \geq 2.$$

Moreover, the eigenspace V_k *with eigenvalue* λ_k *is* $\mathcal{H}_k$.

Considering the Riemannian covering map $\pi : (S^n, g_0) \to (RP^n, g_0)$. Proposition 6.7 implies that proper functions of RP^n are induced from those proper functions of S^n invariant under the antipodal map. Thus, proper functions of RP^n are obtained from harmonic homogeneous polynomials of even degree. Combining this and Proposition 6.8 yields

Proposition 6.9. *The spectrum* $Spec(RP^n), n > 1$ *of the real projective* n*-space of constant curvature one is given by*

$$\lambda_k = 2k(n+2k-1), \quad k \geq 0.$$

The multiplicity $m(\lambda_k)$ *of* λ_k *are given by*

$$m(0) = 1,$$

$$m(\lambda_k) = \frac{(n+2k-2)(n+2k-3)\cdots(n+1)n}{(2k)!}(n+4k-1), \quad k \geq 1.$$

The spectra of the complex projective spaces and quaternion projective spaces are also well-known.

Proposition 6.10. *The spectrum* $Spec(CP^n), n > 1$, *of the complex projective* n*-space of constant holomorphic sectional curvature* 4 *is*

$$\lambda_k = 4k(n+k), \quad k \geq 0.$$

Proposition 6.11. *The spectrum* $Spec(QP^n), n > 1$, *of the quaternion projective* n*-space of constant quaternionic sectional curvature* 4 *is*

$$\lambda_k = 4k(2n+k+1), \quad k \geq 0.$$

6.7 Spectra of flat tori

Consider the flat torus $\mathbb{E}^n/\Lambda$, where Λ is a *lattice* of $\mathbb{E}^n$. Put

$$\Lambda^* = \{u \in \mathbb{E}^n : \langle u, v\rangle \in \mathbb{Z} \ \forall v \in \Lambda\}. \tag{6.40}$$

Then Λ^* is also a lattice, called the *dual lattice* of Λ. We have $(\Lambda^*)^* = \Lambda$. If $v_1, \ldots, v_n$ is a basis of Λ, its dual basis $v_1^*, \ldots, v_n^*$ is a basis for Λ^*.

For each $x \in \Lambda^*$, define a function f_x on $\mathbb{E}^n$ by

$$f_x(y) = e^{2\pi \mathrm{i} x(y)}, \tag{6.41}$$

where $\mathrm{i} = \sqrt{-1}$ and $y \in \mathbb{E}^n$ on the right is regarded a vector. Then f_x defines a function on $\mathbb{E}^n/\Lambda$ also denoted by f_x. If we denote by x_i and y_i the components of x and y with respect to the bases $v_1^*, \ldots, v_n^*$ and $v_1, \ldots, v_n$, respectively, then

$$f_x(y) = e^{2\pi \mathrm{i} \sum_{i=1}^n x_i y_i}. \tag{6.42}$$

By taking differentiation of (6.42) with respect to y_j we find

$$\frac{\partial f_x(y)}{\partial y_j} = 2\pi \mathrm{i} x_j f_x(y), \quad \frac{\partial^2 f_x(y)}{\partial y_j^2} = -4\pi ||x||^2 f_x(y). \tag{6.43}$$

Thus

$$\Delta f_x = 4\pi^2 ||x||^2 f_x, \tag{6.44}$$

which shows that $\lambda = 4\pi^2||x||^2$ is an eigenvalue of Δ with proper function f_x for each $x \in \Lambda^*$. To each eigenvalue λ, the corresponding eigenspace V_λ is generated by the f_x's with $||x||^2 = \frac{\lambda}{4\pi^2}$. The multiplicity $m(\lambda)$ of λ is equal to the number of $x \in \Lambda^*$ such that $||x||^2 = \frac{\lambda}{4\pi^2}$. We summarize this as the following result of [Milnor (1964)].

Proposition 6.12. *Let $(\mathbb{E}^n/\Lambda, g_0)$ be a flat n-torus and Λ^* the dual lattice of Λ. Then the spectrum of $\mathbb{E}^n/\Lambda$ is given by*

$$\{4\pi^2||x||^2 : x \in \Lambda^*\}. \tag{6.45}$$

The multiplicity of $\lambda = 4\pi^2||x||^2$ is equal to the number of $u \in \Lambda^*$ with $||u|| = ||x||$.

It was discovered in [Witt (1941)] that there exist two lattices in $\mathbb{E}^{16}$ not isometric but with the same number of elements of any given norm. By applying this result, J. Milnor showed that there are two 16-dimensional flat tori which are not isometric, but with the same spectrum. Consequently, the information of eigenvalues alone is not enough to determine the isometry class of a Riemannian manifold in general.

6.8 Heat equation, Jacobi's elliptic and theta functions

Let θ be the temperature at time t at any point in a solid material whose conducting properties are uniform and isotropic. If ρ is the material's density, s its specific heat and k its thermal conductivity, then θ satisfies the *heat conduction equation*:

$$\kappa\nabla^2\theta = \frac{\partial\theta}{\partial t}, \tag{6.46}$$

where $\kappa = k/s\rho$ is the diffusivity. In the special case where there is no variation of temperature in the x- and y-directions the heat flow is everywhere parallel to the z-axis and the heat equation reduced to the form:

$$\kappa\frac{\partial^2\theta}{\partial z^2} = \frac{\partial\theta}{\partial t}, \quad \theta = \theta(z,t). \tag{6.47}$$

Consider the boundary conditions: $\theta(0,t) = \theta(\pi,t) = 0$ and the initial condition: $\theta(z,0) = \pi\delta(z-\frac{\pi}{2})$ for $0 < z < \pi$, where $\delta(z)$ is Dirac's unit impulse function. Then the solution of this boundary value problem is

$$\theta(z,t) = 2\sum_{n=0}^{\infty}(-1)^n e^{-(2n+1)^2\kappa t}\sin(2n+1)z. \tag{6.48}$$

By writing $e^{-4\kappa t} = q$, the solution (6.48) assume the form

$$\theta_1(z,q) = 2\sum_{n=0}^{\infty}(-1)^n q^{(n+\frac{1}{2})^2}\sin(2n+1)z, \tag{6.49}$$

which is the first of the four *theta functions*. When the precise value of q is not important we shall suppress the dependence upon q. If one changes the boundary conditions to $\frac{\partial\theta}{\partial z} = 0$ on $z = 0$ and $z = \pi$ with $\theta(z,0) = \pi\delta(z-\frac{\pi}{2})$ for $0 < z < \pi$, then the corresponding solution of the boundary value problem of the heat equation (6.47) is given by

$$\theta_4(z) = \theta_4(z,q) = 1 + 2\sum_{n=1}^{\infty}(-1)^n q^{n^2}\cos 2nz. \tag{6.50}$$

The theta function θ_1(z) of (6.49) is periodic with period 2π. Incrementing z by $\frac{1}{2}\pi$ yields the second theta function:

$$\theta_2(z) = \theta_2(z,q) = \theta_1(z+\tfrac{\pi}{2},q) = 2\sum_{n=1}^{\infty} q^{(n+\frac{1}{2})^2}\cos(2n+1)z. \tag{6.51}$$

Similarly, incrementing z by $\frac{1}{2}\pi$ for θ_4 yields the third theta function:

$$\theta_3(z) = \theta_3(z,q) = \theta_4(z+\tfrac{\pi}{2},q) = 1 + 2\sum_{n=1}^{\infty} q^{n^2}\cos 2nz. \tag{6.52}$$

The four theta functions $\theta_1, \theta_2, \theta_3, \theta_4$ can be extended to complex values for z and q such that $|q| < 1$.

The *Jacobi's elliptic functions* $\mathrm{sn}(u), \mathrm{cn}(u)$ and $\mathrm{dn}(u)$ are defined as ratios of theta functions:

$$\mathrm{sn}(u) = \frac{\theta_3(0)\theta_1(z)}{\theta_2(0)\theta_4(z)}, \ \mathrm{cn}(u) = \frac{\theta_4(0)\theta_2(z)}{\theta_2(0)\theta_4(z)}, \ \mathrm{dn}(u) = \frac{\theta_4(0)\theta_3(z)}{\theta_3(0)\theta_4(z)}, \tag{6.53}$$

where $z = u/\theta_3^2(0)$. Define parameters k and k' by

$$k = \frac{\theta_2^2(0)}{\theta_3^2(0)}, \quad k' = \frac{\theta_4^2(0)}{\theta_3^2(0)},$$

which are called the *modulus* and *complementary modulus*; k and k' satisfy $k^2 + k'^2 = 1$.

Using $\mathrm{cn}(u), \mathrm{dn}(u)$ and $\mathrm{sn}(u)$, one may define minor elliptic functions

$$\mathrm{cd}(u) = \frac{\mathrm{cn}(u)}{\mathrm{dn}(u)}, \ \mathrm{cs}(u) = \frac{\mathrm{cn}(u)}{\mathrm{sn}(u)}, \ \mathrm{nd}(u) = \frac{1}{\mathrm{dn}(u)}, \ \mathrm{ns}(u) = \frac{1}{\mathrm{sn}(u)}, \ \textit{etc.}$$

When it is required to state the modulus explicitly the elliptic functions of Jacobi will be written as $\mathrm{sn}(u,k)$, $\mathrm{cn}(u,k)$, $\mathrm{dn}(u,k)$, etc.

The elliptic functions $\mathrm{sn}\,(u)$, $\mathrm{cn}\,(u)$ and $\mathrm{dn}\,(u)$ satisfy the following relations:

$$\begin{aligned}
&\mathrm{sn}^2(u) + \mathrm{cn}^2(u) = 1, \\
&\mathrm{dn}^2(u) + k^2\mathrm{sn}^2(u) = 1, \\
&k^2\mathrm{cn}^2(u) + k'^2 = \mathrm{dn}^2(u), \\
&\mathrm{sn}'(u) = \mathrm{cn}(u)\mathrm{dn}(u), \\
&\mathrm{cn}'(u) = -\mathrm{sn}(u)\mathrm{dn}(u), \\
&\mathrm{dn}'(u) = -k^2\mathrm{sn}(u)\mathrm{cn}(u).
\end{aligned} \tag{6.54}$$

The Jacobi's *Theta function* $\Theta(u)$ and *Zeta function* $Z(u)$ are defined by

$$\Theta(u) = \theta_4\Big(\frac{\pi u}{2K}\Big), \ Z(u) = \frac{d}{du}(\ln\theta_4), \ K = \frac{\pi}{2}\theta_3^2(0). \tag{6.55}$$

The Zeta and Theta functions satisfy

$$Z(u) = \frac{\Theta'(u)}{\Theta(u)}. \tag{6.56}$$

Jacobi also defined a function $\mathrm{am}(x)$ by means of the equation:

$$\mathrm{am}(x) = \int_0^x \mathrm{dn}(u)du, \tag{6.57}$$

which is known as Jacobi's *amplitude*.

Chapter 7

Submanifolds of Finite Type

The theory of finite type submanifolds began in the late 1970s through author's attempts to find the best possible estimates of the total mean curvature of a compact submanifold of Euclidean space and to find a notion of "degree" for submanifolds of Euclidean space.

The main objects of studies in algebraic geometry are algebraic varieties. Because an algebraic variety is defined by using algebraic equations, one can define the degree of an algebraic variety by its algebraic structure. The concept of degree plays a fundamental role in algebraic geometry.

On the other hand, according to Nash's embedding theorem, every Riemannian manifold can be realized as a Riemannian submanifold of some Euclidean space with sufficiently high codimension. However, one lacks the notion of the degree for Riemannian submanifolds of Euclidean spaces.

Inspired by the above simple observation, the author introduced in the late 1970s the notions of "order" and "type" for submanifolds of Euclidean spaces. By applying these notions, the author was able to obtain some sharp estimates of the total mean curvature for compact submanifolds of Euclidean space in terms of orders. Moreover, by utilizing the notion of order, the notions of finite type submanifolds and of finite type maps were naturally introduced. Although the family of submanifolds of finite type is huge, it contains many important families of submanifolds; including all minimal submanifolds of Euclidean space; all minimal submanifolds of hyperspheres as well as parallel submanifolds and equivariantly immersed compact homogeneous submanifolds. Further, just like minimal submanifolds, submanifolds of finite type can be described by a spectral variation principle, namely as critical points of directional deformations.

On one hand, the notion of finite type submanifolds provides a natural way to combine the spectral geometry with the theory of submanifolds; as well as with maps; in particular, with Gauss map. On the other hand, one can apply the theory of finite type to spectral geometry.

The first results in this subject were collected in [Chen (1984)]. Since then many geometers contributed to the theory. A detailed report on the progress in this subject up to 1996 was presented in [Chen (1996c)].

7.1 Order and type of submanifolds

Let (N, g) be a compact Riemannian n-manifold with Levi-Civita connection ∇. Then the Laplacian Δ of (N, g) acting as an elliptic differential operator on $\mathcal{F}(N)$. The eigenvalues of Δ form a discrete infinite sequence:

$$0 = \lambda_0 < \lambda_1 < \lambda_2 < \ldots \nearrow \infty.$$

Let $V_t = \{f \in \mathcal{F}(N) : \Delta f = \lambda_t f\}$ be the eigenspace of Δ with eigenvalue λ_t. Then V_0 is 1-dimensional and V_t is finite-dimensional for $t \geq 1$. Define an inner product $(\ ,\ )$ on $\mathcal{F}(N)$ by

$$(f, h) = \int_N fh * 1. \tag{7.1}$$

Then $\sum_{t=0}^{\infty} V_t$ is dense in $\mathcal{F}(N)$ (in L^2–sense). If we denote by $\hat{\oplus}_{t=0}^{\infty} V_t$ the completion of $\sum_{t=0}^{\infty} V_t$, we have

$$C^\infty(N) = \hat{\oplus}_k V_k. \tag{7.2}$$

For each $f \in \mathcal{F}(N)$ let f_t denote the projection of f onto the subspace V_t. Then we have the following spectral decomposition:

$$f = \sum\nolimits_{t=0}^{\infty} f_t \quad (\text{in } L^2\text{-sense}). \tag{7.3}$$

Because $\dim V_0 = 1$, for each non-constant $f \in \mathcal{F}(N)$ there is a positive integer $p \geq 1$ such that $f_p \neq 0$ and $f - f_0 = \sum_{t \geq p} f_t$, $f_0 \in V_0$. If there are infinite many nonzero f_t's, we put $q = +\infty$, Otherwise, there is an integer q $(q \geq p)$, such that

$$f_q \neq 0 \ \text{ and } \ f - f_0 = \sum\nolimits_{t=p}^{q} f_t. \tag{7.4}$$

If we allow q to be $+\infty$, we have decomposition (7.4) in general. The set

$$T(f) = \{t > 0 : f_t \neq 0\} \tag{7.5}$$

is called the *order* of f. The smallest element p in $T(f)$ is the *lower order* of f, which is denoted by $l.o.(f)$. The supremum of $T(f)$ is called the *upper order* of f, denoted by $u.o.(f)$. A function f in $\mathcal{F}(N)$ is said to be of *finite type* if $T(f)$ is a finite set, equivalently, if its spectral decomposition (7.3) contains only finitely many nonzero terms. Otherwise f is said to be *of infinite type.* The function f is of *k-type* if $T(f)$ contains k elements.

For an isometric immersion $\mathbf{x} : N \to \mathbb{E}^m_s$ of a compact Riemannian manifold N into a pseudo-Euclidean m-space $\mathbb{E}^m_s$, we put

$$\mathbf{x} = (x_1, \dots, x_m),$$

where x_A is the A-th Euclidean coordinate function of N. For each x_A,

$$x_A - (x_A)_0 = \sum\nolimits_{t=p_A}^{q_A} (x_A)_t, \quad A = 1, \dots, m.$$

For the isometric immersion $\mathbf{x} : M \to \mathbb{E}^m$, we put

$$p = \inf_A \{p_A\}, \quad q = \sup_A \{q_A\},$$

where A ranges among all A with $x_A - (x_A)_0 \neq 0$. Then p and q are well-defined such that p is a positive integer and q is either $+\infty$ or an integer $\geq p$. Consequently, we have the spectral decomposition of $\mathbf{x}$ in vector form:

$$\mathbf{x} = \mathbf{x}_0 + \sum\nolimits_{t=p}^{q} \mathbf{x}_t, \quad \Delta \mathbf{x}_t = \lambda_t \mathbf{x}_t, \quad \mathbf{x}_t \neq 0. \tag{7.6}$$

Since N is compact, the constant vector $\mathbf{x}_0$ in (7.6) is the *center of mass*, of N in $\mathbb{E}^m_s$, i.e., $\mathbf{x}_0 = (\int_N \mathbf{x} * 1)/\mathrm{vol}(N)$.

By applying (7.6), we define the *order* of the immersion $\mathbf{x}$ by

$$T(\mathbf{x}) = \{0 \neq t \in \mathbb{N} : \mathbf{x}_t \neq 0 \text{ in } (7.6)\}.$$

The immersion $\mathbf{x}$ (or the submanifold N) is said to be *of finite type* if $T(\mathbf{x})$ consists of only finite elements; and it is *of k-type* if $T(\mathbf{x})$ contains exactly k elements. It is *of infinite type* if it is not of finite type. Similarly we have the lower order and the upper order of the immersion.

Given two $\mathbb{E}^m_s$-valued functions v, w on N, define the scalar product by

$$(v, w) = \int_N \langle v, w \rangle * 1, \tag{7.7}$$

where $\langle v, w \rangle$ is the pseudo-Euclidean scalar product of v, w.

For $\mathbf{x} : N \to \mathbb{E}^m_s$, the components of the spectral decomposition (7.6) are mutually orthogonal, i.e., $(\mathbf{x}_t, \mathbf{x}_s) = 0$, $t \neq s$.

In general, one cannot make the spectral decomposition of a function on a non-compact Riemannian manifold. However, it remains possible to define the notion of finite type immersions for non-compact manifolds as follows:

An isometric immersion $\mathbf{x} : N \to \mathbb{E}^m_s$ of a non-compact Riemannian manifold N is said to be of *finite type* if the position vector can be decomposed into a finite sum of vector eigenfunctions of the Laplacian of N:

$$\mathbf{x} = c_0 + \mathbf{x}_1 + \cdots + \mathbf{x}_k, \tag{7.8}$$

where c_0 is a constant vector and $\mathbf{x}_1, \ldots, \mathbf{x}_k$ are non-constant vector eigenfunctions of Δ. If all eigenvalues associated with $\mathbf{x}_1, \ldots, \mathbf{x}_k$ are different, the submanifold is said to be of k-type. In particular, if one of the eigenvalues of $\mathbf{x}_1, \ldots, \mathbf{x}_k$ is zero, then N is said to be of *null k-type*. For instance, a *null 2-type* submanifold admits the following spectral decomposition:

$$\mathbf{x} = c_0 + \mathbf{x}_1 + \mathbf{x}_2, \ \ \Delta\mathbf{x}_1 = 0, \ \ \Delta\mathbf{x}_2 = \lambda\mathbf{x}_2, \ \ 0 \neq \lambda \in \mathbf{R}. \tag{7.9}$$

After choosing c_0 to be the origin of $\mathbb{E}^m_s$, (7.9) reduces to

$$\mathbf{x} = \mathbf{x}_1 + \mathbf{x}_2, \ \ \Delta\mathbf{x}_1 = 0, \ \ \Delta\mathbf{x}_2 = \lambda\mathbf{x}_2, \ \ 0 \neq \lambda \in \mathbf{R}. \tag{7.10}$$

The notion of type can be extended to smooth maps, in particular to the Gauss map of a Riemannian submanifold of a pseudo-Euclidean space.

The next lemma is a generalization of Corollary 2.6.

Proposition 7.1. *For integers $k \geq 1$ and $s \geq 0$, every null k-type spacelike submanifold of the pseudo Euclidean m-space $\mathbb{E}^m_s$ is non-compact.*

Proof. Follows from the fact that every harmonic function of a compact Riemannian manifold is constant. □

Obviously, if two Riemannian submanifolds of a pseudo-Euclidean space are congruent, then they have same type number.

Lemma 7.1. *Let $\mathbf{x} : N \to \mathbb{E}^m_s$ be a k-type map of a compact Riemannian manifold N into $\mathbb{E}^m_s$. If $T(\mathbf{x}) = \{t_1, \ldots, t_k\}$ and if $L : \mathbb{E}^m_s \to \mathbb{E}^N_t$ is a linear map, then the composition $L \circ \mathbf{x} : N \to \mathbb{E}^N_t$ is a finite type map. Moreover,*

(1) *the type number of $L \circ \mathbf{x}$ is at most k;*
(2) *the order $T(L \circ \mathbf{x})$ of $L \circ \mathbf{x}$ is a subset of the order $T(\mathbf{x})$ of $\mathbf{x}$.*

Proof. Follows from the fact that a linear combination of finite linear combinations of eigenfunctions of Δ is again a linear combination of eigenfunctions of Δ. □

Lemma 7.2. *Let $\mathbf{x} : N \to \mathbb{E}^m_s$ be a smooth map of a compact Riemannian manifold N into $\mathbb{E}^m_s$ and let $\iota : \mathbb{E}^m_s \subset \mathbb{E}^{m+r}_{s+i}$ be an inclusion map. Then $T(\iota \circ \mathbf{x}) = T(\mathbf{x})$.*

Proof. Follows from Lemma 7.1 and the fact that the inclusion map is a linear map. □

7.2 Minimal polynomial criterion

The following *Minimal Polynomial Criterion* is quite useful to determine whether a submanifold is of finite type.

Theorem 7.1. *Let $\mathbf{x} : N \to \mathbb{E}^m_s$ be an isometric immersion of a compact Riemannian manifold N into a pseudo-Euclidean space $\mathbb{E}^m_s$ and let H be the mean curvature vector of N in $\mathbb{E}^m_s$. Then*

(a) *N is of finite type if and only if there is a nontrivial polynomial $Q(t)$ such that $Q(\Delta)H = 0$;*
(b) *If N is of finite type, there exists a unique monic polynomial $P(t)$ of least degree with $P(\Delta)H = 0$;*
(c) *If N is of finite type, the type number of N is equal to the degree of P.*

The same results holds if H were replaced by $\mathbf{x} - \mathbf{x}_0$.

Proof. Under the hypothesis consider the spectral decomposition (7.6). If $\mathbf{x}$ is of finite type, $q < \infty$. From (7.6) and Beltrami's formula we find

$$-n\Delta^i H = \sum\nolimits_{t=p}^{q} \lambda_t^{i+1}\mathbf{x}_t, \quad i = 0, 1, 2, \cdots. \tag{7.11}$$

Put

$$c_1 = -\sum\nolimits_{t=p}^{q} \lambda_t, \quad c_2 = \sum\nolimits_{t<r} \lambda_t\lambda_r, \quad \cdots,$$
$$c_{q-p+1} = (-1)^{q-p+1}\lambda_p \cdots \lambda_q.$$

Then we have

$$\Delta^{q-p+1}H + c_1\Delta^{q-p}H + \cdots + c_{q-p+1}H = 0. \tag{7.12}$$

Thus, if we put

$$Q(t) = t^{q-p+1} + c_1 t^{q-p} + \cdots + c_{q-p+1}, \tag{7.13}$$

then we get $Q(\Delta)H = 0$.

Conversely, if H satisfies $Q(\Delta)H = 0$ for a nontrivial polynomial $Q(t)$ of degree k, Corollary 2.9 implies $k \geq 1$ since N is compact. Consider spectral decomposition (7.6). From (7.11), (7.13) and $Q(\Delta)H = 0$ we find

$$\sum_{t=1}^{\infty} \lambda_t(\lambda_t^k + c_1\lambda_t^{k-1} + \cdots + c_{k-1}\lambda_t + c_k)\mathbf{x}_t = 0. \tag{7.14}$$

For each integer $r > 0$, (7.14) implies that

$$\sum_{t=1}^{\infty} \lambda_t(\lambda_t^k + c_1\lambda_t^{k-1} + \cdots + c_{k-1}\lambda_t + c_k) \int_N \langle \mathbf{x}_t, \mathbf{x}_r \rangle * 1 = 0.$$

Since $(\mathbf{x}_t, \mathbf{x}_r) = \int_N \langle \mathbf{x}_t, \mathbf{x}_r \rangle * 1 = 0$ for $r \neq t$, we obtain

$$(\lambda_t^k + c_1\lambda_t^{k-1} + \cdots + c_{k-1}\lambda_t + c_k)||\mathbf{x}_r||^2 = 0. \tag{7.15}$$

Thus, for each $\mathbf{x}_t \neq 0$ in (7.6), the corresponding eigenvalue λ_t is a root of the polynomial (7.13). Thus, the decomposition (7.6) contains only finite number of nonzero terms. Hence, $\mathbf{x}$ is of finite type. This proves (a).

If $\mathbf{x}$ is of finite type, it follows from (a) that there exists a nontrivial polynomial $Q(t)$ such that $Q(\Delta)H = 0$. Let $P(t)$ is a monic polynomial of least possible degree satisfying $P(\Delta)H = 0$. Clearly, $P(t)$ is unique. Thus we have (b). (c) follows from the proof of (a).

The remaining part is clear. □

Definition 7.1. The unique monic polynomial P given in Theorem 7.1(b) is called the *minimal polynomial* of the finite type submanifold.

Remark 7.1. Clearly, congruent submanifolds of finite type share the same minimal polynomial. On the other hand, if two finite type submanifolds N_1 and N_2 of $\mathbb{E}_s^m$ have the same minimal polynomial, then

(1) they have the same type number;
(2) the Laplacian of N_1 and N_2 have at least k common eigenvalues given by the roots of the minimal polynomial; and
(3) the orders of N_1, N_2 are the same whenever N_1, N_2 are isospectral.

The isometries of a Riemannian manifold N form a Lie group $I(N)$, which is compact whenever N is compact. Let G be a closed subgroup of $I(N)$. The Riemannian manifold is called a *G-homogeneous* if G acts transitively on N. If the group G is not important, we will simply say that N is a *homogeneous Riemannian manifold.*

Definition 7.2. Let $N = G/H$ be a homogeneous Riemannian manifold and M a Riemannian manifold with group $I(M)$ of isometries. An isometric immersion $\psi : N \to M$ is called *G-equivariant* if there is a homomorphism $\zeta : G \to I(M)$ such that $f(g(p)) = \zeta(g)f(p)$ for each $g \in G$ and each $p \in N$.

An immediate application of Theorem 7.1 is the following.

Proposition 7.2. *Let N be a compact homogeneous Riemannian manifold. If M is equivariantly isometrically immersed in $\mathbb{E}_s^m$, then N is of finite type. Moreover, the type number of N is at most m.*

Proof. Let p be a given point of N. Then $H, \Delta H, \ldots, \Delta^m H$ at p are linearly dependent. Thus, there is a polynomial $Q(t)$ of degree m such that $Q(\Delta)H = 0$ at p. Because N is equivariantly isometrically immersed in $\mathbb{E}^m_s$, $Q(\Delta)H = 0$ holds identically on N. Thus, by Theorem 7.1, N is of finite type. Moreover, since the degree of the minimal polynomial is less than or equal to m, the type number is at most m. □

For non-compact N, the existence of a nontrivial polynomial Q satisfying $Q(\Delta)H = 0$ does not insure finite type of N in general. However, we have the following two results from [Chen and Petrovic (1991)].

Theorem 7.2. *Let $\mathbf{x} : N \to \mathbb{E}^m_s$ be an isometric immersion of a Riemannian n-manifold N into $\mathbb{E}^m_s$. If there is a vector $c \in \mathbb{E}^m_s$ and a nontrivial polynomial P with simple roots such that $P(\Delta)(\mathbf{x} - c) = 0$, then N is of finite type with type number $k \leq \deg P$.*

Proof. Assume that $P(t) = (t - \ell_1) \cdots (t - \ell_k)$ is the polynomial with simple roots satisfying $P(\Delta)(\mathbf{x} - c) = 0$. Consider the following system

$$\begin{aligned} \mathbf{x} - c &= \mathbf{x}_1 + \mathbf{x}_2 + \cdots + \mathbf{x}_k, \\ \Delta\mathbf{x} &= \ell_1\mathbf{x}_1 + \ell_2\mathbf{x}_2 + \cdots + \ell_k\mathbf{x}_k, \\ &\vdots \\ \Delta^{k-1}\mathbf{x} &= \ell_1^{k-1}\mathbf{x}_1 + \ell_2^{k-1}\mathbf{x}_2 + \cdots + \ell_k^{k-1}\mathbf{x}_k \end{aligned} \tag{7.16}$$

for some vector $c \in \mathbb{E}^m_s$. Since $\ell_1, \ldots, \ell_k$ are mutually distinct, we may solve (7.16) for $\mathbf{x}_1, \ldots, \mathbf{x}_k$ in terms of $\mathbf{x} - c, \Delta\mathbf{x}, \ldots, \Delta^{k-1}\mathbf{x}$ to obtain

$$\begin{aligned} \textstyle\prod_{j \neq i}(\ell_j - \ell_i)\mathbf{x}_i &= \sigma_{i,k-1}(\mathbf{x} - c) - \sigma_{i,k-2}\Delta\mathbf{x} + \cdots \\ &\quad + (-1)^{k-2}\sigma_{i,1}\Delta^{k-2}\mathbf{x} + (-1)^{k-1}\Delta^{k-1}\mathbf{x}, \end{aligned} \tag{7.17}$$

where $\sigma_{i,j}$ is the j-th elementary symmetric function of $\ell_1, \ldots, \ell_{i-1}, \ell_{i+1}, \ldots, \ell_k$. In other words, we have

$$\sigma_{i,k-1} = \prod_{j \neq i} \ell_j, \quad \sigma_{i,k-2} = \prod_{\substack{j \neq h \\ j,h \neq i}} \ell_j\ell_h, \quad \ldots, \quad \sigma_{i,1} = \sum_{j \neq i} \ell_j.$$

From the assumption we have

$$\Delta^k\mathbf{x} - \sigma_1\Delta^{k-1}\mathbf{x} + \cdots + (-1)^{k-1}\sigma_{k-1}\Delta\mathbf{x} + (-1)^k\sigma_k(\mathbf{x} - c) = 0, \tag{7.18}$$

where σ_j is the j-th elementary symmetric function of $\ell_1, \ldots, \ell_k$.

It follows from (7.16)-(7.18) that

$$\textstyle\prod_{j \neq i}(\ell_j - \ell_i)\Delta\mathbf{x}_i = \ell_i \prod_{j \neq i}(\ell_j - \ell_i)\mathbf{x}_i, \quad i = 1, \ldots, k. \tag{7.19}$$

Since $\ell_1, \ldots, \ell_k$ are mutually distinct, (7.19) yields $\Delta \mathbf{x}_i = \ell_i \mathbf{x}_i$. Thus, by applying (7.16), we find $\mathbf{x} = c + \mathbf{x}_1 + \cdots + \mathbf{x}_k$. This shows that N is of finite type with type number $k \leq \deg P$. □

Theorem 7.3. *Let $\mathbf{x} : N \to \mathbb{E}^m_s$ be an isometric immersion of a Riemannian n-manifold N into $\mathbb{E}^m_s$. If there exists a nontrivial polynomial Q such that $Q(\Delta)H = 0$, then N is either of infinite type or of finite type with type number $k \leq \deg Q + 1$.*

Proof. Let $Q = t^d + c_1 t^{d-1} + \cdots + c_d$ be a nontrivial polynomial satisfying $Q(\Delta)H = 0$. Suppose that N is of k-type with finite k. Then we have the spectral decomposition

$$\mathbf{x} = \mathbf{c}_0 + \mathbf{x}_1 + \ldots + \mathbf{x}_k, \quad \Delta \mathbf{x}_i = \ell_i \mathbf{x}_i, \tag{7.20}$$

where $\ell_1, \ldots, \ell_k$ are k mutually distinct eigenvalues of Δ. After applying Beltrami's formula, (7.20) implies that

$$-n\Delta^j H = \ell_1^{j+1}\mathbf{x}_1 + \cdots + \ell_k^{j+1}\mathbf{x}_k, \;\; j = 0, 1, 2, \cdots. \tag{7.21}$$

Thus, by using $Q(\Delta)H = 0$, we find that

$$\ell_1 Q(\ell_1)\mathbf{x}_1 + \cdots + \ell_k Q(\ell_k)\mathbf{x}_k = 0. \tag{7.22}$$

Now, by applying Δ^j to (7.22), we find

$$\ell_1^{j+1} Q(\ell_1)\mathbf{x}_1 + \cdots + \ell_k^{j+1} Q(\ell_k)\mathbf{x}_k = 0, \;\; j = 0, 1, 2, \cdots. \tag{7.23}$$

Since $\ell_1, \ldots, \ell_k$ are mutually distinct, (7.23) implies that

$$Q(\ell_1) = \cdots = Q(\ell_k) = 0.$$

Hence, $k \leq d + 1$. □

7.3 A variational minimal principle

Let $\mathbf{x} : N \to \mathbb{E}^m_s$ be an isometric immersion of a compact Riemannian manifold N into a pseudo-Euclidean space $\mathbb{E}^m_s$. Associated to each $\mathbb{E}^m_s$-valued vector field ξ on N, there is a deformation ϕ_t, defined by

$$\phi_t(p) := \mathbf{x}(p) + t\xi(p), \quad p \in M, \quad t \in (-\epsilon, \epsilon), \tag{7.24}$$

where ϵ is a sufficiently small positive number. For each t, ϕ_t gives rise to a submanifold $N_t = \phi_t(N)$. Let $A(t)$ denote the volume of N_t.

Let $\mathfrak{D}$ be the class of all such deformations acting on the submanifold N and $\mathfrak{C}$ a nonempty subclass of $\mathfrak{D}$. Then N is called a *critical point of the area functional* in the class $\mathfrak{C}$ if $A'(0) = 0$ for each deformation in $\mathfrak{C}$; and N is *stable* in the class $\mathfrak{C}$ if $A''(0) \geq 0$ for each deformation in $\mathfrak{C}$.

A compact submanifold N of $\mathbb{E}^m_s$ is said to satisfy the *variational minimal principle in the class* $\mathfrak{C}$ if N minimizes the volume functional for all deformations in $\mathfrak{C}$. In other words, N satisfies the variational minimal principle in the class $\mathfrak{C}$ if N is a critical point of the volume functional and is stable among all deformations of N in the class $\mathfrak{C}$.

Let c be a vector in $\mathbb{E}^m_s$ and $f \in \mathcal{F}(N)$. Then the deformation given by

$$\phi^{fc}_t(p) := \mathbf{x}(p) + tf(p)c, \quad p \in N, \quad t \in (-\epsilon, \epsilon). \tag{7.25}$$

is called a *directional deformation* in the direction c [Voss (1956)]. Such deformations occur when an object moves along a fixed direction.

Let V_i denote the eigenspace of the Laplacian Δ on N associated with i-th eigenvalue λ_i. For each $q \in \mathbb{N}$, let $\mathcal{C}_q$ be the class of all directional deformations defined by smooth functions $f \in \sum_{i \geq q} V_i$ for (7.25). Then

$$\mathcal{C}_0 \supset \mathcal{C}_1 \supset \mathcal{C}_2 \supset \cdots \supset \mathcal{C}_k \supset \cdots,$$

where $\mathcal{C}_0 = \mathcal{F}(N)$. Clearly, if a compact submanifold N of $\mathbb{E}^m_s$ satisfies the variational minimal principle for one class $\mathcal{C}_k$, then it automatically satisfies the variational minimal principle in the class $\mathcal{C}_\ell$ for each $\ell \geq k$.

We have the following variational minimal principle [Chen et al. (1993)].

Theorem 7.4.

(a) *There are no compact Riemannian submanifolds N in $\mathbb{E}^m_s$ which satisfy the variational minimal principle in the classes $\mathcal{C}_0$ and $\mathcal{C}_1$;*
(b) *a compact Riemannian submanifold N of $\mathbb{E}^m_s$ is of finite type if and only if it satisfies the variational minimal principle in the class $\mathcal{C}_q$ for some $q \geq 2$;*
(c) *every compact Riemannian submanifold N of finite type in $\mathbb{E}^m_s$ is stable in the class of $\mathcal{C}_q$ for any $q \geq u.o.(N) + 1$, where $u.o.(N)$ is the upper order of N.*

Proof. Let $c \in \mathbb{E}^m_s$ and $f \in \mathcal{F}(N)$. Consider the directional deformation ϕ^{fc}_t defined by (7.25). Let $e_1, \ldots, e_n$ be an orthonormal local frame field on N. Extended $e_1, \ldots, e_n$ by $(\phi_t)_* e_1, \ldots, (\phi_t)_* e_n$. Put

$$g_{ij}(t) = \langle (\phi_t)_* e_i, (\phi_t)_* e_j \rangle . \tag{7.26}$$

Then

$$A(t) = \int_N \sqrt{det(g_{ij}(t))} * 1 \tag{7.27}$$

On the other hand, it follows from (7.25) that

$$(\phi_t)_* e_i = e_i + t(f_i)c, \quad \frac{\partial}{\partial t} = fc, \tag{7.28}$$

where $f_i = e_i f$. From (7.26) and (7.28) we find

$$g_{ij}(t) = \delta_{ij} + t(f_i \langle c, e_j \rangle + f_j \langle c, e_i \rangle) + t^2 f_i f_j \langle c, c \rangle . \tag{7.29}$$

By using (7.27) and (7.29), we obtain

$$A(t) = \int_N \{1 + 2t \langle c, \nabla f \rangle + t^2 (\langle c^\perp, c^\perp \rangle |\nabla f|^2 + \langle c, \nabla f \rangle^2)\}^{\frac{1}{2}} * 1, \tag{7.30}$$

where $\langle\ ,\ \rangle$ is the scalar product of $\mathbb{E}^m_s$, $c^\perp$ the normal component of c, ∇f the gradient of f, and $*1$ the volume element of N. From (7.30) we obtain

$$A'(0) = \int_N \langle c, \nabla f \rangle * 1, \;\; A''(0) = \int_N \langle c^\perp, c^\perp \rangle |\nabla f|^2 * 1. \tag{7.31}$$

Let c^T be the tangential component of c and $(c^T)^\sharp$ the 1-form on N dual to c^T. Then we have

$$(c^T)^\sharp = dh_c, \tag{7.32}$$

where h_c is the height function of N in $\mathbb{E}^m_s$ with respect to c. By using (7.32) we find

$$\delta (c^T)^\sharp = \Delta h_c = \langle \Delta \mathbf{x}, c \rangle , \tag{7.33}$$

where $\mathbf{x}$ is the position vector field of N in $\mathbb{E}^m_s$.

On the other hand, the Beltrami formula yields

$$\Delta \mathbf{x} = -nH, \;\; n = \dim N. \tag{7.34}$$

Hence, by using the first equation of (7.31), (7.33) and (7.34), we find

$$A'(0) = \int_N H_c f * 1, \tag{7.35}$$

where $H_c = \langle H, c \rangle$. If N is a compact submanifold in $\mathbb{E}^m_s$ which satisfies the variational minimal principle in the class $\mathcal{C}_1$, then (7.35) yields

$$(H_c, f) = \int_N H_c f * 1 = 0. \tag{7.36}$$

Since (7.36) holds for every $f \in \sum_{t=1}^{\infty} V_t$ and $c \in \mathbb{E}^m_s$, we get $H = 0$. But this is impossible since N is compact. Hence, N cannot satisfy the variational minimal principle in the class $\mathcal{C}_1$. Thus, it also does not satisfy the variational minimal principle in the class $\mathcal{C}_0$. This proves (a).

To prove (b) let us assume that N is of finite type. Then we have a finite spectral decomposition:

$$\mathbf{x} = \mathbf{x}_0 + \mathbf{x}_{i_1} + \cdots + \mathbf{x}_{i_k}, \tag{7.37}$$

where $\mathbf{x}_0$ is a constant vector and $\Delta \mathbf{x}_{i_j} = \lambda_{i_j} \mathbf{x}_{i_j}$ with $\lambda_{i_1} < \cdots < \lambda_{i_k}$. From (7.35) and (7.37) we find

$$-nH = \lambda_{i_1} \mathbf{x}_{i_1} + \cdots + \lambda_{i_k} \mathbf{x}_{i_k}, \tag{7.38}$$

which implies that $H_c \in \sum_{j \le \ell} V_j$, where ℓ is the upper order of N. If we put $q = \ell + 1$, then (7.35) implies $A'(0) = 0$ for any deformation in $\mathcal{C}_q$.

Conversely, if N satisfies variational minimal principal in $\mathcal{C}_q$, for some $q \ge 2$, then (7.36) yields $H_c \in \sum_{i<q} V_i$. Since this holds for any $c \in \mathbb{E}^m_s$, H has finite spectral decomposition

$$H = H_0 + H_1 + \cdots + H_{q-1}, \quad \Delta H_j = \lambda_j H_j. \tag{7.39}$$

If we put $P(t) = t \prod_{j=1}^{q-1} (t - \lambda_j)$, we obtain from (7.39) that $P(\Delta)H = 0$. Thus Theorem 7.1 implies that N is of finite type. This proves (b)

It follows from the proof of (b) that a finite type submanifold N satisfies the variational minimal principle in the class of $\mathcal{C}_q$ for $q \ge u.o.(N) + 1$. On the other hand, the second equation in (7.31) implies that $A''(0) \ge 0$. Thus, N is stable in the class $\mathcal{C}_q$. This proves (c). □

7.4 Classification of 1-type submanifolds

The following is the classification of 1-type submanifolds [Chen (1986)].

Theorem 7.5. *Let $\mathbf{x} : N \to \mathbb{E}^{m+1}_s$ be an isometric immersion of a pseudo-Riemannian manifold into $\mathbb{E}^{m+1}_s$. Then N is of 1-type if and only if it is one of the following three types of submanifolds:*

(a) *a minimal submanifold of $\mathbb{E}^{m+1}_s$;*
(b) *a minimal submanifold of a pseudo-sphere $S^m_s(c) \subset \mathbb{E}^{m+1}_s$, $c > 0$;*
(c) *a minimal submanifold of a pseudo-hyperbolic space $H^m_{s-1}(c) \subset \mathbb{E}^{m+1}_s$, $c < 0$.*

Proof. If N is of 1-type in $\mathbb{E}^{m+1}_s$, we have the spectral decomposition:

$$\mathbf{x} = \mathbf{c} + \mathbf{x}_p, \tag{7.40}$$

where $\mathbf{c}$ is vector in $\mathbb{E}^{m+1}_s$ and $\mathbf{x}_p$ is a non-constant vector function satisfying $\Delta \mathbf{x}_\mathbf{p} = \lambda_p \mathbf{x}_p$ for some eigenvalue λ_p. If $\lambda_p = 0$, Beltrami's formula implies that N is minimal in $\mathbb{E}^{m+1}_s$. This gives (a).

If $\lambda_p \neq 0$, Beltrami's formula implies that

$$nH = \lambda_p(\mathbf{c} - \mathbf{x}). \tag{7.41}$$

By differentiating (7.41), we obtain $n\tilde{\nabla}_X H = -\lambda_p X$, $\forall X \in TN$. Hence, $nA_H = \lambda_p I$ and $DH = 0$, which imply that N is a pseudo-umbilical submanifold with parallel mean curvature vector. Thus, by Proposition 3.9, either the immersion is given by (b) or (c), or, up to rigid motion, we have

$$\mathbf{x} = (f, \psi, f), \tag{7.42}$$

where $\psi : N \to \mathbb{E}^{m-1}_{s-1}$ is a minimal immersion and $\Delta f = r$, $0 \neq r \in \mathbf{R}$. If the last case occurs, (7.41), (7.42) and Beltrami's formula imply that $\lambda_p \mathbf{x} = (r, 0, r) + \lambda_p \mathbf{c}$. Thus, $\mathbf{x}$ is a constant vector which is impossible.

The converse can be verified easily. □

Theorem 7.5 implies the following result of [Takahasi (1966)].

Corollary 7.1. *If $\mathbf{x} : N \to \mathbb{E}^{m+1}$ is an isometric immersion, then $\Delta \mathbf{x} = \lambda \mathbf{x}$ for some $\lambda \in \mathbf{R}$ if and only if either N is a minimal submanifold of $\mathbb{E}^{m+1}$ or N is a minimal submanifold of a hypersphere.*

The following conjecture was made in [Chen (1991a, 1996c)].

Conjecture 7.1. *The only compact finite type hypersurfaces of a Euclidean space are hyperspheres.*

7.5 Finite type immersions of compact homogeneous spaces

Let o be a point in a compact Riemannian manifold N. Denote by K the isotropy subgroup at o (i.e., K is the *stabilizer* of o) . Then we have $N = G/K$. The linear isotropy representation is the orthogonal representation of K over the tangent space T_oN at o defined by $K \to O(T_oN) : \phi \mapsto (\phi_*)_o$, where $(\phi_*)_o$ denotes the differential of ϕ at o. A homogeneous Riemannian manifold is called *isotropy-irreducible* if the linear isotropy representation is irreducible. An isotropy-irreducible homogeneous Riemannian manifold is simply called an *irreducible homogeneous Riemannian manifold.*

Let $N = G/K$ be a compact irreducible homogeneous Riemannian manifold. For each eigenvalue λ of the Laplacian on N, we denote by m_λ the multiplicity of the eigenvalue λ. Let $\phi_1, \ldots, \phi_{m_\lambda}$ be an orthonormal basis of the eigenspace V_λ of the Laplacian Δ with eigenvalue λ. Define a map $f_\lambda : N \to \mathbb{E}^{m_\lambda}$ by

$$f_\lambda(u) = c_\lambda(\phi_1(u), \ldots, \phi_{m_\lambda}(u)) \tag{7.43}$$

which is a 1-type isometric immersion for some suitable constant $c_\lambda > 0$. Corollary 7.1 implies that f_λ is a minimal immersion into $S_0^{m_\lambda - 1}(1)$.

If λ_i is the i-th nonzero eigenvalue of Laplacian of N, then $\psi_i = f_{\lambda_i}$ is called the *i-th standard immersion* of $N = G/K$.

Not every 1-type isometric immersion of order $\{k\}$ of an irreducible symmetric space of compact type into Euclidean space is the k-th standard isometric immersion. This is not true even for a Riemannian 3-sphere. In fact, N. Ejiri constructed in [Ejiri (1981)] a 1-type immersion of the 3-sphere $S^3(4)$ of radius 4 into $S^6(1) \subset \mathbb{E}^7$ which is not a standard immersion.

Definition 7.3. Let $\psi_{t_i} : N \to \mathbb{E}^{m_i}$, $i = 1, \ldots, k$, be standard immersions of a compact isotropy-irreducible homogeneous Riemannian manifold N into the Euclidean m_i-space $\mathbb{E}^{m_i}$ and let $c_{t_1}, \ldots, c_{t_k}$ be real numbers such that $c_{t_1}^2 + \ldots + c_{t_k}^2 = 1$. Then

$$D(c_{t_1}, \ldots, c_{t_k}) = (c_{t_1}\psi_{t_1}, \ldots, c_{t_k}\psi_{t_k}) \tag{7.44}$$

defines an isometric immersion of N into $\mathbb{E}^{m_1 + \ldots + m_k}$ which is called a *standard diagonal immersion* of N.

Notice that if the k standard immersions in Definition 7.3 arisen from k distinct eigenvalues of Δ, then $D(c_{t_1}, \ldots, c_{t_k})$ is of k-type.

The following result shows that every finite type immersion of a compact irreducible homogeneous Riemannian manifold is a "screw" diagonal immersion [Deprez (1988); Chen (1996c)].

Theorem 7.6. *Let N be a compact irreducible homogeneous Riemannian manifold. Then*

(1) *for any standard diagonal immersion $D(a_{t_1}, \ldots, a_{t_k}) : N \to \mathbb{E}^m$ and any linear map $L : \mathbb{E}^m \to \mathbb{E}^\ell$, the composition $L \circ D(a_{t_1}, \ldots, a_{t_k})$ is a finite type immersion whose type number is at most k and whose order is a subset of the order $T(D(a_{t_1}, \ldots, a_{t_k}))$ of $D(a_{t_1}, \ldots, a_{t_k})$;*
(2) *if $\mathbf{x} : N \to \mathbb{E}^m$ is a finite type isometric immersion whose center of mass is chosen to be the origin of $\mathbb{E}^m$, then for each standard diagonal immersion $D(c_{t_1}, \ldots, c_{t_k}) : N \to \mathbb{E}^N$ with $T(\mathbf{x}) \subset T(D(c_{t_1}, \ldots, c_{t_k}))$, there exists a linear map $L : \mathbb{E}^N \to \mathbb{E}^m$ such that $\mathbf{x} = L \circ D(c_{t_1}, \ldots, c_{t_k})$.*

Proof. Statement (1) follows from Lemma 7.1. Now, assume that $\mathbf{x} : N \to \mathbb{E}^m$ is a finite type isometric immersion of a compact irreducible homogeneous Riemannian manifold N whose center of mass is chosen to be the origin of $\mathbb{E}^m$. Let $T(\mathbf{x}) = \{t_1, \ldots, t_\ell\}$. For each t_i, let $\{\phi_1^{t_i}, \ldots, \phi_{m_{t_i}}^{t_i}\}$ be an orthonormal basis of the eigenspace V_{t_i} of Δ with eigenvalue λ_{t_i}. Then

$$\mathbf{x} = \sum\nolimits_{i=1}^{\ell} \left(a_1^i \phi_1^{t_i} + \cdots + a_{m_{t_i}}^i \phi_{m_{t_i}}^{t_i}\right), \tag{7.45}$$

for some vectors $a_1^1, \ldots, a_{m_{t_1}}^1, \ldots, a_1^\ell, \ldots, a_{m_{t_\ell}}^\ell \in \mathbb{E}^m$. Consider a standard diagonal immersion $D(c_{t_1}, \ldots, c_{t_k}) : N \to \mathbb{E}^N$ of order $\{t_1, t_2, \ldots, t_k\}$ with $k \geq \ell$. Then $D(c_{t_1}, \ldots, c_{t_k})$ is of the form:

$$\left(c_{t_1} b_{t_1} \phi_1^{t_1}, \ldots, c_{t_1} b_{t_1} \phi_{m_{t_1}}^{t_1}, \ \ldots, c_{t_k} b_{t_k} \phi_1^{t_k}, \ldots, c_{t_k} b_{t_k} \phi_{m_{t_k}}^{t_k}\right) \tag{7.46}$$

for some positive numbers $b_{t_1}, \ldots, b_{t_k}$. Let L be the linear map from $\mathbb{E}^N$ into $\mathbb{E}^m$ defined by

$$\begin{aligned} &L(c_{t_1} b_{t_1} e_1) = a_1^1, \ \ldots, \ L(c_{t_1} b_{t_1} e_{m_{t_1}}) = a_{m_{t_1}}^1, \\ &\quad \vdots \\ &L(c_{t_\ell} b_{t_\ell} e_{m_{t_1}+\cdots+m_{t_{\ell-1}}+1}) = a_1^\ell, \ \ldots, \\ &L(c_{t_\ell} b_{t_\ell} e_{m_{t_1}+\cdots+m_{t_\ell}}) = a_{m_{t_\ell}}^\ell, \\ &L(e_{m_{t_1}+\cdots+m_{t_\ell}+1}) = \cdots = L(e_N) = 0, \end{aligned} \tag{7.47}$$

where $\{e_1, \ldots, e_N\}$ is a standard basis of $\mathbb{E}^N$. Then $\mathbf{x} = L \circ D(c_{t_1}, \ldots, c_{t_k})$, which implies statement (2). □

Remark 7.2. When N is a circle, Theorem 7.6 implies that every finite type closed curve in Euclidean space is the composition of a W-curve and a linear map of Euclidean spaces.

7.6 Submanifolds of $\mathbb{E}_s^m$ satisfying $\Delta H = \lambda H$

Theorem 7.3 implies that a pseudo-Riemannian submanifold of $\mathbb{E}_s^m$ is 1-type if and only if, up to translations, the position vector satisfies $\Delta\mathbf{x} = \lambda\mathbf{x}$.

The next result determines pseudo-Riemannian submanifolds of $\mathbb{E}_s^m$ which satisfy $\Delta H = \lambda H$ for some $\lambda \in \mathbf{R}$ [Chen (1988a)].

Theorem 7.7. *If $\mathbf{x} : N \to \mathbb{E}_s^m$ is an isometric immersion of a pseudo-Riemannian manifold into $\mathbb{E}_s^m$, then the mean curvature vector H is an eigenvector of Δ if and only if it is one of the following:*

(a) *a biharmonic submanifold;*
(b) *a 1-type submanifold;*
(c) *a null 2-type submanifold.*

Proof. Under the hypothesis, assume that $\Delta H = \lambda H$ holds for some real number λ. If $\lambda = 0$, then N is a biharmonic submanifold, which gives (a).

Now, assume that $\Delta H = \lambda H$ with $\lambda \neq 0$. Let us put

$$\mathbf{x}_p = \frac{1}{\lambda}\Delta \mathbf{x} \text{ and } \mathbf{x}_0 = \mathbf{x} - \mathbf{x}_p. \tag{7.48}$$

Then we find

$$\Delta \mathbf{x}_p = -\frac{n}{\lambda}\Delta H = \Delta \mathbf{x} = \lambda \mathbf{x}_p, \;\; \Delta \mathbf{x}_0 = \Delta \mathbf{x} - \Delta \mathbf{x}_p = 0. \tag{7.49}$$

Hence, N is either of 1-type or of null 2-type, depending on $\mathbf{x}_0$ is a constant or non-constant. The converse is easy to verify. □

For spherical submanifolds satisfying $\Delta H = \lambda H$ we have the following.

Proposition 7.3. *Let $\mathbf{x} : N \to S^{m-1}(1) \subset \mathbb{E}^m$ be an isometric immersion of a Riemannian manifold into the unit hypersphere $S^{m-1}(1)$ such that the mean curvature vector H of N in $\mathbb{E}^m$ satisfies $\Delta H = \lambda H$ for some $\lambda \in \mathbf{R}$. Then N is either a minimal submanifold of $S^{m-1}(1)$ or a minimal submanifold of a totally umbilical hypersurface of $S^{m-1}(1)$. In both cases, M is of 1-type.*

Proof. Under the hypothesis, by comparing $\Delta H = \lambda H$ using formula (2.45) of ΔH, we find

$$\lambda = n + ||A_\xi||^2 + ||D'\xi||^2, \;\; \lambda = n||H||^2, \tag{7.50}$$

where ξ is a unit normal vector in the direction of the mean curvature vector H' of N in $S^{m-1}(1)$ and D' is the normal connection of N in S^{m-1}. These two expressions of λ imply that N is a pseudo-umbilical submanifold of S^{m-1} with parallel mean curvature vector. Thus, by applying Proposition 3.10, we obtain the desired result. □

Theorem 7.7 and Proposition 7.3 imply the following.

Corollary 7.2. *We have*

(1) *every null 2-type submanifold N of $\mathbb{E}^m$ is non-spherical, i.e., N does not lie in any hypersphere of $\mathbb{E}^m$;*
(2) *every biharmonic submanifolds in a Euclidean space is non-spherical.*

Remark 7.3. Conformally flat hypersurface in $\mathbb{E}^4$ satisfying $\Delta H = \lambda H$ was classified in [Garay (1994)]. Also, it was proved in [Arvanitoyeorgos et al. (2009)] that if a Lorentz hypersurface in $\mathbb{E}^4_1$ satisfies $\Delta H = \lambda H$ for some constant λ, then it has constant mean curvature.

7.7 Submanifolds of $H^m(-1)$ satisfying $\Delta H = \lambda H$

For submanifolds of $H^m(1) \subset \mathbb{E}_1^{m+1}$, we have [Chen (1994a)].

Proposition 7.4. *Let $\mathbf{x} : N \to H^m(-1) \subset \mathbb{E}_1^{m+1}$ be an isometric immersion of a Riemannian manifold into the hyperbolic space $H^m(-1)$. Then the mean curvature vector of N in $\mathbb{E}_1^{m+1}$ is an eigenvector of Δ if and only if one of the following two cases occurs:*

(a) *N is a minimal submanifold of $H^m(-1)$;*
(b) *N lies in a totally umbilical hypersurface of $H^m(-1)$ as minimal submanifold.*

Proof. Under the hypothesis, choose an orthonormal frame $\{e_1, \ldots, e_n, e_{n+1}, \ldots, e_{m+1}\}$ such that $e_1, \ldots, e_n$ are tangent to N and $e_{m+1} = \mathbf{x}$. Then

$$H = H' + \mathbf{x}, \quad A_{\mathbf{x}} = -I. \tag{7.51}$$

Thus, after applying Lemma 3.5, formula (2.45) reduces to

$$\begin{aligned} \Delta H = \Delta^D H + \frac{n}{2} \nabla \langle H, H \rangle + 2 \operatorname{trace} A_{DH} - nH' \\ + n(\langle H', H' \rangle - 1)\mathbf{x} + \sum\nolimits_{r=n+1}^{m} \operatorname{trace}(A_{H'} A_r) e_r, \end{aligned} \tag{7.52}$$

where H' is the mean curvature vector of N in $H^m(-1)$. Assume that $\Delta H = \lambda H$ for some constant λ. Then, (7.51) and (7.52) give

$$\langle H', H' \rangle = 1 + \frac{\lambda}{n}, \quad \langle H, H \rangle = \frac{\lambda}{n}. \tag{7.53}$$

Thus, N has constant mean curvature in $H^{m-1}(-1)$ and as well as in $\mathbb{E}_1^m$.

Case (1): $H' = 0$. We find $H = \mathbf{x}$. Hence, Beltrami's formula yields $\Delta H = -nH$. Thus, H is an eigenvector of Δ with eigenvalue $\lambda = -n$.

Case (2): If $H' \neq 0$. Let us choose e_{n+1} such that

$$H' = a e_{n+1}, \quad a = \sqrt{1 + \lambda/n}, \quad \lambda > -n. \tag{7.54}$$

Put $De_r = \sum_{t=n+1}^{m+1} \omega_r^t e_t$. From $e_{m+1} = \mathbf{x}$ we find $\omega_{n+1}^{m+1} = \cdots = \omega_m^{m+1} = 0$. Hence

$$\begin{aligned} \Delta^D H = a \sum_{r=n+2}^{m} \Big\{ \sum_{i=1}^{n} \sum_{t=n+1}^{m} \omega_{n+1}^t(e_i) \omega_r^t(e_i) - \operatorname{trace}(\nabla \omega_{n+2}^r) \Big\} e_r \\ + ||De_{n+1}||^2 H'. \end{aligned} \tag{7.55}$$

Also, $\operatorname{trace}(A_{n+1}^2) = n + \lambda - ||De_{n+1}||^2$ holds. From this and (7.53) we find

$$n \operatorname{trace}(A_{n+1}^2) - (\operatorname{trace}(A_{n+1}))^2 = -n ||De_{n+1}||^2. \tag{7.56}$$

On the other hand, we also have

$$n\,\mathrm{trace}\,(A_{n+2}^2) - (\mathrm{trace}\,(A_{n+2}))^2 = \sum\nolimits_{i<j} (\kappa_i - \kappa_j)^2, \tag{7.57}$$

where $\kappa_1, \ldots, \kappa_n$ are the eigenvalues of A_{n+1}. Hence, we conclude from (7.54), (7.56) and (7.57) that

$$A_{n+1} = aI, \quad De_{n+1} = DH' = DH = 0, \tag{7.58}$$

which show that N is a pseudo-umbilical submanifold of $H^m(-1)$ with parallel nonzero mean curvature vector. Thus, by Proposition 3.11, N lies in a totally umbilical hypersurface of $H^m(-1)$ as a minimal submanifold.

Conversely, if M is minimal submanifold of a totally umbilical hypersurface of $H^m(-1)$, then N is a pseudo-umbilical submanifold with parallel nonzero mean curvature vector. Thus, (7.58) holds for some nonzero real number a. Hence, we find from (7.55) that $\Delta^D H = 0$. Also, (7.58) yields

$$\sum\nolimits_{r=n+1}^{m} \mathrm{trace}(A_{H'} A_r) e_r = n \langle H', H' \rangle\, e_{n+1}. \tag{7.59}$$

Thus, by (7.55), (7.58) and (7.59), we find $\Delta H = \lambda H$, $\lambda = n(a^2 - 1)$. This completes the proof of the Proposition. □

A complete flat totally umbilical hypersurface of a hyperbolic $(n+1)$-space H^{n+1} is called a *horosphere*.

For submanifolds of $H^m(1) \subset \mathbb{E}_1^{m+1}$, we also have [Chen (1994a)].

Proposition 7.5. *If N is a non-minimal submanifold of $H^m(-1) \subset \mathbb{E}_1^{m+1}$, then the following four statements are equivalent:*

(1) *N is a biharmonic submanifold of $\mathbb{E}_1^{m+1}$;*
(2) *the mean curvature vector of N in $\mathbb{E}_1^{m+1}$ is a lightlike constant vector;*
(3) *N is a pseudo-umbilical submanifold with unit parallel mean curvature vector in $H^m(-1)$;*
(4) *N is a minimal submanifold of a horosphere of $H^m(-1)$.*

Proof. Assume that N is a non-minimal submanifold of the hyperbolic space $H^{m-1}(-1)$.

(1) ⇒ (2) & (3): If N is biharmonic, then $\Delta H = 0$. Thus, (7.52) yields $\langle H, H \rangle = 0$ and $\langle H', H' \rangle = 1$. Since $H = H' + \mathbf{x}$ is nonzero, H is lightlike. Moreover, by Proposition 7.4, N is a pseudo-umbilical submanifold with parallel mean curvature vector in $H^m(-1)$. Thus (3) holds. Moreover, it follows from (7.51), (7.53), (7.58) and $\Delta H = 0$ that $A_H = 0$. Hence, we conclude from Weingarten's formula and (7.58) that H is a constant vector in $\mathbb{E}_1^{m+1}$. This proves (2).

(2) $\Rightarrow$ (1) is obvious.

(3) $\Rightarrow$ (1): If N is a pseudo-umbilical submanifold with unit parallel mean curvature vector in $H^{m-1}(-1)$, then, by Proposition 7.6, N satisfies $\Delta H = \lambda H$ for some $\lambda \in \mathbf{R}$. On the other hand, (7.53) and $\langle H', H' \rangle = 1$ yield $\lambda = 0$. Thus, N is biharmonic.

(3) $\Rightarrow$ (4): If N is pseudo-umbilical with unit parallel mean curvature vector in $H^m(-1)$, then $H = H' + \mathbf{x}$ is a lightlike constant vector by (2). Thus, $\langle c, \mathbf{x} \rangle = -1$. Let N be the hyperplane section of $H^m(-1)$ given by $N = \{x \in \mathbb{E}^{m+1}_1 : \langle x, x \rangle = -1 \ \langle c, x \rangle = -1\}$.

If we put $\xi = c - x$, then $\langle \xi, x \rangle = 0$ and $\langle \xi, \xi \rangle = 1$. Thus, ξ is a unit normal vector field of N in $H^m(-1)$. Moreover, from the definition of ξ, it is clear that N is a totally umbilical hypersurface of $H^m(-1)$ with constant mean curvature one. Thus, by Gauss' equation, N is flat. Because the mean curvature vector H of N in $\mathbb{E}^{m+1}_1$ is c which is equal to $\xi - x$, H is also normal to N. Therefore, N lies in a horosphere of $H^m(-1)$ as a minimal submanifold. This proves (4).

(4) $\Rightarrow$ (3): If N lies in a horosphere N' of $H^{m-1}(-1)$ as a minimal submanifold, then N is a pseudo-umbilical submanifold with parallel mean curvature vector in $H^m(-1)$. Because N' is flat and $H^m(-1)$ has constant curvature -1, Gauss' equation implies that N has constant mean curvature one in $H^m(-1)$. This proves (3). □

When $m = n + 1$, Proposition 7.5 implies the following.

Corollary 7.3. *Let N be a hypersurface of $H^{n+1}(-1) \subset \mathbb{E}^{n+1}_1$. Then N is an open part of a horosphere if and only if N is biharmonic in $\mathbb{E}^{n+1}_1$.*

Remark 7.4. Another simple characterization of horosphere in terms of ΔH can be found in [Dimitric (2009)].

7.8 Submanifolds of $S^m_1(1)$ satisfying $\Delta H = \lambda H$

For submanifolds of the de Sitter spacetime $S^m_1(1)$ we have [Chen (1995a)].

Proposition 7.6. *Let N be an n-dimensional spacelike submanifold of the de Sitter spacetime $S^{m-1}_1(1) \subset \mathbb{E}^m_1$. Then the mean curvature vector of N in $\mathbb{E}^m_1$ satisfies $\Delta H = \lambda H$ for some real number $\lambda < n$ if and only if N lies in a spacelike, non-totally geodesic, totally umbilical hypersurface of $S^{m-1}_1(1)$ as a minimal submanifold.*

Proposition 7.7. *Let $x : N \to S_1^{m-1}(1) \subset \mathbb{E}_1^m$ be an isometric immersion of a Riemannian n-manifold into $S_1^{m-1}(1)$. Then the mean curvature vector of N in $\mathbb{E}_1^m$ is an eigenvector of Δ with eigenvalue n if and only if ether*

(a) *N is a minimal submanifold of $S_1^{m-1}(1)$ or*
(b) *$\mathbf{x} : N \to \mathbb{E}_1^m$ is congruent to*

$$(f + \varphi, x_3, \ldots, x_m, f + \varphi),$$

where φ is a harmonic function, f is an eigenfunction of Δ with eigenvalue n, and $\mathbf{y} = (x_3, \ldots, x_m) : N \to S^{m-3}(1) \subset \mathbb{E}^{m-2}$ is an isometric minimal immersion of N into the unit hypersphere $S^{m-3}(1)$.

Proposition 7.8. *Let λ be a real number $> n$, c a spacelike vector in $\mathbb{E}_1^m$ satisfying $\langle c, c \rangle = 1 - n/\lambda > 0$, and $N_1(c)$ the hypersurface of the de Sitter spacetime S_1^{m-1} defined by $N_1(c) = \{x \in S_1^{m-1}(1) : \langle c, x \rangle = \langle c, c \rangle\}$. Then*

(i) *$N_1(c)$ is a non-totally geodesic totally umbilical hypersurface of $S_1^{m-1}(1)$;*
(ii) *if N is an n-dimensional spacelike minimal submanifold of $N_1(c)$, then*
 (i.1) *N is a pseudo-umbilical submanifold of $S_1^{m-1}(1)$;*
 (ii.2) *the mean curvature vector of N in $S_1^{m-1}(1)$ is a nonzero parallel normal vector field; and*
 (ii.3) *the mean curvature vector H of N in $\mathbb{E}_1^m$ satisfying $\Delta H = \lambda H$ for some real number $\lambda > n$.*

Conversely, if $\mathbf{x} : N \to S_1^{m-1}(1) \subset \mathbb{E}_1^m$ is an isometric immersion of a Riemannian n-manifold such that the mean curvature vector of N in $\mathbb{E}_1^m$ satisfies (ii.2) *and* (ii.3), *then N lies in $N_1(c)$ as a minimal submanifold, where c is a spacelike vector in $\mathbb{E}_1^m$ with $\langle c, c \rangle = 1 - n/\lambda$.*

7.9 Biharmonic submanifolds

Minimal submanifold of $\mathbb{E}_s^m$ are trivially biharmonic, i.e., $\Delta H = 0$. So the real problem is if there are other submanifolds besides minimal that are biharmonic. The study of biharmonic submanifolds was initiated by the author in the middle of 1980s. In fact, he proved in 1985 that every biharmonic surface in $\mathbb{E}^3$ is minimal (unpublished then). This result was the starting point of I. Dimitric work on his doctoral thesis at Michigan State University [Dimitric (1989)]. In particular, Dimitric extended this result on biharmonic surface in $\mathbb{E}^3$ in his thesis to that if N is a biharmonic hypersurface of $\mathbb{E}^m$ with at most two distinct principal curvatures, then N is minimal (cf. [Dimitric (1989, 1992)]). Since conformally flat hypersurfaces of $\mathbb{E}^m$ with $m \geq 5$ have at most two distinct principal curvatures, Dimitric's result implies that biharmonic conformally flat hypersurfaces of $\mathbb{E}^m$ with $m \geq 5$ are minimal. The nonexistent result on biharmonic surfaces in $\mathbb{E}^3$ was also extended by the author and Ishikawa to surfaces in pseudo-Euclidean 3-space $\mathbb{E}^3_s$ while Ishikawa was working at MSU for his doctoral thesis during the period 1988–1990 (cf. [Chen and Ishikawa (1991, 1998)]). Another extension of the result on biharmonic surfaces in $\mathbb{E}^3$ was done in [Hasanis-Vlachos (1995)], in which biharmonic hypersurfaces of $\mathbb{E}^4$ are shown to be minimal (for an alternative proof, see [Defever (1998)]).

I. Dimitric also proved the following [Dimitric (1989, 1992)].

Proposition 7.9. *Every finite type biharmonic pseudo-Riemannian submanifold in $\mathbb{E}^m_s$ are minimal.*

Proof. Let N be a finite type biharmonic pseudo-Riemannian submanifold of $\mathbb{E}^m_s$. Then it has finite spectral decomposition

$$\mathbf{x} = \mathbf{x}_0 + \mathbf{x}_{t_1} + \cdots + \mathbf{x}_{t_k} \tag{7.60}$$

with $\Delta \mathbf{x}_0 = 0$, $\Delta \mathbf{x}_{t_i} = \lambda_{t_i} \mathbf{x}_{t_i}$ for nonzero distinct eigenvalues $\lambda_{t_1}, \ldots, \lambda_{t_k}$ of Δ. By taking Δ^r of (7.60) we find

$$0 = \Delta^r \mathbf{x} = \lambda^r_{t_1} \mathbf{x}_{t_1} + \cdots + \lambda^r_{t_k} \mathbf{x}_{t_k}, \quad r = 2, 3, \cdots . \tag{7.61}$$

Since $\lambda_{t_1}, \ldots, \lambda_{t_k}$ are distinct eigenvalues of Δ, system (7.61) is impossible unless $k = 0$. Thus, $\mathbf{x} = \mathbf{x}_0$, which implies that N is minimal. □

Corollary 7.4. *Every biharmonic pseudo-Riemannian submanifold of $\mathbb{E}^m_s$ is either minimal or of infinite type.*

Remark 7.5. There exist many infinite type pseudo-Riemannian biharmonic submanifolds in pseudo-Euclidean spaces[Chen and Ishikawa (1991)]. One simple example is the non-minimal spacelike curve (s^3, s, s^3) in $\mathbb{E}^3_1$.

Proposition 7.10. *Every compact biharmonic Riemannian submanifold N of a pseudo-Euclidean m-space $\mathbb{E}^m_s$ satisfies*

$$\int_N \langle H, H \rangle * 1 = 0. \tag{7.62}$$

Hence, every biharmonic Riemannian submanifold of $\mathbb{E}^m$ is non-compact.

Proof. If N is compact, Beltrami's formula and Proposition 6.1 yield

$$\int_N \langle \Delta H, \mathbf{x} \rangle * 1 = -n \int_N \langle H, H \rangle * 1,$$

which gives (7.62). In particular, if $\mathbb{E}^m_s$ is Euclidean, (7.62) implies that $H = 0$, which contradicts to Corollary 2.6. □

The following *biharmonic conjecture* was made in [Chen (1991a)].

Conjecture 7.2. *The only biharmonic submanifolds in Euclidean spaces are the minimal ones.*

Remark 7.6. A *biharmonic map* is a map $\phi : (N, h) \to (M, g)$ between Riemannian manifolds that is a critical point of the bienergy functional

$$E^2(\phi, D) = \frac{1}{2} \int_D ||\tau_\phi||^2 * 1 \tag{7.63}$$

for every compact subset D of N, where $\tau_\phi = \text{trace}_h \nabla d\phi$ is the tension field ϕ. The Euler-Lagrange equation of this functional gives the biharmonic map equation [Jiang (1986)]

$$\tau_\phi^2 := \text{trace}_h(\nabla^\phi \nabla^\phi - \nabla^\phi_{\nabla^N})\tau_\phi - \text{trace}_h R^M(d\phi, \tau_\phi)d\phi = 0, \tag{7.64}$$

where R^M is the curvature tensor of (M, g). Equation (7.64) states that ϕ is a biharmonic map if and only if its bi-tension field τ_ϕ^2 vanishes identically.

Let N be an n-dimensional submanifold of a Euclidean m-space $\mathbb{E}^m$. If we denote by $\iota : N \to \mathbb{E}^m$ the inclusion map of the submanifold, then the tension field of the inclusion map is given by $\tau_\iota = -\Delta \iota = -nH$ according to Beltrami's formula. Thus, N is a biharmonic submanifold if and only if $n\Delta H = -\Delta^2 \iota = -\tau_\iota^2 = 0$, i.e., the inclusion map ι is a biharmonic map.

Remark 7.7. Caddeo, Montaldo and Oniciuc showed in [Caddeo et al. (2002)] that any biharmonic submanifold in the hyperbolic 3-space $H^3(-1)$ is minimal, and that any n-dimensional pseudo-umbilical biharmonic submanifold of hyperbolic m-space is minimal if $n \neq 4$. Based on these, Caddeo, Montaldo and Oniciuc extended Conjecture 7.2 to the following.

Generalized Chen Conjecture: *Any biharmonic submanifold in a Riemannian manifold of non-positive sectional curvature is minimal.*

Recently, it is proved in [Ou and Tang (2010)] that this generalized conjecture is false. However, Conjecture 7.2 is still open.

Remark 7.8. During the last two decades, biharmonic submanifolds in various Riemannian manifolds have been investigated by many authors. As results, many examples of non-minimal biharmonic submanifolds in non-flat Riemannian manifolds have been produced (see, for instance, [Ou (2010)] and the survey article [Montaldo and Oniciuc (2006)]).

7.10 Null 2-type submanifolds

The first result on null 2-type submanifolds is the following [Chen (1988b)].

Theorem 7.8. *A surface N in $\mathbb{E}^3$ is of null 2-type if and only if it is an open portion of a circular cylinder.*

Proof. Let N be a surface in $\mathbb{E}^3$, ξ a unit local normal vector field, and $\{e_1, e_2\}$ an orthonormal local frame of N. If we put $H = \alpha\xi$ and $A = A_\xi$, then it follows (2.50) that $\mathrm{trace}(A_{DH}) = A(\nabla\alpha)$. Also, we have

$$\Delta^D H = (\Delta\alpha)\xi, \ \sum\nolimits_{i=1}^{2} h(A_H e_i, e_i) = \alpha||A||^2\xi. \tag{7.65}$$

Thus, formula (2.45) reduces to

$$\Delta H = \{\Delta\alpha + \alpha||A||^2\}\xi + \nabla\alpha^2 + 2\,A(\nabla\alpha). \tag{7.66}$$

Assume that N is of null 2-type. Then Proposition 7.3 implies that $\Delta H = \lambda H$ for some constant $\lambda \neq 0$. Comparing this with (7.66) gives

$$\Delta\alpha = \alpha(\lambda - ||A||^2), \ \ A(\nabla\alpha^2) = -\alpha\,\nabla\alpha^2. \tag{7.67}$$

If we put $U = \{p \in M : \nabla\alpha^2 \neq 0 \text{ at } p\}$, then $\nabla\alpha^2$ is an eigenvector of A with eigenvalue $-\alpha$ on U. The other eigenvalue is 3α. Let us choose e_1 in the direction of $\nabla\alpha^2$. Then $e_2\alpha = 0$. Also, we have

$$h(e_1, e_1) = -\alpha\xi, \ \ h(e_1, e_2) = 0, \ \ h(e_2, e_2) = 3\alpha\xi. \tag{7.68}$$

It follows from (7.68) and Codazzi's equation that

$$3e_1\alpha = -4\alpha\omega_1^2(e_2), \ \ \omega_1^2(e_1) = 0 \ \text{ on } U. \tag{7.69}$$

Let ω^1, ω^2 be the dual 1-forms of e_1, e_2. It follows from $\omega_1^2(e_1) = 0$ and Cartan's structure equation that $d\omega^1 = 0$. Thus, by Poincaré lemma, we find $\omega^1 = du$ for some local function u. Combining this with $d\alpha = (e_1\alpha)\omega^1$ yields $d\alpha \wedge du = 0$. Thus α is a function of u. Hence, by (7.69), we find

$$d\alpha = \alpha'(u)du, \ \ 4\alpha\omega_1^2 = -3\alpha'(u)\omega^1. \tag{7.70}$$

Exterior differentiating the second equation in (7.70) gives

$$4\alpha\alpha'' - 7(\alpha')^2 + 16\alpha^4 = 0. \tag{7.71}$$

Let $y = (\alpha')^2$. Then we obtain from (7.71) that

$$y = (\alpha')^2 = c\alpha^{\frac{7}{2}} - 16\alpha^4, \quad c \in \mathbf{R}. \tag{7.72}$$

On the other hand, it follows from (7.67) and (7.68) that

$$\alpha\Delta\alpha = (\lambda - 10\alpha^2)\alpha^2. \tag{7.73}$$

By applying (7.69) we obtain from (7.73) that

$$4\alpha\alpha'' - 3(\alpha')^2 + 4(\lambda - 10\alpha^2)\alpha^2 = 0. \tag{7.74}$$

From (7.71) and (7.74) we find $(\alpha')^2 = 14\alpha^4 - \lambda\alpha^2$. By combining this with (7.72) we conclude that α^2 is constant on U. So, U is empty. Thus N has constant mean curvature. Now, we find from (7.67) that $||A||^2$ is the constant λ. Thus, N has constant mean curvature and Gauss curvature. Since N is of null 2-type, N is an open portion of a circular cylinder.

The converse is trivial. □

By combining Theorem 7.7, Theorem 7.8 and the minimality of biharmonic surfaces in $\mathbb{E}^3$, we obtain the following.

Theorem 7.9. *Let N be a surface in $\mathbb{E}^3$. Then the mean curvature vector of N is an eigenvector of Δ if and only if either N is a minimal surface or N is an open portion of a circular cylinder.*

The following result was proved in [Chen (1992)].

Theorem 7.10. *Let N be a 2-type pseudo-Riemannian submanifold of $\mathbb{E}^m_s$. If N has parallel mean curvature vector, i.e., $DH = 0$, then either*

(a) *N is of null 2-type;*
(b) *or N is non-null and, up to translations and dilations, N lies in the pseudo-hypersphere $S^{m-1}_s(1)$, or in the pseudo-hyperbolic $(m-1)$-space $H^{m-1}_{s-1}(-1)$, or in the light cone $\mathcal{LC}$.*

Proof. Let N be a pseudo-Riemannian submanifold of $\mathbb{E}^m_s$ with parallel mean curvature vector. Then Corollary 2.8 gives

$$\Delta H = \sum\nolimits_{r=n+1}^{m} \epsilon_r \text{trace}(A_H A_{e_r}) e_r, \tag{7.75}$$

where $e_{n+1}, \ldots, e_m$ is an orthonormal frame of the normal bundle of N.

If N is of 2-type, then up to translation, the position vector $\mathbf{x}$ of N takes the following spectral decomposition

$$\mathbf{x} = \mathbf{x}_p + \mathbf{x}_q, \quad \Delta\mathbf{x}_p = \lambda_p \mathbf{x}_p, \quad \Delta\mathbf{x}_q = \lambda_q \mathbf{x}_q$$

with $\lambda_p < \lambda_q$. Thus, we find

$$n\Delta H = n(\lambda_p + \lambda_q)H + \lambda_p \lambda_q \mathbf{x}. \tag{7.76}$$

It follows from (7.75) and (7.76) that either (i) N is of null 2-type or (ii) N is non-null and $\mathbf{x}$ is normal to N each point. Case (ii) implies that $\langle \mathbf{x}, \mathbf{x} \rangle = c$, which is constant. Consequently, up to dilations, N lies in $S_s^{m-1}(1)$, or in $H_{s-1}^{m-1}(-1)$, or in $\mathcal{LC}$ according to $c > 0$, $c < 0$ or $c = 0$. □

The following is an immediate consequence of Theorem 7.10.

Corollary 7.5. *Every 2-type pseudo-Riemannian hypersurface N of $\mathbb{E}_s^{n+1}$ with parallel mean curvature vector is of null 2-type.*

Proof. Due to the fact that pseudo-hyperspheres and pseudo-hyperbolic hyperpanes are of 1-type and the light-cone $\mathcal{LC}$ are degenerated. □

For 2-type surface with parallel mean curvature vector, we have the following result from [Chen and Lue (1988)].

Theorem 7.11. *Let N be a surface of a Euclidean m-space $\mathbb{E}^m$ with parallel mean curvature vector. Then N is of 2-type if and only if N is an open portion of one of the following two surfaces:*

(a) *a circular cylinder $\mathbf{R} \times S^1(r) \subset \mathbb{E}^3 \subset \mathbb{E}^m$, which is of null 2-type;*
(b) *the direct product of two plane circles $S^1(a) \times S^1(b) \subset \mathbb{E}^4 \subset \mathbb{E}^m$ with different radii.*

Proof. Let N be a surface with $DH = 0$ in $\mathbb{E}^m$. Then N lies either in a totally geodesic $\mathbb{E}^3$ or in a hypersphere S^3 of a totally geodesic $\mathbb{E}^4$ (cf. page 106 of [Chen (1973b)]). Now, assume that N is of 2-type. If lies in a totally geodesic $\mathbb{E}^3$, then N is of null 2-type by Theorem 7.10. Thus, by Theorem 7.8 we get case (a). If N lies in S^3, we obtain case (b) according to Theorem 7.20. □

The following results were proved in [Chen and Song (1989a,b)].

Theorem 7.12. *Let N be a spacelike surface in the Minkowski spacetime $\mathbb{E}_1^4$. Then N is of null 2-type with constant mean curvature if and only if, up to rigid motions, N is an open part of one of the following surfaces:*

(1) *a helical cylinder of first kind defined by*

$$\left(\frac{bv}{c}, a\cos\left(\frac{v}{c}\right), a\sin\left(\frac{v}{c}\right), u\right), \quad c = \sqrt{a^2 - b^2},$$

where a, b are real numbers with $a^2 > b^2$;

(2) *a helical cylinder of second kind defined by*

$$\left(a\cosh\left(\frac{v}{c}\right), a\sinh\left(\frac{v}{c}\right), \frac{bv}{c}, u\right), \quad c = \sqrt{a^2 + b^2},$$

where a, b are real numbers with $a > 0$.

Theorem 7.13. *Let N be a spacelike surface in $\mathbb{E}^4_2$. Then N is of null 2-type with constant mean curvature if and only if, up to rigid motions, N is an open part of one of the following surfaces:*

(1) *A helical cylinder of first kind in $\mathbb{E}^4_2$ defined by*

$$\left(b\sin\left(\frac{v}{c}\right), b\cos\left(\frac{v}{c}\right), \frac{av}{c}, u\right), \quad c = \sqrt{a^2 - b^2},$$

where a, b are real numbers with $a^2 > b^2 > 0$;

(2) *a helical cylinder of second kind in $\mathbb{E}^4_2$ defined by*

$$\left(\frac{bv}{c}, a\cosh\left(\frac{v}{c}\right), a\sinh\left(\frac{v}{c}\right), u\right), \quad c = \sqrt{a^2 - b^2},$$

where a, b are real numbers with $a^2 > b^2 > 0$.

Theorem 7.14. *Let N be a marginally trapped surface in $\mathbb{E}^4_1$. Then N is of null 2-type surface if and only if N is a flat surface such that, up to rigid motions, N is an open portion of the surface defined by*

$$(f(u,v) + \varphi(u,v), f(u,v) + \varphi(u,v), u, v)$$

with f is an eigenfunction of the Laplacian Δ with nonzero eigenvalue and φ is harmonic.

Theorem 7.14 implies immediately the following.

Corollary 7.6. *There exist no marginally trapped null 2-type surfaces in $\mathbb{E}^4_1$ which lies either in the de Sitter spacetime $S^3_1(1)$ or in the hyperbolic $H^3(-1)$.*

Remark 7.9. Due to simplicity it is very natural and interesting to classify all null 2-type submanifolds of $\mathbb{E}^m_s$. So far only partial results have been obtained by the author, U. Dursun, A. Ferrández, C.-S. Houh, D.-S. Kim, Y.-H. Kim, S.-J. Li, P. Lucas, H.-Z. Song, among others.

The following conjecture was made in [Chen (1991a, 1996c)].

Conjecture 7.3. *The only complete non-compact hypersurfaces of finite type in a Euclidean space are either minimal or of null 2-type.*

Remark 7.10. Conjectures 7.1, 7.2 and 7.3 stay open.

7.11 Spherical 2-type submanifolds

Definition 7.4. An isometric immersion $\phi : N \to S^{m-1}(1) \subset \mathbb{E}^m$ of a compact Riemannian manifold N into a hypersphere $S^{m-1}(1)$ is called *mass-symmetric* in $S^{m-1}(1)$ if the center of mass is the center of the hypersphere.

For simplicity we assume that the center of $S^{m-1}(1)$ is the origin of $\mathbb{E}^m$.

Lemma 7.3. *Every compact minimal submanifold N of a hypersphere $S^{m-1}(1) \subset \mathbb{E}^m$ is mass-symmetric.*

Proof. By Beltrami's formula and Hopf's lemma, we find $\int_N H * 1 = 0$, where H is the mean curvature vector of N in $\mathbb{E}^m$. Since N is a minimal in $S^{m-1}(1)$ (centered at the origin of $\mathbb{E}^m$), $H = -\mathbf{x}$, where $\mathbf{x}$ is the position vector field of N in $\mathbb{E}^m$. Then $0 = \int_N \mathbf{x} * 1 / \int_N *1$. This shows that the hypersurface is mass-symmetric. □

For spherical hypersurfaces, we have [Chen (1979c)].

Theorem 7.15. *If N is a compact, mass-symmetric, 2-type hypersurface of $S^{n+1}(1) \subset \mathbb{E}^{n+2}$, then*

(a) *the mean curvature vector H' of N in $S^{n+1}(1)$ satisfies*

$$n^2 ||H'||^2 = (n - \lambda_p)(\lambda_q - n);$$

(b) *the scalar curvature τ of N is given by*

$$\tau = \frac{1}{n}(\lambda_p + \lambda_q) - \frac{\lambda_p \lambda_q}{n(n-1)};$$

(c) $(\Delta H)^T = 0$, *where $(\Delta H)^T$ is the tangential component of ΔH.*

Proof. Under the hypothesis, the position vector $\mathbf{x}$ satisfies the spectral decomposition $\mathbf{x} = \mathbf{x}_p + \mathbf{x}_q$, $\Delta \mathbf{x}_p = \lambda_p \mathbf{x}_p$, $\Delta \mathbf{x}_q = \lambda_q \mathbf{x}_q$ with $\lambda_p < \lambda_q$. Thus

$$\Delta^2 \mathbf{x} - (\lambda_p + \lambda_q) \Delta \mathbf{x} + \lambda_p \lambda_q \mathbf{x} = 0.$$

Hence, by Beltrami's formula,

$$n\Delta H = n(\lambda_p + \lambda_q)H + \lambda_p\lambda_q \mathbf{x}. \tag{7.77}$$

Hence, by using (2.45), (3.11), (7.77), and by comparing the $\mathbf{x}$-component and the tangential component of (7.77), we obtain (a) and (c).

For (b), let ξ be a unit normal vector field of N in $S^{n+1}(1)$. Then the ξ-component of (7.77) yields

$$||h||^2 = ||h'||^2 + n = \lambda_p + \lambda_q, \tag{7.78}$$

where h' is the second fundamental form of N in $S^{n+1}(1)$. Now, (b) follows from (a), (7.78) and Proposition 2.4. □

Corollary 2.8 and Theorem 7.2 imply easily the following.

Theorem 7.16. *Let N be a hypersurface of $S^{n+1}(1) \subset \mathbb{E}^{n+2}$. If N has nonzero constant mean curvature and constant scalar curvature, then N is of 1-type or of 2-type. Moreover, when N is compact, N is mass-symmetric unless N is totally umbilical.*

Proof. Under the hypothesis, we find $DH = 0$. Thus, Corollary 2.8 gives

$$\Delta H = \alpha' ||h||^2 \xi - n||H||^2 \mathbf{x} \tag{7.79}$$

where α' is the mean curvature of N in $S^{n+1}(1)$. Because $H = H' - \mathbf{x}$, (7.79) implies that

$$\Delta H - ||h||^2 H + (n||H||^2 - ||h||^2)\mathbf{x} = 0, \tag{7.80}$$

where $||h||$ and $||H||$ are constant. Hence, we conclude from Theorem 7.2 that N is of 1-type or of 2-type.

If we further assume that N is compact, then it follows from (7.80), Hopf's lemma and Beltrami's formula that $(n||H||^2 - ||h||^2) \int_N \mathbf{x} * 1 = 0$. Since N is of 2-type, it cannot be a totally umbilical hypersurface of $S^{n+1}(1)$. Thus, $n||H||^2 \neq ||h||^2$ holds. Hence, N is mass-symmetric. □

Definition 7.5. A hypersurface of an $(n+1)$-sphere S^{n+1} is called an *isoparametric hypersurface* if it has constant principal curvatures.

Corollary 7.7. *Every isoparametric hypersurface of $S^{n+1}(1) \subset \mathbb{E}^{n+2}$ is either of 1-type or of 2-type.*

Conversely, we have the following result from [Hasanis-Vlachos (1991)].

Theorem 7.17. *Let N be a 2-type hypersurface of $S^{n+1}(1) \subset \mathbb{E}^{n+2}$. Then N has nonzero constant mean curvature and constant scalar curvature.*

For spherical 2-type hypersurfaces we also have the following [Chen (1991c)].

Theorem 7.18. *Let N be a hypersurface of $S^{n+1}(1) \subset \mathbb{E}^{n+2}$ with at most two distinct principal curvatures. Then N is of 2-type if and only if N is an open portion of the direct product of two spheres $S^j \times S^{n-j}$ such that the radii r_1 and r_2 of S^j and S^{n-j}, respectively, satisfying $r_1^2 + r_2^2 = 1$ and $(r_1, r_2) \neq (\frac{\sqrt{j}}{\sqrt{n}}, \frac{\sqrt{n-j}}{\sqrt{n}})$.*

Theorem 7.19. *Let N be a hypersurface of $S^{n+1}(1) \subset \mathbb{E}^{n+2}$ with $n \geq 4$. Then N is conformally flat and of 2-type if and only if N is an open portion of $S^1 \times S^{n-1}$ such that the radii r_1 and r_2 of S^1 and S^{n-1}, respectively, satisfying $r_1^2 + r_2^2 = 1$ and $(r_1, r_2) \neq (\frac{1}{\sqrt{n}}, \frac{\sqrt{n-1}}{\sqrt{n}})$.*

An application of Theorems 7.16 and 7.17 is the following.

Corollary 7.8. *Every compact 2-type hypersurface of $S^{n+1}(1) \subset \mathbb{E}^{n+2}$ is mass-symmetric.*

When $n = 2$, we have [Chen (1979c); Barros and Garay (1987)].

Theorem 7.20. *Let N be a surface of $S^3(1) \subset \mathbb{E}^4$. Then N is of 2-type if and only if N is open part of the product of two plane circles of different radii, i.e., $N \subset S^1(a) \times S^1(b)$ with $a \neq b$ and $a^2 + b^2 = 1$.*

Proof. Follows from Theorem 7.17 and that the only surface in $S^3(1)$ with constant scalar curvature and nonzero mean curvature are open part of the product of two plane circles of different radii (cf. Remark 20.5). □

Let x, y, z be the natural coordinates of $\mathbb{E}^3$ and $u_1, \ldots, u_5$ that of $\mathbb{E}^5$. The mapping defined by

$$u_1 = \frac{yz}{\sqrt{3}}, \; u_2 = \frac{xz}{\sqrt{3}}, \; u_3 = \frac{xy}{\sqrt{3}}, \; u_4 = \frac{x^2 - y^2}{2\sqrt{3}},$$
$$u_5 = \frac{1}{6}(x^2 + y^2 - 2z^2)$$

gives rise to an isometric immersion of $S^2(\frac{1}{3})$ of curvature $\frac{1}{3}$ into $S^4(1)$. Two points (x, y, z) and $(-x, -y, -z)$ of $S^2(\frac{1}{3})$ are mapped into the same point. Thus, this mapping defines an embedding of the real projective plane $RP^2(\frac{1}{3})$ into $S^4(1)$, known as the *Veronese surface.* The Veronese immersion is the second standard immersion of the 2-sphere $S^2(\frac{1}{3})$.

Definition 7.6. Let H^ℓ be a topologically embedded ℓ-dimensional submanifold in a Riemannian m-manifold M^m ($\ell < m-1$). Denote by $\nu_1(H^\ell)$

the unit normal subbundle of the normal bundle $T^{\perp}H^{\ell}$ of H^{ℓ} in M^m. Then, for a sufficiently small $r > 0$, the mapping

$$\psi : \nu_1(H^{\ell}) \to M^m : (p, e) \mapsto \exp_p(re)$$

is an immersion which is called the *tube* (or *tubular hypersurface*) with radius r around H^{ℓ}. We denote it by $T^r(H^{\ell})$.

For 2-type compact hypersurfaces in $S^4(1)$, we have [Chen (1993b)].

Theorem 7.21. *A compact hypersurface N of $S^4(1) \subset \mathbb{E}^5$ is of 2-type if and only if it is congruent to one of the following two hypersurfaces:*

(a) *a standard embedding $S^1 \times S^2 \subset S^4(1) \subset \mathbb{E}^5$ such that the radii r_1 of S^1 and r_2 of S^2 satisfying $r_1^2 + r_2^2 = 1$ and $(r_1, r_2) \neq (\frac{1}{\sqrt{3}}, \frac{\sqrt{2}}{\sqrt{3}})$;*
(b) *a tube $T^r(V^2)$ with radius $r \neq \frac{\pi}{2}$ over the Veronese surface V^2 in $S^4(1)$.*

Proof. If N is a hypersurface of $S^4(1)$ given by (a) or (b), then N is an isoparametric hypersurface with constant scalar curvature and nonzero constant mean curvature. Thus, by Theorem 7.16, N is either of 1-type or of 2-type. Clearly, if N is given by (a), then it is of 2-type. Next, assume that N is given by (b), then it follows from a direct computation that, at the point $\psi(p, e) = \exp_p(re)$ corresponding to a point $(p, e) \in T^r(V^2)$, the principal curvatures of the tube in $S^4(1)$ are given by

$$\frac{\cot r - \sqrt{3}}{1 + \sqrt{3}\cot r}, \quad \frac{\cot r + \sqrt{3}}{1 - \sqrt{3}\cot r}, \quad \cot r. \tag{7.81}$$

It is direct to verify from (7.81) that the tube is of 1-type if and only if $r = \frac{\pi}{2}$. But this case was excluded.

Conversely, if N is a 2-type compact hypersurface of $S^4(1) \subset \mathbb{E}^5$, then it follows from Theorem 7.17 that N has nonzero constant mean curvature and constant scalar curvature. Thus, a result of [Chang (1993)] implies that N is an isoparametric hypersurface of $S^4(1)$. Since N is assumed to be of 2-type, the number q of distinct principal curvatures must 2 or 3.

If $q = 2$, then Theorem 1 of [Chen (1991c)] implies that N is the direct product of a circle and a 2-sphere as described in case (a).

If $q = 3$, N lies in the family of parallel hypersurfaces of Cartan's hypersurface. It is known that the Cartan hypersurface is the only minimal one in this family of 3-dimensional isoparametric hypersurfaces. It is also know that such hypersurfaces can be described as tubes of constant radius r about the minimal Veronese embedding of $RP^2(\frac{1}{3})$ into $S^4(1)$ (cf. Remark 20.5). Therefore, if we denote such a tubular hypersurface by $T^r(RP^2)$, then the three principal curvatures are given by (7.81). As we already pointing out above, the tube $T^r(RP^2)$ is of 1-type if and only if $r = \frac{\pi}{2}$. Hence, for $q = 3$, we have case (b), since N is assumed to be of 2-type. $\square$

7.12 2-type hypersurfaces in hyperbolic spaces

Lemma 7.4. *Every spacelike null k-type submanifold of a pseudo-Euclidean space* $\mathbb{E}^m_s$ *is non-compact.*

Assume that N is a hypersurface of the hyperbolic space $H^{n+1}(-1)$, imbedded standardly in the Minkowski spacetime $\mathbb{E}^{n+2}_1$, i.e.,

$$H^{n+1}(-1) = \{\mathbf{x} = (x_1, \ldots, x_{n+2}) \in \mathbb{E}^{n+2}_1 : \langle \mathbf{x}, \mathbf{x} \rangle = -1,\ x_1 > 0\}. \tag{7.82}$$

Then mean curvature vector H of N in $\mathbb{E}^{n+2}_1$ is given by

$$H = \beta\xi + \mathbf{x}, \tag{7.83}$$

where ξ is a unit normal vector field and β is the mean curvature of N in $H^{n+1}(-1)$. Thus, by applying Proposition 2.7, we find

$$\begin{aligned} \Delta H = &(\Delta\beta + \beta||A_\xi||^2 - n\beta)\xi + (n\beta^2 - n)\mathbf{x} \\ &+ \frac{n}{2}\nabla\beta^2 + 2A_\xi(\nabla\beta). \end{aligned} \tag{7.84}$$

If N is of 2-type, we have the spectral decomposition:

$$\mathbf{x} = \mathbf{c} + \mathbf{x}_p + \mathbf{x}_q,\ \ \Delta\mathbf{x}_p = \lambda_p\mathbf{x}_p,\ \ \Delta\mathbf{x}_q = \lambda_q\mathbf{x}_q,\ \ \lambda_p < \lambda_q, \tag{7.85}$$

for some vector $\mathbf{c} \in \mathbb{E}^{n+2}_1$. Thus, by applying Beltrami's formula, we find

$$n\Delta H = n(\lambda_p + \lambda_q)H + \lambda_p\lambda_q(\mathbf{x} - \mathbf{c}). \tag{7.86}$$

Because ξ and $\mathbf{x}$ form an orthonormal normal frame of the normal bundle of N in $\mathbb{E}^{n+2}_1$, we have

$$\mathbf{c} = \mathbf{c}_T + \langle \mathbf{c}, \xi\rangle\,\xi - \langle \mathbf{c}, \mathbf{x}\rangle\,\mathbf{x}, \tag{7.87}$$

where $\mathbf{c}_T$ denotes the tangential component of $\mathbf{c}$. Thus, we have

$$\begin{aligned} n\Delta H = &\{n(\lambda_p + \lambda_q)\beta - \lambda_p\lambda_q\langle \mathbf{c}, \xi\rangle\}\xi - \lambda_p\lambda_q\mathbf{c}_T \\ &+ \{n(\lambda_p + \lambda_q) + \lambda_p\lambda_q(1 + \langle \mathbf{c}, \mathbf{x}\rangle)\}\mathbf{x}. \end{aligned} \tag{7.88}$$

From (7.84) and (7.88) we find

$$4nA_\xi(\nabla\beta) + n^2\nabla\beta^2 = -2\lambda_p\lambda_q\mathbf{c}_T, \tag{7.89}$$

$$n^2\beta^2 - n^2 = n(\lambda_p + \lambda_q) + \lambda_p\lambda_q(1 + \langle \mathbf{c}, \mathbf{x}\rangle), \tag{7.90}$$

$$n\Delta\beta - n^2\beta + n\beta||A_\xi||^2 = n(\lambda_p + \lambda_q)\beta - \lambda_p\lambda_q\langle \mathbf{c}, \xi\rangle. \tag{7.91}$$

By taking covariant derivative of (7.87), we obtain

$$\nabla \langle \mathbf{c}, \mathbf{x} \rangle = \mathbf{c}_T, \quad \nabla \langle \mathbf{c}, \xi \rangle = -A_\xi(\mathbf{c}_T), \tag{7.92}$$

$$\nabla_X \mathbf{c}_T = \langle \mathbf{c}, \xi \rangle A_\xi X + \langle \mathbf{c}, \mathbf{x} \rangle X, \tag{7.93}$$

for $X \in TN$. From (7.89), (7.90), and (7.92), we derive that

$$2n^2 \nabla \beta^2 = \lambda_p \lambda_q \mathbf{c}_T, \tag{7.94}$$

$$4A_\xi(\nabla \beta) = -3n \nabla \beta^2. \tag{7.95}$$

It follows from (7.92), (7.94) and (7.95) that

$$\lambda_p \lambda_q \langle \mathbf{c}, \xi \rangle = (n\beta)^3 + k \tag{7.96}$$

for a constant k. By combining (7.87), (7.90), (7.94) and (7.96), we find

$$\lambda_p^2 \lambda_q^2 ||\mathbf{c}||^2 = 4n^4 \beta^2 ||\nabla \beta||^2 + \{(n\beta)^3 + k\}^2 - \{(n\beta)^2 + b\}^2, \tag{7.97}$$

where $b = -n^2 - n(\lambda_p + \lambda_q) - \lambda_p \lambda_q$ which is a constant.

The following result was obtained in [Chen (1992)].

Theorem 7.22. *Every 2-type hypersurface of the hyperbolic $(n+1)$-space $H^{n+1}(-1) \subset \mathbb{E}_1^{n+2}$ has nonzero constant mean curvature and constant scalar curvature.*

Proof. Let N be a 2-type hypersurface of the hyperbolic $(n+1)$-space $H^{n+1}(-1) \subset \mathbb{E}_1^{n+2}$. Put $U = \{p \in N : \nabla \beta^2(p) \neq 0\}$. Assume that U is not empty. Let V be a connected component of U. Then V is non-empty open submanifold of N. Now, let us study only on open submanifold V unless mentioned otherwise. On V, we choose an orthonormal frame $\{e_1, \ldots, e_n\}$ such that e_1 is the unit vector field in the direction of $\nabla \beta$. It follows from (7.95) that e_1 is an eigenvector of A_ξ with eigenvalue $\kappa_1 = -\frac{3n}{2}\beta$. For simplicity we choose $e_2, \ldots, e_n$ to be eigenvectors of A_ξ with eigenvalues given by $\kappa_2, \ldots, \kappa_n$, respectively. Then we have

$$\kappa_1 = -\frac{3n}{2}\beta, \quad \kappa_2 + \cdots + \kappa_n = \frac{5n}{2}\beta. \tag{7.98}$$

By taking the covariant derivative of (7.94) with respect to a tangent vector $X \in TN$ and by applying (7.93), we find

$$2n^2\{(X\beta)\nabla\beta + \beta \nabla_X(\nabla \beta)\} = \lambda_p \lambda_q (\langle \mathbf{c}, \xi \rangle A_\xi X + \langle \mathbf{c}, \mathbf{x} \rangle X). \tag{7.99}$$

In particular, if we choose $X = e_1$, then (7.99) implies that $\nabla_{e_1} e_1 = 0$, which shows that the integral curves of e_1 are geodesics of N. Hence, the connection forms with respect to $\{e_1, \ldots, e_n\}$ satisfy

$$\omega_1^j(e_1) = 0, \quad j = 1, \ldots, n. \tag{7.100}$$

Furthermore, since e_1 is in the direction of $\nabla\beta$, we also have

$$d\beta = (e_1\beta)\omega^1. \tag{7.101}$$

If we choose $X = e_i, i = 2, \ldots, n$, then (7.99) implies that

$$2n^2\beta\,||\nabla\beta||\,\omega_1^i(e_i) = \lambda_p\lambda_q(\kappa_i\,\langle \mathbf{c}, \xi\rangle + \langle \mathbf{c}, \mathbf{x}\rangle). \tag{7.102}$$

On the other hand, from Codazzi's equation, we find

$$e_1\kappa_i = (\kappa_1 - \kappa_i)\omega_1^i(e_i), \quad i > 1. \tag{7.103}$$

By using (7.98), (7.102) and (7.103) we have

$$\begin{aligned}\lambda_p\lambda_q\{2\,\langle \mathbf{c}, \xi\rangle\,||A_\xi||^2 + 3n^2\beta^2\,\langle \mathbf{c}, \xi\rangle + (3n+2)n\beta\,\langle \mathbf{c}, \mathbf{x}\rangle\} \\ = -10n^3\beta||\nabla\beta||^2.\end{aligned} \tag{7.104}$$

Now, by combining (7.90), (7.96) and (7.104), we get

$$\begin{aligned}2(n^3\beta^3 + k)||A_\xi||^2 + 3n^2\beta^2(n^3\beta^3 + k) \\ + (3n+2)n\beta(n^2\beta^2 + b) = -10n^3\beta||\nabla\beta||^2.\end{aligned} \tag{7.105}$$

Differentiating (7.97) with respect to e_1 and using (7.101), we find

$$4n^2||\nabla\beta||^2 + 4n^2\beta e_1e_1\beta + 3n\beta(n^3\beta^3 + k) = 2n^2\beta^2 + 2b. \tag{7.106}$$

From (7.98), (7.100), (7.101), (7.102) and the definition of Δ, we find

$$\Delta\beta = -e_1e_1\beta - \lambda_p\lambda_q\left\{\frac{5}{4n}\,\langle \mathbf{c}, \xi\rangle + \frac{n-1}{2n^2\beta}\,\langle \mathbf{c}, \mathbf{x}\rangle\right\}. \tag{7.107}$$

On the other hand, by using (7.91) and (7.96), we find

$$\Delta\beta = n\beta - \beta||A_\xi||^2 + (\lambda_p + \lambda_q)\beta - n^2\beta^3 - \frac{k}{n}. \tag{7.108}$$

By combining (7.107) and (7.108) and using (7.90) and (7.96), we obtain

$$\begin{aligned}4ne_1e_1\beta + 5(n^3\beta^3 + k) + \frac{2(n-1)}{n\beta}(n^2\beta^2 + b) + 4n^2\beta \\ = 4n^3\beta^3 + 4k + 4n\beta||A_\xi||^2 - 4n\beta(\lambda_p + \lambda_q).\end{aligned} \tag{7.109}$$

From (7.105) and (7.109) we find

$$\begin{aligned}(n^3\beta^3 + k)\{4n^2\beta e_1e_1\beta + 7n^4\beta^4 + (4\lambda_p + 4\lambda_q + 6n - 2)n^2\beta^2 \\ + nk\beta + 2(n-1)b\} + 20n^5\beta^3||\nabla\beta||^2 \\ + 2(3n+2)n^3\beta^3(n^2\beta^2 + b) = 0.\end{aligned} \tag{7.110}$$

Now, by combining (7.103) and (7.110), we obtain

$$\begin{aligned}2n(k - 4n^3\beta^3)||\nabla\beta||^2 - (3n+2)n^2\beta^3(n^2\beta^2 + b) \\ = (n^3\beta^3 + k)\{2n^3\beta^4 + (2\lambda_p + 2\lambda_q + 3n)n\beta^2 - k\beta + b\}.\end{aligned} \tag{7.111}$$

Therefore, we conclude from (7.97) and (7.111) that the mean curvature β is constant on the open submanifold $V \subset N$, which is a contradiction. Consequently, N has constant mean curvature β in $H^{n+1}(-1)$. Since N is of 2-type, the mean curvature β must be nonzero. Since N has nonzero constant mean curvature β in $H^{n+1}(-1)$, it follows from from (7.89) that

$$\lambda_p \lambda_q \mathbf{c}_T = 0. \tag{7.112}$$

Now, we find from (7.92) and (7.112) that $\lambda_p \lambda_q \langle \mathbf{c}, \xi \rangle$ is a constant. Thus, $||A_\xi||$ is a constant by (7.91). Since the scalar curvature τ of N satisfies $\tau = n^2\beta^2 - ||A_\xi||^2 - n(n-1)$, N has constant scalar curvature as well. □

Conversely, we have the following result from [Chen (1992)].

Theorem 7.23. *Let N be a hypersurface of $H^{n+1}(-1) \subset \mathbb{E}_1^{n+2}$. If N has constant mean curvature and constant scalar curvature, then N is either of 1-type or of 2-type.*

Proof. Under the hypothesis, we find $DH = 0$. Thus, by Corollary 2.8,

$$\Delta H = \beta ||h||^2 \xi - n||H||^2 \mathbf{x} \tag{7.113}$$

where β is the mean curvature of N in $H^{n+1}(-1)$. Because $H = \beta\xi + \mathbf{x}$, (7.113) implies

$$\Delta H - ||h||^2 H + (n||H||^2 + ||h||^2)\mathbf{x} = 0,$$

where $||h||$ and $||H||$ are constant. Hence, after applying Theorem 7.2, we conclude that N is either of 1-type or of 2-type. □

Theorem 7.24. *Every 2-type hypersurface in $H^{n+1}(-1) \subset \mathbb{E}_1^{n+2}$ is non-compact.*

Proof. Assume that N is a compact 2-type hypersurface of $H^{n+1}(-1) \subset \mathbb{E}_1^{n+2}$, then N has constant scalar curvature and nonzero constant mean curvature in $H^{n+1}(-1)$ by Theorem 7.22 . Thus, Corollary 2.8 implies

$$\Delta H = c_1 H + c_2 \mathbf{x}, \tag{7.114}$$

where c_1, c_2 are constants satisfying $c_1 = ||A_\xi||^2 - n, \ \ c_2 = n\beta^2 - ||A_\xi||^2$. Since $\Delta\mathbf{x} = -nH$ by Beltrami's formula and N is compact, Hopf's Lemma and (7.114) imply that

$$c_2 \int_M \mathbf{x} * 1 = 0, \tag{7.115}$$

In particular, this gives $c_2 \int_M x_1 * 1 = 0$, where x_1 is the first coordinate function of N in $\mathbb{E}_1^{n+2}$. Since the hypersurface lies in the hyperbolic space $H^{n+1}(-1) \subset \mathbb{E}_1^{n+2}$, x_1 is always positive. Thus, by (7.115), we must have $c_2 = 0$. This together with (7.114), shows that N is of 1-type in $\mathbb{E}_1^{n+2}$ according to Theorem 7.1. But this is an contradiction. □

Example 7.1. For each positive number r and each integer k, $(2 \le k \le n)$, let $N^n_{k,r}$ denote the n-dimensional submanifold of $\mathbb{E}^{n+2}_1$ defined by

$$\Big\{(x_1, x_2, \ldots, x_{n+2}) : x_1^2 - x_2^2 - \cdots - x_k^2 = 1 + r^2,\ x_{k+1}^2 + \ldots + x_{n+2}^2 = r^2\Big\}.$$

Then $N^n_{k,r}$ is a complete non-compact 2-type hypersurface of $H^{n+1}(-1)$. Geometrically, $N^n_{k,r}$ is the direct product of a hyperbolic space and a sphere.

Theorem 7.25. *Let N be a hypersurface of $H^{n+1}(-1) \subset \mathbb{E}^{n+2}_1$. If N has at most two distinct principal curvatures. Then N is of 2-type if and only if N is an open portion of $N^n_{k,r}$ for some positive integer k, $2 \le k \le n$, and some positive number r.*

Proof. Let N be a 2-type hypersurface of $H^{n+1}(-1) \subset \mathbb{E}^{n+2}_1$. Then Theorem 7.22 implies that N has constant mean curvature and constant scalar curvature. Thus, if N has at most two distinct principal curvature, then N is an an open portion of $N^n_{k,r}$ for some positive integer $k, 2 \le k \le n$ according a result of Cartan. (cf. Remark 20.4) □

If $n = 2$, this implies the following.

Corollary 7.9. *Let M be a surface in the hyperbolic 3-space $H^3(-1) \subset \mathbb{E}^4_1$. Then N is of 2-type if and only if N is a flat surface which is an open portion of $N^2_{2,r}$ for some positive number r.*

Remark 7.11. Similarly, for 2-type hypersurfaces in the Lorentzian space forms, we have the following results from [Chen (1992)].

Theorem 7.26. *We have*

(a) *every spacelike 2-type hypersurface of the de Sitter spacetime $S^{n+1}_1 \subset \mathbb{E}^{n+2}_2$ has nonzero constant mean curvature and constant scalar curvature;*
(b) *every spacelike 2-type hypersurface of the anti-de Sitter spacetime $H^{n+1}_1(-1)$ has nonzero constant mean curvature and constant scalar curvature;*
(c) *every spacelike 2-type hypersurface of the Minkowski spacetime $\mathbb{E}^{n+1}_1$ is of null 2-type if it has constant mean curvature;*
(d) *hyperplanes, hyperbolic hypersurfaces and null 2-type hypersurfaces are the only spacelike hypersurfaces in the Minkowski spacetime $\mathbb{E}^{n+1}_1$ which have constant mean curvature and constant length of second fundamental form.*

Chapter 8

Total Mean Curvature

The two most important invariants of a surface in a Euclidean 3-space $\mathbb{E}^3$ are the (intrinsic) Gauss curvature G and the (extrinsic) squared mean curvature $||H||^2$.

According to Gauss' Theorema Egregium, Gauss curvature is an intrinsic invariant and the integral of the Gaussian curvature over a closed surface gives the well-known Gauss-Bonnet formula

$$\int_N G *1 = 2\pi\chi(M), \tag{8.1}$$

where $\chi(N)$ denotes the Euler number of N. For the Gauss curvature of a closed surface N in $\mathbb{E}^3$, there is an inequality due to [Chern and Lashof (1958)]

$$\int_N |G| *1 \geq 4\pi(1+g), \tag{8.2}$$

where g is the genus of N.

The study of the total mean curvature

$$w(f) = \int_N ||H||^2 *1 \tag{8.3}$$

of a surface in $\mathbb{E}^3$ goes back at least to Blaschke's school in the 1920s.[1] Among others, G. Thomsen in [Thomsen (1923)] studied the first variations of (8.3) and proved that the Euler-Lagrange equation[2] of $w(f)$ is

$$\Delta H + 2H(H^2 - K) = 0. \tag{8.4}$$

[1]R. Kusner pointed out that as early as 1810, Marie-Sophie Germain (1776-1831), a French mathematician, physicist, and philosopher, already proposed $\int H^2 *1$ as the "virtual work" in her study of vibrating curved plates; whether she carried out her program is unclear.

[2]Thomsen attributes the Euler-Lagrange equation (8.4) to W. Schadow.

W. Blaschke proved in his book [Blaschke (1929)] that $\int_N H^2 *1$ of a closed surface N in $\mathbb{E}^3$ is a conformal invariant, i.e, it keeps the same value under conformal mappings of $\mathbb{E}^3$.

About 40 years after Thomsen's work, Willmore reintroduced in [Willmore (1968)] the problem. He observed that, by combining the well-known inequality $H^2 \geq G$, Gauss-Bonnet's formula and Chern-Lashof's inequality, one obtains

$$w(f) = \int_N H^2 *1 \geq 4\pi, \tag{8.5}$$

for a closed surface N in $\mathbb{E}^3$, with equality holding if and only if N is a round sphere.

8.1 Total mean curvature of tori in $\mathbb{E}^3$

For tubes in $\mathbb{E}^3$ one has the following [Shiohama and Tagaki (1970); Willmore (1971)].

Theorem 8.1. *Let N be a torus embedded in $\mathbb{E}^3$ such that the embedded surface is generated by carrying a small circle around a closed curve so that the center moves along the curve and plane of the circle is in the normal plane to the curve at each point. Then*

$$\int_N ||H||^2 *1 \geq 2\pi^2. \tag{8.6}$$

The equality sign holds if and only if the embedded surface is congruent to the anchor ring defined by

$$a\left((\sqrt{2}+\cos u)\cos v, (\sqrt{2}+\cos u)\sin v, \sin u\right),$$

where a is a positive number.

Proof. Let $\gamma(s)$ be the closed curve of length ℓ described in the theorem. Denote by κ and τ the curvature and torsion of γ. Then the position vector of N is given by

$$\mathbf{x}(s,v) = \gamma(s) + r(\cos v)\mathbf{N} + r(\sin v)\mathbf{B},$$

where $\mathbf{N}, \mathbf{B}$ are the principal and binormal vector fields of γ. A direct computation shows that the two principal curvatures of N are given by

$$\kappa_1 = \frac{1}{r}, \quad \kappa_2 = \frac{\kappa \cos v}{\kappa r \cos v - 1}.$$

Thus the squared mean curvature is given by

$$||H||^2 = \left(\frac{1 - 2\kappa r \cos v}{2r(1 - \kappa r \cos v)}\right)^2.$$

Hence

$$\begin{aligned}\int_M ||H||^2 * 1 &= \int_0^\ell \int_0^{2\pi} \left(\frac{1 - 2\kappa r \cos v}{2r(1 - \kappa r \cos v)}\right)^2 *1 \\ &= \frac{\pi}{2r}\int_0^\ell \frac{ds}{\sqrt{1 - \kappa^2 r^2}} \\ &\geq \int_0^\ell |\kappa| ds \geq 2\pi^2,\end{aligned} \tag{8.7}$$

where we use the fact that, for any real variable x, $x\sqrt{1 - x^2}$ takes its maximum value $\frac{1}{2}$ at $x = \frac{1}{\sqrt{2}}$; and also we apply a well-known *inequality of Fenchel-Borsuk* which states that, for a closed curve γ in a Euclidean m-space, the total absolute curvature satisfies

$$\int_\gamma |\kappa| ds \geq 2\pi, \tag{8.8}$$

with equality holding if and only if γ is a convex planar curve.

If the equality sign of (8.7) holds, then the inequality in (8.7) is an equality. Thus, after applying Fenchel-Borsuk's theorem, we conclude that γ is a convex planar curve. In this case, we have $\kappa = \frac{1}{\sqrt{2}r}$. Hence γ is a circle of radius $\sqrt{2}r$.

The converse is easy to verify. □

Willmore's Conjecture: Willmore conjectured that inequality (8.6) holds for all tori in $\mathbb{E}^3$.

Remark 8.1. It was proved in [Hertrich-Jeromin and Pinkall (1992)] that inequality (8.6) is also true if the closed circular tube N in Theorem 8.1 were replaced by a closed tube in $\mathbb{E}^3$ generated by a small ellipse. (A special case of this was proved by I. van de Woestijne and L. Verstraelen in 1990 (cf. [Verstraelen (1990)]).

Remark 8.2. So far, only partial answers on Willmore's conjecture have been obtained by many geometers, but the conjecture remains open. (See, for instance, pages 367–372 of the survey article [Chen (2000c)] for more details.)

8.2 Total mean curvature and conformal invariants

Let (M, g) be a Riemannian m-manifold and ρ be a positive function on M. We put

$$\tilde{g} = \rho^2 g. \tag{8.9}$$

Then $\tilde{g}$ is called a *conformal change of the metric* g. Denote by $\tilde{\nabla}$ and ∇ the Levi-Civita connections of $\tilde{g}$ and g, respectively. Then $\tilde{\nabla}$ and ∇ are related by

$$\tilde{\nabla}_{\tilde{X}}\tilde{Y} - \nabla_{\tilde{X}}\tilde{Y} = (\tilde{X}\ln\rho)\tilde{Y} + (\tilde{Y}\ln\rho)\tilde{X} - g(\tilde{X}, \tilde{Y})U, \tag{8.10}$$

where $U = (d\rho)_{\#}$ is the vector field associated with $d\rho$.

Let N be an n-dimensional submanifold of M. Denote by g_N and $\tilde{g}_N$ the metrics on N induced from g and $\tilde{g}$, respectively. Then, for each normal vector field ξ of N in M, we have

$$\tilde{\nabla}_X\xi - \nabla_X\xi = (X\ln\rho)\xi - (\xi\ln\rho)X, \quad X \in TN. \tag{8.11}$$

Thus, after applying Weingarten's formula, we find

$$\tilde{D}_X\xi - D_X\xi = (X\ln\rho)\xi, \tag{8.12}$$

where $\tilde{D}$ and D are the normal connections of N with respect to $\tilde{\nabla}$ and ∇, respectively. Hence

$$\tilde{D}_X - D_X = (X\ln\rho)I. \tag{8.13}$$

Therefore, the normal curvature tensors $R^{\tilde{D}}$ and R^D are related by

$$\begin{aligned} R^{\tilde{D}}(X,Y) = R^D(X,Y) &+ D_X((Y\ln\rho)I) + (X\ln\rho)D_Y \\ &- D_Y((X\ln\rho)I) - (Y\ln\rho)D_X - ([X,Y]\ln\rho)I. \end{aligned} \tag{8.14}$$

Consequently, by the definition of Lie bracket, we obtain

$$R^{\tilde{D}}(X,Y) = R^D(X,Y), \quad \forall X, Y \in \mathcal{X}(N). \tag{8.15}$$

This implies the following [Chen (1974a)].

Proposition 8.1. *Let N be a submanifold of a Riemannian manifold M. Then the normal curvature tensor R^D of N is a conformal invariant.*

Let h, A and $\tilde{h}, \tilde{A}$ be the second fundamental form and shape operator of N in (M, g) and $(M, \tilde{g})$, respectively. Then it follows from (8.10) that

$$g(\tilde{A}_\xi X, Y) = g(A_\xi X, Y) + g(X,Y)g(U,\xi). \tag{8.16}$$

Let $e_1, \ldots, e_n$ be the eigenvectors of A_ξ with respect to the metric on N induced from g. Then

$$\rho^{-1}e_1, \ldots, \rho^{-1}e_n \tag{8.17}$$

form an orthonormal frame of N with respect to the metric induced from $\tilde{g}$; and they are eigenvectors of $\tilde{A}_\xi$. Denote by $\kappa_1(\xi), \ldots, \kappa_n(\xi)$ the eigenvalues of A_ξ and by $\tilde{\kappa}_1(\xi), \ldots, \tilde{\kappa}_n(\xi)$ that of $\tilde{A}_\xi$. Then (8.16) implies that

$$\tilde{\kappa}_i(\xi) = \kappa_i(\xi) + \lambda_\xi, \quad \lambda_\xi = g(U, \xi). \tag{8.18}$$

Since $\tilde{A}_\xi = \rho \tilde{A}_{\tilde{\xi}}$ and $\tilde{\xi} = \rho^{-1}\xi$ is a unit normal vector with respect to $\tilde{g}$, we find from (8.18) that

$$\rho(\tilde{\kappa}_i(\tilde{\xi}) - \tilde{\kappa}_j(\tilde{\xi})) = \kappa_i(\xi) - \kappa_j(\xi). \tag{8.19}$$

Now, let $\xi_{n+1}, \ldots, \xi_m$ be an orthonormal normal frame of N with respect to g. Then the mean curvature vector H of N in (M, g) is given by

$$H = \frac{1}{n} \sum_r \sum_i \kappa_i(\xi_r)\xi_r. \tag{8.20}$$

Let us put

$$\tau_{ext} = \frac{2}{n(n-1)} \sum_r \sum_{i<j} \kappa_i(\xi_r)\kappa_j(\xi_r). \tag{8.21}$$

Then τ_{ext} is well-defined. We call τ_{ext} the *extrinsic scalar curvature* with respect to g. In particular, if N is of constant curvature c, then Gauss' equation implies that

$$\tau_{ext} = \tau - c, \tag{8.22}$$

where τ is the scalar curvature.

By using (8.19)-(8.21), we find the following [Chen (1974a)].

Proposition 8.2. *If N is a submanifold of a Riemannian manifold, then*

$$(||H||^2 - \tau_{ext})g \tag{8.23}$$

is invariant under any conformal change of metric.

When N is compact, this implies immediately the following.

Proposition 8.3. *If N is an n-dimensional compact submanifold of a Riemannian manifold, then*

$$\int_N (||H||^2 - \tau_{ext})^{\frac{n}{2}} *1 \tag{8.24}$$

is a conformal invariant.

When N is 2-dimensional, Proposition 8.3 becomes the following.

Proposition 8.4. *If N is a closed surface in a Riemannian m-manifold M, then*

$$\int_N (||H||^2 + \tilde{K}(N)) *1 \tag{8.25}$$

is a conformal invariant, where $\tilde{K}(N)$ denotes the sectional curvature of M restricted to the tangent planes of N.

If M is $\mathbb{E}^m$, Proposition 8.4 implies the following [Chen (1973c)].

Corollary 8.1. *Let N be a closed surface in $\mathbb{E}^m$ and ψ is a diffeomorphism of $\mathbb{E}^m$ which induces a conformal change of metric on $\mathbb{E}^m$. Then*

$$\int_N ||H||^2 dS_N = \int_{\psi(N)} ||H_\psi||^2 dS_{\psi(N)}, \tag{8.26}$$

where dS_N is the area element of N.

When $m = 3$, this corollary is due to [Blaschke (1929)].

Remark 8.3. Since the functional $w(f) = \int_N ||H||^2 dA$ is compatible with conformal geometry, the problem of finding the infimum of $\int_N ||H||^2 dA$ in $\mathbb{E}^3$ can be stated as a problem in the unit 3-sphere S^3. In fact, if N is viewed, via the stereographic projection, as a surface in S^3, then the functional above transforms into

$$\int_N (1 + ||H_S||^2) *1, \tag{8.27}$$

where H_S is the mean curvature vector of N in S^3. The above relation shows that the area of a minimal surface $N \subset S^3$ equals the total mean curvature $\int_{\sigma(N)} H^2 dS_{\sigma(N)}$ of the stereographic projection of N in $\mathbb{E}^3$.

Many closed minimal surfaces in S^3 were discovered in [Lawson (1970a)]. Lawson proved that every closed surface, but the real projective plane, can be minimally immersed into S^3. Every closed orientable surface can be embedded minimally in S^3. All Lawson's minimal surfaces are constructed by starting with classical solutions to the Plateau problem for specific piecewise geodesic curves in S^3 and extending these solutions to complete minimal surfaces of the desired type using a reflection principle.

A *Hopf torus* of S^3 is the inverse image of a closed spherical curve in S^2 via the Hopf map $\pi : S^3 \to S^2$. Pinkall used Hopf tori in S^3 to show that there exists surfaces in $\mathbb{E}^3$ satisfying (8.4) which cannot be obtained by stereographic projections of minimal surfaces in S^3 [Pinkall (1985)].

Definition 8.1. A surface N in $\mathbb{E}^m$ is said to be *conformally equivalent* to another surface $\bar{N}$ in $\mathbb{E}^m$ if N can be obtained from $\bar{N}$ via a conformal mapping of $\mathbb{E}^m$.

For surfaces in a Euclidean space we have the following [Chen (1973a, 1981e)].

Theorem 8.2. *Let N be a compact surface in $\mathbb{E}^m$. If N is conformally equivalent to a flat surface in $\mathbb{E}^m$, then*

$$\int_N ||H||^2 *1 \geq 2\pi^2. \tag{8.28}$$

The equality sign holds if and only if N is a conformal Clifford torus, i.e., N is conformally equivalent to a square torus $S^1(r) \times S^1(r) \subset \mathbb{E}^4 \subset \mathbb{E}^m$.

8.3 Total mean curvature for arbitrary submanifolds

According to Nash's embedding theorem, every compact Riemannian n-manifold can be isometrically embedded in $\mathbb{E}^n$ with $n = \frac{1}{2}n(3n+11)$. On the other hand, most compact Riemannian n-manifold cannot be isometrically immersed in $\mathbb{E}^{n+1}$ as a hypersurface. Thus, the theory of submanifolds with arbitrary codimension is far richer than the theory of hypersurfaces; in particular, much richer than surfaces in $\mathbb{E}^3$.

For the total mean curvature of a compact submanifold in an arbitrary Euclidean space, we have the following [Chen (1971)].

Theorem 8.3. *Let N be an n-dimensional compact submanifold of a Euclidean m-space $\mathbb{E}^m$. Then*

$$\int_N ||H||^n *1 \geq c_n, \tag{8.29}$$

where c_n is the volume of unit n-sphere $S^n(1)$.

The equality sign of (8.29) *holds if and only if N is embedded as an ordinary n-sphere in a totally geodesic $(n+1)$-subspace $\mathbb{E}^{n+1}$ when $n > 1$; and as a convex planar curve when $n = 1$.*

Proof. Let $\mathbf{x} : N \to \mathbb{E}^m$ be an isometric immersion of a compact Riemannian n-manifold into $\mathbb{E}^m$. Let B denote the bundle space consisting of all frame $\{p, \mathbf{x}(p), e_1, \ldots, e_n, e_{n+1}, \ldots, e_m\}$ such that $e_1, \ldots, e_n$ are orthonormal vectors in T_pN, $p \in N$, and $e_{n+1}, \ldots, e_m$ are orthonormal vectors in

$T_p^{\perp}N$. Let us choose the frame $\{p, \mathbf{x}(p), e_1, \ldots, e_n, \bar{e}_{n+1}, \ldots, \bar{e}_m\}$ in B such that $\bar{e}_m$ is parallel to the mean curvature vector H of N at p. Then

$$||H|| = \frac{1}{n}(\bar{h}_{11}^m + \cdots + \bar{h}_{nn}^m), \tag{8.30}$$

$$\sum_{i=1}^{n} \bar{h}_{ii}^r = 0, \quad r = n+1, \ldots, m-1, \tag{8.31}$$

where $\bar{h}_{ij}^s = \langle h(e_i, e_j), \bar{e}_s \rangle$.

On the other hand, for each (p, e_m) in the unit normal bundle B_1, we may put

$$e_m = \sum_{r=n+1}^{m} \cos\theta_r \bar{e}_r, \tag{8.32}$$

where θ_r is the angle between e_m and $\bar{e}_r$. For each $(p, e) \in B_1$, we put

$$m(p, e) = \frac{1}{n}\text{trace}\,(A_e). \tag{8.33}$$

From (8.30)-(8.33) we find

$$m(p, e) = \sum_{r=n+1}^{m} (\cos\theta_r) m(p, \bar{e}_r) = ||H(p)|| \cos\theta_m. \tag{8.34}$$

Hence

$$\begin{aligned} \int_{B_1} |m(p, e_m)|^n \, dV \wedge d\sigma &= \int_{B_1} ||H||^n \, |\cos^n \theta_m| \, dV \wedge d\sigma \\ &= \frac{2c_{m-1}}{c_n} \int_N ||H||^n *1, \end{aligned} \tag{8.35}$$

where dV is the volume element of N and $d\sigma$ is the volume element of fibers.

Let e be a unit vector in the unit sphere $S^{m-1}(1)$. Consider the height function $h_e = \langle e, \mathbf{x}(p) \rangle$ on N. Then $Xh_e = \langle e, X \rangle$ for $X \in \mathcal{X}(N)$. Thus

$$XYh_e = \langle e, \nabla_X Y + h(X, Y) \rangle \tag{8.36}$$

for $X, Y \in \mathcal{X}(N)$. Since h_e is continuous on N, h_e has at least one maximum and one minimum, say at q and q', respectively. Since e is normal to N at q and q', we obtain from (8.36) that

$$XYh_e = \langle A_e X, Y \rangle \tag{8.37}$$

at q, q'. This implies that the shape operator A is either non-negative or non-positive definite at (q, e) and at (q', e).

Let U denote the set consisting of all elements in B_1 such that the eigenvalues of $\kappa_1(p,e),\dots,\kappa_n(p,e)$ of A_e have the same sign. Then $S^{m-1}(1)$ is covered by U at least twice under the map $\nu : B_1 \to S^{m-1}(1)$ defined by $\nu(p,e) = e$. This shows that

$$\int_U \nu^* d\Sigma \geq 2c_{m-1}, \tag{8.38}$$

where $d\Sigma$ is the volume element of B_1.

Since $\kappa_1(p,e),\dots,\kappa_n(p,e)$ have the same sign on U, we have

$$\begin{aligned} |m(p,e)|^n &= |\tfrac{1}{n}(\kappa_1(p,e)+\cdots+\kappa_n(p,e))|^n \\ &\geq |\kappa_1(p,e)\cdots\kappa_n(p,e)| = |K(p,e)|, \end{aligned} \tag{8.39}$$

where $K(p,e) = \kappa_1(p,e)\cdots\kappa_n(p,e)$ is called the *Liptschitz-Killing curvature* at (p,e). Hence, by using (8.35), (8.38), (8.39) and a result of [Chern and Lashof (1958)], we obtain

$$\int_N ||H||^n *1 = \frac{c_n}{2c_{m-1}}\int_{B_1} |m(p,e)|^n dV \wedge d\sigma \geq \frac{c_n}{2c_{m-1}}\int_U \nu^* d\Sigma \geq c_n.$$

This proves (8.29).

Now, assume that the equality case of (8.29) holds. Consider the map

$$\mathbf{y} : B_1 \to \mathbb{E}^m : (p,e) \mapsto \mathbf{x}(p) + re, \tag{8.40}$$

where r is a sufficiently small positive number such that (8.40) is an immersion. In this way, we may regard B_1 as a hypersurface in $\mathbb{E}^m$. Because

$$\langle e, d\mathbf{y}\rangle = \langle e, d\mathbf{x}\rangle + r\,\langle e, de\rangle = 0,$$

e is a unit normal vector field of B_1 in $\mathbb{E}^m$ at (p,e). Thus, $e_1,\dots,e_{m-1}$ form an orthonormal basis of $T_{(p,e)}B_1$. A direct computation shows that the principal curvatures of B_1 in $\mathbb{E}^m$ at (p,e) are given by

$$\begin{aligned} \bar\kappa_i(p,e) &= \frac{\kappa_i(p,e)}{1 + r\kappa_i(p,e)}, \quad i = 1,\dots,n, \\ \bar\kappa_r(p,e) &= \frac{1}{r}, \quad r = n+1,\dots,m-1. \end{aligned} \tag{8.41}$$

Let $\bar U = \{(p,e) : \kappa_1(p,e) = \cdots = \kappa_n(p,e) \neq 0\}$ and $\bar V = B_1 - \bar U$. Under the hypothesis that the equality sign of (8.29) holds, we may prove that $m(p,e) = 0$ identically on $\bar V$.

Let $\pi : B_1 \to N$ be the projection defined by $\pi(p,e) = p$. Then $U = \pi(\bar U)$ is totally umbilical in $\mathbb{E}^m$. Hence, $m(p,e)$ is a nonzero constant on each connected component. Now, by applying the continuity of $m(p,e)$, we obtain $\pi(\bar U) = N$. Thus, N is totally umbilical in $\mathbb{E}^m$. Consequently, if $n \geq 2$, N is embedded as an ordinary hypersphere in a totally geodesic $(n+1)$-subspace of $\mathbb{E}^m$. If $n = 1$, then N is a convex planar curve according to Fenchel-Borsuk's theorem.

The converse is trivial. □

An immediate consequence of Theorem 8.3 is the following.

Corollary 8.2. *If N is an n-dimensional compact minimal submanifold of a unit m-sphere $S^m(1)$, then*

$$\mathrm{vol}\,(N) \geq c_n,$$

with the equality holding if and only if N is a great n-sphere of $S^m(1)$.

Proof. We may regard $S^m(1)$ as an ordinary hypersphere of $\mathbb{E}^{m+1}$. Then it follows from the minimality of N in $S^m(1)$ and Theorem 8.3 that

$$\mathrm{vol}\,(N) = \int_N ||H||^2 *1 \geq c_n,$$

where H is the mean curvature vector N in $\mathbb{E}^{m+1}$.

The remaining part follows immediately from Theorem 8.3. □

Similarly, we also have the following from Theorem 8.3.

Corollary 8.3. *If N is an n-dimensional compact minimal submanifold of the real projective m-space $RP^m(1)$ of constant curvature 1, then*

$$\mathrm{vol}\,(N) \geq \frac{c_n}{2},$$

with the equality holding if and only if N is a totally geodesic submanifold of RP^m.

Corollary 8.4. *If N is an n-dimensional compact minimal submanifold of the complex projective m-space $CP^m(4)$ of constant holomorphic sectional curvature 4, then*

$$\mathrm{vol}\,(N) \geq \frac{c_{n+1}}{2\pi},$$

with the equality holding if and only if $N = CP^k, n = 2k$, which is embedded as a totally geodesic complex submanifold of $CP^m(4)$.

Corollary 8.5. *If N is an n-dimensional compact minimal submanifold of the quaternionic projective m-space $HP^m(4)$ of constant quaternionic sectional curvature 4, then*

$$\mathrm{vol}\,(N) \geq \frac{c_{n+3}}{2\pi^2},$$

with the equality holding if and only if $N = HP^k, n = 4k$, which is embedded as a totally geodesic quaternionic submanifold of $HP^m(4)$.

Corollary 8.6. *If N is an n-dimensional compact minimal submanifold of the Cayley plane $\mathcal{O}P^2(4)$ of maximal sectional curvature 4, then*

$$\mathrm{vol}\,(N) \geq \frac{c_n}{2^n}.$$

The proofs of Corollaries 8.3-8.6 base on Theorem 8.1 and the Hopf fibration from spheres onto projective spaces (see [Chen (1984)] for details).

8.4 Total mean curvature and order of submanifolds

The following problem was proposed in [Chen (1975)].

Problem 8.1. *Let (N, g) be a compact Riemannian manifold. What are the relationship between the total mean curvature of an isometric immersion $\mathbf{x} : N \to \mathbb{E}^m$ and the Riemannian structure of (N, g)?*

We need the following.

Lemma 8.1. *Let $\mathbf{x} : N \to \mathbb{E}^m$ be an isometric immersion of a compact oriented Riemannian n-manifold N into $\mathbb{E}^m$. Then $\langle d\mathbf{x}, d\mathbf{x}\rangle = n$.*

Proof. Let $e_1, \ldots, e_n$ be an orthonormal basis of T_pN, $p \in N$. Then

$$\langle d\mathbf{x}, d\mathbf{x}\rangle = \sum_{i=1}^{n} \langle d\mathbf{x}(e_i), d\mathbf{x}(e_i)\rangle = \sum_{i=1}^{n} \langle e_i, e_i\rangle = n.$$

□

The following is known as Minkowski-Hsiung's formula.

Proposition 8.5. *Let $\mathbf{x} : N \to \mathbb{E}^m$ be an isometric immersion of a compact oriented Riemannian n-manifold N into $\mathbb{E}^m$. Then*

$$\int_N \{1 + \langle \mathbf{x}, H\rangle\} *1 = 0. \tag{8.42}$$

Proof. From the formula of Beltrami, we have $\Delta\mathbf{x} = -nH$. Thus, by Lemma 8.1 and Proposition 6.1, we find

$$n\int_N \langle \mathbf{x}, H\rangle *1 = -(\mathbf{x}, \Delta\mathbf{x}) = -(d\mathbf{x}, d\mathbf{x}) = -n\int_N *1.$$

□

Next, we give some sharp relations between total mean curvature and the order of submanifolds which provide some solutions to Problem 8.1.

Theorem 8.4. *Let* $\mathbf{x} : N \to \mathbb{E}^m$ *be an isometric immersion of a compact Riemannian n-manifold N into* $\mathbb{E}^m$. *Then*

$$\int_N ||H||^k *1 \geq \left(\frac{\lambda_p}{n}\right)^{k/2} \mathrm{vol}\,(N), \quad k = 2, 3, \ldots, n, \tag{8.43}$$

where p is the lower order of N.

The equality sign of (8.43) *holds for some k if and only if N is of 1-type.*

Proof. Let p and q be the lower and upper order of N in $\mathbb{E}^m$. Then we have the spectral decomposition:

$$\mathbf{x} = \mathbf{x}_0 + \sum_{t=p}^{q} \mathbf{x}_t, \quad \Delta \mathbf{x}_t = \lambda_t \mathbf{x}_t. \tag{8.44}$$

Thus

$$n^2 \int_N ||H||^2 *1 = n^2(H, H) = (\Delta \mathbf{x}, \Delta \mathbf{x}) = \sum_{t=p}^{q} \lambda_t^2 ||\mathbf{x}_t||^2. \tag{8.45}$$

On the other hand, it follows from (8.42) and (8.44) that

$$n \int_N *1 = -n(\mathbf{x}, H) = (\mathbf{x}, \Delta \mathbf{x}) = \sum_{t=p}^{q} \lambda_t ||\mathbf{x}_t||^2. \tag{8.46}$$

Hence, by combining (8.45) and (8.46), we find

$$n^2 \int_N ||H||^2 *1 - n\lambda_p \int_N *1 = \sum_{t=p+1}^{q} \lambda_t(\lambda_t - \lambda_p)||\mathbf{x}_t||^2 \geq 0.$$

Therefore we obtain

$$\int_N ||H||^2 *1 \geq \left(\frac{\lambda_p}{n}\right) \mathrm{vol}(N), \tag{8.47}$$

with the equality holding if and only if N is of 1-type.

Now, by using Hölder's inequality, we find

$$\left(\frac{\lambda_p}{n}\right) \mathrm{vol}(N) \leq \int_N ||H||^2 *1 \leq \left(\int_N ||H||^{2r} *1\right)^{1/r} \left(\int_N *1\right)^{1/s},$$

with $r^{-1} + s^{-1} = 1$, $r, s > 1$. By choosing $r = k/2$, we obtain (8.43). The remaining part is clear. □

Remark 8.4. Theorem 8.4 generalizes a result of [Reilly (1977)] who proved the inequality

$$\int_N ||H||^2 *1 \geq \frac{\lambda_1}{n} \mathrm{vol}(N), \tag{8.48}$$

with the equality holding if and only if N is a minimal submanifold of a hypersphere.

The following result provides a sharp relationship between total mean curvature and the upper order of the submanifold.

Theorem 8.5. *Let* $\mathbf{x}: N \to \mathbb{E}^m$ *be an isometric immersion of a compact Riemannian n-manifold* N *into* $\mathbb{E}^m$*. Then*

$$\int_N ||H||^k *1 \leq \left(\frac{\lambda_q}{n}\right)^{k/2} \operatorname{vol}(N), \quad k = 1, 2, 3, 4, \tag{8.49}$$

where q *is the upper order of* N*.*

The equality sign of (8.49) *holds for some* k *if and only if* N *is of 1-type.*

Proof. From Proposition 2.7 we have

$$\Delta H = \Delta^D H + \sum_{i=1}^{n} h(A_H e_i, e_i) + \frac{n}{2}\nabla\langle H, H\rangle + 2\operatorname{trace} A_{DH}.$$

Thus

$$\begin{aligned}\langle \Delta H, H\rangle &= \langle \Delta^D H, H\rangle + \sum_{i=1}^{n} \langle h(A_H e_i, e_i), H\rangle \\ &= \langle \Delta^D H, H\rangle + ||A_H||^2.\end{aligned} \tag{8.50}$$

Moreover, it follows from (8.44)-(8.46) that

$$n\int_N *1 = \sum_{t=p}^{q} \lambda_t\, ||\mathbf{x}_t||^2, \tag{8.51}$$

$$n^2\int_N ||H||^2 *1 = \sum_{t=p}^{q} \lambda_t^2\, ||\mathbf{x}_t||^2, \tag{8.52}$$

$$n^2\int_N \langle H, \Delta H\rangle *1 = \sum_{t=p}^{q} \lambda_t^3\, ||\mathbf{x}_t||^2. \tag{8.53}$$

If $q = \infty$, (8.49) is trivial. Thus we may assume that $q < \infty$. It we put

$$\begin{aligned}\Lambda =& n^2\int_N \langle H, \Delta H\rangle *1 - n^2(\lambda_p + \lambda_q)\int_N ||H||^2 *1 \\ &+ n\lambda_p\lambda_q\int_N *1,\end{aligned} \tag{8.54}$$

then we find from (8.51)-(8.53) that

$$\Lambda = \sum_{t=p+1}^{q-1} \lambda_t(\lambda_t - \lambda_q)(\lambda_t - \lambda_q)||\mathbf{x}_t||^2 \leq 0, \tag{8.55}$$

with the equality holding if and only if N is of 1-type or 2-type. Combining (8.50), (8.54) and (8.55) gives

$$\begin{aligned} n^2\int_N \langle H, \Delta^D H\rangle *1 + n^2\int_N ||A_H||^2 *1 \\ - n^2(\lambda_p+\lambda_q)\int_N ||H||^2 *1 + n\lambda_p\lambda_q\int_N *1 = 0. \end{aligned} \tag{8.56}$$

Since N is compact, Hopf's lemma implies that

$$\int_N \langle H, \Delta H\rangle *1 = \int_N ||DH||^2 *1. \tag{8.57}$$

Let $\kappa_1, \ldots, \kappa_n$ denote the eigenvalues of A_H. Then it is easy to verify that

$$||A_H||^2 = n||H||^4 + \frac{1}{n}\sum_{i<j}(\kappa_i - \kappa_j)^2. \tag{8.58}$$

By combining (8.56), (8.57) and (8.58), and applying Schwartz's inequality, we find

$$\begin{aligned} 0 &\geq n^2\int_N ||DH||^2 *1 + n^3\int_N ||H||^4 *1 + n\sum_{i<j}\int_N (\kappa_i-\kappa_j)^2 *1 \\ &\quad - n^2(\lambda_p+\lambda_q)\int_N ||H||^2 *1 + n\lambda_p\lambda_q\int_N *1 \\ &\geq n^2\int_N ||DH||^2 *1 + \frac{n^3\left(\int_N ||H||^2 * 1\right)^2}{\int_N *1} + n\lambda_p\lambda_q\int_N *1 \\ &\quad + n\sum_{i<j}\int_N (\kappa_i-\kappa_j)^2 *1 - n^2(\lambda_p+\lambda_q)\int_N ||H||^2 *1. \end{aligned} \tag{8.59}$$

Thus

$$\begin{aligned} 0 &\geq n\,\mathrm{vol}(N)\int_N ||DH||^2 *1 + \mathrm{vol}(N)\sum_{i<j}\int_N (\kappa_i-\kappa_j)^2 *1 \\ &\quad + \left(n\int_N ||H||^2 *1 - \lambda_p\mathrm{vol}(N)\right)\left(n\int_N ||H||^2 *1 - \lambda_q\mathrm{vol}(N)\right). \end{aligned} \tag{8.60}$$

Now, by combining Theorem 8.4 and (8.60), we have

$$\int_N ||H||^2 *1 \leq \left(\frac{\lambda_q}{n}\right)\mathrm{vol}\,(N). \tag{8.61}$$

Substituting (8.61) into the first inequality of (8.59), we find (8.49) for $k = 4$. Thus, by using Hölder's inequality, we derive that

$$\int_N ||H||^k *1 \leq \left(\int_N ||H||^4 *1\right)^{k/4} (\mathrm{vol}(N))^{1-k/4}$$

for $k < 4$. Hence, by applying inequality (8.49) with $k = 4$, we get (8.49).

If the equality sign of (8.49) holds for some $k \in \{1, 2, 3, 4\}$, then all of the inequalities in (8.55)-(8.61) become equalities. Therefore, we conclude that N is pseudo-umbilical with parallel mean curvature vector. Hence, N is of 1-type according to Proposition 3.9 and Theorem 7.5.

The converse is easy to verify. □

By combining Theorems 8.4 and 8.5 we have the following.

Theorem 8.6. *Let $\mathbf{x} : N \to \mathbb{E}^m$ be an isometric immersion of a compact Riemannian n-manifold N into $\mathbb{E}^m$. Then*

$$\left(\frac{\lambda_p}{n}\right) \text{vol}(N) \leq \int_N ||H||^2 *1 \leq \left(\frac{\lambda_q}{n}\right) \text{vol}(N), \tag{8.62}$$

where p and q are the lower and the upper orders of N. Either equality sign in (8.62) *holds if and only if N is of 1-type.*

An immediate consequence of Theorem 8.6 is the following.

Corollary 8.7. *Let N be an n-dimensional compact submanifold of a Euclidean m-space. If N has constant mean curvature, then*

$$\lambda_p \leq n\,||H||^2 \leq \lambda_q,$$

with either equality holding if and only if N is of 1-type.

8.5 Conformal property of $\lambda_1 \text{vol}(M)$

A 1-type immersion $\mathbf{x} : N \to \mathbb{E}^m$ with order $\{p\}$ is simply called an *immersion of order p*.

The following *conformal property of* $\lambda_1 \text{vol}(N)$ as an application of Reilly's inequality (8.48) was first proved in [Chen (1979a)].

Theorem 8.7. *Let N be a compact Riemannian surface which admits an order* 1 *isometric embedding into $\mathbb{E}^m$. Then, for any compact surface $\overline{N}$ in $\mathbb{E}^m$ which is conformally equivalent to N, we have*

$$\lambda_1 \text{vol}(N) \geq \bar{\lambda}_1 \text{vol}(\overline{N}), \tag{8.63}$$

with equality sign of (8.63) *holding if and only if $\overline{N}$ is also of order* 1.

Proof. Let $\mathbf{x} : N \to \mathbb{E}^m$ be an order 1 isometric embedding of a compact Riemannian surface. Then, by Reilly's result, we have

$$\int_N ||H||^2 dS_N = \frac{\lambda_1}{2} \text{vol}(N). \tag{8.64}$$

Since the total mean curvature is a conformal invariant, we find

$$\int_{\bar{N}} ||\bar{H}||^2 dS_{\bar{N}} = \int_N ||H||^2 dS_N. \tag{8.65}$$

where the left-hand-side is the corresponding quantity for $\bar{N}$.

On the other hand, we also have

$$\int_{\bar{N}} ||\bar{H}||^2 dS_{\bar{N}} \geq \bar{\lambda}_1 \mathrm{vol}(\bar{N}). \tag{8.66}$$

Thus, after combining (8.64), (8.65) and (8.66), we obtain (8.63).

If the equality sign of (8.63) holds, then the equality sign of (8.66) holds. Hence, $\bar{N}$ is of order 1. The converse is clear. □

Let $T_a^2 = S^1(a) \times S^1(a)$ be a square torus. Then

$$\lambda_1 \mathrm{vol}\,(T_a^2) = 4\pi^2. \tag{8.67}$$

The standard embedding of T_a^2 into $\mathbb{E}^4 \subset \mathbb{E}^m$ is of order 1.

A compact surface M in $\mathbb{E}^m$ is called a *conformal square torus* if it can be obtained from the standard embedding of T_a^2 via some conformal mappings of Euclidean space.

Theorem 8.7 implies immediately the following [Chen (1979a)].

Corollary 8.8. *If N is a conformal square torus in $\mathbb{E}^m$, then*

$$\lambda_1 \mathrm{vol}(N) \leq 4\pi^2, \tag{8.68}$$

with the equality holding if and only if N admits an order 1 isometric embedding.

8.6 Total mean curvature and λ_1, λ_2

Let $\mathbf{x} : N \to \mathbb{E}^m$ be an isometric immersion of a compact Riemannian manifold N into the Euclidean m-space$\mathbb{E}^m$ with center of mass at $\mathbf{x}_0$. The *moment* of N is given by

$$\mathcal{M} = \int_N ||\mathbf{x} - \mathbf{x}_0||^2 * 1. \tag{8.69}$$

The next result provides a sharp relationship between the total mean curvature and the first and second nonzero eigenvalues λ_1, λ_2 of Δ [Chen (1987a)].

Proposition 8.6. *Let* $\mathbf{x} : N \to \mathbb{E}^m$ *be an isometric immersion of a compact Riemannian n-manifold* N *into* $\mathbb{E}^m$. *Then*

$$\int_N ||H||^2 *1 \geq \frac{1}{n}(\lambda_1 + \lambda_2)\mathrm{vol}(N) - \frac{\lambda_1\lambda_2}{n^2}\mathcal{M}, \tag{8.70}$$

where $\mathcal{M}$ *is the moment of* N.

The equality sign of (8.70) *holds if and only if* N *is either of 1-type with order* 1 *or order* 2, *or it is of 2-type with order* $\{1, 2\}$.

Proof. Let $\mathbf{x} : N \to \mathbb{E}^m$ be an isometric immersion of a compact Riemannian n-manifold N into $\mathbb{E}^m$. Then we have the spectral decomposition (8.44) of $\mathbf{x}$. Thus

$$\mathcal{M} = \langle \mathbf{x} - \mathbf{x}_0, \mathbf{x} - \mathbf{x}_0 \rangle = \sum_{t=p}^{q} (\mathbf{x}_t, \mathbf{x}_t). \tag{8.71}$$

From Corollary 8.1 and (8.44), we also have

$$n\,\mathrm{vol}(N) = (d\mathbf{x}, d\mathbf{x}) = (\mathbf{x}, \delta d\mathbf{x}) = (\mathbf{x}, \Delta\mathbf{x}) = \sum_{t=p}^{q} \lambda_t(\mathbf{x}_t, \mathbf{x}_t). \tag{8.72}$$

Moreover, we have

$$n^2 \int_N ||H||^2 *1 = (\Delta\mathbf{x}, \Delta\mathbf{x}) = \sum_{t=p}^{q} \lambda_t^2(\mathbf{x}_t, \mathbf{x}_t). \tag{8.73}$$

By combining (8.71), (8.72) and (8.73), we find

$$\begin{aligned} n^2 \int_N ||H||^2 *1 - n(\lambda_1 + \lambda_2)\mathrm{vol}(N) + \lambda_1\lambda_2\mathcal{M} \\ = \sum_{t=p}^{q} (\lambda_t - \lambda_1)(\lambda_t - \lambda_2)(\mathbf{x}_t, \mathbf{x}_t) \geq 0. \end{aligned} \tag{8.74}$$

This proves (8.70). If the equality sign of (8.70) holds, inequality (8.74) becomes equality. Thus N must be of order 1, 2 or $\{1, 2\}$.

The converse is clear. □

Proposition 8.7. *Let* $\mathbf{x} : N \to S^{m-1}(1) \subset \mathbb{E}^m$ *be an isometric immersion of a compact Riemannian n-manifold into the unit hypersphere* $S^{m-1}(1) \subset \mathbb{E}^m$. *If* N *is mass-symmetric in* $S^{m-1}(1)$, *then*

$$\int_N ||H||^2 *1 \geq \frac{1}{n^2}\{n(\lambda_1 + \lambda_2) - \lambda_1\lambda_2\}\mathrm{vol}(N), \tag{8.75}$$

with the equality holding if and only if N *is of order* 1, 2 *or* $\{1, 2\}$.

Proof. Under the hypothesis, the moment $\mathcal{M}$ is nothing but the volume of N. Hence, the proposition follows from Proposition 8.6. □

By applying Proposition 8.7 we have the following following relationship between λ_1, λ_2 of compact minimal submanifolds in projective spaces [Chen (1987a)].

Theorem 8.8. *Let N be a compact minimal Riemannian n-manifold. If N admits an isometric minimal immersion into a projective space FP^m ($FP^m = RP^m(1)$, $CP^m(4)$ or $HP^m(4)$), then*

$$\frac{m}{2(m+1)}\lambda_1\lambda_2 \geq n(\lambda_1+\lambda_2-2n-2d), \quad d = \dim F. \tag{8.76}$$

Moreover, we have

(1) *If $FP^m = HP^m(4)$, then the equality sign of* (8.76) *holds if and only if $n = 4m$ and $N = HP^m(4)$;*
(2) *If $FP^m = CP^m(4)$, then the equality holding if and only if N is one of the compact Hermitian symmetric spaces:*

$$CP^k(4),\ CP^k(2),\ SO(2+k)/SO(2)\times SO(k),$$
$$CP^k(4)\times CP^k(4),\ U(2+k)/U(2)\times U(k),\ (k>2),$$
$$SO(10)/U(5) \text{ and } E_6/\mathrm{Spin}(10)\times T,$$

with an appropriate metric, where m is given respectively by

$$k,\ \frac{k(k+3)}{2},\ k+1,\ k(k+2),\ \frac{k(k+3)}{2},\ 15 \text{ and } 26.$$

For the proof of this theorem, see [Chen (1987a)]. For similar results on Kähler submanifolds in $CP^m(4)$ and their applications, see [Ros (1984)].

8.7 Total mean curvature and circumscribed radii

Let $B_o(r)$ denote the open ball in $\mathbb{E}^m$ with radius r centered at the origin o and put $c = ||\mathbf{x}_0||$, where $\mathbf{x}_0$ is the center of mass of N in $\mathbb{E}^m$.

The total mean curvature can also be estimated in terms of the order of immersions and circumscribed radii [Chen and Jiang (1995)].

Theorem 8.9. *Let N be an n-dimensional compact submanifold of $\mathbb{E}^m$.*

(a) *If N lies in a closed ball $\overline{B_o(R)}$ with radius R, then*

$$\int_N ||H||^k *1 \geq \frac{\mathrm{vol}(N)}{(R^2-c^2)^{k/2}}, \quad k = 2,3,\cdots,n, \tag{8.77}$$

with any one of the equalities holding if and only if N is of 1-type and N lies in the boundary $\partial(B_o(R))$ of $B_o(R)$;

(b) *Moreover, if N also lies in $\mathbb{E}^m - B_o(r)$, then*

$$\left(\frac{\lambda_p}{n}\right)^2 (r^2-c^2)\mathrm{vol}(N) \le \int_N ||H||^2 *1 \le \left(\frac{\lambda_q}{n}\right)^2 (R^2-c^2)\mathrm{vol}(N), \quad (8.78)$$

where p and q are the lower and the upper orders of N in $\mathbb{E}^m$. Either equality of (8.78) holding if and only if N is of type 1 and it lies in the boundary $\partial(B_o(r))$.

Proof. For a given integer $\nu \ge p$, we put

$$u^\nu = (||\mathbf{x}_p||, ||\mathbf{x}_{p+1}||, \dots, ||\mathbf{x}_\nu||),$$

$$v^\nu = (\lambda_p||\mathbf{x}_p||, \lambda_{p+1}||\mathbf{x}_{p+1}||, \dots, \lambda_\nu||\mathbf{x}_\nu||).$$

Then

$$\langle u^\nu, u^\nu\rangle = \sum_{t=p}^{\nu} ||\mathbf{x}_t||^2, \ \ \langle u^\nu, v^\nu\rangle = \sum_{t=p}^{\nu} \lambda_t||\mathbf{x}_t||^2, \ \ \langle v^\nu, v^\nu\rangle = \sum_{t=p}^{\nu} \lambda_t^2||\mathbf{x}_t||^2.$$

Thus, by the Schwartz inequality, we find

$$\left(\sum_{t=p}^{\nu} ||\mathbf{x}_t||^2\right)\left(\sum_{t=p}^{\nu} \lambda_t^2||\mathbf{x}_t||^2\right) \ge \left(\sum_{t=p}^{\nu} \lambda_t||\mathbf{x}_t||^2\right)^2. \quad (8.79)$$

On the other hand, we have

$$\langle u^\nu, u^\nu\rangle \to (\mathbf{x},\mathbf{x}) - c^2\,\mathrm{vol}(N),$$
$$\langle v^\nu, v^\nu\rangle \to n^2\int_N ||H||^2 *1,$$
$$\langle u^\nu, v^\nu\rangle \to -n\int_N \langle \mathbf{x}, H\rangle *1 = n\,\mathrm{vol}(N)$$

as $\nu \to \infty$. Thus (8.79) yields

$$((\mathbf{x},\mathbf{x}) - c^2\,\mathrm{vol}(N))\int_N ||H||^2 *1 \ge \mathrm{vol}(N)^2. \quad (8.80)$$

Since $(\mathbf{x},\mathbf{x}) \le R^2\,\mathrm{vol}(N)$, (8.80) implies (8.77) for $k = 2$.

Now, by using Hölder's inequality, we find

$$\left(\int_N ||H||^{2r} *1\right)^{1/r}\left(\int_N *1\right)^{1/s} \ge \int_N ||H||^2 *1 \ge \frac{\mathrm{vol}(N)}{R^2 - c^2}$$

with $r^{-1} + s^{-1} = 1, r, s > 1$. If we choose $r = \frac{k}{2}$, we obtain (8.77).

It is easy to verify that the equality sign of (8.77) holds for some k if and only if N is of 1-type and it lies in the boundary $\partial(B_o(R))$. This proves statement (a).

For statement (b) we consider

$$\begin{aligned}
&((\mathbf{x},\mathbf{x}) - c^2\mathrm{vol}(N))n^2 \int_N ||H||^2 *1 \\
&\geq n^2\mathrm{vol}(N)^2 = \left(\sum_{t\geq p} \lambda_t ||\mathbf{x}_t||^2\right)^2 \\
&\geq \lambda_p^2 \left(\sum_{t\geq p} ||\mathbf{x}_t||^2\right)^2 = \lambda_p^2((\mathbf{x},\mathbf{x}) - c^2\mathrm{vol}(N))^2,
\end{aligned}$$

which implies the first inequality of (8.78). The second inequality of (8.78) can be proved in a similar way.

The remaining part can be verified easily. □

Theorem 8.9 implies immediately the following.

Corollary 8.9. *Let N be an n-dimensional compact submanifold of $\mathbb{E}^m$.*

(1) *If N is contained in the closed ball $\overline{B_o(R)}$, then*

$$\max ||H||^2 \geq \frac{1}{R^2 - c^2},$$

with equality holding if and only if N is a minimal submanifold of a hypersphere of $\mathbb{E}^m$;

(2) *If N has constant mean curvature and N lies in $\mathbb{E}^m - B_o(r)$, then*

$$\left(\frac{\lambda_p}{n}\right)^2 (r^2 - c^2) \leq ||H||^2 \leq \left(\frac{\lambda_q}{n}\right)^2 (R^2 - c^2),$$

with either equality holding if and only if N is of 1-type.

The following result provides a sharp estimate of λ_1 for compact submanifolds of Euclidean space in term of circumscribed radii.

Proposition 8.8. *Let N be an n-dimensional compact submanifold of $\mathbb{E}^m$.*

(1) *If N is contained in $\mathbb{E}^m - B_o(r)$, then the first nonzero eigenvalue of the Laplacian of N satisfies*

$$\lambda_1 \leq \frac{n}{r^2 - c^2},$$

with equality sign holding if and only if N is of 1-type and it lies in the boundary $\partial(B_o(r))$ of $B_o(r)$;

(2) *If N is contained in the closed ball $\overline{B_o(R)}$, we have*

$$\lambda_p(r^2 - c^2) \leq n \leq \lambda_q(R^2 - c^2),$$

where p and q are the lower and upper orders of N. Either equality sign holding if and only if N is of 1-type and it lies in the boundary $\partial(B_o(R))$ of $B_o(R)$.

Proof. This proposition follows from the following two equations:

$$\begin{aligned} n\,\mathrm{vol}(N) &= \sum_{t \geq p} \lambda_t ||\mathbf{x}_t||^2 \geq \lambda_p((\mathbf{x}, \mathbf{x}) - c^2 \mathrm{vol}(N)) \\ &\geq \lambda_1(r^2 - c^2)\mathrm{vol}(N) \end{aligned}$$

and

$$\begin{aligned} n\,\mathrm{vol}(N) &= \sum_{t \geq p} \lambda_t ||\mathbf{x}_t||^2 \leq \lambda_q((\mathbf{x}, \mathbf{x}) - c^2 \mathrm{vol}(N)) \\ &\leq \lambda_q(R^2 - c^2)\mathrm{vol}(N). \end{aligned}$$

□

Proposition 8.8 implies immediately the following estimates on λ_1 for ellipsoids.

Corollary 8.10. *Let N be the ellipsoid in $\mathbb{E}^3$ defined by*

$$\frac{x^2}{a^2} + \frac{y^2}{b^2} + \frac{z^2}{c^2} = 1, \quad a \leq b \leq c.$$

Then the first nonzero eigenvalue of the Laplacian of N satisfies

$$\lambda_1 \leq \frac{2}{a^2},$$

with equality sign holding if and only if N is a sphere, i.e., $a = b = c$.

Proposition 8.8 can also be applied to obtain the following estimate of λ_1 for closed tubes [Chen and Jiang (1995)].

Proposition 8.9. *Let σ be a closed curve in $\mathbb{E}^3$. Denote by σ_0 the center of mass of σ and by $T^\epsilon(\sigma)$ the tube around σ with a sufficiently small radius ϵ. If σ is contained in $\mathbb{E}^3 - B_{\sigma_0}(r)$ and $\epsilon < r$, then the first nonzero eigenvalue λ_1 of the Laplacian of the tube $T^\epsilon(\sigma)$ satisfies*

$$\lambda_1 < \frac{2}{(r - \epsilon)^2}. \tag{8.81}$$

Proof. Without loss of generality, we may assume that $\sigma : [0, \ell] \to \mathbb{E}^3$ is a unit speed closed curve whose center of mass is the origin. Denote by $\{\mathbf{T}, \mathbf{N}, \mathbf{B}\}$ the Frenet frame of σ. Then the tube $T^\epsilon(\sigma)$ is given by

$$\mathbf{x}(s, \theta) = \sigma(s) + \epsilon \cos\theta\, \mathbf{N} + \epsilon \sin\theta\, \mathbf{B}. \tag{8.82}$$

From (8.82) we find

$$*1 = \epsilon(1 - \epsilon\kappa\cos\theta)d\theta \wedge ds, \tag{8.83}$$

where κ is the curvature function of σ.

By using (8.82) and (8.83) we have

$$\begin{aligned} \mathbf{x}_0 &= \frac{\epsilon}{\text{vol}(M)} \int_0^\ell \int_0^{2\pi} \mathbf{x}(s, \theta)(1 - \epsilon\kappa\cos\theta)d\theta \wedge ds \\ &= \frac{\epsilon}{\text{vol}(M)} \left(2\pi \int_0^\ell \sigma(s)ds - \epsilon\pi \int_0^\ell H_\sigma ds \right), \end{aligned} \tag{8.84}$$

where H_σ is the mean curvature vector of σ. Since $\int_0^\ell H_\sigma ds = 0$, (8.84) implies that the center of mass of the tube coincides with the center of mass of σ. Therefore, by applying Proposition 8.8, we obtain $(r - \epsilon)^2\lambda_1 \le 2$. If $(r - \epsilon)^2\lambda_1 = 2$, then the tube is of 1-type which is impossible. Thus, we obtain (8.81). □

The following result is a generalization of Proposition 8.7.

Theorem 8.10. *Let N be an n-dimensional compact submanifold of $\mathbb{E}^m$ and $c = ||\mathbf{x}_0||$ with $\mathbf{x}_0$ being the center of mass. Then*

(1) *If N is contained in the closed ball $\overline{B_o(R)}$, then*

$$\int_N ||H||^2 *1 \ge \frac{1}{n^2}\{n(\lambda_1 + \lambda_2) + (c^2 - R^2)\lambda_1\lambda_2\}\text{vol}(N),$$

with equality sign holding if and only if N is contained in the boundary $\partial(B_o(R))$ and, moreover, either N is of 1-type with order 1 or 2, or M is of 2-type and of order $\{1, 2\}$;

(2) *If N is contained in $\mathbb{E}^m - B_o(r)$ and N is of finite type with lower order p and upper order q, then*

$$\int_N ||H||^2 *1 \le \frac{1}{n^2}\{n(\lambda_p + \lambda_q) + (c^2 - r^2)\lambda_p\lambda_q\}\text{vol}(N),$$

with the equality holding if and only if N lies in the boundary $\partial(B_o(r))$ and M is either of 1-type or of 2-type.

This can be proved in the same way as Proposition 8.7.

Chapter 9

Pseudo-Kähler Manifolds

Algebraic geometry studies algebraic varieties which are defined locally as the common zero sets of polynomials. Descartes's idea of coordinates is crucial to algebraic geometry, but it has undergone some significant transformations beginning in the early 19th century. Before then, the coordinates were made of real numbers, but this changed with complex numbers, and then with an arbitrary field. Also, homogeneous coordinates of projective geometry provide another extension of the coordinate system, which enriched complex algebraic geometry. During 19th and the first half of 20th century many great mathematicians contributed to the important developments of algebraic geometry; included among the principal investigators are Cayley, Laguerre, Klein, Riemann, Picard, Enriques, Castelnuovo, C. Segre, Severi, Shafarevich, van der Waeden, Zariski, and Weil.

It was Schouten, van Dantzig and Kähler who around 1929–1932 first tried to study complex manifolds from the viewpoint of Riemannian geometry. Consequently, a Hermitian space with the so-called symmetric unitary connection was introduced. Hermitian spaces with such a connection are now known today as Kähler manifolds, which include all projective non-singular algebraic varieties in complex algebraic geometry.

It was Weil who pointed out in [Weil (1947)] that there is on a complex manifold a $(1,1)$-tensor field J whose square is the negative of the identity transformation, i.e, $J^2 = -I$. In the same year, Ehresmann introduced the notion of an almost complex manifold as an even-dimensional manifold which admits such a $(1,1)$-tensor J.

A Kähler manifold M is a $2n$-dimensional manifold with a complex structure J, $J^2 = -I$, and a compatible Riemannian metric g; namely,

$$g(Ju, Jv) = g(u, v), \quad \nabla J = 0, \quad \forall u, v \in TM.$$

Thus, a Kähler manifold admits a $U(n)$-structure satisfying an integrability condition. If we define a 2-form Ω on M by $\Omega(u,v) = g(u, Jv)$, then Ω is a *symplectic structure*, i.e., Ω is a non-degenerate closed 2-form. Hence, a Kähler manifold is a Riemannian manifold, a complex manifold, and a symplectic manifold, with these three structures all mutually compatible. This threefold structure corresponds to the presentation of the unitary group as an intersection: $U(n) = O(2n) \cap GL(n, \mathbf{C}) \cap Sp(2n)$. Without any integrability conditions, the corresponding notion is an almost Hermitian manifold. If the Sp-structure is integrable but the complex structure need not be, the notion is an almost Kähler manifold; if the complex structure is integrable but the Sp-structure need not be, the notion is a Hermitian manifold. When g is a pseudo-Riemannian metric, the corresponding manifold is pseudo-Kählerian.

Kähler manifolds are important in algebraic and differential geometry and in mathematical physics, especially in string theory.

9.1 Pseudo-Kähler manifolds

Let M be a complex manifold of complex dimension n and $\{z_1, \ldots, z_n\}$ a local complex coordinate system on a coordinate neighborhood U. If

$$z_j = x_j + \mathrm{i}y_j, \quad \mathrm{i} = \sqrt{-1}, \quad j = 1, \ldots, n,$$

then $\{x_1, \ldots, x_n, y_1, \ldots, y_n\}$ forms a system of local coordinates on U.

Put $X_j = \partial/\partial x_j, Y_j = \partial/\partial y_j$. Then $X_1, \ldots, X_n, Y_1, \ldots, Y_n$ form a local frame of TM. Let J be the endomorphism of TU defined by

$$JX_j = Y_j, \quad JY_j = -X_j, \quad j = 1, \ldots, n.$$

Then $J^2 = -I$. It is easy to verify that J does not depend on the choice of $(z, \ldots, z_n)$. J is called the *complex structure* of the complex manifold M.

Definition 9.1. A pseudo-Riemannian metric g on a complex manifold M is called *pseudo-Hermitian* if g and J are compatible, i.e.,

$$g(JX, JY) = g(X, Y), \quad X, Y \in T_pM, \; p \in M. \tag{9.1}$$

A *pseudo-Hermitian manifold* is a complex manifold equipped with a pseudo-Hermitian metric. If the pseudo-Hermitian metric is of index zero, M is called a *Hermitian manifold.* If the index of the metric is positive, M is called an *indefinite Hermitian manifold.*

It follows from (9.1) that the index of g is an even integer $2s$ with $0 \leq s \leq n$, $n = \dim_{\mathbb{C}} M$. The integer s is called the *complex index*. The *fundamental 2-form* Ω of a pseudo-Hermitian manifold (M, g) is defined by

$$\Omega(X, Y) = g(X, JY), \quad X, Y \in TM. \tag{9.2}$$

Definition 9.2. A pseudo-Hermitian manifold (resp. an indefinite Hermitian manifold) is called *pseudo-Kähler* (resp. *indefinite Kähler*) if its fundamental 2-form Ω is closed, i.e., $d\Omega = 0$. The corresponding metric is called *pseudo-Kähler* (resp. *indefinite Kähler*). A pseudo-Kähler manifold with complex index one is called a *Lorentzian Kähler manifold.*

Proposition 9.1. *A pseudo-Hermitian manifold is pseudo-Kählerian if and only if J is parallel, i.e., $\nabla J = 0$.*

Proof. Let (M, g, J) be a pseudo-Hermitian manifold. Then

$$3d\Omega(X, Y, X) = g(X, (\nabla_Z J)Y) - g(Y, (\nabla_X J)Z) + g(Z, (\nabla_Y J)X)$$

for X, Y, Z tangent to M. Thus, $\nabla J = 0$ implies $d\Omega = 0$.

Conversely, because

$$2g((\nabla_X J)Y, Z) = d\Omega(X, JY, JZ) - d\Omega(X, Y, Z),$$

$d\Omega = 0$ implies $\nabla J = 0$. □

Lemma 9.1. *The Riemann curvature tensor R of a pseudo-Kähler manifold satisfies*

$$R(X, Y) \circ JZ = J \circ R(X, Y)Z, \tag{9.3}$$

$$R(JX, JY)Z = R(X, Y)Z. \tag{9.4}$$

Proof. Since J is parallel, $\nabla_X JY = J\nabla_X Y$. Thus

$$R(X, Y) \circ JZ = [\nabla_X, \nabla_Y]JZ - \nabla_{[X,Y]}JZ = J(R(X, Y)Z),$$

which gives (9.3). By applying (1.6) and (1.9) we have

$$\begin{aligned} g(R(JX, JY)V, U) &= g(R(U, V)JY, JX) = g(J(R(U, V)Y), JX) \\ &= g(R(U, V)Y, X) = g(R(X, Y)V, U). \end{aligned}$$

Thus we obtain (9.4). □

A plane section on a pseudo-Kähler manifold is called *holomorphic* if it is spanned by $\{v, Jv\}$ for some non-null vector $v \in TM$. The sectional curvature $K(v \wedge Jv)$ of a holomorphic section is called the *holomorphic sectional curvature* at v, which is denoted by $H(v)$. The holomorphic sectional curvature $H(v)$ is independent of the choice of v in $\mathrm{Span}\{v, Jv\}$.

Definition 9.3. A Kähler manifold (resp. *indefinite Kähler manifold*) is called a *complex space form* (resp. *indefinite complex space form*) if it has constant holomorphic sectional curvature. An indefinite complex space form with complex index one is called a *Lorentzian complex space form.*

The following proposition and examples were given in [Barros and Romero (1982)].

Proposition 9.2. *The curvature tensor of an (indefinite) complex space form $M^m_s(4c)$ of constant holomorphic sectional curvature* $4c$ *satisfies*

$$\begin{aligned} R(X,Y)Z = &c\{g(Y,Z)X - g(X,Z)Y + g(JY,Z)JX \\ &- g(JX,Z)JY + 2g(X,JY)JZ\}. \end{aligned} \tag{9.5}$$

Conversely, if the curvature tensor of an (indefinite) Kähler manifold satisfies (9.5)*, then it is an (indefinite) complex space form of constant holomorphic sectional curvature* $4c$.

Proof. Let R^o be a $(1,3)$ tensor on M defined by

$$\begin{aligned} R^o(X,Y)Z = &c\{g(Y,Z)X - g(X,Z)Y + g(JY,Z)JX \\ &- g(JX,Z)JY + 2g(X,JY)JZ\}. \end{aligned} \tag{9.6}$$

Consider $\hat{R} = R - R^o$. Then $g(\hat{R}(v,Jv)Jv,v) = 0$ for $v \in TM$ with $g(v,v) \neq 0$. Since $\{v \in T_pM : g(v,v) \neq 0\}$ is dense in T_pM, $g(\hat{R}(v,Jv)Jv,v) = 0$ holds in general. Thus, by applying polarization, we find $\hat{R} = 0$.

The converse is trivial. □

Example 9.1. Let n and s be integers such that $n \geq 1$ and $0 \leq s \leq n$. The complex manifold $\mathbf{C}^n$ endowed with the real part of the Hermitian form

$$b_{s,n}(z,w) = -\sum_{j=1}^{s} \bar{z}_j w_j + \sum_{j=s+1}^{n} \bar{z}_j w_j \ \text{ for } z,w \in \mathbf{C}^n \tag{9.7}$$

defines a flat (indefinite) complex space form of index $2s$, denote by $\mathbf{C}^n_s$.

Example 9.2. For a positive number c and an integer $s \geq 0$, there exists an (indefinite) complex space form $CP^n_s(4c)$ with complex dimension n, complex index s and of constant holomorphic sectional curvature $4c$. $CP^n_s(4c)$ is homotopy equivalent to the standard complex projective space CP^{n-s}; and therefore, CP^n_s is simply-connected. The underlying complex manifold of $CP^n_s(4c)$ is the open submanifold

$$\{z \in \mathbf{C}^{n+1} : b_{s,n+1}(z,z) > 0\}/\mathbf{C}^*$$

of $CP^n = (\mathbf{C}^{n+1} - \{0\})/\mathbf{C}^*$, where $\mathbf{C}^* = \mathbf{C} - \{0\}$.

Consider the pseudo hypersphere of curvature c defined by

$$S^{2n+1}_{2s}(c)=\Big\{z\in \mathbf{C}^{n+1}_s : b_{s,n+1}(z,z)=\frac{1}{c}\Big\}. \tag{9.8}$$

Then $\pi : S^{2n+1}_{2s}(c)\to CP^n_s;\ z\mapsto z\cdot \mathbf{C}^*$ is a submersion. There is a unique pseudo-Kähler metric of index $2s$ on CP^n_s which make CP^n_s an (indefinite) complex space form of constant holomorphic sectional curvature $4c$ such that π is a pseudo-Riemannian submersion.

Example 9.3. For a negative number c and a non-negative integer s, the (indefinite) complex hyperbolic n-space $CH^n_s(4c)$ is obtained from $CP^n_{n-s}(-4c)$ by replacing the metric of $CP^n_{n-s}(-4c)$ by its negative. In particular, $CP^n = CP^n_0$ and $CH^n = CH^n_0$ are called *complex projective n-space* and *complex hyperbolic n-space*, respectively.

Remark 9.1. Every complete simply-connected pseudo-Kähler manifold of complex dimension n, of complex index s and of constant holomorphic sectional curvature $4c$ is holomorphically isometric to $CP^n_s(4c), \mathbf{C}^n_s$ or $CH^n_s(4c)$ according to $c>0, c=0$ or $c<0$, respectively.

9.2 Pseudo-Kähler submanifolds

Based on the behavior of the tangent bundle of pseudo-Riemannian submanifolds under the action of the complex structure J of a pseudo-Kähler manifold, there are several typical interesting families of submanifolds; namely, complex, purely real, totally real, CR and slant submanifolds.

Definition 9.4. A pseudo-Riemannian submanifold N of a pseudo-Kähler manifold M is called a *complex submanifold* if each of its tangent spaces is invariant under the action of the complex structure J of M.

The study of complex submanifolds of Kähler manifolds from differential geometric point of view (i.e., with emphasis on the Riemannian metric) was initiated in [Calabi (1953)] and studied extensively during the second half of the last century by many geometers (cf. [Ogiue (1974); Chen (2000c)]).

Proposition 9.3. *A complex submanifold of a pseudo-Kähler manifold is pseudo-Kählerian with respect to its induced structures. Moreover, its second fundamental form h and shape operator A satisfy*

$$h(JX,Y)=h(X,JY)=Jh(X,Y), \tag{9.9}$$

$$A_{J\xi}=JA_\xi,\ \ JA_\xi=-A_\xi J, \tag{9.10}$$

for X,Y tangent to N and ξ normal to N.

Proof. Let N be a complex submanifold of a pseudo-Kähler manifold M with complex structure J and pseudo-Kähler metric g. Obviously, it follows from (9.1) that N is a pseudo-Hermitian manifold with respect to the induced metric g' and the induced complex structure, also denoted by J. For any vector fields X, Y tangent to N we have

$$\nabla_X(JY) = \nabla'_X(JY) + h(X, JY), \tag{9.11}$$

where ∇' is the induced connection.

On the other hand, since M is pseudo-Kählerian, $\nabla J = 0$. Thus

$$\nabla_X(JY) = J(\nabla_X Y) = J\nabla'_X Y + Jh(X, Y). \tag{9.12}$$

Comparing (9.11) and (9.12) gives $\nabla' J = 0$ and $h(X, JY) = Jh(X, Y)$. Hence, N is a pseudo-Kählerian manifold. Further, by symmetry of h we have (9.9). It is easy to verify that (9.10) follows from (2.16) and (9.9). □

By a *pseudo-Kähler submanifold* we mean a complex submanifold of a pseudo-Kähler manifold with its induced pseudo-Kählerian structure.

Example 9.4. (Segre embedding) A point of $CP^n_s(4)$ can be represented by $[(z, w)]$, where $z = (z_1, \ldots, z_s) \in \mathbf{C}^s, w = (w_1, \ldots, w_{n-s+1}) \in \mathbf{C}^{n-s+1}$, $(z, w) \in S^{2n+1}_{2s}(1) \subset \mathbf{C}^{n+1}_s$ and $[(z, w)]$ denotes the class $(z, w) \cdot S^1(1)$.

Consider a mapping

$$\psi_{pqst} : CP^p_s(4) \times CP^q_t(4) \to CP^{p+q+pq}_r(4),$$

defined by

$$\psi_{pqst}([(z, w)], [(x, y)]) = [(z_i y_\alpha, w_k x_a, z_j x_b, w_\ell y_\beta)],$$

with $r = s(q - t) + t(p - s) + s + t$. Then ψ_{pqst} is an isometric holomorphic embedding. Thus, it defines a pseudo-Kähler submanifold of $CP^{p+q+pq}_r(4)$ [Ikawa et al. (1988)].

Definition 9.5. A pseudo-Riemannian submanifold N of a pseudo-Riemannian manifold M is called *austere* if there exists a local orthonormal frame $\{e_1, \ldots, e_n, e_{1^*}, \ldots, e_{n^*}\}$ on N such that the second fundamental form of N satisfies [Harvey and Lawson (1982)]

$$\epsilon_{i^*} h(e_{i^*}, e_{i^*}) + \epsilon_i h(e_i, e_i) = 0, \quad i = 1, \ldots, n. \tag{9.13}$$

Obviously, every austere submanifold is minimal.

Proposition 9.4. *Every pseudo-Kähler submanifold of a pseudo-Kähler manifold is austere.*

Proof. Assume that N is a pseudo-Kähler submanifold of a pseudo-Kähler manifold M with $n = \dim_{\mathbf{C}} N$. Then, according to (9.1) we may choose a local orthonormal frame $\{e_1, \ldots, e_n, e_{1^*}, \ldots, e_{n^*}\}$ on N such that $e_{i^*} = Je_i, i = 1, \ldots, n$. Then

$$\epsilon_i = \langle e_i, e_i \rangle = \langle e_{i^*}, e_{i^*} \rangle = \epsilon_{i^*}. \tag{9.14}$$

Thus, it follows from (9.9) that (9.13) holds. Hence N is austere. □

Definition 9.6. A pseudo-Riemannian submanifold N of a pseudo-Riemannian manifold is called *isotropic* at a point $p \in N$ if $\langle h(v,v), h(v,v) \rangle$ is independent of the choice of the unit vector $v \in T_pN$ at p, where h is the second fundamental form of N. N is called *isotropic* if it is isotropic at each point. Moreover, N is called *null-isotropic* if $\langle h(v,v), h(v,v) \rangle = 0$ for any unit tangent vector v of N.

Proposition 9.5. *A pseudo-Kähler submanifold of an indefinite complex space form $M_s^m(4c)$ has constant holomorphic sectional curvature $4c$ if and only if N is null-isotropic.*

Proof. Let N be a pseudo-Kähler submanifold of an indefinite complex space form $M_s^m(c)$. For a unit vector v of N, it follows from Gauss' equation and Proposition 9.3 that the holomorphic sectional curvature of N satisfies

$$H(v) = 4c - 2 \langle h(v,v), h(v,v) \rangle . \tag{9.15}$$

Thus N has constant holomorphic sectional curvature $4c$ if and only if $\langle h(v,v), h(v,v) \rangle = 0$ for any unit vector v. □

It also follows from Gauss' equation and Propositions 9.3 and 9.4 that the scalar curvature τ of a pseudo-Kähler submanifold N of $M_s^m(c)$ satisfies

$$2\tau = 4n(n+1)c - S_h, \tag{9.16}$$

where $S_h = \sum_{i,j=1}^{2n} \epsilon_i \epsilon_j \langle h(e_i, e_j), h(e_i, e_j) \rangle$ and $e_1, \ldots, e_{2n}$ is an orthonormal basis of N. Hence, we also have the following.

Proposition 9.6. *The scalar curvature of a pseudo-Kähler submanifold of an indefinite complex space form $M_s^m(4c)$ is equal to $2n(n+1)c$ if and only if $S_h = 0$.*

Remark 9.2. Contrast to Kählerian case, null-isotropic pseudo-Kähler submanifolds are not necessary totally geodesic. A simple example is the flat pseudo-Kähler submanifold $\mathbf{C}_t^n$ embedded in $\mathbf{C}_{t+1}^{n+2}$ defined by

$$\left(\sum_{j=1}^n z_j^2, z_1, \ldots, z_n, \sum_{j=1}^n z_j^2 \right), \quad z_j = x_j + \mathrm{i} y_j.$$

This non-totally geodesic example satisfies $S_h = 0$ identically.

9.3 Purely real submanifolds of pseudo-Kähler manifolds

Let N be a pseudo-Riemannian submanifold of a pseudo-Kähler manifold. For each $X \in TN$ we put

$$JX = PX + FX, \tag{9.17}$$

where PX and FX are the tangential and the normal components of JX. Then P is an endomorphism of TN and F is a $T^\perp N$-valued 1-form. Similarly, for each normal vector ξ of N, we put

$$J\xi = t\xi + f\xi, \tag{9.18}$$

where $t\xi$ and $f\xi$ are the tangential and the normal components of $J\xi$. Then f is an endomorphism of $T^\perp N$ and t is a TN-valued 1-form on $T^\perp N$.

It follows from (9.1) and (9.17) that

$$\langle PX, Y\rangle = -\langle X, PY\rangle \tag{9.19}$$

for X, Y tangent to N. Thus

$$\left\langle P^2X, Y\right\rangle = \left\langle X, P^2Y\right\rangle = -\langle PX, PY\rangle. \tag{9.20}$$

Define $\nabla' P$ and $\nabla' F$ by

$$(\nabla'_X P)Y = \nabla'_X(PY) - P(\nabla'_X Y), \tag{9.21}$$

$$(\nabla'_X F)Y = D_X(FY) - F(\nabla'_X Y), \tag{9.22}$$

where ∇' is the Levi-Civita connection of the purely real submanifold N.

Proposition 9.7. *Let N be a pseudo-Riemannian submanifold of a pseudo-Kähler manifold. Then $\nabla' P = 0$ if and only if the shape operator satisfies*

$$A_{FY}Z = A_{FZ}Y \tag{9.23}$$

for vectors Y, Z tangent to N.

Proof. It follows from (2.6), (2.15), (9.17), (9.18) and $\nabla J = 0$ that

$$\begin{aligned} 0 &= \nabla_X(JY) - J\nabla_X Y \\ &= \nabla'_X(PY) + h(X, PY) - A_{FY}X + D_X(FY) \\ &\quad - P(\nabla'_X Y) - F(\nabla'_X Y) - th(X,Y) - fh(X,Y) \end{aligned} \tag{9.24}$$

for X, Y tangent to N. The tangential components of (9.24) yield

$$(\nabla'_X P)Y = A_{FY}X + th(X,Y).$$

Thus, $\nabla' P = 0$ holds identically if and only if $A_{FY}X + th(X,Y) = 0$, which is equivalent to $A_{FY}Z = A_{FZ}Y$. □

Proposition 9.8. *Let N be a pseudo-Riemannian submanifold of a pseudo-Kähler manifold. Then the following three statements are equivalent:*

(i) $\nabla' F = 0$;
(ii) $F\nabla'_X Y = D_X FY$ *for* X, Y *tangent to* N;
(iii) $h(X, PY) = fh(X, Y)$ *for* X, Y *tangent to* N.

Proof. The equivalence of (i) and (ii) is a consequence of (9.22). By comparing the normal components of (9.24) we find

$$h(X, PY) + D_X(FY) - F(\nabla'_X Y) - fh(X, Y) = 0. \tag{9.25}$$

If $\nabla' F = 0$, then by (ii) and (9.25) we have (iii). Conversely, if (iii) holds, then (9.25) reduces to (ii). Thus, $\nabla' F = 0$ holds. □

Definition 9.7. A pseudo-Riemannian submanifold N of a pseudo-Kähler manifold M is called *purely real* if the complex structure J on M carries the tangent bundle of N into a transversal bundle, i.e., $J(T_pN) \cap T_pN = \{0\}$ for each $p \in N$.

Remark 9.3. A purely real submanifold N of a pseudo-Kähler manifold contains no complex points. When $\dim N = 2$, the converse is also true.

Next, we present a basic property of purely real surfaces in an indefinite Kähler manifold. Suppose that N is a Lorentz surface in an indefinite Kähler manifold M. Let us choose a local frame $\{e_1, e_2\}$ on N such that

$$\langle e_1, e_1\rangle = \langle e_2, e_2\rangle = 0, \ \langle e_1, e_2\rangle = -1. \tag{9.26}$$

We call a frame a *pseudo-orthonormal frame* of N.

It follows from (9.1) and (9.17) that $\langle Pu, u\rangle = 0$ for any $u \in TN$. Thus, by using (9.26), we know that there is a function α such that

$$Pe_1 = (\sinh\alpha)e_1, \ \ Pe_2 = -(\sinh\alpha)e_2. \tag{9.27}$$

If we put

$$e_3 = (\operatorname{sech}\alpha)Fe_1, \ \ e_4 = (\operatorname{sech}\alpha)Fe_2, \tag{9.28}$$

then we derive from (9.27) and (9.28) that

$$Je_1 = \sinh\alpha e_1 + \cosh\alpha e_3, \ \ Je_2 = -\sinh\alpha e_2 + \cosh\alpha e_4. \tag{9.29}$$

By applying $J^2 = -I$, (9.26) and (9.29), we find

$$Je_3 = -\cosh\alpha e_1 - \sinh\alpha e_3, \ \ Je_4 = -\cosh\alpha e_2 + \sinh\alpha e_4, \tag{9.30}$$

$$\langle e_3, e_3\rangle = \langle e_4, e_4\rangle = 0, \ \ \langle e_3, e_4\rangle = -1. \tag{9.31}$$

Since $\cosh\alpha \geq 1$, we conclude from (9.29) the following general result.

Proposition 9.9. *Every Lorentz surface in any indefinite Kähler manifold is purely real.*

When N is a Lorentz surface in a Lorentzian Kähler surface, we call a frame $\{e_1, e_2, e_3, e_4\}$ chosen as above an *adapted pseudo-orthonormal frame.*

9.4 Dependence of fundamental equations for Lorentz surfaces

Let N be a Lorentz surface in a Lorentzian Kähler surface M^2_1. Let ∇' and ∇ be the Levi-Civita connection of N and M^2_1, respectively. We may assume that locally N is equipped with the Lorentzian metric:

$$g = -m^2(x, y)(dx \otimes dy + dy \otimes dx)$$

for some positive function m. The Levi-Civita connection of g satisfies

$$\nabla'_{\frac{\partial}{\partial x}} \frac{\partial}{\partial x} = \frac{2m_x}{m} \frac{\partial}{\partial x}, \; \nabla'_{\frac{\partial}{\partial x}} \frac{\partial}{\partial y} = 0, \; \nabla'_{\frac{\partial}{\partial y}} \frac{\partial}{\partial y} = \frac{2m_y}{m} \frac{\partial}{\partial y} \tag{9.32}$$

and the Gaussian curvature K is given by

$$K = \frac{2mm_{xy} - 2m_x m_y}{m^4}.$$

If we put

$$e_1 = \frac{1}{m}\frac{\partial}{\partial x}, \quad e_2 = \frac{1}{m}\frac{\partial}{\partial y}, \tag{9.33}$$

then $\{e_1, e_2\}$ is a pseudo-orthonormal frame satisfying

$$\langle e_1, e_1 \rangle = \langle e_2, e_2 \rangle = 0, \; \langle e_1, e_2 \rangle = -1. \tag{9.34}$$

From (9.32) and (9.33) we find

$$\begin{aligned} \nabla'_{e_1} e_1 &= \frac{m_x}{m^2} e_1, \; \nabla'_{e_2} e_1 = -\frac{m_y}{m^2} e_1, \\ \nabla'_{e_1} e_2 &= -\frac{m_x}{m^2} e_2, \; \nabla'_{e_2} e_2 = \frac{m_y}{m^2} e_2. \end{aligned} \tag{9.35}$$

It follows from (9.1), (9.17) and (9.34) that

$$Je_1 = \sinh \alpha e_1 + \cosh \alpha e_3, \qquad Je_2 = -\sinh \alpha e_2 + \cosh \alpha e_4, \tag{9.36}$$

$$Je_3 = -\cosh \alpha e_1 - \sinh \alpha e_3, \;\; Je_4 = -\cosh \alpha e_2 + \sinh \alpha e_4, \tag{9.37}$$

$$\langle e_3, e_3 \rangle = \langle e_4, e_4 \rangle = 0, \;\; \langle e_3, e_4 \rangle = -1 \tag{9.38}$$

for some function α. If we put $\nabla'_X e_j = \sum_{k=1}^2 \omega_j^k(X) e_k$, $j = 1, 2$, we deduce from (9.34) that

$$\nabla'_X e_1 = \omega(X) e_1, \;\; \nabla'_X e_2 = -\omega(X) e_2, \;\; \omega = \omega_1^1. \tag{9.39}$$

Similarly, if we put $D_X e_r = \sum_{s=3}^{4} \omega_r^s(X) e_s$; $r = 3, 4$, then (9.38) gives

$$D_X e_3 = \psi(X) e_3, \quad D_X e_4 = -\psi(X) e_4, \quad \psi = \omega_3^3. \tag{9.40}$$

Let us put $h(e_i, e_j) = h_{ij}^3 e_3 + h_{ij}^4 e_4$. By applying $\nabla_X(JY) = J\nabla_X Y$ and (9.36)-(9.40), we find

$$A_{e_3} e_j = h_{j2}^4 e_1 + h_{1j}^4 e_2, \quad A_{e_4} e_j = h_{j2}^3 e_1 + h_{1j}^3 e_2, \tag{9.41}$$

$$e_j \alpha = (\omega_j - \psi_j) \coth \alpha - 2h_{1j}^3, \tag{9.42}$$

$$e_1 \alpha = h_{12}^4 - h_{11}^3, \quad e_2 \alpha = h_{22}^4 - h_{12}^3, \tag{9.43}$$

$$\omega_j - \psi_j = (h_{1j}^3 + h_{j2}^4) \tanh \alpha, \tag{9.44}$$

where $\omega_j = \omega(e_j)$ and $\psi_j = \psi(e_j)$ for $j = 1, 2$. For simplicity let us put

$$\begin{aligned} h(e_1, e_1) &= \beta e_3 + \gamma e_4, \\ h(e_1, e_2) &= \delta e_3 + \varphi e_4, \\ h(e_2, e_2) &= \lambda e_3 + \mu e_4. \end{aligned} \tag{9.45}$$

In view of (9.38) and (9.45), the equation of Gauss can be expressed as

$$\gamma\lambda + \beta\mu - 2\delta\varphi = \frac{2(m m_{xy} - m_x m_y)}{m^4} - \tilde{K}, \tag{9.46}$$

where $\tilde{K}$ is the sectional curvature of M_1^2 with respect to the plane section spanned by e_1, e_2.

The equations of Gauss, Codazzi and Ricci are independent in general. However, for Lorentz surfaces we have the following [Chen (2009a)].

Theorem 9.1. *The equation of Ricci is a consequence of the equations of Gauss and Codazzi for a Lorentz surface in any Lorentzian Kähler surface.*

Proof. Let N be a Lorentz surface in a Lorentzian Kähler surface M_1^2. By using (9.35), (9.37) and (9.44) we find

$$\begin{aligned} D_{e_1} e_3 &= \left(\frac{m_x}{m^2} - (\beta + \varphi) \tanh \alpha\right) e_3, \\ D_{e_2} e_3 &= -\left(\frac{m_y}{m^2} + (\delta + \mu) \tanh \alpha\right) e_3, \\ D_{e_1} e_4 &= \left((\beta + \varphi) \tanh \alpha - \frac{m_x}{m^2}\right) e_4, \\ D_{e_2} e_4 &= \left(\frac{m_y}{m^2} + (\delta + \mu) \tanh \alpha\right) e_4. \end{aligned} \tag{9.47}$$

Thus, it follows from (9.35), (9.45) and (9.47) that

$$\begin{aligned}
(\bar{\nabla}_{e_1}h)(e_1,e_2) &= \Big(\frac{\delta_x}{m}+\frac{\delta m_x}{m^2}-\delta(\beta+\varphi)\tanh\alpha\Big)\,e_3\\
&\quad+\Big(\frac{\varphi_x}{m}-\frac{\varphi m_x}{m^2}+\varphi(\beta+\varphi)\tanh\alpha\Big)\,e_4,\\
(\bar{\nabla}_{e_2}h)(e_1,e_1) &= \Big(\frac{\beta_y}{m}+\frac{\beta m_y}{m^2}-\beta(\delta+\mu)\tanh\alpha\Big)\,e_3\\
&\quad+\Big(\frac{\gamma_y}{m}+\frac{3\gamma m_y}{m^2}+\gamma(\delta+\mu)\tanh\alpha\Big)\,e_4,\\
(\bar{\nabla}_{e_1}h)(e_2,e_2) &= \Big(\frac{\lambda_x}{m}+\frac{3\lambda m_x}{m^2}-\lambda(\beta+\varphi)\tanh\alpha\Big)\,e_3\\
&\quad+\Big(\frac{\mu_x}{m}+\frac{\mu m_x}{m^2}+\mu(\beta+\varphi)\tanh\alpha\Big)\,e_4,\\
(\bar{\nabla}_{e_2}h)(e_1,e_2) &= \Big(\frac{\delta_y}{m}-\frac{\delta m_y}{m^2}-\delta(\delta+\mu)\tanh\alpha\Big)\,e_3\\
&\quad+\Big(\frac{\varphi_y}{m}+\frac{\varphi m_y}{m^2}+\varphi(\delta+\mu)\tanh\alpha\Big)\,e_4.
\end{aligned}\tag{9.48}$$

On the other hand, from (9.36) we also have

$$\begin{aligned}
(\tilde{R}(e_1,e_2)e_2)^{\perp} &= -\operatorname{sech}\alpha\tilde{R}(e_1,e_2;e_2,Je_2)e_3\\
&\quad-\{\tanh\alpha\tilde{K}+\operatorname{sech}\alpha\tilde{R}(e_1,e_2;e_2,Je_1)\}e_4,\\
(\tilde{R}(e_2,e_1)e_1)^{\perp} &= \{\tanh\alpha\tilde{K}-\operatorname{sech}\alpha\tilde{R}(e_2,e_1;e_1,Je_2)\}e_3\\
&\quad-\operatorname{sech}\alpha\tilde{R}(e_2,e_1;e_1,Je_1)e_4.
\end{aligned}\tag{9.49}$$

From (9.33), (9.38), (9.48), (9.49) and Codazzi's equation we find

$$\begin{aligned}
\lambda_x-\delta_y &= (\lambda\beta+\lambda\varphi-\delta^2-\delta\mu)m\tanh\alpha-\frac{\delta m_y+3\lambda m_x}{m}\\
&\quad-m\operatorname{sech}\alpha\tilde{R}(e_1,e_2;e_2,Je_2),\\
\mu_x-\varphi_y &= (\delta\varphi-\beta\mu)m\tanh\alpha+\frac{\varphi m_y-\mu m_x}{m}\\
&\quad-m\operatorname{sech}\alpha\tilde{R}(e_1,e_2;e_2,Je_1)-m(\tanh\alpha)\tilde{K},\\
\beta_y-\delta_x &= (\beta\mu-\delta\varphi)m\tanh\alpha+\frac{\delta m_x-\beta m_y}{m}\\
&\quad-m\operatorname{sech}\alpha\tilde{R}(e_2,e_1;e_1,Je_2)+m(\tanh\alpha)\tilde{K},\\
\gamma_y-\varphi_x &= (\beta\varphi+\varphi^2-\delta\gamma-\gamma\mu)m\tanh\alpha-\frac{\varphi m_x+3\gamma m_y}{m}\\
&\quad-m\operatorname{sech}\alpha\tilde{R}(e_2,e_1;e_1,Je_1).
\end{aligned}\tag{9.50}$$

Also, from (9.33), (9.34), (9.41), (9.43) and (9.45) we have

$$A_{e_3}=\begin{pmatrix}\varphi & \mu\\ \gamma & \varphi\end{pmatrix},\quad A_{e_4}=\begin{pmatrix}\delta & \lambda\\ \beta & \delta\end{pmatrix},\tag{9.51}$$

$$\alpha_x=m(\varphi-\beta),\quad \alpha_y=m(\mu-\delta).\tag{9.52}$$

By applying (9.36), (9.37) and (9.51) we derive that

$$\tilde{R}(e_1, e_2; e_3, e_4) = (\operatorname{sech}^2\alpha - \tanh^2\alpha)\tilde{K} \\ - 2\operatorname{sech}\alpha\tanh\alpha\tilde{R}(e_1, e_2; e_2, Je_1), \tag{9.53}$$

$$\langle [A_{e_3}, A_{e_4}]e_1, e_2\rangle = \gamma\lambda - \beta\mu. \tag{9.54}$$

Now, we find from (9.35), (9.47), and (9.54) that

$$\tilde{g}(R^D(e_1, e_2)e_3, e_4) = \frac{2mm_{xy} - 2m_x m_y}{m^4} \\ + \{(\delta + \mu)\alpha_x - (\beta + \varphi)\alpha_y\}\frac{\operatorname{sech}^2\alpha}{m} \\ + \{(\delta + \mu)m_x - (\beta + \varphi)m_y + m(\delta_x + \mu_x - \beta_y - \varphi_y)\}\frac{\tanh\alpha}{m^2}. \tag{9.55}$$

Therefore, the equation of Ricci is given by

$$\beta\mu - \gamma\lambda + \frac{2mm_{xy} - 2m_x m_y}{m^4} + \{(\delta + \mu)\alpha_x - (\beta + \varphi)\alpha_y\}\frac{\operatorname{sech}^2\alpha}{m} \\ + \{(\delta + \mu)m_x - (\beta + \varphi)m_y + m(\delta_x + \mu_x - \beta_y - \varphi_y)\}\frac{\tanh\alpha}{m^2} \tag{9.56} \\ = (\operatorname{sech}^2\alpha - \tanh^2\alpha)\tilde{K} - 2\operatorname{sech}\alpha\tanh\alpha\tilde{R}(e_1, e_2; e_2, Je_1).$$

On the other hand, using (9.33) and (9.43) we find

$$(\delta + \mu)\alpha_x - (\beta + \varphi)\alpha_y = 2m(\delta\varphi - \beta\mu). \tag{9.57}$$

Also, by applying (9.50), we have

$$(\delta + \mu)m_x - (\beta + \varphi)m_y + m(\delta_x + \mu_x - \beta_y - \varphi_y) \\ = 2(\delta\varphi - \beta\mu)m^2\tanh\alpha - 2m^2\tanh\alpha\tilde{K} \\ + m^2\operatorname{sech}\alpha\big\{R(e_2, e_1; e_1, Je_2) - \tilde{R}(e_1, e_2; e_2, Je_1)\big\}. \tag{9.58}$$

Substituting (9.57) and (9.58) into equation (9.56) gives

$$\gamma\lambda + \beta\mu - 2\delta\varphi = \frac{2mm_{xy} - 2m_x m_y}{m^4} - \tilde{K} \\ - \tanh\alpha\operatorname{sech}\alpha\big\{\tilde{R}(e_2, e_1; e_1, Je_2) + \tilde{R}(e_1, e_2; e_2, Je_1)\big\}. \tag{9.59}$$

On the other hand, by applying (1.9) and (9.4), we find

$$\tilde{R}(e_2, e_1; e_1, Je_2) = -\tilde{R}(e_1, e_2; e_2, Je_1).$$

Combining this with (9.59) shows that (9.59) is nothing but equation (9.46) of Gauss. Consequently, Ricci's equation is a consequence of equations of Gauss and Codazzi. □

Remark 9.4. Similarly, one may also prove that Gauss' equation is a consequence of equations of Codazzi and Ricci. The same phenomena occurs for purely real surfaces in Kähler surfaces as well [Chen (2010f)].

9.5 Totally real and Lagrangian submanifolds

The study of totally real and Lagrangian submanifolds of Kähler manifolds from differential geometric point of view was initiated in the early 1970s. Such submanifolds have been studied extensively by many geometers during the last three decades. (For a detailed survey on Lagrangian submanifolds of Kähler manifolds from Riemannian view point, see [Chen (2001e)].)

Definition 9.8. A *totally real submanifold* N of a pseudo-Kähler manifold M is a pseudo-Riemannian submanifold such that the complex structure J of M carries each tangent space of N into the corresponding normal space, i.e., $J(T_pN) \subset T_p^{\perp}N$, $p \in N$. A totally real submanifold N of M is called *Lagrangian* if $J(T_pN) = T_p^{\perp}N$ for each $p \in N$.

Since curves in a pseudo-Kähler manifold are totally real, we only consider totally real submanifolds of dimension ≥ 2. A well known result of [Gromov (1970)] states that any embedded compact Lagrangian submanifold in $\mathbf{C}_s^n$ is not simply-connected (for a complete proof of this see [Sikorav (1986)]). Gromov's result is false when the Lagrangian submanifolds were immersed but not embedded. The simplest example is the following.

Example 9.5. Consider the *Whitney n-sphere* defined by

$$w(y_0, y_1, \dots, y_n) = \frac{1 + \mathrm{i}\, y_0}{1 + y_0^2}(y_1, \dots, y_n), \tag{9.60}$$

with $y_0^2 + y_1^2 + \cdots + y_n^2 = 1$. The Whitney's immersion $w : S^n \to \mathbf{C}^n$ is a Lagrangian immersion of an n-sphere S^n into $\mathbf{C}^n$ which has a unique self-intersection point at $w(-1, 0, \dots, 0) = w(1, 0, \dots, 0)$.

Some basic properties of Lagrangian submanifolds are the following.

Lemma 9.2. *Let N be a Lagrangian submanifold of a pseudo-Kähler manifold M. If $\sigma = Jh$, then for X, Y, Z tangent to N we have*

(1) *$g(\sigma(X,Y), Z)$ is totally symmetric, i.e.,*

$$g(\sigma(X,Y), Z) = g(\sigma(Y,Z), X) = g(\sigma(Z,X), Y);$$

(2) *the induced Levi-Civita connection ∇' and the normal connection D of N satisfy $D_X(JY) = J\nabla'_X Y$;*
(3) *$J(R(X,Y)Z) = R^D(X,Y)JZ$;*
(4) *N is flat if and only N has flat normal connection;*

(5) *if M is an (indefinite) complex space form, we have*

$$(\bar{\nabla}\sigma)(X,Y,Z) := \nabla'_X\sigma(Y,Z) - \sigma(\nabla'_XY,Z) - \sigma(Y,\nabla'_XZ) \tag{9.61}$$

is totally symmetric.

Proof. Let X,Y,Z be tangent to N. Then

$$J\nabla'_XY + Jh(X,Y) = J\nabla_XY = \nabla_X(JY) = -A_{JY}X + D_X(JY),$$

which implies (2) and $\sigma(X,Y) = -A_{JY}X$. Thus, after applying (2.16) and symmetry of h we obtain (1).

(3) follows from (2); and (4) is an immediate consequence of (3).

If N is a Lagrangian submanifold of an (indefinite) complex space form $M^n_s(c)$, then Codazzi's equation implies that $(\bar{\nabla}_Xh)(Y,Z) = (\bar{\nabla}_Yh)(X,Z)$. Hence, by applying (2) and $\sigma = Jh$, we have (9.61). □

The equations of Gauss, Codazzi and Ricci for Lagrangian submanifolds N in an (indefinite) complex space form $M^m_s(4c)$ are given respectively by

$$R(X,Y;Z,W) = \langle A_{h(Y,Z)}X,W\rangle - \langle A_{h(X,Z)}Y,W\rangle + c\,(\langle X,W\rangle\langle Y,Z\rangle - \langle X,Z\rangle\langle Y,W\rangle), \tag{9.62}$$

$$(\bar{\nabla}\sigma)(X,Y,Z) = (\bar{\nabla}\sigma)(Y,X,Z), \tag{9.63}$$

$$R^D(X,Y;JZ,JW) = \langle[A_{JZ},A_{JW}]X,Y\rangle + c\,(\langle X,W\rangle\langle Y,Z\rangle - \langle X,Z\rangle\langle Y,W\rangle) \tag{9.64}$$

for X,Y,Z,W tangent to N.

Due to Lemma 9.2(2), equation (9.64) of Ricci is nothing but equation (9.62) of Gauss. Consequently, after applying a result of [Eschenburg and Tribuzy (1993)] we obtain the following existence and uniqueness theorems for Lagrangian submanifolds in (indefinite) complex space forms.

Theorem 9.2. *Let (N^n_s,g) be a simply-connected pseudo-Riemannian n-manifold with index $s \geq 0$. If σ is a TN^n_s-valued symmetric bilinear form on N^n_s such that*

(a) *$g(\sigma(X,Y),Z)$ is totally symmetric,*
(b) *$(\bar{\nabla}\sigma)(X,Y,Z)$ is totally symmetric,*
(c) *$R(X,Y)Z = cg(Y,Z)X - cg(X,Z)Y + \sigma(\sigma(Y,Z),X) - \sigma(\sigma(X,Z),Y)$,*

then there is a Lagrangian isometric immersion $L : N^n_s \to M^n_s(4c)$ of N^n_s into a complete simply-connected (indefinite) complex space form $M^n_s(4c)$ whose second fundamental form h is given by $h(X,Y) = J\sigma(X,Y)$.

Theorem 9.3. *Let $L_1, L_2 : N_s^n \to M_s^n(4c)$ be two Lagrangian isometric immersions of a pseudo-Riemannian manifold N_s^n with second fundamental forms h^1 and h^2. If $\langle h^1(X,Y), JL_{1\star}Z\rangle = \langle h^2(X,Y), JL_{2\star}Z\rangle$, for all vector fields X, Y, Z tangent to N_s^n, then there exists an isometry Φ of $M_s^n(4c)$ such that $L_1 = \Phi \circ L_2$.*

Remark 9.5. There exist essential difference between Lagrangian submanifolds of Kähler manifolds and of pseudo-Kähler manifolds. For instance, the only minimal Lagrangian submanifolds of constant curvature c in a complex space form of constant holomorphic curvature $4c$ are the totally geodesic ones [Chen and Ogiue (1974a); Ejiri (1982)]. In contrast it is shown in [Chen and Vrancken (2002a); Vrancken (2001)] that there exist many non-totally geodesic minimal Lagrangian Lorentzian submanifolds of constant curvature c in a Lorentzian complex space form $M_1^n(4c)$.

9.6 CR-submanifolds of pseudo-Kähler manifolds

Let N be an n-manifold and $T^{\mathbf{C}}N$ its complexified tangent bundle, i.e.,

$$T_p^{\mathbf{C}}N = T_pN \otimes_{\mathbf{R}} \mathbf{C} \equiv T_pN \oplus \mathrm{i}(T_pN).$$

Let $\mathcal{H}$ be a complex subbundle of complex dimension h. A *CR-manifold* of real dimension n and CR-dimension h is a pair $(N, \mathcal{H})$ such that $\mathcal{H}$ is involutive and $\mathcal{H}_p \cap \bar{\mathcal{H}}_p = 0$ [Greenfield (1968)].

If $(N, \mathcal{H})$ is a CR-manifold, then there exists a unique subbundle $\mathcal{D}$ of TN such that $\mathcal{D}^{\mathbf{C}} = \mathcal{H} \oplus \bar{\mathcal{H}}$, and a unique bundle map $\mathcal{J} : \mathcal{D} \to \mathcal{D}$ such that $\mathcal{J}^2 = -I$ and $\mathcal{H} = \{X - \mathrm{i}\mathcal{J}X : X \in \mathcal{D}\}$.

Let N be a pseudo-Riemannian submanifold of a pseudo-Hermitian manifold (M, g, J). For each point $p \in N$, let $\mathcal{D}_p$ denote the maximal holomorphic subspace of T_pN, i.e., $\mathcal{D}_p = T_pN \cap J(T_pN)$. When $\dim \mathcal{D}_p$ is the same for all p, $\{\mathcal{D}_p : p \in N\}$ defines a *holomorphic distribution* $\mathcal{D}$.

Definition 9.9. A pseudo-Riemannian submanifold N of an almost-Hermitian manifold M, equipped with a pseudo-Riemannian metric, is called a *CR-submanifold* if there exist a holomorphic distribution $\mathcal{D}$ and a totally real distribution $\mathcal{D}^\perp$ on N, $J(\mathcal{D}_p^\perp) \subset T_p^\perp M$, such that $TN = \mathcal{D} \oplus \mathcal{D}^\perp$.

A CR-submanifold is called *proper* if $\operatorname{rank} \mathcal{D} > 0$ and $\operatorname{rank} \mathcal{D}^\perp > 0$.

The notion of CR-submanifolds was introduced in [Bejancu (1978)]. Such submanifolds in Kähler manifolds have been studied extensively since late 1970s. For surveys of main results on CR-submanifolds of Kähler manifolds see [Yano and Kon (1983); Bejancu (1986); Chen (2000c)].

If N is a CR-submanifold of a pseudo-Hermitian manifold, let P and Q denote the natural projections of TN to $\mathcal{D}$ and $\mathcal{D}^\perp$, respectively. Since the holomorphic distribution $\mathcal{D}$ is J-invariant, we may put $\mathcal{J} = J \circ P$ and define a complex subbundle $\mathcal{H}$ on $T^{\mathbf{C}}N$ by $\mathcal{H} = \{X - \mathrm{i}\mathcal{J}X : X \in \mathcal{D}\}$.

Lemma 9.3. $\mathcal{J}(X - \mathrm{i}\mathcal{J}X) = J(PX) - \mathrm{i}\mathcal{J}(JPX)$.

Proof. Follows easily from definitions. □

Lemma 9.4. *Let N be a CR-submanifold of a pseudo-Hermitian manifold (M, J, g). Then $Q([JX, Y] + [X, JY]) = 0$ for $X, Y \in \mathcal{D}$.*

Proof. Since M is pseudo-Hermitian, $[J, J] = 0$. Hence

$$0 = [J, J](JX, Y) = J([X, Y] - [JX, JY]) - [JX, Y] - [X, JY],$$

but $[X, Y]$ and $[JX, JY]$ are tangent to N, thus $J([X, Y] - [JX, JY])$ has no $\mathcal{D}^\perp$-component. So $[JX, Y] + [X, JY]$ has no $\mathcal{D}^\perp$-component. □

For CR-submanifolds we have the following [Blair and Chen (1979)].

Theorem 9.4. *Every CR-submanifold of a pseudo-Hermitian manifold is a CR-manifold.*

Proof. Let $X, Y \in \mathcal{D}$. Then, by $[J, J] = 0$ and Lemma 9.4, we find

$$\begin{aligned}[X - \mathrm{i}\mathcal{J}X, Y - \mathrm{i}\mathcal{J}Y] &= [X, Y] - [JX, JY] - \mathrm{i}[JX, Y] - \mathrm{i}[X, JY] \\ &= -J[JX, Y] - J[X, JY] - \mathrm{i}P[JX, Y] - \mathrm{i}P[X, JY].\end{aligned}$$

Continuing this computation using Lemmas 9.3 and 9.4, we obtain

$$\begin{aligned}[X - \mathrm{i}\mathcal{J}X, Y - \mathrm{i}\mathcal{J}Y] &= \mathrm{i}\mathcal{J}^2([JX, Y] + [X, JY]) - \mathcal{J}([JX, Y] + [X, JY]) \\ &= -\mathcal{J}([JX, Y] - \mathrm{i}\mathcal{J}[JX, Y]) - \mathcal{J}([X, JY] - \mathrm{i}\mathcal{J}[X, JY]) \in \mathcal{H}.\end{aligned}$$

□

A fundamental properties of CR-submanifolds of pseudo-Kähler manifolds is the following.

Proposition 9.10. *The totally real distribution of a CR-submanifold N of a pseudo-Kähler manifold M is a nondegenerate integrable distribution.*

Proof. Let N be a CR-submanifold of a pseudo-Kähler manifold M. If a vector $v \in \mathcal{D}_p^\perp$ is orthogonal to every vector in $\mathcal{D}_p^\perp$, then it is orthogonal to every vector in T_pN. Thus $v = 0$, so $\mathcal{D}^\perp$ is nondegenerate.

Next, assume that X is a vector field in $\mathcal{D}$ and Z, W in $\mathcal{D}^\perp$. Then we find from (9.2) that $\Omega(X,Z) = \Omega(X,W) = \Omega(Z,W) = 0$. Since M is pseudo-Kählerian, $d\Omega = 0$. Hence

$$\begin{aligned} 0 &= 3d\Phi(X,Z,W) \\ &= X\Omega(Z,W) - Z\Omega(X,W) + W\Omega(X,Z) - \Omega([X,Z],W) \\ &\qquad - \Omega([W,X],Z) - \Omega([Z,W],X) \\ &= \Omega(J[Z,W],X), \end{aligned}$$

which implies that $\mathcal{D}^\perp$ is an integrable distribution. □

Corollary 9.1. *Every CR-submanifold N of a pseudo-Kähler manifold M is foliated by totally real pseudo-Riemannian submanifolds of M.*

Proof. Follows immediately from Proposition 9.10. □

Let $\mathcal{K}$ be a nondegenerate distribution of a pseudo-Riemannian manifold N and let $\mathcal{K}^\perp$ denote the orthogonal complementary distribution of $\mathcal{K}$. Put

$$\mathring{h}(X,Y) = (\nabla'_X Y)^\perp \tag{9.65}$$

for vector fields X, Y in $\mathcal{K}$, where $(\nabla'_X Y)^\perp$ is the $\mathcal{K}^\perp$-component of $\nabla'_X Y$. Then $\mathring{h}$ is a well-defined $\mathcal{K}^\perp$-valued $(0,2)$-tensor field. Moreover, it follows from Frobenius' theorem that $\mathcal{K}$ is integrable if and only if $\mathring{h}$ is symmetric.

Let $e_1, \ldots, e_k$ be an orthonormal basis of $\mathcal{K}$. If we put

$$\mathring{H} = \frac{1}{k}\text{trace}\,\mathring{h} = \frac{1}{k}\sum_{j=1}^{k} \epsilon_j \mathring{h}(e_j,e_j), \tag{9.66}$$

then, up to sign, $\mathring{H}$ is a well-defined vector field, which is called the *mean curvature vector of* $\mathcal{K}$. If $\mathring{H} = 0$, $\mathcal{K}$ is called a *minimal distribution*. In particular, if $\mathring{h} = 0$, then $\mathcal{K}$ is called a *totally geodesic distribution*.

Every totally geodesic distribution on N is an integrable distribution whose leaves are totally geodesic submanifolds of N.

Another fundamental properties of CR-submanifolds is the following.

Proposition 9.11. *If N is a CR-submanifold of a pseudo-Kähler manifold, then*

(a) *the holomorphic distribution $\mathcal{D}$ of N is a nondegenerate minimal distribution;*
(b) *$\mathcal{D}$ is integrable if and only if $\mathring{h}(JX,Y) = \mathring{h}(X,JY)$ for X, Y in $\mathcal{D}$.*

Proof. Let N be a CR-submanifold of a pseudo-Kähler manifold M. Denote by ∇ the Levi-Civita connection of M. Since $\mathcal{D}^{\perp}$ is nondegenerate, its orthogonal distribution $\mathcal{D}$ is also nondegenerate.

For each vector field X in $\mathcal{D}$ and Z in $\mathcal{D}^{\perp}$ we have

$$\langle Z, \nabla'_X X\rangle = \langle JZ, \nabla_X JX\rangle = -\langle \nabla_X JZ, JX\rangle = \langle A_{JZ}X, JX\rangle .$$

Replacing X by JX yields

$$\langle Z, \nabla'_{JX} JX\rangle = -\langle A_{JZ}X, X\rangle = -\langle A_{JZ}X, JX\rangle .$$

Combining both equations gives $\langle Z, \nabla'_X X + \nabla'_{JX} JX\rangle = 0$. Consequently, we have $\mathring{H} = 0$, which gives (a).

For vector fields X, Y in $\mathcal{D}$ and Z in $\mathcal{D}^{\perp}$ we have $\langle [X,Y], JZ\rangle = 0$. Thus

$$\begin{aligned}\langle \mathring{h}(X, JY), Z\rangle &= \langle \nabla'_X JY, Z\rangle = \langle \nabla_X JY, Z\rangle = -\langle \nabla_X Y, JZ\rangle \\ &= -\langle \nabla_Y X, JZ\rangle = \langle \nabla'_Y JX, Z\rangle .\end{aligned}$$

Combining this with $\langle \mathring{h}(JX, Y), Z\rangle = \langle \nabla'_{JX} Y, Z\rangle$ yields

$$\langle \mathring{h}(JX, Y) - \mathring{h}(X, JY), Z\rangle = \langle [X, JY], Z\rangle .$$

Hence we have (b). □

The pseudo-sphere $S^{2n-1}_{2s}(1)$ and pseudo-hyperbolic space $H^{2n-1}_{2s-1}(-1)$ of $\mathbf{C}^n_s$ are totally umbilical proper CR-submanifolds of flat indefinite complex space forms. In contrast, we have the following.

Proposition 9.12. *There are no totally umbilical proper CR-submanifolds in any non-flat indefinite complex space form.*

Proof. If N is a totally umbilical CR-submanifold of a pseudo-Kähler manifold M. Then it follows from (2.26) and (2.56) that

$$(\bar{\nabla}_X h)(Y, Z) = \langle Y, Z\rangle D_X H$$

for X, Y, Z tangent to N. Thus, Codazzi's equation yields

$$\tilde{R}(X, Z; JX, JZ) = \langle Z, JX\rangle \langle D_X H, JZ\rangle - \langle X, JX\rangle \langle D_Z H, JZ\rangle = 0$$

for vectors X in $\mathcal{D}$ and Z in $\mathcal{D}^{\perp}$.

On the other hand, since M is pseudo-Kählerian, we also have

$$\tilde{R}(X, Z; JZ, JX) = \tilde{R}(X, Z, Z, X) = \tilde{K}(X, Z).$$

Hence, if M is an indefinite complex space form, it must be flat. □

Definition 9.10. A pseudo-Riemannian submanifold of a pseudo-Kähler manifold is called a *CR-warped product* if it is a warped product $N_T \times_f N_\perp$ of an invariant submanifold N_T and a totally real submanifold $N_\perp$.

For arbitrary CR-warped products in an arbitrary Kähler manifold, we have the following general result from [Chen (2001b)].

Theorem 9.5. *Let $N = N_T \times_f N_\perp$ be a CR-warped product in an arbitrary Kähler manifold M. Then*

$$||h||^2 \geq 2p\,||\nabla(\ln f)||^2, \tag{9.67}$$

where $\nabla \ln f$ is the gradient of $\ln f$ and p is the dimension of $N_\perp$.

If the equality sign of (9.67) holds identically, then N_T is a totally geodesic submanifold and $N_\perp$ is a totally umbilical submanifold of M. Moreover, N is a minimal submanifold in M.

Remark 9.6. CR-warped products satisfying the equality case of (9.67) in complex space forms were completely classified in [Chen (2001b,c)]. For further results on CR-warped products in Kähler manifolds, see the survey article [Chen (2008a)].

9.7 Slant submanifolds of pseudo-Kähler manifolds

Another important class of pseudo-Riemannian submanifolds of pseudo-Kähler manifolds is the family of slant submanifolds. Slant submanifolds in Kähler manifolds have been studied extensively since 1990 by many geometers (see [Chen (2000c)]). In this section we provide some basic properties of slant submanifolds in pseudo-Kähler manifolds.

Definition 9.11. A pseudo-Riemannian submanifold N of a pseudo-Kähler manifold is called *slant* if the squared norm $||Pu||^2$ is independent of the choice of unit vector $u \in TN$, where P is defined by (9.17).

Clearly, if N is a spacelike slant submanifold of a pseudo-Kähler manifold, then $\langle Pu, Pu \rangle = k \geq 0$ for each u in the unit tangent bundle T^1N.

Lemma 9.5. *Let N be a pseudo-Riemannian manifold with indefinite metric. Assume that N is isometrically immersed in a pseudo-Kähler manifold M as a slant submanifold.*

1. *If $\langle Pu, Pu \rangle = 0$ for spacelike unit vectors $u \in TN$, then N is totally real;*
2. *If $\langle Pu, Pu \rangle < 0$ for spacelike unit vectors $u \in TN$, then N is even-dimensional and N has neutral metric.*

Proof. Under the hypothesis, if $\langle Pu, Pu \rangle = 0$ holds for each spacelike unit vector $u \in TN$, then $\langle Pu, Pu \rangle = 0$ holds for any unit tangent vector u in TN. Thus $\langle P^2u, u \rangle = 0$ hold identically on TN. Hence, by polarization we get $P^2 = 0$. Now, by using (9.19) and $P^2 = 0$ we find $P = 0$, which shows that N is totally real.

If $\langle Pu, Pu \rangle = k < 0$ for spacelike unit tangent vectors u, then P maps spacelike vectors into timelike vectors. On the other hand, since N is slant, we have $\langle Pv, Pv \rangle = \pm k$ for unit timelike vectors v. If $\langle Pv_o, Pv_o \rangle = -k$ for some timelike vector v_o, then by continuity $\langle Pv, Pv \rangle = -k < 0$ for all unit timelike vectors v. Thus, P maps timelike vectors into timelike vector. But this is impossible since, for a unit spacelike vector u, Pu is timelike and $\langle P^2u, u \rangle = k > 0$. Hence, $\langle Pv, Pv \rangle = k$ for every unit timelike vector v. Therefore, P maps timelike vectors into spacelike vectors. Consequently, $\dim N$ is even, say $2n$, and the metric is neutral. □

Lemma 9.6. *Let N be a slant submanifold of a pseudo-Kähler manifold. If $\langle Pu, Pu \rangle = k < 0$ for spacelike unit vectors $u \in TN$, then there exist a positive real number θ and an orthogonal decomposition $TN = \mathcal{D}_s \oplus \mathcal{D}_t$ with $\dim \mathcal{D}_s = \dim \mathcal{D}_t$ such that*

(1) $P^2 = (\sinh^2 \theta)I$;
(2) *$\mathcal{D}_s$ is a space-like distribution and $\mathcal{D}_t$ is a time-like distribution;*
(3) *$P(\mathcal{D}_s) = \mathcal{D}_t$ and $P(\mathcal{D}_t) = \mathcal{D}_s$,*

Proof. Let N be a slant submanifold of a pseudo-Kähler manifold. If $\langle Pu, Pu \rangle = k < 0$ for unit spacelike vectors u, we may put $k = -\sinh^2 \theta$ for some positive number θ. Then $\langle Pu, Pu \rangle = -\sinh^2 \theta \langle u, u \rangle$. Thus

$$\langle P^2u, u \rangle = \sinh^2 \theta \langle u, u \rangle \quad \text{for each spacelike vector } u. \tag{9.68}$$

Similarly, we have $\langle P^2v, v \rangle = \sinh^2 \theta \langle v, v \rangle$ for timelike vector v. Hence, by continuity, we find $\langle P^2w, w \rangle = \sinh^2 \theta \langle w, w \rangle$ for any $w \in TN$. Now, by applying polarization we have $\langle P^2u, v \rangle = \sinh^2 \theta \langle u, v \rangle$. Since N is pseudo-Riemannian, this gives (1).

Since the dimension of N is even, say $2n$, and the metric has signature (n, n), we may choose an orthonormal frame $\{e_1, \ldots, e_n, e_{n+1}, \ldots, e_{2n}\}$ of N such that $\epsilon_1 = \cdots = \epsilon_n = -1$ and $\epsilon_{n+1} = \cdots = \epsilon_{2n} = 1$. Now, if we put $\mathcal{D}_t = \text{Span}\{e_1, \ldots, e_n\}, \mathcal{D}_s = \text{Span}\{e_{n+1}, \ldots, e_{2n}\}$, we find (2) and (3). □

The following two results can be found in [Arslan et al. (2010)].

Proposition 9.13. *Let N be a Lorentz surface in a Lorentzian Kähler surface. Then N is a slant surface if and only if $\nabla' P = 0$ holds identically.*

Proof. Let N be a Lorentz surface in a Lorentzian Kähler surface. Then N is purely real by Proposition 9.9. Let us choose an orthonormal frame $\{e_1, e_2\}$ on N such that $\langle e_1, e_1\rangle = -1, \langle e_1, e_2\rangle = 0, \langle e_2, e_2\rangle = 1$. If we put

$$Pe_1 = \sinh \alpha e_2,\ Fe_1 = \cosh \alpha e_3,\ Fe_2 = \cosh \alpha e_4, \tag{9.69}$$

then

$$\begin{aligned} &Je_1 = \sinh \alpha e_2 + \cosh \alpha e_3, \quad Je_2 = \sinh \alpha e_1 + \cos \alpha e_4, \\ &\langle e_3, e_3\rangle = -1,\ \langle e_3, e_4\rangle = 0,\ \langle e_4, e_4\rangle = 1. \end{aligned} \tag{9.70}$$

Let us put

$$\nabla'_X e_i = \sum_{j=1}^{2} \omega_i^j(X) e_j,\ \ D_X e_r = \sum_{s=3}^{4} \omega_r^s(X) e_s.$$

Then

$$\omega_1^1 = \omega_2^2 = \omega_3^3 = \omega_4^4 = 0,\ \ \omega_1^2 = \omega_1^2,\ \ \omega_3^4 = \omega_4^3. \tag{9.71}$$

Hence, we find from (9.21), (9.69) and (9.71) that

$$(\nabla'_X P)e_1 = (X\alpha)(\cosh \alpha)e_2,\ \ (\nabla'_X P)e_2 = (X\alpha)(\cosh \alpha)e_1,$$

which imply that α is constant if and only if $\nabla' P \equiv 0$. Consequently, N is slant if and only if $\nabla' P \equiv 0$. □

Proposition 9.14. *Let N be a Lorentz surface in a Lorentzian Kähler surface. If N is non-Lagrangian, then N is a minimal slant surface if and only if $\nabla' F = 0$ holds identically on N.*

Proof. Let N be a Lorentz surface in a Lorentzian Kähler surface. Then N is purely real. Let us choose e_1, e_2, e_3, e_4 as in the proof of Proposition 9.13. Then it follows from (9.22), (9.70) and (9.71) that

$$\begin{aligned} &(\nabla'_X F)e_1 = (X\alpha)(\sinh \alpha)e_3 + (\cosh \alpha)(\omega_3^4(X) - \omega_1^2(X))e_4, \\ &(\nabla'_X F)e_2 = (X\alpha)(\sinh \alpha)e_4 + (\cosh \alpha)(\omega_3^4(X) - \omega_1^2(X))e_3. \end{aligned} \tag{9.72}$$

If $\nabla' F \equiv 0$, then (9.72) gives $X\alpha = 0$ and $\omega_3^4 = \omega_1^2$. Thus N is a slant surface. Also from Proposition 9.8(iii) we find

$$h(X, PY) = fh(X, Y). \tag{9.73}$$

Thus $h(e_1, Pe_2) = h(Pe_1, e_2)$. Combining this with (9.70) yields

$$(h(e_1, e_1) - h(e_2, e_2)) \sinh \alpha = 0.$$

Since N is non-Lagrangian, this shows that $\operatorname{trace} h = 0$, i.e., N is minimal.

Conversely, if N is minimal and slant, then Proposition 9.7 and Proposition 9.13 imply that $A_{FX}Y = A_{FY}X$. Since N is minimal, we also have

$$h(e_1, e_1) = h(e_2, e_2). \tag{9.74}$$

It follows from $A_{FX}Y = A_{FY}X$ and (9.74) that

$$h(e_1, e_1) = h(e_2, e_2) = \beta e_3 + \gamma e_4,\ h(e_1, e_2) = -\gamma e_3 - \beta e_4 \tag{9.75}$$

for some functions β and γ. Now, by applying (9.70) and (9.75) we obtain (9.73). Consequently, $\nabla' F = 0$ according to Proposition 9.8. □

Chapter 10

Para-Kähler Manifolds

Para-complex numbers were introduced in [Graves (1845)] as generalization of complex numbers. Such numbers have the expression $x + y\mathrm{j}$, where x, y are real numbers and j satisfies $\mathrm{j}^2 = 1$, $\mathrm{j} \neq 1$. J. T. Graves applied para-complex numbers to solve some questions in number theory.

Para-Kähler geometry is the geometry related to the algebra of para-complex numbers. Properties of para-Kähler manifolds were first studied in [Rashevskij (1948)], in which P.K. Rashevski considered a neutral metric of signature (n, n) defined from a potential function on a locally product $2n$-manifold. He called such manifolds *stratified space*. Para-Kähler manifolds were explicitly defined in [Rozenfeld (1949)]. B. A. Rozenfeld compared Rashevskij's definition with Kähler's definition in the complex case and established the analogy between Kähler and para-Kähler ones. Such manifolds were defined independently in [Ruse (1949)]; and studied in [Libermann (1954)] in the context of G-structures.

The Levi-Civita connection of a para-Kähler manifold $(M, g, \mathcal{P})$ preserves the almost product structure $\mathcal{P}$; or equivalently, its holonomy group Hol_x preserves the eigenspace decomposition $T_xM = T_x^+ \oplus T_x^-$. The parallel eigendistributions $T^\pm$ of $\mathcal{P}$ are g-isotropic integrable distributions. Moreover, they are Lagrangian distributions with respect to the para-Kähler form $\omega = g \circ \mathcal{P}$, which is parallel and hence closed. The leaves of these distributions are totally geodesic submanifolds. Thus, from the standpoint of symplectic manifolds, a para-Kähler structure can be regarded as a pair of complementary integrable Lagrangian distributions (T^+, T^-) on a symplectic manifold (M, ω). Such a structure is often called a *bi-Lagrangian structure* or a *Lagrangian 2-web*.

Para-Kähler manifolds have been applied to supersymmetric field theories as well as to string theory in recent years.

10.1 Para-Kähler manifolds

Definition 10.1. An *almost para-Hermitian manifold* is a manifold M endowed with an almost product structure $\mathcal{P} \neq \pm I$ and a pseudo-Riemannian metric g such that

$$\mathcal{P}^2 = I, \quad g(\mathcal{P}v, \mathcal{P}w) = -g(v, w) \tag{10.1}$$

for vectors $v, w \in T_p(M)$, $p \in M$, where I is the identity map. The *fundamental 2-form* Ω of M is defined by

$$\Omega(X, Y) = g(X, \mathcal{P}Y), \quad X, Y \in TM.$$

Definition 10.1 shows that the dimension of an almost para-Hermitian manifold M is even and the metric is neutral.

Definition 10.2. An almost para-Hermitian manifold $(M, \mathcal{P}, g)$ is called *para-Kähler* (or *hyperbolic Kähler*) if it satisfies $\nabla\mathcal{P} = 0$ identically, where ∇ is the Levi-Civita connection of M.

Definition 10.3. Two para-Kähler manifolds $(M, \mathcal{P}, g)$ and $(M', \mathcal{P}', g')$ are *$\mathcal{P}$-isometric* if there is an isometric $\psi : M \to M'$ such that $\psi_* \circ \mathcal{P} = \mathcal{P}' \circ \psi_*$.

There are many para-Kähler manifolds, for instance, a homogeneous manifold $M = G/H$ of a semisimple Lie group G admits an invariant para-Kähler structure $(g, \mathcal{P})$ if and only if it is a covering of the adjoint orbit $\mathrm{Ad}_G h$ of a semisimple element h [Hou et al. (1997)].

The simplest example of para-Kähler manifold is $(\mathbb{E}^{2n}_n, \mathcal{P}, g_0)$ consisting of the pseudo-Euclidean $2n$-space $\mathbb{E}^{2n}_n$, the standard flat neutral metric

$$g_0 = -\sum\nolimits_{j=1}^{n} dx_j^2 + \sum\nolimits_{j=1}^{n} dy_j^2, \tag{10.2}$$

and the almost product structure

$$\mathcal{P} = \sum\nolimits_{j=1}^{n} \frac{\partial}{\partial y_j} \otimes dx_j + \sum\nolimits_{j=1}^{n} \frac{\partial}{\partial x_j} \otimes dy_j.$$

We simply called $(\mathbb{E}^{2n}_n, \mathcal{P}, g_0)$ the *para-Kähler n-plane*, denoted by $\mathcal{P}K^n$.

Lemma 10.1. *The curvature tensor and the Ricci tensor of a para-Kähler manifold satisfy*

$$R(X, Y) \circ \mathcal{P} = \mathcal{P} \circ R(X, Y); \tag{10.3}$$

$$R(\mathcal{P}X, \mathcal{P}Y) = -R(X, Y), \quad R(X, \mathcal{P}Y) = -R(\mathcal{P}X, Y); \tag{10.4}$$

$$Ric(\mathcal{P}X, \mathcal{P}Y) = -Ric(X, Y). \tag{10.5}$$

Proof. The proof of (10.3) is the same as that of (8.3). (10.4) follows from (1.9), (10.1), (10.3) and $\mathcal{P}^2 = I$. (10.5) follows from (10.4). $\square$

From (10.1) we find

$$g(\mathcal{P}v, w) + g(v, \mathcal{P}w) = 0, \quad v, w \in T_pM, \ p \in M. \tag{10.6}$$

Thus $g(v, \mathcal{P}v) = 0$. When $\{v, \mathcal{P}v\}$ determines a nondegenerate plane, the sectional curvature $H^{\mathcal{P}}(v) = K(v \wedge \mathcal{P}v)$ is called the *para-holomorphic sectional curvature* of the $\mathcal{P}$-plane section spanned by v.

10.2 Para-Kähler space forms

Definition 10.4. A *para-Kähler space form* is a para-Kähler manifold of constant para-holomorphic sectional curvature.

Proposition 10.1. *Let M be a para-Kähler manifold of constant para-holomorphic sectional curvature c. Then the Riemann curvature tensor of M satisfies*

$$\begin{aligned} R(X,Y;Z,W) = \frac{c}{4}\{ &\langle X,W\rangle \langle Y,Z\rangle - \langle X,Z\rangle \langle Y,W\rangle + \langle \mathcal{P}X,W\rangle \langle Y,\mathcal{P}Z\rangle \\ &- \langle X,\mathcal{P}Z\rangle \langle \mathcal{P}Y,W\rangle - 2\langle X,\mathcal{P}Y\rangle \langle \mathcal{P}Z,W\rangle\}. \end{aligned}$$

Proof. This can be done in the same way as Proposition 9.2. $\square$

The model of a nonflat para-Kähler space form was constructed in [Gadea et al. (1989)] as follows: Let B be $\mathbf{R}^2$ with the product

$$(a,b)(a',b') = (aa', bb'),$$

then B is a commutative algebra. If we define the conjugate $\bar{w}$ of an element $w = (a,b) \in B$ by $\bar{w} = (b,a)$, then an element $w \in B$ is called *real* if $w = \bar{w}$, and is invertible if $w\bar{w} \neq 0$.

We put $B_+ = \{(a,b) \in B : a > 0, b > 0\}$. Then B_+ is a Lie group. Let

$$B_0^{r+1} = \{z = (z_j) \in B^{r+1} : \langle z, \bar{z}\rangle > 0\}, \quad \langle z, \bar{z}\rangle = \sum\nolimits_{j=0}^{r} z_j \bar{z}_j.$$

Denote by $\mathfrak{gl}(B; r+1)$ the algebra of $(r+1)\times(r+1)$-matrices with elements in B. Then $\mathfrak{gl}(B, r+1) = \mathfrak{gl}(\mathbf{R}; r+1) \times \mathfrak{gl}(\mathbf{R}; r+1)$. We have the Lie group

$$U(B; r+1) = \{Z \in \mathfrak{gl}(B; r+1) : \langle Zz, \bar{Z}\bar{z}\rangle = \langle z, \bar{z}\rangle \ \forall z \in B^{r+1}\}.$$

Let $P^r(B)$ be the quotient of B_0^{r+1} under the equivalence given by $(z_j) = (qz_j)$, $q \in B_+$. Then if $\pi : B_0^{r+1} \to P^r(B)$ is the natural projection, we can identify $\pi(z)$ with the unique element $w = qz$ such that $\langle w, \bar{w}\rangle = 1$, $\langle w, w\rangle = \langle \bar{w}, \bar{w}\rangle$, where $q = (a, b) \in B_+$. Indeed, if $z = (z_j) = ((u_j, v_j))$, we have

$$\langle w, \bar{w}\rangle = (ab\,\langle u, v\rangle, ab\,\langle u, v\rangle), \quad \langle w, w\rangle = (a^2\,\langle u, u\rangle, b^2\,\langle v, v\rangle),$$
$$\langle \bar{w}, \bar{w}\rangle = (b^2\,\langle v, v\rangle, a^2\,\langle u, u\rangle)$$

with

$$a = \frac{\langle v, v\rangle^{\frac{1}{4}}}{\langle u, u\rangle^{\frac{1}{4}}\,\langle u, v\rangle^{\frac{1}{2}}}, \quad b = \frac{\langle u, u\rangle^{\frac{1}{4}}}{\langle v, v\rangle^{\frac{1}{4}}\,\langle u, v\rangle^{\frac{1}{2}}}.$$

Thus

$$P^r(B) \cong \{(u, v) \in \mathbf{R}^{r+1} \times \mathbf{R}^{r+1} : \langle u, u\rangle = \langle v, v\rangle\,, \langle u, v\rangle = 1\}. \tag{10.7}$$

Since $Z(qz) = qZ(z)$ for all $Z \in U(B, r+1)$, $z \in B_0^{r+1}$, $q \in B_+$, the action of $U(B; r+1)$ pass to the quotient $P^r(B)$.

For a nonzero real number c, consider on B_0^{r+1} the tensor field

$$\begin{aligned}\tilde{g}_c = \frac{2}{c\,\langle u, v\rangle}\Bigg\{ &\sum_{j=0}^{r} du_j \otimes dv_j + \sum_{j=0}^{r} dv_j \otimes du_j \\ &- \frac{1}{\langle u, v\rangle} \sum_{i,j=0}^{r} u_i v_j (dv_i \otimes du_j + du_j \otimes dv_i)\Bigg\}.\end{aligned}$$

Then $\tilde{g}_c$ is invariant by $U(B, r+1)$. If $\iota : P^r(B) \to B_0^{r+1}$ is the inclusion, we have $(\iota \circ \pi)^*\tilde{g}_c = \tilde{g}_c$. Hence the tensor field $g_c = \iota^*\tilde{g}_c$ is a pseudo-Riemannian metric on $P^r(B)$, which is also invariant by $U(B; r+1)$. For $P^r(B)$ we have the coordinate charts $(U_j^{\pm}, \varphi_j)$, where

$$U_j^{+} = \{(u, v)\in P^r(B) : u_j, v_j > 0\}, \ U_j^{-} = \{(u, v)\in P^r(B) : u_j, v_j < 0\},$$
$$\varphi_j(u, v) = \left(\frac{u_0}{u_j}, \ldots, \widehat{\frac{u_j}{u_j}}, \ldots, \frac{u_r}{u_j}; \frac{v_0}{v_j}, \ldots, \widehat{\frac{v_j}{v_j}}, \ldots, \frac{v_r}{v_j}\right),$$

and $\widehat{\cdot}$ denotes the missing term. If we call (x_k, y_k) to the coordinates of any one of these charts, say $x_k = u_k/u_0, y_k = v_k/v_0$, then

$$\begin{aligned}g_c = \frac{2}{c(1 + \langle x, y\rangle)}\Bigg\{ &\sum_j dx_j \otimes dy_j + \sum_j dy_j \otimes dx_j \\ &- \frac{1}{1 + \langle x, y\rangle} \sum_{i,j} x_i y_j (dy_i \otimes dx_j + dx_j \otimes dy_i)\Bigg\}.\end{aligned} \tag{10.8}$$

Also, we have the almost-product structure on B_0^{r+1} given by

$$\hat{\mathcal{P}} = \sum_j \frac{\partial}{\partial u_j} \otimes du_j - \sum_j \frac{\partial}{\partial v_j} \otimes dv_j,$$

and it defines an almost-product structure $\mathcal{P}$ on $P^r(B)$ by the relation $\pi_* \circ \hat{\mathcal{P}} = \mathcal{P} \circ \pi_*$. Thus in the same chart $\mathcal{P}$ is given by

$$\mathcal{P} = \sum_j \frac{\partial}{\partial x_j} \otimes dx_j - \sum_j \frac{\partial}{\partial y_j} \otimes dy_j. \tag{10.9}$$

The $P^r(B)$, together with the metric (10.8) and the almost product structure $\mathcal{P}$ given by (10.9), is a para-Kähler manifold of constant para-holomorphic sectional curvature $c \neq 0$. Moreover, if $r > 1$, then $P^r(B)$ is complete, connected, simply connected para-Kähler space form. To obtain a complete simply-connected model of para-Kähler space form for $r = 1$, it is enough to extend the above structure on $P^1(B) = S^1 \times \mathbf{R}$ to its universal covering, which is $\mathbf{R}^2$. We denote by $P^r_c(B)$ the triple $(P^r(B), \mathcal{P}, g_c)$.

Two complete simply-connected para-Kähler space forms of equal para-holomorphic sectional curvature are $\mathcal{P}$-isometric [Gadea et al. (1989)].

10.3 Invariant submanifolds of para-Kähler manifolds

Definition 10.5. A pseudo-Riemannian submanifold N of a para-Kähler manifold $(M, \mathcal{P}, g)$ is called an *invariant submanifold* if each of its tangent spaces is invariant under the action of $\mathcal{P}$, i.e., $\mathcal{P}(T_pN) = T_pN$, $p \in N$.

Example 10.1. For two integers $r, p \geq 1$, we consider the mapping

$$\phi_{r,p} : P^r_c(B) \times P^p_c(B) \to \mathbf{R}^{r+p+rp+1} \times \mathbf{R}^{r+p+rp+1}$$

defined by

$$P^r_c(B) \times P^p_c(B) \ni (((u_j, v_j)), ((\hat{u}_\alpha, \hat{v}_\alpha))) \mapsto ((u_j\hat{u}_\alpha, v_j\hat{v}_\alpha)).$$

It is obvious that conditions $\langle u, u \rangle = \langle v, v \rangle, \langle \hat{u}, \hat{u} \rangle = \langle \hat{v}, \hat{v} \rangle$ and $\langle u, v \rangle = \langle \hat{u}, \hat{v} \rangle = 1$ imply that

$$\sum_{j=0}^{r} \sum_{\alpha=0}^{p} u_j^2 \hat{u}_\alpha^2 = \sum_{j=0}^{r} \sum_{\alpha=0}^{p} v_j^2 \hat{v}_\alpha^2, \quad \sum_{j=0}^{r} \sum_{\alpha=0}^{p} u_j v_j \hat{u}_\alpha \hat{v}_\alpha = 1.$$

Thus, according to (10.7), $\phi_{r,p}$ gives rise to a mapping

$$\varphi_{r,p} : P^r_c(B) \times P^p_c(B) \to P^{r+h+rp}_c(B). \tag{10.10}$$

It is direct to verify that $\varphi_{r,p}$ gives rise to an invariant submanifolds of $P^{r+h+rp}_c(B)$.

Lemma 10.2. *An invariant submanifold of a para-Kähler manifold is para-Kählerian with respect to its induced structures. Moreover, its second fundamental form satisfies*

$$h(\mathcal{P}X,Y)=h(X,\mathcal{P}Y)=\mathcal{P}h(X,Y),\quad X,Y\in TN. \tag{10.11}$$

Proof. If N is an invariant submanifold of a para-Kähler manifold M, (10.1) implies that N is an almost para-Hermitian manifold with respect to the induced metric g' and the induced para-structure, also denoted by $\mathcal{P}$. For any vector fields X,Y tangent to N we have

$$\nabla_X(\mathcal{P}Y)=\nabla'_X(\mathcal{P}Y)+h(X,\mathcal{P}Y). \tag{10.12}$$

On the other hand, since M is para-Kählerian, $\nabla\mathcal{P}=0$. Thus

$$\nabla_X(\mathcal{P}Y)=\mathcal{P}\nabla'_XY+\mathcal{P}h(X,Y). \tag{10.13}$$

Comparing (10.12) and (10.13) gives $\nabla'\mathcal{P}=0$ and $h(X,\mathcal{P}Y)=\mathcal{P}h(X,Y)$. Thus N is para-Kählerian. Further, by symmetry of h we have (10.11). □

Proposition 10.2. *Every invariant submanifold of a para-Kähler manifold is austere.*

Proof. Assume that N is an invariant submanifold of a para-Kähler manifold M. Then, by (10.1) and Lemma 10.2, we may choose a local orthonormal frame $\{e_1,\ldots,e_n,e_{1^*},\ldots,e_{n^*}\}$ on N such that $e_{i^*}=\mathcal{P}e_i$. Hence we have $\epsilon_{i^*}=\langle e_{i^*},e_{i^*}\rangle=-\langle e_i,e_i\rangle=-\epsilon_i$. On the other hand, we find from (10.11) that $h(e_{i^*},e_{i^*})=h(e_i,e_i)$. Thus, N is austere. □

Proposition 10.3. *An invariant submanifold of a para-Kähler space form $M^{2m}_m(c)$ has constant para-holomorphic sectional curvature c if and only if N is null-isotropic.*

Proof. Let N be an invariant submanifold of $M^{2m}_m(c)$. For any $v\in T^1N$, it follows from Gauss' equation, Proposition 10.1 and Proposition 10.2 that the para-holomorphic sectional curvature $H^{\mathcal{P}}(v)$ of $v\in T^1N$ satisfies

$$H^{\mathcal{P}}(v)=c-2\langle h(v,v),h(v,v)\rangle. \tag{10.14}$$

Hence, N has constant para-holomorphic sectional curvature c if and only if N is null-isotropic. □

Similarly, we also have the following.

Proposition 10.4. *The scalar curvature of an invariant submanifold N^{2n}_n of a para-Kähler space form $M^{2m}_m(c)$ is equal to $\frac{1}{2}n(n+1)c$ if and only if $S_h=0$, where S_h is defined by (2.32).*

Remark 10.1. Some simple examples of non-totally geodesic null-isotropic invariant submanifolds of para-Kähler manifolds are given by the isometric immersions of $(\mathbb{E}^{2n}_n, g_0, \mathcal{P})$ into $(\mathbb{E}^{2n+2}_{n+1}, g_0, \mathcal{P})$ defined by

$$\left(\sum_{j=1}^{n} a_i(x_i+y_i)^2, x_1, \ldots, x_n, \sum_{j=1}^{n} a_i(x_i+y_i)^2, y_1, \ldots, y_n\right),$$

where $a_1, \ldots, a_n$ are real numbers, not all zero. Such null-isotropic invariant submanifolds satisfy $S_h = 0$.

10.4 Lagrangian submanifolds of para-Kähler manifolds

Analogous to totally real submanifolds of an almost Hermitian manifold, a pseudo-Riemannian submanifold N of an almost para-Hermitian manifold is called *anti-invariant* if $\mathcal{P}(T_pN) \subset T_p^{\perp}N$ for each $p \in N$. In particular, N is called *Lagrangian* if $\mathcal{P}(T_pN) = T_p^{\perp}N$ for each $p \in N$.

Lemma 10.3. *Let N be a Lagrangian submanifold of a para-Kähler manifold M^{2n}_n. Then*

(a) $\mathcal{P}(\nabla'_X Y) = D_X(\mathcal{P}Y)$;
(b) $A_{\mathcal{P}X}Y = -\mathcal{P}(h(X,Y))$;
(c) $\langle h(X,Y), \mathcal{P}Z\rangle = \langle h(Y,Z), \mathcal{P}X\rangle = \langle h(Z,X), \mathcal{P}Y\rangle$;
(d) $\mathcal{P}(R'(X,Y)Z) = R^D(X,Y)\mathcal{P}Z$

for X, Y, Z tangent to N, where ∇' and R' are the Levi-Civita connection and curvature tensor of N.

Proof. Let X, Y, Z be vector fields tangent to N. Then it follows from (2.6) and (2.15) that

$$\mathcal{P}(\nabla'_X Y) + \mathcal{P}(h(X,Y)) = \mathcal{P}(\nabla_X Y) = \nabla_X(\mathcal{P}Y) = D_X(\mathcal{P}Y) - A_{\mathcal{P}Y}X,$$

which implies (a) and (b). From (2.16) and (b), we find

$$\langle h(Z,X), \mathcal{P}Y\rangle = -\langle \mathcal{P}(h(X,Z)), Y\rangle = \langle A_{\mathcal{P}X}Z, Y\rangle = \langle h(Y,Z), \mathcal{P}X\rangle.$$

Similarly, we also have $\langle h(X,Y), \mathcal{P}Z\rangle = \langle h(Y,Z), \mathcal{P}X\rangle$.

Statement (d) follows immediately from (a) and (2.27). □

The equations of Gauss and Codazzi of a Lagrangian submanifold N of a para-Kähler space form $M^{2n}_n(4c)$ are given respectively by

$$R(X,Y;Z,W) = \left\langle A_{h(Y,Z)}X, W\right\rangle - \left\langle A_{h(X,Z)}Y, W\right\rangle + c\,(\langle X,W\rangle\langle Y,Z\rangle - \langle X,Z\rangle\langle Y,W\rangle), \tag{10.15}$$

$$(\bar{\nabla}_X h)(Y,Z) = (\bar{\nabla}_Y h)(X,Z) \tag{10.16}$$

for X, Y, Z, W tangent to N. Lemma 10.3(b) implies that Ricci's equation is a consequence of Gauss for such submanifolds.

If we put $h = \mathcal{P} \circ \sigma$, then (10.1) and Lemma 10.3(c) imply that

$$\begin{aligned}\left\langle A_{h(Y,Z)}X, W\right\rangle &= \langle h(X,W), h(Y,Z)\rangle = \langle h(X,W), \mathcal{P}\sigma(Y,Z)\rangle \\ &= \langle h(\sigma(Y,Z),X), \mathcal{P}W\rangle = -\langle \sigma(\sigma(Y,Z),X), W\rangle .\end{aligned}$$

Therefore, equation (10.11) can be rephrased as

$$R(X,Y)Z = \sigma(\sigma(X,Z),Y) - \sigma(\sigma(Y,Z),X) + c\langle Y,Z\rangle X - c\langle X,Z\rangle Y.$$

Consequently, after applying a result of [Eschenburg and Tribuzy (1993)], we have the following fundamental existence and uniqueness theorems for Lagrangian submanifolds in para-Kähler space forms.

Theorem 10.1. *Let N be a simply-connected Riemannian n-manifold. If σ is a TN-valued symmetric bilinear form on N such that*

(a) *$g(\sigma(X,Y),Z)$ is totally symmetric;*
(b) *$(\bar{\nabla}\sigma)(X,Y,Z)$ is totally symmetric;*
(c) *$R(X,Y)Z = cg(Y,Z)X - cg(X,Z)Y + \sigma(\sigma(X,Z),Y) - \sigma(\sigma(Y,Z),X)$,*

then there exists a Lagrangian immersion L of N into a complete simply-connected para-Kähler space form $M^{2n}_n(4c)$ whose second fundamental form is given by $h = \mathcal{P} \circ \sigma$.

Theorem 10.2. *Let $L_1, L_2 : N \to M^{2n}_n(4c)$ be two Lagrangian immersions of a Riemannian n-manifold N into a complete simply-connected para-Kähler space form $M^{2n}_n(4c)$ with second fundamental forms h^1 and h^2, respectively. If*

$$g(h^1(X,Y), \mathcal{P}L_{1\star}Z) = g(h^2(X,Y), \mathcal{P}L_{2\star}Z)$$

for vectors X, Y, Z tangent to N, then there exists a $\mathcal{P}$-isometry Φ of $M^{2n}_n(4c)$ such that $L_1 = \Phi \circ L_2$.

Proposition 10.5. *The only totally umbilical Lagrangian submanifold N of a para-Kähler space form $M^{2n}_n(4c)$ with $n \geq 2$ is the totally geodesic one.*

Proof. Let N be a totally umbilical Lagrangian submanifold of $M^{2n}_n(4c)$ with $n \geq 2$. Then the second fundamental form satisfies

$$h(X,Y) = \langle X,Y\rangle H, \quad X, Y \in TN. \tag{10.17}$$

If $H = 0$, then N is totally geodesic. Thus, we may assume that $H \neq 0$. It follows from (10.17) that $(\bar{\nabla}_X h)(Y,Z) = \langle Y, Z\rangle D_X H$. After applying Codazzi's equation, we find

$$\langle Y, Z\rangle D_X H = \langle X, Z\rangle D_Y H, \quad X, Y, Z \in \mathcal{X}(N). \tag{10.18}$$

For any $X \in TN$, by choosing $0 \neq Y = Z \perp X$ we find $DH = 0$. Thus $\langle H, H\rangle$ is constant. Hence Gauss's equation and (10.17) imply that N is of constant curvature $c - ||H||^2$, where $||H|| = \sqrt{-\langle H, H\rangle}$.

Let us put $Z = \mathcal{P}H$. Then Lemma 10.3(i) yields $\nabla' Z = 0$. Thus, Z is a nonzero parallel vector field on N, which implies that N is flat. Therefore $c = ||H||^2 > 0$.

From totally umbilicity of N, we also have $[A_H, A_\xi] = 0$ for any normal vector ξ. Hence, by using $DH = 0$ we find from Ricci's equation that

$$g(R(Z,Y)H, \mathcal{P}Y) = 0, \quad Y, Z \in TN. \tag{10.19}$$

On the other hand, by (10.1) and Proposition 10.1 we find

$$g(R(Z,Y)H, \mathcal{P}Y) = c\{g(Z, \mathcal{P}H)g(Y,Y) - g(Y, \mathcal{P}H)g(Y,Z)\}.$$

Hence, after choosing Y, Z in such way that $Z = \mathcal{P}H$ and $g(Y,Z) = 0$, we find $g(H,H) = 0$, which is a contradiction. □

Consider a *graph* in the para-Kähler n-plane $(\mathbb{E}^{2n}_n, \mathcal{P}, g_0,)$ given by

$$\left(f^1(y_1,\ldots,y_n),\ldots,f^n(y_1,\ldots,y_n),y_1,\ldots,y_n\right), \tag{10.20}$$

where $f^1,\ldots,f^n$ are functions defined on an open domain $\mathcal{D}$ of $\mathbb{E}^n$. Then

$$L_{y_i} = \big(f^1_{y_i},\ldots,f^n_{y_i},0,\ldots,0,\overbrace{1}^{(n+i)-th},0,\ldots,0\big), \quad i = 1,\ldots,n. \tag{10.21}$$

Thus the graph is a Lagrangian submanifold of $(\mathbb{E}^{2n}_n, g_0, \mathcal{P})$ if and only if $\left(\frac{\partial f^i}{\partial x_j}\right)$ is a symmetric matrix. In particular, when $\mathcal{D}$ is simply-connected, there exists a function $F : \mathcal{D} \to \mathbf{R}$ with $(f^1,\ldots,f^n) = \nabla F$.

Lemma 10.4. *Let $\mathcal{D}$ be an open domain of $\mathbb{E}^n$ and let $F : \mathcal{D} \to \mathbf{R}$ be a smooth function. Then*

(i) $L(y_1,\ldots,y_n) = (F_{y_1},\ldots,F_{y_n},y_1,\ldots,y_n)$ *defines a Lagrangian graph of the para-Kähler n-plane $(\mathbb{E}^{2n}_n, g_0, \mathcal{P})$;*
(ii) *the second fundamental form h of $L : \mathcal{D} \to \mathbb{E}^{2n}_n$ satisfies*

$$\left\langle h\left(\frac{\partial}{\partial y_i}, \frac{\partial}{\partial y_j}\right), \mathcal{P}\left(\frac{\partial}{\partial y_k}\right)\right\rangle = -\frac{\partial^3 F}{\partial y_i \partial y_j \partial y_k}, \quad i,j,k = 1,\ldots,n.$$

Proof. (i) follows immediately from the observation made just before the Lemma. (ii) follows from

$$\nabla_{\partial/\partial y_i}\frac{\partial}{\partial y_j} = \left(F_{y_1y_iy_j},\ldots,F_{y_ny_iy_j},0,\ldots,0\right),$$

$$\mathcal{P}\Big(\frac{\partial}{\partial y_k}\Big) = \big(0,\ldots,0,\overbrace{1}^{k-th},0,\ldots,0,F_{y_1y_k},\ldots,F_{y_ny_k}\big).$$

□

10.5 Scalar curvature of Lagrangian submanifolds

Definition 10.6. A Lagrangian submanifold N of a para-Kähler manifold is called a *Lagrangian H-umbilical* if the second fundamental form takes the following simple form:

$$\begin{aligned} &h(e_1,e_1)=\lambda\mathcal{P}e_1,\ h(e_2,e_2)=\cdots=h(e_n,e_n)=\mu\mathcal{P}e_1,\\ &h(e_1,e_j)=\mu\mathcal{P}e_j,\ h(e_j,e_k)=0,\ 2\le j\ne k\le n, \end{aligned} \tag{10.22}$$

for some suitable functions λ and μ with respect to some orthonormal local frame $\{e_1,\ldots,e_n\}$.

Theorem 10.3. *If N is a Lagrangian spacelike submanifold of a para-Kähler space form $M^{2n}_n(4c)$ of constant para-holomorphic sectional curvature $4c$, then*

$$\tau \ge \frac{n^2(n-1)}{2(n+2)}\langle H,H\rangle + \frac{n(n-1)}{2}c. \tag{10.23}$$

The equality sign of (10.23) *holds identically on N if and only if N is a Lagrangian H-umbilical submanifold with $\lambda = 3\mu$.*

Proof. Under the hypothesis let $h^i_{jk} = \langle h(e_j,e_k),\mathcal{P}e_i\rangle$ with respect to an orthonormal frame $\{e_1,\ldots,e_n\}$ of N. Then, by Lemma 10.3(iii), we have

$$h^i_{jk} = h^j_{ik} = h^k_{ij},\quad i,j,k=1,\ldots,n. \tag{10.24}$$

From the definition of mean curvature vector we find

$$n^2\langle H,H\rangle = -\sum_i\Big(\sum_j (h^i_{jj})^2 + 2\sum_{j<k} h^i_{jj}h^i_{kk}\Big). \tag{10.25}$$

Also, from Gauss's equation we have

$$2\tau = n(n-1)c + n^2\langle H,H\rangle + \sum_{i,j,k=1}^n (h^i_{jk})^2. \tag{10.26}$$

Thus, by applying (10.24), (10.25) and (10.26), we obtain

$$\tau = \frac{n(n-1)c}{2} + \sum_{i\neq j}(h^i_{jj})^2 + 3\sum_{i<j<k}(h^i_{jk})^2 - \sum_i\sum_{j<k} h^i_{jj}h^i_{kk}. \tag{10.27}$$

If we put $m = \frac{n+2}{n-1}$, then it follows from (10.25) and (10.27) that

$$\begin{aligned} m(2\tau - n(n-1)c) - n^2\langle H,H\rangle &= \sum_i (h^i_{ii})^2 + (1+2m)\sum_{i\neq j}(h^i_{jj})^2 \\ &\quad + 6m\sum_{i<j<k}(h^i_{jk})^2 - 2(m-1)\sum_i\sum_{j<k}h^i_{jj}h^i_{kk}. \\ &= \sum_i (h^i_{ii})^2 + 6m\sum_{i<j<k}(h^i_{jk})^2 + (m-1)\sum_{i\neq j,k}\sum_{j<k}(h^i_{jj}-h^i_{kk})^2 \\ &\quad + (1+2m-(n-2)(m-1))\sum_{j\neq i}(h^i_{jj})^2 - 2(m-1)\sum_{j\neq i}h^i_{ii}h^i_{jj} \\ &= 6m\sum_{i<j<k}(h^i_{jk})^2 + (m-1)\sum_{i\neq j,k}\sum_{j<k}(h^i_{jj}-h^i_{kk})^2 \\ &\quad + \frac{1}{n-1}\sum_{j\neq i}(h^i_{ii} - (n-1)(m-1)h^i_{jj})^2 \geq 0, \end{aligned} \tag{10.28}$$

which implies (10.23). It follows from (10.28) that the equality sign of (10.23) holds if and only if $h^i_{ii} = 3h^i_{jj}$ and $h^i_{jk} = 0$ for distinct i,j,k.

Now, assume that N satisfies the equality sign of (10.23). If we choose $e_1,\ldots,e_n$ such that $\mathcal{P}e_1$ is parallel to the mean curvature vector H, then $h^j_{kk} = 0$ for $j > 1$, $k = 1,\ldots,n$. Hence we have (10.22) with $\lambda = 3\mu$.

The converse is easy to verify. □

Remark 10.2. It was proved in [Chen (2010h)] that a spacelike Lagrangian H-submanifold with $\lambda = 3\mu$ in $(\mathbb{E}^{2n}_n, \mathcal{P}, g_0)$ is congruent to an open part of

$$\left(\frac{\mathrm{dn}(\sqrt{2}as)\cosh(a^2t)}{a\,\mathrm{cn}^2(\sqrt{2}as)}, \frac{\mathrm{sn}(\sqrt{2}as)\sinh(a^2t)}{\sqrt{2}a\,\mathrm{cn}^2(\sqrt{2}as)}z_2, \ldots, \frac{\mathrm{sn}(\sqrt{2}as)\sinh(a^2t)}{\sqrt{2}a\,\mathrm{cn}^2(\sqrt{2}as)}z_n,\right.$$
$$\left.\frac{\mathrm{sn}(\sqrt{2}as)\cosh(a^2t)}{\sqrt{2}a\,\mathrm{cn}^2(\sqrt{2}as)}, \frac{\mathrm{dn}(\sqrt{2}as)\sinh(a^2t)}{a\,\mathrm{cn}^2(\sqrt{2}as)}z_2, \ldots, \frac{\mathrm{dn}(\sqrt{2}as)\sinh(a^2t)}{a\,\mathrm{cn}^2(\sqrt{2}as)}z_n\right)$$

with $a > 0$ and $z_2^2 + z_3^2 + \cdots + z_n^2 = 1$, where dn, cn, sn are Jacobi's elliptic functions with modulus $k = \frac{1}{\sqrt{2}}$.

10.6 Ricci curvature of Lagrangian submanifolds

The following provides an optimal result on Ricci curvature for Lagrangian submanifolds of para-Kähler space forms [Chen (2010h)].

Theorem 10.4. *If N is a Lagrangian spacelike submanifold of a para-Kähler space form $M^{2n}_n(4c)$, then*

$$Ric(u) \geq \frac{n(n-1)}{4}\langle H,H\rangle + (n-1)c, \quad u \in T^1N. \tag{10.29}$$

The equality sign of (10.29) *holds identically if and only if either*

(a) *N is totally geodesic or*
(b) *$n = 2$ and N is a Lagrangian H-umbilical surface $\lambda = 3\mu$.*

Proof. Let $u \in T^1N$. We choose an orthonormal basis $\{e_1, \cdots, e_n\}$ such that $e_1 = u$. By applying Gauss's equation and (10.24), we have

$$\begin{aligned} Ric(u) - (n-1)c &= \sum_{k=1}^{n}\sum_{j=2}^{n}\left((h^k_{1j})^2 - h^k_{11}h^k_{jj}\right) \\ &\geq \sum_{k=2}^{n}(h^1_{1k})^2 + \sum_{k=2}^{n}(h^k_{1k})^2 - \sum_{k=1}^{n}\sum_{j=2}^{n} h^k_{11}h^k_{jj} \\ &= \sum_{k=2}^{n}(h^k_{11})^2 + \sum_{k=2}^{n}(h^1_{kk})^2 - \sum_{k=1}^{n}\sum_{j=2}^{n} h^k_{11}h^k_{jj} \\ &= \sum_{k=2}^{n}(h^1_{kk})^2 - h^1_{11}\sum_{k=2}^{n} h^1_{kk} + \sum_{k=2}^{n}\left((h^k_{11})^2 - h^k_{11}\sum_{j=2}^{n} h^k_{jj}\right). \end{aligned} \tag{10.30}$$

On the other hand, if we put $H_k = \frac{1}{n}\sum_{j=1}^{n} h^k_{jj}$, it follows from Cauchy-Schwarz's inequality that

$$\left(h^1_{11} - \frac{n}{2}H_1\right)^2 + \sum_{j=2}^{n}(h^1_{jj})^2 \geq \frac{nH_1^2}{4}, \tag{10.31}$$

or equivalently

$$\sum_{k=2}^{n}(h^1_{kk})^2 - h^1_{11}\sum_{k=2}^{n} h^1_{kk} \geq \frac{n(1-n)}{4}H_1^2. \tag{10.32}$$

Similarly, by applying Cauchy-Schwarz's inequality, we also have

$$(h^k_{11})^2 - h^k_{11}\sum_{j=2}^{n} h^k_{jj} \geq -\frac{n^2}{8}H_k^2. \tag{10.33}$$

By combining (10.30), (10.32) and (10.33), we find

$$\begin{aligned} Ric(X) &\geq \frac{n(1-n)}{4} H_1^2 - \frac{n^2}{8} \sum_{k=2}^{n} H_k^2 \\ &\geq \frac{n(1-n)}{4} \left(H_1^2 + \sum_{k=2}^{n} H_k^2 \right) \\ &= \frac{n(n-1)}{4} \langle H, H \rangle , \end{aligned} \tag{10.34}$$

which gives inequality (10.29).

Now, assume that the equality sign of (10.29) holds identically, then the inequalities in (10.30), (10.31) and (10.33) become equalities. Thus

$$2h_{11}^1 - nH_1 = 2h_{22}^1 = \cdots = 2h_{nn}^1, \tag{10.35}$$

$$4h_{11}^k = nH_k, \ \ h_{jk}^1 = 0, \ \ 2 \leq j \neq k \leq n. \tag{10.36}$$

The equality sign of (10.34) implies that either $n = 2$ or $H_2 = \cdots = H_n = 0$.

If $n = 2$, either N is totally geodesic or N is a Lagrangian H-umbilical surface with $\lambda = 3\mu$ according to (10.35) and (10.36).

If $n \geq 3$, we have $H_2 = \cdots = H_n = 0$. Thus the mean curvature vector H is parallel to $\mathcal{P}e_1$. Moreover, it follows from (10.35) and (10.36) that

$$\begin{aligned} &h_{11}^1 = (n+1)h_{22}^1, \ \ h_{22}^1 = \cdots = h_{nn}^1, \\ &h_{11}^k = h_{1k}^1 = h_{1k}^j = h_{jk}^1 = 0, \ \ 2 \leq j \neq k \leq n. \end{aligned} \tag{10.37}$$

Now, if we choose $X = e_2$ and apply the same argument as for $X = e_1$, we find $H_1 = 0$. Combining this with (10.37) yields $A_{\mathcal{P}e_1} = 0$. Similarly, we also have $A_{\mathcal{P}e_2} = \cdots = A_{\mathcal{P}e_n} = 0$. Consequently, M is totally geodesic.

The converse can be verified by a direct computation. □

By combining Remark 10.2 and Theorem 10.4 we have the following result of [Chen (2010h)].

Theorem 10.5. *Let N be a Lagrangian spacelike submanifold of the para-Kähler n-plane $(\mathbb{E}_n^{2n}, \mathcal{P}, g_0)$. Then*

$$Ric(u) \geq \frac{n(n-1)}{4} \langle H, H \rangle \tag{10.38}$$

for any unit vector u tangent to N. The equality sign of (10.38) holds identically if and only if one of the following two cases occurs:

(a) *N is congruent to an open portion of a spacelike n-plane defined by $(0, \ldots, 0, y_1, \ldots, y_n)$;*

(b) *$n = 2$ and N is congruent to a space-like surface defined by*

$$\left(\frac{\mathrm{dn}(\sqrt{2}as)\cosh(a^2t)}{a\mathrm{cn}^2(\sqrt{2}as)}, \frac{\mathrm{sn}(\sqrt{2}as)\sinh(a^2t)}{\sqrt{2}a\mathrm{cn}^2(\sqrt{2}as)}, \frac{\mathrm{sn}(\sqrt{2}as)\cosh(a^2t)}{\sqrt{2}a\mathrm{cn}^2(\sqrt{2}as)}, \frac{\mathrm{dn}(\sqrt{2}as)\sinh(a^2t)}{a\mathrm{cn}^2(\sqrt{2}as)}\right)$$

with $a > 0$.

10.7 Lagrangian H-umbilical submanifolds

Lagrangian H-umbilical submanifolds are the simplest Lagrangian submanifolds next to totally geodesic ones. The next result shows that there exist many non-totally geodesic Lagrangian H-umbilical submanifolds.

Proposition 10.6. *Let $\gamma = (\gamma_1, \gamma_2) : I \to \mathbb{E}^2_1$ be a unit speed spacelike curve satisfying $\langle \gamma, \gamma \rangle < 0$. Define $L : I \times \mathbf{R} \times S^{n-2}(1) \to \mathbb{E}^{2n}_n$ by*

$$L(s,t,z) = \big(\gamma_1(s)\cosh t, \gamma_2(s) z \sinh t, \gamma_2(s)\cosh t, \gamma_1(s) z \sinh t\big),$$

$z \in \{(z_2, \ldots, z_n) \in \mathbb{E}^{n-1} : z_2^2 + z_3^2 + \cdots + z_n^2 = 1\}$. Then L defines a Lagrangian H-umbilical submanifold of $(\mathbb{E}^{2n}_n, \mathcal{P}, g_0)$.

Proof. It follows from hypothesis that

$$L_s = (\gamma_1' \cosh t, \gamma_2' z \sinh t, \gamma_2' \cosh t, \gamma_1' z \sinh t), \tag{10.39}$$

$$L_t = (\gamma_1 \sinh t, \gamma_2 z \cosh t, \gamma_2 \sinh t, \gamma_1 z \cosh t), \tag{10.40}$$

$$XL = (0, \gamma_2(\sinh t)X, 0, \gamma_1(\sinh t)X), \tag{10.41}$$

$$L_{ss} = (\gamma_1'' \cosh t, \gamma_2'' z \sinh t, \gamma_2'' \cosh t, \gamma_1'' z \sinh t), \tag{10.42}$$

$$L_{st} = (\gamma_1' \sinh t, \gamma_2' z \cosh t, \gamma_2' \sinh t, \gamma_1' z \cosh t), \tag{10.43}$$

$$XL_s = (0, \gamma_2'(\sinh t)X, 0, \gamma_1'(\sinh t)X), \tag{10.44}$$

$$XL_t = (0, \gamma_2(\cosh t)X, 0, \gamma_1(\cosh t)X), \tag{10.45}$$

$$\begin{aligned} XYL &= (0, \gamma_2(\cosh t)\nabla'_X Y, 0, \gamma_1(\cosh t)\nabla'_X Y) \\ &\quad - (0, \langle X, Y\rangle \gamma_2 z \cosh t, 0, \langle X, Y\rangle \gamma_1 z \cosh t) \end{aligned} \tag{10.46}$$

for X, Y tangent to $S^{n-2}(1)$. From (10.39)-(10.41) we get

$$\mathcal{P}(L_s) = (\gamma_2' \cosh t, \gamma_1' z \sinh t, \gamma_1' \cosh t, \gamma_2' z \sinh t), \tag{10.47}$$

$$\mathcal{P}(L_t) = (\gamma_2 \sinh t, \gamma_1 z \cosh t, \gamma_1 \sinh t, \gamma_2 z \cosh t), \tag{10.48}$$

$$\mathcal{P}(XL) = (0, \gamma_1(\sinh t)X, 0, \gamma_2(\sinh t)X). \tag{10.49}$$

Since $\gamma = (\gamma_1, \gamma_2)$ is a unit speed space-like curve in $\mathbb{E}^2_1$, (10.39)-(10.41) imply that the induced metric via L is given by

$$g = ds^2 + ||\gamma||^2(dt^2 + \sinh^2 t g_1), \tag{10.50}$$

where g_1 is the metric of $S^{n-2}(1)$. It follows from (10.39)-(10.41) and (10.47)-(10.49) that L is Lagrangian. Since γ is unit speed and spacelike,

$$(\gamma_1''(s), \gamma_2''(s)) = \kappa(s)(\gamma_2'(s), \gamma_1'(s)) \tag{10.51}$$

for some function κ. Thus, by (10.42)-(10.51) and $\langle z, X\rangle = 0$ for $X \in TN$, we obtain (10.22) with

$$\lambda = \kappa, \quad \mu = \frac{\gamma_1'\gamma_2 - \gamma_1\gamma_2'}{||\gamma||^2}. \tag{10.52}$$

Consequently, L defines a Lagrangian H-umbilical submanifold with the desired properties. This completes the proof of the proposition. □

Similarly, we also have the following.

Proposition 10.7. *Let $\gamma = (\gamma_1, \gamma_2) : I \to \mathbb{E}^2_1$ be a unit speed spacelike curve satisfying $\langle \gamma, \gamma\rangle > 0$. Define $L : I \times \mathbf{R} \times S^{n-2}(1) \to \mathbb{E}^{2n}_n$ by*

$$\big(\gamma_1(s)\sin t, \gamma_1(s) z \cos t, \gamma_2(s)\sin t, \gamma_2(s) z\cos t\big),$$

where $z = (z_2, \ldots, z_n) \in \mathbb{E}^{n-1}$ satisfies $\langle z, z\rangle = 1$. Then L defines a Lagrangian H-umbilical submanifold of $(\mathbb{E}^{2n}_n, \mathcal{P}, g_0)$.

Proof. This can be proved in the same way as Proposition 10.6. □

The following classification theorem of Lagrangian H-umbilical submanifolds of $(\mathbb{E}^{2n}_n, g_0, \mathcal{P})$ was proved in [Chen (2010h)].

Theorem 10.6. *Let $L : N \to (\mathbb{E}^{2n}_n, \mathcal{P}, g_0)$ be a Lagrangian H-umbilical immersion of a Riemannian n-manifold N into the para-Kähler n-plane with $n \geq 3$. Then*

(i) *if N is of constant curvature, then either N is flat or L is congruent to an open portion of*

$$\frac{1}{2b}\Big(2\cosh^2(bs)\cosh t, z\sinh(2bs)\sinh t, \sinh(2bs)\cosh t, \\ 2z\cosh^2(bs)\sinh t\Big)$$

with $b \neq 0$, where $z = (z_2, \ldots, z_n) \in \mathbb{E}^{n-1}$ satisfies $z_2^2 + z_3^2 + \cdots + z_n^2 = 1$;

(ii) *if N contains no open subset of constant sectional curvature, then L is locally congruent to one of the following three types of submanifolds:*

(ii.1) *a Lagrangian submanifold defined by*

$$\Bigg(a^2\sum_{j=2}^n x_j^2 + \frac{e^{2r}}{8} - \frac{e^{-2r}}{2r'^2} - \int^s \frac{2r'^2 + r''}{e^{2r}r'^3}ds, \frac{1 - a^2e^{2r}}{2}x_2, \ldots, \frac{1 - a^2e^{2r}}{2}x_n, \\ a^2\sum_{j=2}^n x_j^2 - \frac{e^{2r}}{8} - \frac{e^{-2r}}{2r'^2} - \int^s \frac{2r'^2 + r''}{e^{2r}r'^3}ds, \frac{1 + a^2e^{2r}}{2}x_2, \ldots, \frac{1 + a^2e^{2r}}{2}x_n\Bigg),$$

where $r = r(s)$ is a non-constant function and a is positive number;

(ii.2) *a Lagrangian submanifold defined by*

$$\frac{1}{2}\left(\left(\frac{e^{\int^s \lambda ds}}{\mu+\varphi}+\frac{e^{-\int^s \lambda ds}}{\mu-\varphi}\right)\sin t, \left(\frac{e^{\int^s \lambda ds}}{\mu+\varphi}+\frac{e^{-\int^s \lambda ds}}{\mu-\varphi}\right) z\cos t,\right.$$
$$\left.\left(\frac{e^{\int^s \lambda ds}}{\mu+\varphi}-\frac{e^{-\int^s \lambda ds}}{\mu-\varphi}\right)\sin t, \left(\frac{e^{\int^s \lambda ds}}{\mu+\varphi}-\frac{e^{-\int^s \lambda ds}}{\mu-\varphi}\right) z\cos t\right),$$

where $\mu(s), \varphi(s)$ are nonzero functions satisfies $\varphi\varphi' - \mu\mu' = (\mu^2-\varphi^2)\varphi$, $\lambda = 2\mu + \mu\varphi^{-1}$ and $z = (z_2, \dots, z_n) \in \mathbb{E}^{n-1}$ satisfies $\langle z, z\rangle = 1$;

(ii.3) *a Lagrangian submanifold defined by*

$$\frac{1}{2}\left(\left(\frac{e^{\int^s \lambda ds}}{\mu+\varphi}+\frac{e^{-\int^s \lambda ds}}{\mu-\varphi}\right)\cosh t, \left(\frac{e^{\int^s \lambda ds}}{\mu+\varphi}-\frac{e^{-\int^s \lambda ds}}{\mu-\varphi}\right) z\sinh t,\right.$$
$$\left.\left(\frac{e^{\int^s \lambda ds}}{\mu+\varphi}-\frac{e^{-\int^s \lambda ds}}{\mu-\varphi}\right)\cosh t, \left(\frac{e^{\int^s \lambda ds}}{\mu+\varphi}+\frac{e^{-\int^s \lambda ds}}{\mu-\varphi}\right) z\sinh t\right),$$

where $\mu(s), \varphi(s)$ are nonzero functions satisfies $\varphi\varphi' - \mu\mu' = (\mu^2-\varphi^2)\varphi$, $\lambda = 2\mu + \mu\varphi^{-1}$ and $z = (z_2, \dots, z_n) \in \mathbb{E}^{n-1}$ satisfies $\langle z, z\rangle = 1$.

A unit speed curve $z : I \to S^{n-1}_n(1) \subset (\mathbb{E}^{2n}_n, \mathcal{P}, g_0)$ is called a *para-Legendre curve* if $\langle z'(t), Pz(t)\rangle = 0$ for each $t \in I$. For a unit speed space-like para-Legendre curve $z : I \to S^{n-1}_n(1) \subset (\mathbb{E}^{2n}_n, \mathcal{P}, g_0)$, we have

$$\langle z, z\rangle = \langle z', z'\rangle = 1, \;\; \langle z, z'\rangle = \langle z, Pz\rangle = \langle z', Pz\rangle = 0. \tag{10.53}$$

Thus we may extend z, Pz, z', Pz' to an orthonormal frame z, Pz, z', Pz', $w_3, Pw_3, \dots, w_n, Pw_n$ along the curve. From (10.53) we find

$$\langle z'', Pz\rangle = \langle z'', z'\rangle = 0, \;\; \langle z'', z\rangle = -1. \tag{10.54}$$

Hence, it follows from (10.53)-(10.54) that

$$z''(t) = \mu(t)Pz'(t) - z(t) - \sum_{j=3}^{n} a_j(t)w_j(t) + \sum_{j=3}^{n} b_j(t)Pw_j(t) \tag{10.55}$$

for some functions $\mu, a_3, \dots, a_n, b_3, \dots, b_n$.

Definition 10.7. A unit speed space-like para-Legendre curve $z : I \to S^{n-1}_n(1) \subset (\mathbb{E}^{2n}_n, \mathcal{P}, g_0)$ is called *special para-Legendre* if (10.55) reduces to

$$z''(t) = -z(t) + \mu(t)Pz'(t) - \sum_{j=3}^{n} a_j(t)w_j(t) \tag{10.56}$$

for some parallel normal vector fields $w_3, \dots, w_n$.

Flat Lagrangian H-umbilical submanifolds of $(\mathbb{E}_n^{2n}, \mathcal{P}, g_0)$ are classified in [Chen (2011b)].

Theorem 10.7. *Let $L : N \to (\mathbb{E}_n^{2n}, \mathcal{P}, g_0), n \geq 2$, be a Lagrangian H-umbilical immersion of a flat Riemannian n-manifold into the para-Kähler n-plane. Then locally L is congruent to one of the following two types of submanifolds:*

(a) *a Lagrangian submanifolds defined by*

$$L(t, u_2, \ldots, u_n) = (\gamma_1(t), 0, \ldots, 0, \gamma_{n+1}(t), u_2, \ldots, u_n), \tag{10.57}$$

where $(\gamma_1(t), \gamma_{n+1}(t))$ is a space-like curve in $\mathbb{E}_1^2$;

(b) *a Lagrangian submanifold defined by*

$$L(t, u_2, \ldots, u_n) = u_2 z(t) + \sum_{j=3}^{n} u_j w_j(t) + \int^t b(t) z'(t) dt \tag{10.58}$$

for some function b, where $z : I \to S_n^{2n-1} \subset (\mathbb{E}_n^{2n}, g_0, P)$ is a unit speed space-like special para-Legendre curve satisfying

$$z''(t) = -z(t) + \varphi(t) P z'(t) - \sum_{j=3}^{n} a_j(t) w_j(t) \tag{10.59}$$

for a function $\varphi \neq 0$ and parallel normal vector fields $w_3, \ldots, w_n$ along z.

10.8 $\mathcal{PR}$-submanifolds of para-Kähler manifolds

Definition 10.8. A pseudo-Riemannian submanifold N of a para-Kähler manifold $(M, \mathcal{P}, g)$ is called a *$\mathcal{PR}$-submanifold* if there exist an *invariant distribution* $\mathcal{D}$, $\mathcal{P}(\mathcal{D}) = \mathcal{D}$, and an *anti-invariant distribution* $\mathcal{D}^\perp$ on N, $\mathcal{P}(\mathcal{D}_p^\perp) \subset T_p^\perp N$, such that $TN = \mathcal{D} \oplus \mathcal{D}^\perp$. A $\mathcal{PR}$-submanifold is called a *$\mathcal{PR}$-warped product* if it is a warped product $N_T \times_f N_\perp$ of an invariant submanifold N_T and a totally real submanifold $N_\perp$.

Proposition 10.8. *The anti-invariant distribution of a $\mathcal{PR}$-submanifold of a para-Kähler manifold is a nondegenerate integrable distribution.*

Proof. This can be proved in the same way as Proposition 9.10. □

An immediate consequence of Proposition 10.8 is the following.

Corollary 10.1. *Every $\mathcal{PR}$-submanifold N of a para-Kähler manifold M is foliated by anti-invariant pseudo-Riemannian submanifolds of M.*

Proposition 10.9. *If N is a $\mathcal{P}R$-submanifold of a para-Kähler manifold, then*

(a) *the invariant distribution $\mathcal{D}$ is a nondegenerate minimal distribution;*
(b) *the distribution $\mathcal{D}$ is integrable if and only if the second fundamental form $\mathring{h}$ of $\mathcal{D}$ satisfies $\mathring{h}(\mathcal{P}X, Y) = \mathring{h}(X, \mathcal{P}Y)$;*
(c) $\langle A_{\mathcal{P}Z}U, \mathcal{P}X\rangle = \langle \nabla'_U Z, X\rangle$;
(d) $A_{\mathcal{P}Z}W = A_{\mathcal{P}W}Z$, $A_{\mathcal{P}\xi}X = -A_\xi \mathcal{P}X$

for $X, Y \in \mathcal{D}$; $Z, W \in \mathcal{D}^\perp$; $U \in N$ and normal vector $\xi \perp \mathcal{P}\mathcal{D}^\perp$.

Proof. (a) and (b) can be proved in the same way as Proposition 9.11.

Now, consider the following identity

$$\begin{aligned} \mathcal{P}\nabla'_U Z + \mathcal{P}h(U, Z) &= \nabla_U \mathcal{P}Z \\ &= -A_{\mathcal{P}Z}U + D_U \mathcal{P}Z. \end{aligned} \tag{10.60}$$

By taking scalar product of (10.60) with $\mathcal{P}X \in \mathcal{D}$, we find (c). Similarly, by taking the scalar product of (10.60) with $W \in \mathcal{D}^\perp$, we have the first equation of (d). The second equation of (d) follows from the fact that

$$\begin{aligned} \langle A_\xi \mathcal{P}X, U\rangle &= \langle h(\mathcal{P}X, U), \xi\rangle = \langle \nabla_U \mathcal{P}X, \xi\rangle \\ &= \langle \mathcal{P}\nabla_U X, \xi\rangle = -\langle \nabla_U X, \mathcal{P}\xi\rangle \\ &= -\langle h(X, U), \mathcal{P}\xi\rangle = -\langle A_{\mathcal{P}\xi}X, U\rangle. \end{aligned}$$ □

Proposition 10.10. *There does not exist a totally umbilical proper $\mathcal{P}R$-submanifold in any non-flat para-Kähler space form.*

Proof. If N is a totally umbilical proper $\mathcal{P}R$-submanifold of a para-Kähler manifold M, then it follows from (2.26) and (2.56) that

$$(\bar{\nabla}_X h)(Y, Z) = \langle Y, Z\rangle\, D_X H$$

for X, Y, Z tangent to N. Thus, by Codazzi's equation, we find

$$\begin{aligned} \tilde{R}(X, W; \mathcal{P}X, \mathcal{P}W) &= \langle W, \mathcal{P}X\rangle \langle D_X H, \mathcal{P}W\rangle - \langle X, \mathcal{P}X\rangle \langle D_W H, \mathcal{P}W\rangle \\ &= 0 \end{aligned}$$

for vectors X in $\mathcal{D}$ and W in $\mathcal{D}^\perp$.

On the other hand, since M is para-Kählerian, we also have

$$\tilde{R}(X, W; \mathcal{P}X, \mathcal{P}W) = \tilde{R}(X, W; W, X) = \tilde{K}(X, W).$$

Hence, if M is a para-Kähler space form, it must be flat by Proposition 10.1. □

Definition 10.9. A $\mathcal{P}R$-submanifold of a para-Kähler manifold M is called a *$\mathcal{P}R$-product* if it is locally a direct product $N^T \times N^\perp$ of an invariant submanifold N^T and an anti-invariant submanifold $N^\perp$ of M.

Proposition 10.11. *Let N be a $\mathcal{P}R$-product of a para-Kähler manifold M. Then for any unit vector fields $X \in \mathcal{D}$ and $Z \in \mathcal{D}^\perp$ we have*

$$2\langle h(X,Z), h(X,Z)\rangle = \epsilon_X \epsilon_Z (\tilde{K}(X,Z) + \tilde{K}(X,\mathcal{P}Z)), \tag{10.61}$$

where $\tilde{K}$ is the sectional curvature of M, $\epsilon_X = \langle X, X\rangle$ and $\epsilon_Z = \langle Z, Z\rangle$.

Proof. Under the hypothesis, $\nabla'_X Z = \nabla'_Z X = 0$. So, Proposition 10.9(c) yields

$$h(\mathcal{D}, \mathcal{D}^\perp) \perp \mathcal{P}\mathcal{D}^\perp. \tag{10.62}$$

Thus, by applying Codazzi's equation, we find

$$\tilde{R}(X, \mathcal{P}X; Z, \mathcal{P}Z) = \langle D_X h(\mathcal{P}X, Z), \mathcal{P}Z\rangle - \langle D_{\mathcal{P}X} h(X,Z), \mathcal{P}Z\rangle. \tag{10.63}$$

Since N is a $\mathcal{P}R$-product, Proposition 10.9(c) implies that $A_{\mathcal{P}Z}X = 0$. Hence we find from (10.62) and (10.63) that

$$\begin{aligned} \tilde{R}(X, \mathcal{P}X; Z, \mathcal{P}Z) &= \langle h(X,Z), D_{\mathcal{P}X}\mathcal{P}Z\rangle - \langle h(\mathcal{P}X, Z), D_X \mathcal{P}Z\rangle \\ &= \langle h(X,Z), \mathcal{P}\nabla_{\mathcal{P}X} Z\rangle - \langle h(\mathcal{P}X, Z), \mathcal{P}\nabla_X Z\rangle \\ &= -2\langle h(\mathcal{P}X, Z), \mathcal{P}h(X,Z)\rangle \\ &= -2\left\langle A_{\mathcal{P}h(X,Z)}\mathcal{P}X, Z\right\rangle \\ &= 2\left\langle A_{h(X,Z)}X, Z\right\rangle \\ &= 2\langle h(X,Z), h(X,Z)\rangle, \end{aligned} \tag{10.64}$$

where we have applied 10.9(d). On the other hand, it follows from the first Bianchi identity, (10.1) and Lemma 10.1 that

$$\begin{aligned} \tilde{R}(X, \mathcal{P}X; Z, \mathcal{P}Z) &= -\tilde{R}(Z, X; \mathcal{P}X, \mathcal{P}Z) - \tilde{R}(\mathcal{P}X, Z; X, \mathcal{P}Z) \\ &= \epsilon_X \epsilon_Z (\tilde{K}(X,Z) + \tilde{K}(X, \mathcal{P}Z)). \end{aligned}$$

Combining this with (10.64) gives (10.61). □

Corollary 10.2. *Let N be a $\mathcal{P}R$-product in a para-Kähler space form $M^{2n}_n(4c)$. Then*

$$\langle h(X,Z), h(X,Z)\rangle = c\langle X, X\rangle\langle X, Z\rangle \tag{10.65}$$

for any nonzero vector fields $X \in \mathcal{D}$ and $Z \in \mathcal{D}^\perp$. In particular, if $c \neq 0$ and X, Z are non-null, then $h(X,Z) \neq 0$.

Proof. It follows from Proposition 10.1 that

$$\tilde{K}(X,Z) = \tilde{K}(X,\mathcal{P}Z) = c$$

for unit vectors X, Z satisfying $\langle X, Z\rangle = \langle X, \mathcal{P}Z\rangle = 0$. Hence, by applying Proposition 10.11 we obtain the corollary. □

Example 10.2. The simplest examples of $\mathcal{P}R$-products in para-Kähler manifolds are the direct products of the form:

$$\mathcal{P}K^r \times N^p \subset \mathcal{P}K^r \times \mathcal{P}K^p = \mathcal{P}K^{r+p}, \tag{10.66}$$

where $\mathcal{P}K^r = (\mathbb{E}^{2r}_r, \mathcal{P}, g_0)$ is the para-Kähler r-plane and N^p is either a spacelike or timelike Lagrangian submanifold of $\mathcal{P}K^p$.

Proposition 10.12. *Let $N^T \times N^\perp$ be a $\mathcal{P}R$-product of the para-Kähler $(r+p)$-plane $(\mathbb{E}^{2r+2p}_{r+p}, \mathcal{P}, g_0)$ with $r = \frac{1}{2}\dim N^T$ and $p = \dim N^\perp$. If $N^\perp$ is either spacelike or timelike, then the $\mathcal{P}R$-product is given by a direct product immersion as described in* (10.66).

Proof. Let $N^T \times N^\perp$ be a $\mathcal{P}R$-product in a para-Kähler m-plane $\mathcal{P}K^m$. Then it follows from Corollary 10.2 that $\langle h(X,Z), h(X,Z)\rangle = 0$ for any vectors $X \in \mathcal{D}$ and $Z \in \mathcal{D}^\perp$. Thus, if $m = \dim N^\perp + \frac{1}{2}\dim N^T$ and if $N^\perp$ is either spacelike or timelike, then we have $h(X,Z) = 0$. Thus, the immersion is mixed totally geodesic. Hence, according to Lemma 4.3 of Moore, the immersion is a direct product immersion. Therefore, after applying the assumption on the dimension, we conclude that N^T is an open portion of a para-Kähler r-plane with $2r = \dim N^T$ and the immersion is given by (10.66). □

Example 10.3. Some other important examples of $\mathcal{P}R$-products can be obtained via the embedding

$$\varphi_{r,p} : P^r_c(B) \times P^p_c(B) \longrightarrow P^{r+p+rp}_c(B)$$

introduced in Example 10.1 of Section 10.3. In fact, if $N^T = P^r_c(B)$ and $N^\perp$ is a Lagrangian submanifold of $P^p_c(B)$, then

$$N^T \times N^\perp \subset P^r_c(B) \times P^p_c(B) \xrightarrow{\varphi_{r,p}} P^{r+p+rp}_c(B)$$

is a $\mathcal{P}R$-product of $P^{r+p+rp}_c(B)$ such that N^T is immersed as a totally geodesic invariant submanifold in $P^{r+p+rp}_c(B)$. Clearly, the codimension of such $\mathcal{P}R$-products is $p + 2rp$.

The next result shows that the codimension of a $\mathcal{P}R$-product in any non-flat para-Kähler space form is at least $p + 2rp$. Moreover, when the codimension is $p + 2rp$, then the $\mathcal{P}R$-product is obtained in the way described in Example 10.3.

Theorem 10.8. *Let $N = N^T \times N^\perp$ be a $\mathcal{P}R$-product of a para-Kähler space form $M^{2m}_m(4c)$ with $c \neq 0$, $\dim N^T = 2r$ and $\dim N^\perp = p$. Then $m \geq r + p + rp$.*

Moreover, if $m = r + p + rp$, then N^T is immersed in $M^{2m}_m(4c)$ as a totally geodesic invariant submanifold.

Proof. Let $N^T \times N^\perp$ be a $\mathcal{P}R$-product in a para-Kähler space form $M^{2m}_m(4c)$, $c \neq 0$. Let $u_1, \ldots, u_{2r}$ be an orthonormal basis of $\mathcal{D}$ and $v_1, \ldots, v_p$ orthonormal of $\mathcal{D}^\perp$. Then (10.65) and the linearity of h imply that

$$\langle h(u_i, v_\alpha), h(u_j, v_\alpha)\rangle = 0 \tag{10.67}$$

for $1 \leq i \neq j \leq 2r$ and $\alpha = 1, \ldots, p$.

If $p = 1$, then (10.62), (10.65) and (10.67) imply that the codimension of the $\mathcal{P}R$-product in $M^{2m}_m(4c)$ is at least $p + 2rp$. Thus $m \geq r + p + rp$.

If $p \geq 2$, then (10.67) and the linearity of h imply that

$$\langle h(u_i, v_\alpha), h(u_j, v_\beta)\rangle + \langle h(u_i, v_\beta), h(u_j, v_\alpha)\rangle = 0 \tag{10.68}$$

for $1 \leq i \neq j \leq 2r$ and $1 \leq \alpha \neq \beta \leq p$.

On the other hand, Gauss's equation and Proposition 10.1 imply that the curvature tensor R' and the second fundamental form h of N satisfy

$$\begin{aligned} 0 &= R'(u_i, u_j; v_\beta, v_\alpha) \\ &= \langle h(u_i, v_\alpha), h(u_j, v_\beta)\rangle - \langle h(u_i, v_\beta), h(u_j, v_\alpha)\rangle . \end{aligned}$$

Combining this with (10.68) gives

$$\langle h(u_i, v_\alpha), h(u_j, v_\beta)\rangle = 0, \quad 1 \leq i \neq j \leq 2r; \quad 1 \leq \alpha \neq \beta \leq p. \tag{10.69}$$

Thus, by (10.62), (10.65) and (10.69), we find $m \geq r + p + rp$.

Now, let us assume that $m = r+p+rp$. Then $h(\mathcal{D}, \mathcal{D}^\perp)$ is the orthogonal complementary subbundle of $\mathcal{P}\mathcal{D}$ in $T^\perp N$. It follows from Proposition 10.9(c) that

$$h(\mathcal{D}, \mathcal{D}) \perp \mathcal{D}^\perp. \tag{10.70}$$

Also, from Gauss's equation and Proposition 10.1, we find

$$\langle h(X, Y), h(\mathcal{P}X, W)\rangle = \langle h(X, W), h(\mathcal{P}X, Y)\rangle \tag{10.71}$$

for X, Y in $\mathcal{D}$ and $W \in \mathcal{D}^\perp$. Further, it follows from (10.62) and statements (b) and (e) of Proposition 10.9 that

$$\begin{aligned} \langle h(X,W), h(\mathcal{P}X,Y)\rangle &= \left\langle A_{h(X,W)}\mathcal{P}X, Y\right\rangle \\ &= -\left\langle A_{\mathcal{P}h(X,W)}X, Y\right\rangle \\ &= -\left\langle h(X,Y), \mathcal{P}h(X,W)\right\rangle. \end{aligned} \tag{10.72}$$

Moreover, from (10.6) and (10.70) we also have

$$\begin{aligned} \langle h(X,Y), \mathcal{P}h(X,W)\rangle &= -\left\langle \mathcal{P}h(X,Y), h(X,W)\right\rangle \\ &= -\left\langle \nabla_W X, \mathcal{P}h(X,Y)\right\rangle \\ &= \left\langle \nabla_W(\mathcal{P}X), h(X,Y)\right\rangle \\ &= \left\langle h(X,Y), h(\mathcal{P}X,W)\right\rangle. \end{aligned} \tag{10.73}$$

Therefore, by combining (10.71), (10.72) and (10.73), we obtain

$$\langle h(X,Y), h(\mathcal{P}X,W)\rangle = \langle h(X,W), h(\mathcal{P}X,Y)\rangle = 0. \tag{10.74}$$

Now, by applying linearity of h, we obtain

$$h(\mathcal{D},\mathcal{D}) \perp h(\mathcal{D},\mathcal{D}^\perp).$$

Since $h(\mathcal{D},\mathcal{D}^\perp)$ is the orthogonal complementary normal subbundle of $\mathcal{P}\mathcal{D}$. Thus, by applying (10.70), we find $h(X,Y) = 0$, $X, Y \in \mathcal{D}$. Combining this with the total geodesy of N^T in $N^T \times N^\perp$, we conclude that N^T is immersed as a totally geodesic submanifold of $M^{2m}_m(4c)$. □

An immediate application of Theorem 10.8 is the following.

Corollary 10.3. *If the direct product $R^{2r}_r \times P^{2p}_p$ of two para-Kähler manifolds can be isometrically immersed in a non-flat para-Kähler space form $M^{2(r+p+rp)}_{r+p+rp}(4c)$ as an invariant submanifold, then both R^{2r}_r and P^{2p}_p are of constant para-holomorphic sectional curvature $4c$. Furthermore, both R^{2r}_r and P^{2p}_p are immersed as totally geodesic invariant submanifolds of $M^{2(r+p+rp)}_{r+p+rp}(4c)$.*

Proof. Under the hypothesis of the corollary, let us consider the $\mathcal{P}R$-submanifold $R^{2r}_r \times N^\perp$ of $M^{2(r+p+rp)}_{r+p+rp}(4c)$, where $N^\perp$ is a Lagrangian submanifold of P^{2p}_p. Then, according to Theorem 10.8, R^{2r}_r is totally geodesic invariant submanifold of $M^{2(r+p+rp)}_{r+p+rp}(4c)$. Thus, R^{2r}_r is of constant para-holomorphic sectional curvature $4c$. Similarly, P^{2p}_p is also a totally geodesic submanifold of $M^{2(r+p+rp)}_{r+p+rp}(4c)$. Therefore, P^{2p}_p is also of constant para-holomorphic sectional curvature $4c$. □

Chapter 11

Pseudo-Riemannian Submersions

A submersion $\pi : M \to B$ is a differentiable map between differentiable manifolds whose differential is everywhere surjective. The notion of submersions is a fundamental concept in differential topology. The bundle map $\pi : E \to B$ of a fiber bundle is an example of a submersion. Submersions appear often in classical differential geometry, e.g. a surface of revolution or a family of parallel surfaces in $\mathbb{E}^3$ each leads in an obvious way to a submersion.

Riemannian submersions are submersions equipped with compatible Riemannian metrics. The theory of Riemannian submersions arose from [Hermann (1960); Nagano (1960); O'Neill (1966); Gray (1967)]. Riemannian submersions are natural generalizations of warped products, which occur widely in geometry. For instance, when a Lie group acts isometrically, freely and properly on a Riemannian manifold M, the projection $\pi : M \to B$ to the quotient space $B = M/G$ equipped with quotient metric is a Riemannian submersion.

The fundamental equations of Riemannian submersions were derived in [O'Neill (1966)], which are well-known as the O'Neill equations. One important consequence of O'Neill's equations is that, for a Riemannian submersion $\pi : M \to B$, the lower bound for the sectional curvature of B is at least as big as the lower bound for the sectional curvature of M.

Pseudo-Riemannian submersions were first introduced in O'Neill's book [O'Neill (1983)]. The notion can be regarded as the "dual" of pseudo-Riemannian immersions. The most interesting and important family of pseudo-Riemannian submersions consists of those with totally geodesic fibers (or more generally, with minimal fibers). The Hopf fibration and the generalized Hopf fibrations are typical examples of pseudo-Riemannian submersions with totally geodesic fibers.

11.1 Pseudo-Riemannian submersions

Let $\pi : M \to B$ be a pseudo-Riemannian submersion with $\dim M > \dim B$. Denote by $\mathcal{V}$ and $\mathcal{H}$ the vertical and horizontal distributions as well as for the orthogonal projections of TM on its horizontal and vertical subspaces, respectively. A vector field $\bar{X}$ on M is said to be *basic* if $\bar{X}$ is horizontal and π-related to a vector field X on B. Notice that every vector field X on B has a unique horizontal lift $\bar{X}$ to M which is basic.

Since π is a submersion, π_* gives a linear isomorphism $\mathcal{H}_p \approx T_bB$. Moreover, it follows from (S3) of Definition 6.5 that it is a linear isometry.

Lemma 11.1. *For $X, Y \in \mathcal{X}(B)$, we have*

(1) $\langle \bar{X}, \bar{Y} \rangle = \langle X, Y \rangle \circ \pi$;
(2) $\mathcal{H}[\bar{X}, \bar{Y}] = [X, Y]^-$;
(3) $\mathcal{H}\nabla_{\bar{X}}\bar{Y} = (\nabla'_X Y)^-$, *where ∇' is the Levi-Civita connection of B.*

Proof. (1) follows immediately form (S3). (2) follows from the fact that $[\bar{X}, \bar{Y}]$ is π-related to $[X, Y]$; hence $\mathcal{H}[\bar{X}, \bar{Y}]$ is also π-related to $[X, Y]$.

Clearly, (3) holds if both sides have the same scalar product with every horizontal vector field, or merely with every horizontal lift $\bar{Z}$. Thus, by (1), it suffices to show $\langle \nabla_{\bar{X}}\bar{Y}, \bar{Z} \rangle = \langle \nabla'_X Y, Z \rangle \circ \pi$, which follows by expanding both sides in the Koszul formula. This is due to the fact that, using (1) and (2), we have

$$\bar{X}\langle \bar{Y}, \bar{Z} \rangle = \bar{X}(\langle Y, Z \rangle \circ \pi) = (\pi_*(\bar{X}))\langle Y, Z \rangle = X\langle Y, Z \rangle \circ \pi,$$
$$\langle \bar{X}, [\bar{Y}, \bar{Z}] \rangle = \langle \bar{X}, [X, Y]^- \rangle = \langle X, [Y, Z] \rangle \circ \pi.$$ □

B. O'Neill characterized the geometry of a Riemannian submersion in terms of the two *fundamental tensors* T, A defined by

$$A_E F = \mathcal{H}\nabla_{\mathcal{H}E}\mathcal{V}F + \mathcal{V}\nabla_{\mathcal{H}E}\mathcal{H}F, \tag{11.1}$$

$$T_E F = \mathcal{H}\nabla_{\mathcal{V}E}\mathcal{V}F + \mathcal{V}\nabla_{\mathcal{V}E}\mathcal{H}F \tag{11.2}$$

for vector fields $E, F \in \mathcal{X}(M)$. Here ∇ is the Levi-Civita connection of g.

The letters U, V, W will always denote vertical vector fields, X, Y, Z horizontal vector fields. Notice that $T_U V$ is the second fundamental form of each fiber.

The following result is due to O'Neill.

Theorem 11.1. *Let $\pi_1, \pi_2 : (M, g) \to (B, g_B)$ be Riemannian submersions. If π_1, π_2 have the same fundamental tensors A and T and ${\pi_1}_*(p) = {\pi_2}_*(p)$ at a point $p \in M$, then $\pi_1 = \pi_2$.*

11.2 O'Neill integrability tensor and O'Neill's equations

For a pseudo-Riemannian submersion $\pi : M \to B\,(\dim B \geq 2)$, leaves not necessary exist, even locally. In view of Frobenius' theorem, this failure can be measured by the vector field $\mathcal{V}[X,Y] = A_X Y - A_Y X$. Thus, the tensor A provides a natural obstruction to the integrability of horizontal distribution $\mathcal{H}$. For this reason the tensor A is called the *O'Neill integrability tensor*.

O'Neill's integrability tensor satisfies the following [O'Neill (1966)].

Lemma 11.2. *Let X, Y be horizontal vector fields and E, F be vector fields on M. Then each of the following holds:*

(a) $A_X Y = -A_Y X$, *or equivalently,* $A_X Y = \frac{1}{2}\mathcal{V}[X,Y]$;
(b) $A_{\mathcal{H}E} F = A_E F$;
(c) A_E *maps the horizontal subspace into the vertical one and the vertical subspace into the horizontal one;*
(d) $g(A_X E, F) = -g(E, A_X F)$;
(e) *If X is basic, then* $A_X V = \mathcal{H}\nabla_V X$ *for every vertical vector field V;*
(f) $g((\nabla_Y A)_X E, F) = g(E, (\nabla_Y A)_X F)$.

Let g and g_B be the metric tensors of M and B, respectively and g_F the induced metric on fiber $\pi^{-1}(\pi(p))$, $p \in M$. Denote by R, R^B and R^F the Riemann curvature tensors of the metrics g, g_B and g_F respectively.

The following equations, known as *O'Neill's equations*, characterize the geometry of a pseudo-Riemannian submersion [O'Neill (1966)].

Proposition 11.1. *For vertical vector fields U, V, W, W' and horizontal vector fields X, Y, Z, Z', we have the following formulas:*

(1) $R(U,V;W,W') = R^F(U,V;W,W') + g(T_U W, T_V W') - g(T_V W, T_U W')$;

(2) $R(U,V,W,X) = g((\nabla_U T)_V W, X) - g((\nabla_V T)_U W, X)$;

(3) $R(X,U;V,Y) = g((\nabla_X T)_U V, Y) - g(T_U X, T_V Y) + g((\nabla_U A)_X Y, V) + g(A_X U, A_Y V)$;

(4) $R(U,V;Y,X) = g((\nabla_U A)_X Y, V) - g((\nabla_V A)_X Y, U) + g(A_X U, A_Y V) - g(A_X V, A_Y U) - g(T_U X, T_V Y) + g(T_V X, T_U Y)$;

(5) $R(Y,X;Z,U) = g((\nabla_Z A)_X Y, U) + g(A_X Y, T_U Z) - g(A_Y Z, T_U X) - g(A_Z X, T_U Y)$;

(6) $R(X,Y;Z,Z') = R^B(\pi_* X, \pi_* Y; \pi_* Z, \pi_* Z') + 2g(A_X Y, A_Z Z') - g(A_Y Z, A_X Z') + g(A_X Z, A_Y Z')$.

The following follows from O'Neill's equations and Lemma 11.2(a).

Lemma 11.3. *Let $\pi : M \to B$ be a pseudo-Riemannian submersion. Then*

$$R^B(\pi_* X, \pi_* Y; \pi_* Y, \pi_* X) = R(X, Y; Y, X) + 3g(A_X Y, A_X Y). \tag{11.3}$$

Moreover, if π has totally geodesic fibers, then we also have

(1) $R(U, V; V, U) = R^F(U, V; V, U)$;
(2) $R(X, U; U, X) = g(A_X U, A_X U)$;

Formula (11.3) and Lemma 11.2(a) imply the following [O'Neill (1966)].

Corollary 11.1. *For a Riemannian submersion $\pi : M \to B$, we have*

$$K_B(X, Y) = K(\bar{X}, \bar{Y}) + \frac{3}{4} ||\mathcal{V}[\bar{X}, \bar{Y}]||^2 \tag{11.4}$$

for orthonormal vector fields X, Y on B. Consequently, the lower bound for the sectional curvature of B is at least as big as the lower bound for the sectional curvature of M.

11.3 Submersions with totally geodesic fibers

The following lemma is obvious.

Lemma 11.4. *Let $\pi : M \to B$ be a Riemannian submersion with totally geodesic fibers. If B' is a submanifold of B, then $\pi|_{\pi^{-1}(B')} : \pi^{-1}(B') \to B'$ is a Riemannian submersion with totally geodesic fibers.*

Many explicit examples of pseudo-Riemannian submersions with totally geodesic fibers can be obtained from the following standard construction [Besse (1987)]: Let G be a Lie group and let K and H be two compact Lie subgroups of G with $K \subset H$. Let $\pi : G/K \to G/H$ be the associated bundle with fiber H/K to the H-principal bundle $p : G \to G/H$. Let $\mathfrak{g}$ be the Lie algebra of G and $\mathfrak{k} \subset \mathfrak{h}$ the corresponding Lie subalgebras of K and H. We choose an $ad(H)$-invariant complement $\mathfrak{m}$ to $\mathfrak{h}$ in $\mathfrak{g}$, and an $ad(K)$-invariant complement $\mathfrak{p}$ to $\mathfrak{k}$ in $\mathfrak{h}$. An $ad(H)$-invariant nondegenerate bilinear symmetric form on $\mathfrak{m}$ defines a G-invariant pseudo-Riemannian metric g' on G/H and an $ad(K)$-invariant nondegenerate bilinear symmetric form on $\mathfrak{p}$ defines a H-invariant pseudo-Riemannian metric $\hat{g}$ on H/K. The orthogonal direct sum for these nondegenerate bilinear symmetric forms on $\mathfrak{p} \oplus \mathfrak{m}$ defines a G-invariant pseudo-Riemannian metric g on G/K.

The following proposition was proved in [Besse (1987)].

Proposition 11.2. *The mapping $\pi : (G/K, g) \to (G/H, g')$ is a pseudo-Riemannian submersion with totally geodesic fibers.*

By applying this proposition we have the following examples.

Example 11.1. Let $G = SU(m+1), H = S(U(1)\cdot U(m))$ and $K = SU(m)$. The corresponding mapping $\pi : (G/K, g) \to (G/H, g')$ is the Riemannian submersion $\pi : S^{2m+1}(1) \to CP^m(4)$, where we choose the canonical metric g on $S^{2m+1}(1)$ which is $SU(m+1)$-invariant. π is a submersion with totally geodesic fibers onto the complex projective m-space $CP^m(4)$. This Riemannian submersion is known as the *Hopf fibration.*

Example 11.2. Let $G = Sp(m+1)$, $H = Sp(1)\cdot Sp(m)$, $K = Sp(m)$. Then we get $\pi : S^{4m+3}(1) \to HP^m(4)$ with totally geodesic fibers $S^3(1)$, where $HP^m(4)$ is the quaternion projective m-space with constant quaternionic sectional curvature 4.

Example 11.3. Let $G = Sp(m+1)$, $H = Sp(1)\cdot Sp(m)$, $K = U(1)Sp(m)$. Then $\pi : CP^{2m+1} \to HP^m(4)$ is Riemannian submersion with totally geodesic fibers S^2

Example 11.4. Let $G = Spin(9)$, $H = Spin(8)$ and $K = Spin(7)$. We have the Riemannian submersion with totally geodesic fibers:

$$S^{15}(1) = Spin(9)/Spin(7) \to S^8(4) = Spin(9)/Spin(8).$$

Example 11.5. Let $G = SU(t+1, n-t)$, $H = S(U(1)U(t, n-t))$ and $K = SU(t, n-t)$, $0 \le t \le n$. We have the pseudo-Riemannian submersion

$$\begin{aligned} H^{2n+1}_{2t+1} &= SU(t+1, n-t)/SU(t, n-t) \to \\ CH^n_t &= SU(t+1, n-t)/S(U(1)U(t, n-t)). \end{aligned}$$

Example 11.6. Let $G = Sp(t+1, n-t)$, $H = Sp(1)Sp(t, n-t)$ and $K = U(1)Sp(t, n-t)$. We obtain the pseudo-Riemannian submersion

$$\begin{aligned} CH^{2n+1}_{2t+1} &= Sp(t+1, n-t)/U(1)Sp(t, n-t) \to \\ QH^n_t &= Sp(t+1, n-t)/Sp(1)Sp(t, n-t). \end{aligned}$$

Example 11.7. Let $G = Sp(t+1, n-t)$, $H = Sp(1)Sp(t, n-t)$ and $K = Sp(t, n-t)$, $0 \le t \le n$. We have the pseudo-Riemannian submersion

$$\begin{aligned} H^{4n+3}_{4t+3} &= Sp(t+1, n-t)/Sp(t, n-t) \to \\ QH^n_t &= Sp(t+1, n-t)/Sp(1)Sp(t, n-t). \end{aligned}$$

Example 11.8. Let $G = Spin(1,8)$, $H = Spin(8)$ and $K = Spin(7)$. We have the pseudo-Riemannian submersion

$$H_7^{15} = Spin(1,8)/Spin(7) \to H^8(-4) = Spin(1,8)/Spin(8).$$

Examples 11.2-11.8 are called *generalized Hopf fibration.*

More examples of submersions are the following.

Example 11.9. The frame bundle $F(B)$ of a b-dimensional Riemannian manifold B is a principal bundle over B with structure group $O(b)$. There exists a natural Riemannian structure on $F(B)$ such that the projection: $\pi : F(B) \to B$ is a Riemannian submersion with totally geodesic fibers.

Example 11.10. If (M, J, g) is an almost Hermitian manifold, its tangent bundle TM is also an almost Hermitian manifold with almost Hermitian structure (J^H, g_s), where J^H is the horizontal lift of J and g_s is the Sasaki metric given by: $g_s(X^H, Y^H) = g_s(X^V, Y^V) = (g(X,Y))^V$ and $g_s(X^H, Y^V) = 0$, where X^H and Y^V are the horizontal and vertical lifts of X and Y, respectively. (cf. [Yano and Ishihara (1973)].) The projection π: $(TM, J^H, g_s) \to (M, J, g)$ is an almost Hermitian submersion with totally geodesic fibers.

Example 11.11. On an oriented Riemannian 4-manifold N, there exists an S^2-bundle Z, called the *twistor space* of N, whose fiber over any point $p \in N$ consists of all almost complex structures on T_pN that are compatible with the metric and the orientation. It is known that there is one-parameter family of metrics g^t on Z, making the projection $Z \to N$ into a Riemannian submersion with totally geodesic fibers.

The following four results were obtained in [Băditoiu (2004)].

Theorem 11.2. *Let $\pi : H_{s+r'}^{n+r} \to B_s^n$ be a pseudo-Riemannian submersion from a pseudo-hyperbolic space onto a pseudo-Riemannian manifold. If π has totally geodesic fibers and the dimension of the fibers is ≤ 3, then π is equivalent to one of the two canonical pseudo-Riemannian submersions: (1) $\pi : H_{2t+1}^{2k+1} \to CH_t^k$, $0 \leq t \leq k$ or (2) $\pi : H_{4t+3}^{4k+3} \to QH_t^k$, $0 \leq t \leq k$.*

Theorem 11.3. *Let $\pi : H_{s+r'}^{n+r} \to B_s^n$ be a pseudo-Riemannian submersion of a pseudo-hyperbolic space onto a pseudo-Riemannian manifold. If π has totally geodesic fibers and one of the following two conditions is satisfied: (i) B is an isotropic pseudo-Riemannian manifold, i.e., for any $p \in B_s^n$ and any real number t, the group of isometries $I(B_s^n, g')$ preserving p acts transitively on the set of all nonzero tangent vectors v at p for which $g'(v,v) = t$,* *or* (ii) $\text{index}(B) \in \{0, \dim B\}$, *then* π *is equivalent to one of the following canonical pseudo-Riemannian submersions:*

(a) $H^{2k+1}_{2t+1} \to CH^k_t$, $0 \le t \le k$;
(b) $H^{4k+3}_{4t+3} \to QH^k_t$, $0 \le t \le k$;
(c) $H^{15}_{7+8t} \to H^8_{8t}(-4)$, $t \in [0, 1]$.

Theorem 11.4. *Let* $\pi : CH^n_s \to B$ *be a pseudo-Riemannian submersion of a complex pseudo-hyperbolic space onto a pseudo-Riemannian manifold. If* π *has complex totally geodesic submanifolds and if one of the following conditions is satisfies:*

(1) *the real dimension of the fibers is* $r \le 2$;
(2) B *is an isotropic pseudo-Riemannian manifold;*
(3) $\text{index}(B) \in \{0, \dim B\}$,

then π *is equivalent to the canonical pseudo-Riemannian submersion* $\pi : CH^{2k+1}_{2t+1} \to QH^k_t$, $0 \le t \le k$.

Theorem 11.5. *There do not exist pseudo-Riemannian submersions* $\pi : QH^n_s \to B$ *with quaternionic fibers from a quaternionic pseudo-hyperbolic space onto an isotropic pseudo-Riemannian manifold or onto a pseudo-Riemannian manifold of* $\text{index}(B) \in [0, \dim(B)]$.

Theorems 11.2-11.4 are pseudo-Riemannian versions of the following result of [Escobales (1975); Ranjan (1985)].

Theorem 11.6. *Let* $\pi : S^m(1) \to B$ *be a Riemannian submersion with totally geodesic fibers. Then, up to equivalence by a fiber-preserving isometry of the total space,* π *is one following standard ones:*

(1) $\pi : S^{2n+1}(1) \to CP^n(4)$ *with* S^1 *as fibers;*
(2) $\pi : S^{4n+3}(1) \to HP^n(4)$ *with* S^3 *as fibers;*
(3) $\pi : S^{15}(1) \to S^8(4)$ *with* S^7 *as fibers.*

Here $CP^n(4)$ *and* $HP^n(4)$ *denote the complex projective space and quaternionic projective space of real dimensions* $2n$ *and* $4n$, *respectively, taking sectional curvature in the interval* $[1, 4]$.

R. Escobales also obtained the following [Escobales (1978)].

Theorem 11.7. *A Riemannian submersion* $\pi : CP^r(4) \to B$, *with complex totally geodesic fibers and* $2 \le \dim fiber \le 2r - 2$, *is one of the following:*

(1) $\pi : CP^{2n+1}(4) \to QP^n(4)$ *with* S^2 *as fibers;*
(2) $\pi : CP^7(4) \to S^8(4)$ *with* $CP^3(4)$ *as fibers.*

Escobales did not know whether case (2) in Theorem 11.7 was empty or not. By applying homotopy theory it was shown in [Ucci (1983)] that case (2) cannot occur.

Two Riemannian submersions $\pi_i : S^m \to B_i \, (i = 1, 2)$ are said to be *metric congruent* if there are isometric diffeomorphisms $\iota_1 : S^m \to S^m$ and $\iota_2 : B_1 \to B_2$ such that

$$\begin{array}{ccc} S^m & \xrightarrow{\iota_1} & S^m \\ {\scriptstyle \pi_1}\downarrow & & \downarrow{\scriptstyle \pi_2} \\ B & \xrightarrow{\iota_2} & B' \end{array}$$

commutes.

The following result of [Gromoll and Grove (1988); Wilking (2001)] generalizes completely the earlier result of Escobales and Ranjan, without the assumption of totally geodesic fibers.

Theorem 11.8. *Let* $\pi : S^m(1) \to B$ *be a Riemannian submersion with connected fibers of positive dimension. Then* $\pi : S^m(1) \to B$ *is metrically congruent to one of the following fibrations:*

(1) $\pi : S^{2n+1}(1) \to CP^n(4)$ *with* S^1 *as fibers;*
(2) $\pi : S^{4n+3}(1) \to HP^n(4)$ *with* S^3 *as fibers;*
(3) $\pi : S^{15}(1) \to S^8(4)$ *with* S^7 *as fibers.*

11.4 Submersions with minimal fibers

Let $\pi : (M, g) \to (B, g_B)$ be a map from a Riemannian manifold (M, g) into another Riemannian manifold (B, g_B). The *second fundamental form* h_π of π is defined by

$$h_\pi(X, Y) = \tilde{\nabla}_X \pi_*(Y) - \pi_*(\nabla_X Y), \tag{11.5}$$

for vector fields X, Y tangent to M, where ∇ is the Levi-Civita connection of M and $\tilde{\nabla}$ is the pullback of the connection ∇' of B to the induced vector bundle $\pi^{-1}(TB)$. The *tension field* τ_π of π is the trace of h_π, i.e.,

$$\tau_\pi = \sum\nolimits_{j=1}^{m} \epsilon_j h_\pi(e_j, e_j), \tag{11.6}$$

where $\{e_1, \ldots, e_m\}$ is a local orthonormal frame of M. A map π is called *harmonic* if τ_π vanishes identically.

The following result is well-known [Eells and Sampson (1964)].

Theorem 11.9. *A pseudo-Riemannian submersion $\pi : (M, g) \to (B, g_B)$ is harmonic if and only if each fiber of π is a minimal submanifold.*

Proof. Let $\pi : (M, g) \to (B, g_B)$ be a pseudo-Riemannian submersion. Let us choose a local orthonormal frame on M such that $e_1, \ldots, e_b$ are basic and $e_{b+1}, \ldots, e_m$ are vertical. Then

$$\tilde{\nabla}_{e_r} \pi_* e_r = \pi_*(\nabla_{e_r} e_r), \quad r = 1, \ldots, b,$$
$$\tilde{\nabla}_{e_\alpha} \pi_* e_\alpha = 0, \quad \alpha = b+1, \ldots, m.$$

Hence

$$-\tau_\pi = \sum_{\alpha=b+1}^{m} \epsilon_\alpha \pi_*(\nabla_{e_\alpha} e_\alpha) = (m-b)\pi_* \hat{H},$$

where $\hat{H}$ is the mean curvature vector of fibers. Therefore, the submersion is harmonic if and only if fibers are minimal. □

The following results were obtained in [Kwon and Suh (1997)].

Theorem 11.10. *Let (M, g) be a pseudo-Riemannian $(b+s)$-manifold with index s and constant sectional curvature $c \neq 0$. If (B, g_B) is a Riemannian b-manifold, then there do not exist pseudo-Riemannian submersions $\pi : (M, g) \to (B, g_B)$ with minimal fibers.*

Theorem 11.11. *Let $\pi : (M, g) \to (B, g_B)$ be a pseudo-Riemannian submersion with minimal fibers such that (M, g) is a pseudo-Riemannian $(b+s)$-manifold with index s and let (B, g_B) be a Riemannian b-manifold. If the Ricci tensor Ric^F of fibers and Ricci tensor Ric^M of M satisfy*

$$Ric^F(u, u) \geq Ric^M(u, u)$$

for u tangent to fibers, then the horizontal distribution is integrable.

The next two theorems were proved in [Chen (2007b)].

Theorem 11.12. *Let $\pi : M \to B$ be a Riemannian submersion with minimal fibers. If M is positively curved, then the horizontal distribution of the Riemannian submersion is a non-totally geodesic distribution.*

Proof. Let $\pi : M \to B$ be a Riemannian submersion with minimal fibers. Assume $n = \dim M > \dim B = b > 0$. We use the following ranges of indices:

$$1 \leq i, j, k \leq n-b; \quad 1 \leq r, s, t \leq b; \quad 1 \leq \alpha, \beta, \gamma \leq n,$$

unless mentioned otherwise. Assume that $v_1, \ldots, v_{n-b}$ are orthonormal vertical vector fields and $X_1, \ldots, X_b$ are orthonormal horizontal vector fields of the submersion. Then it follows from Proposition 11.1(3) that

$$K(X_i \wedge v_r) = \langle (\nabla_{X_i} T)_{v_r} v_r, X_i \rangle + ||A_{X_i} v_r||^2 - ||T_{v_r} X_i||^2. \tag{11.7}$$

From the definition of ∇T, we have

$$(\nabla_{X_i} T)_{v_r} v_r = \nabla_{X_i}(T_{v_r} v_r) - T_{\nabla_{X_i} v_r} v_r - T_{v_r}(\nabla_{X_i} v_r). \tag{11.8}$$

Thus, by applying the minimality of fibers, we find

$$\begin{aligned} \sum_{i,r} \langle \nabla_{X_i}(T_{v_r} v_r), X_i \rangle &= \sum_{i,r} \langle \nabla_{X_i}(\mathcal{H}\nabla_{v_r} v_r), X_i \rangle \\ &= \sum_{i,r} X_i \langle h_F(v_r, v_r), X_i \rangle = 0. \end{aligned} \tag{11.9}$$

If we put

$$\nabla_X v_r = \sum_s \omega_r^s(X) v_s + \sum_i \omega_r^i(X) X_i, \tag{11.10}$$

then

$$\begin{aligned} &\sum_{i,r} \big\langle T_{\nabla_{X_i} v_r} v_r + T_{v_r}(\nabla_{X_i} v_r), X_i \big\rangle \\ &= \sum_{i,r} \big\langle \nabla_{\mathcal{V}\nabla_{X_i} v_r} v_r + \nabla_{v_r}(\mathcal{V}\nabla_{X_i} v_r), X_i \big\rangle \\ &= \sum_{i,r,s} \omega_r^s(X_i) \langle \nabla_{v_s} v_r + \nabla_{v_r} v_s, X_i \rangle \\ &= \sum_{i,r,s} \omega_r^s(X_i)(\omega_r^i(v_s) + \omega_s^i(v_r)) \\ &= \sum_{i,r,s} \omega_r^s(X_i)\omega_r^i(v_s) + \sum_{i,r,s} \omega_s^r(X_i)\omega_r^i(v_s) \\ &= 0. \end{aligned} \tag{11.11}$$

Consequently, we find

$$\sum_{i,r} \langle (\nabla_{X_i} T)_{v_r} v_r), X_i \rangle = 0. \tag{11.12}$$

By combining (11.7) and (11.12), we obtain

$$\sum_{i,r} K(X_i \wedge v_r) = \breve{A}_\pi - ||h_F||^2, \tag{11.13}$$

where $\breve{A}_\pi$ is defined by

$$\breve{A}_\pi = \sum_{i=1}^{b} \sum_{r=b+1}^{n} \langle A_{X_i} v_r, A_{X_i} v_r \rangle, \tag{11.14}$$

and h_F is the second fundamental form of the fibers in M. Consequently, the theorem follows from (11.13). □

Theorem 11.13. *Let $\pi : M \to B$ be a Riemannian submersion with minimal fibers. If M is non-positively curved (resp. negatively curved), then fibers have non-positive (resp. negative) scalar curvature. Further, if M is non-positively curved and fibers have zero scalar curvature, then the horizontal distribution is totally geodesic; and M, B, F are flat. Moreover, $\pi : M \to B$ is trivial, i.e, M is locally the direct product of B and F.*

Proof. If $K^F(v_r \wedge v_s)$ denotes the sectional curvature of a fiber F associated with a plane section $v_r \wedge v_s$, then Gauss's equation gives

$$K^F(v_r \wedge v_s) = K(v_r \wedge v_s) + \langle h_F(v_r, v_r), h_F(v_s, v_s)\rangle - ||h_F(v_r, v_s)||^2. \quad (11.15)$$

Since π has minimal fibers, (11.15) shows that the scalar curvature of F satisfies

$$2\hat{\tau}_F = \sum_{r \neq s} K(v_r \wedge v_s) - ||h_F||^2, \quad (11.16)$$

where $\hat{\tau}_F = \sum_{r<s} K_F(v_r \wedge v_s)$. Combining (11.13) and (11.16) gives

$$\sum_{i,r} K(X_i \wedge v_r) + \sum_{r \neq s} K(v_r \wedge v_s) = 2\hat{\tau}_F + \breve{A}_\pi. \quad (11.17)$$

Thus, if M is non-positively curved (resp. negatively curved), the fibers have non-positive (resp. negative) scalar curvature. In particular, if M is non-positively curved and the fibers have zero scalar curvature, then we derive from (11.17) that

$$K(X_i \wedge v_r) = K(v_r \wedge v_s) = \breve{A}_\pi = 0. \quad (11.18)$$

Thus, the horizontal distribution $\mathcal{H}$ is integrable and its integrable submanifolds are totally geodesic. Also, from (11.16) and (11.18) we find

$$K(v_r \wedge v_s) = ||h_F||^2 = 0. \quad (11.19)$$

Thus, π has totally geodesic fibers as well. Hence, M is locally the direct product of a fiber F and the base manifold B. The flatness of M, F and B follows from (11.18) and (11.19). □

11.5 A cohomology class for Riemannian submersion

Define a cohomology class $c_\pi(M)$ associated with a Riemannian submersion $\pi : M \to B$ with orientable base manifold B as follows: Let $b = \dim B, n = \dim M$, and let $e_1, \ldots, e_n$ be a local orthonormal frame on M satisfying the following two conditions:

(i) $e_{b+1}, \dots, e_n$ are vertical vector fields and

(ii) $e_1, \dots, e_b$ are basic horizontal vector fields such that $\pi_* e_1, \dots, \pi_* e_b$ gives rise to the positive orientation of B.

Let $\omega^1, \dots, \omega^n$ be the dual frame of $e_1, \dots, e_n$. Consider the b-form

$$\omega = \omega^1 \wedge \cdots \wedge \omega^b. \tag{11.20}$$

Then $d\omega = 0$, since ω is the pull back of the volume form of B. Thus ω defines a cohomology class $c_\pi(M) = [\omega] \in H^b(M; \mathbb{R})$.

Theorem 11.14. *Let $\pi : M \to B$ be a Riemannian submersion from a closed manifold M onto an orientable base manifold B. Then the pull back of the volume element of B is harmonic if and only if the horizontal distribution $\mathcal{H}$ is integrable and fibers are minimal.*

Proof. Let $e_1, \dots, e_n$ and $\omega^1, \dots, \omega^n$ be defined as above. Then

$$\omega^j(e_s) = 0, \;\; \omega^i(e_j) = \delta_{ij} \tag{11.21}$$

for $1 \leq i, j \leq b$; $b + 1 \leq s \leq n$. Let us put

$$\omega^\perp = \omega^{b+1} \wedge \cdots \wedge \omega^n. \tag{11.22}$$

The we find

$$d\omega^\perp = \sum_{s=b+1}^{n} (-1)^{s-b} \omega^{b+1} \wedge \cdots \wedge d\omega^s \wedge \cdots \wedge \omega^n. \tag{11.23}$$

It follows from (11.21) and (11.23) that $d\omega^\perp = 0$ if and only if the following two conditions hold:

$$d\omega^\perp(X, Y, V_1, \dots, V_{n-b-1}) = 0, \tag{11.24}$$

$$d\omega^\perp(e_i, e_{b+1}, \dots, e_n) = 0, \;\; i = 1, \dots, b \tag{11.25}$$

for horizontal vector fields X, Y and vertical vector fields $V_1, \dots, V_{n-b-1}$.

From (11.21) and (11.22) we find

$$d\omega^\perp(X, Y, V_1, \dots, V_{n-b-1}) = \omega^\perp([X, Y], V_1, \dots, V_{n-b-1}).$$

Thus (11.24) holds if and only if the horizontal distribution $\mathcal{H}$ of the Riemannian submersion is integrable. From (11.21) and (11.22) we also find

$$\begin{aligned} & d\omega^\perp(e_i, e_{b+1}, \dots, e_n) \\ & = \sum_{s=b+1}^{n} (-1)^{1+s-b} \omega^\perp([e_i, e_s], e_{b+1}, \dots, \widehat{e}_s, \dots, e_n) \\ & = \sum_{s=b+1}^{n} \{\omega^s(\nabla_{e_i} e_s) - \omega^s(\nabla_{e_s} e_i)\} = - \sum_{s=b+1}^{n} \langle \nabla_{e_s} e_i, e_s \rangle \\ & = \sum_{s=b+1}^{n} \langle A_{e_s} e_s, e_i \rangle \end{aligned}$$

for $i \in \{1, \ldots, b\}$. Thus (11.25) holds if and only if fibers are minimal. Hence the pull back of the volume element of B is co-closed, equivalently, $d\omega^{\perp} = 0$, if and only if $\mathcal{H}$ is integrable and fibers are minimal. □

Theorem 11.15. *Let $\pi : M \to B$ be a Riemannian submersion with minimal fibers and orientable base manifold B. If M is a closed manifold satisfying $H^b(M; \mathbb{R}) = 0$ with $b = \dim B$, then the horizontal distribution $\mathcal{H}$ of π is never integrable. Consequently, the submersion is always non-trivial.*

Proof. Since each nonzero harmonic form of M represents a non-trivial cohomology class, this theorem follows easily from Theorem 11.14. □

11.6 Geometry of horizontal immersions

In differential geometry there are important examples of manifolds which occur as base space B of a pseudo-Riemannian submersion $\pi : M \to B$ and whose geometric structure is obtained by projection the structure of the bundle space M. For example, the geometric structure of $CP^n(4)$ arises from the Sasakian structure of $S^{2n+1}(1)$ via Hopf's fibration.

In this section we present the close relation between immersed submanifolds of the bundle space and the base manifold B of a pseudo-Riemannian submersion. The general reference of this section is [Reckziegel (1985)].

For a pseudo-Riemannian submersion $\pi : M \to B$, the *curvature form of* the horizontal distribution $\mathcal{H}$ is the skew-symmetric bilinear form defined by

$$\Omega(X, Y) = \mathcal{V}[X, Y], \quad X, Y \in \mathcal{H}. \tag{11.26}$$

Clearly, distribution $\mathcal{H}$ is integrable if and only if $\Omega = 0$.

Lemma 11.5. *Let $\pi : M \to B$ be a pseudo-Riemannian submersion. If $X \in \mathcal{X}(N)$ such that $\phi_* X$ is horizontal and if $\phi : N \to M$ is a smooth map, then*

$$2\mathcal{V}(\nabla_X \eta) = \Omega(\phi_* X, \eta), \tag{11.27}$$

$$\pi_*(\nabla_X \eta) = \nabla'_X \pi_* \eta \tag{11.28}$$

for horizontal vector field η along ϕ, where ∇ and ∇' are the Levi-Civita connection of M and B, respectively.

Proof. From (11.1), (11.26) and Lemma 11.2(a) we find

$$\Omega(\phi_* X, \eta) = \mathcal{V}[\phi_* X, \eta] = 2A_{\phi_* X}\eta = 2\mathcal{V}\nabla_X \eta.$$

This gives (11.27). Equation (11.28) follows from Lemma 11.1(3). □

Let $\pi : M \to B$ be a pseudo-Riemannian submersion. An immersion $\phi : N \to M$ is called *horizontal* if $\phi_*(T_pN) \subset \mathcal{H}_{\phi(p)}$ for all $p \in N$.

The following result is due to [Reckziegel (1985)].

Theorem 11.16. *Let $\pi : M \to B$ be a pseudo-Riemannian submersion and $\phi : N \to M$ be an horizontal isometric immersion from a pseudo-Riemannian manifold N into M. Then*

(a) *O'Neill's integrability tensor A vanishes on vector fields tangent along ϕ, i.e., $\Omega(\phi_*(T_pN), \phi_*(T_pN)) = 0$ for each $p \in N$;*
(b) *$\hat{\phi} = \pi \circ \phi$ is an isometric immersion $N \to B$;*
(c) *the second fundamental form h^ϕ of ϕ takes its values in $\mathcal{H}$ and $\pi_* h^\phi$ equals the second fundamental form $h^{\hat{\phi}}$ of $\hat{\phi} : N \to B$;*
(d) *for every normal vector field η of N in M, $\pi_*\eta$ is a normal vector field of $\hat{\phi}$; in case that η is horizontal, we have*

$$\mathcal{V}(D_X\eta) = A_X\eta, \tag{11.29}$$

$$\pi_*(D_X\eta) = D'_X(\pi_*\eta), \tag{11.30}$$

where D and D' are the normal connection of ϕ and $\hat{\phi}$ and X is a tangent vector field of N.

Proof. Let $X, Y \in \mathcal{X}(N)$. By combining Gauss's formula and (11.27), we find $\Omega(\phi_*X, \phi_*Y) = 2\mathcal{V}h^\phi(X,Y)$. As the term on the right (resp. left) is symmetric (resp. skew-symmetric), both terms vanish. Thus we have (a) and the first half of (c).

(b) follows trivially from the definition of pseudo-Riemannian submersions. For the second half of (c) we only need to substitute Gauss's formula in (11.28). Formula (11.29) is immediate from (11.27) because $\nabla_X\eta - D_X\eta$ is tangent to N according to Weingarten's formula; hence it is horizontal. For the proof of (11.30) we use shape operators A^ϕ and $A^{\hat{\phi}}$ of ϕ and $\hat{\phi}$. From (c) we find $A^{\hat{\phi}}_{\pi_*\eta} = A^\phi_\eta$. Hence

$$\pi_*(\nabla_X\eta - D_X\eta) = \nabla'_X(\pi_*\eta) - D'_X(\pi_*\eta)$$

by means of Weingarten's formula. Now (11.30) is an easy consequence of (11.28). □

The following is an immediate consequence of Theorem 11.16.

Corollary 11.2. *Under the hypothesis of Theorem 11.16, the immersion $\hat{\phi} = \pi \circ \phi : N \to B$ is totally geodesic (resp. minimal, parallel, totally umbilical, or pseudo-umbilical) if and only if ϕ is totally geodesic (resp. minimal, parallel, totally umbilical, or pseudo-umbilical).*

Chapter 12

Contact Metric Manifolds and Submanifolds

Contact geometry is an odd-dimensional counterpart of symplectic geometry. Both contact and symplectic geometry are motivated by the mathematical formalism of classical mechanics, where one can consider either the even-dimensional phase space of a mechanical system or the odd-dimensional extended phase space including the time variable.

Contact geometry studies a geometric structure on manifolds given by a hyperplane distribution in the tangent bundle and specified by a 1-form satisfying a *maximum nondegeneracy condition.* In view of the Frobenius theorem, one may recognize the condition as the opposite of the condition that the distribution be determined by a codimension one foliation on the manifold

A contact metric manifold is a contact manifold equipped with an associated metric, called a contact metric. An important class of contact metric manifolds is the class of Sasakian manifolds. Sasakian manifolds have an associated vector field, called the characteristic vector field which generates a one-dimensional foliation. If the leaves of this foliation are compact, then the space of leaves is a Kähler "orbifold"[1].

Sasakian manifolds with Riemannian metric were introduced in 1960 by Shigeo Sasaki (1912–1987) in [Sasaki (1960)]. The notion was extended by T. Takahashi in 1969 to manifolds with pseudo-Riemannian metrics in [Takahasi (1969)].

Contact geometry has broad applications in physics; including classical mechanics, dynamics, geometrical optics and control theory.

For the general references of contact metric and Sasakian manifolds we refer to [Blair (2010); Takahasi (1969); Calvaruso and Perrone (2010)].

[1]Orbifolds are generalization of manifolds. Roughly speaking, an orbifold is the object obtained by identifying any two points of a map which are equivalent under some symmetry of the map group.

12.1 Contact pseudo-Riemannian metric manifolds

Definition 12.1. Let M be a $(2n+1)$-manifold and ϕ, ξ and η tensor fields of type $(1,1)$, $(1,0)$ and $(0,1)$ on M, respectively. The triple (ϕ, ξ, η) is called an *almost contact structure* if the following are satisfied:

$$\eta(\xi) = 1, \quad \eta(\phi X) = 0, \quad \phi^2(X) = -X + \eta(X)\xi \tag{12.1}$$

for $X \in TM$. The vector field ξ is called the *characteristic vector field.* (ξ is also called the *Reeb vector field* if the almost contact structure is contact.)

Definition 12.2. Let g be a pseudo-Riemannian metric on M. Then $(\phi, \xi, \eta, g, \epsilon)$ is called an *almost contact metric structure* on M if (ϕ, ξ, η) is an almost contact structure such that

$$g(\xi, \xi) = \epsilon, \quad \eta(X) = \epsilon g(\xi, X), \quad X \in TM, \quad \epsilon = 1 \text{ or } -1, \tag{12.2}$$

$$\eta(\phi(X), \phi(Y)) = g(X, Y) - \epsilon\eta(X)\eta(Y), \quad X, Y \in TM. \tag{12.3}$$

An almost contact metric structure is called a *contact metric structure* if it satisfies $d\eta(X, Y) = g(X, \phi Y)$. A pseudo-Riemannian manifold with a contact metric structure is called a *contact metric manifold.*

Definition 12.3. A contact metric manifold $(M, \phi, \xi, \eta, g, \epsilon)$ is said to be *K-contact* if its characteristic vector field ξ is a Killing vector field.

A plane section of a contact metric manifold $(M, \phi, \xi, \eta, g, \epsilon)$ is called a *ϕ-section* if it is spanned by $\{v, \phi v\}$ for some non-null vector v. The section curvature of a ϕ-section is called a *ϕ-sectional curvature.*

12.2 Sasakian manifolds

Definition 12.4. A contact metric structure $(\phi, \xi, \eta, g, \epsilon)$ on M is called a *normal contact metric structure* if it satisfies

$$(\nabla_X \phi)Y = g(X, Y)\xi - \epsilon\eta(Y)X, \quad X, Y \in TM, \tag{12.4}$$

where ∇ is the Levi-Civita connection of g. A manifold M endowed with a normal contact metric structure $(\phi, \xi, \eta, g, \epsilon)$ is called a *Sasakian manifold.*

The following result is due to [Takahasi (1969)].

Proposition 12.1. *For an almost contact metric manifold (M, ϕ, ξ, η, g), condition* (12.4) *implies* (1) $\nabla_X \xi = \phi(X)$, (2) *ξ is a Killing vector field, and* (3) $d\eta(X, Y) = g(X, \phi Y)$.

Remark 12.1. For an almost contact metric structure $(\phi, \xi, \eta, g, \epsilon)$ on M, if we put $\bar{g} = -g, \bar{\xi} = -\xi, \bar{\eta} = -\eta, \bar{\phi} = \phi$, then $(\bar{\phi}, \bar{\xi}, \bar{\eta}, \bar{g}, -\epsilon)$ is an almost contact metric structure on M. Hence, we may assume that $\epsilon = 1$. In this case, we simply denote $(\phi, \xi, \eta, g, \epsilon)$ by (ϕ, ξ, η, g).

If (M, ϕ, ξ, η, g) is Sasakian and α is a nonzero constant, we put

$$D_p = \{v \in T_pM : \eta(v) = 0\},\ p \in M,$$
$$\bar{g} = \alpha g + (\alpha^2 - \alpha)\eta \otimes \eta,$$
$$\bar{\xi} = \alpha^{-1}\xi,\ \bar{\eta} = \alpha\eta,\ \bar{\phi} = \phi.$$

Then $(\bar{\phi}, \bar{\xi}, \bar{\eta}, \bar{g})$ is also a Sasakian structure on M. (M, ϕ, ξ, η, g) is said to be *D-homothetic to* $(M, \bar{\phi}, \bar{\xi}, \bar{\eta}, \bar{g})$ and D is called the *contact distribution.*

For a Sasakian manifold, the sectional curvature of the plane section spanned by $\{X, \phi X\}$ with $X \perp \xi$ is called a *ϕ-sectional curvature.*

The following results can be found in [Takahasi (1969)].

Proposition 12.2. *A Sasakian manifold of constant ϕ-sectional curvature $c \neq -3$ is D-homothetic to a Sasakian manifold of constant curvature one.*

Proof. If (M, g) is of constant ϕ-sectional curvature c, we have

$$\bar{K} = \bar{K}(X, \bar{\phi}X) = \frac{c - 3(\alpha - 1)}{\alpha}$$

for any non-lightlike vector $X \in D_p$, $p \in M$. Thus $(M, \bar{g})$ is of constant ϕ-sectional curvature $\{c - 3(\alpha - 1)\}/\alpha$. Hence if $c \neq -3$ and if we take $\alpha = (c + 3)/4$, then $(M, \bar{g})$ is of constant ϕ-sectional curvature one, and therefore of constant curvature one. □

Proposition 12.3. *Let $M_i = (M_i, \phi_i, \xi_i, \eta_i, g_i)$, $i = 1, 2$, be complete, simply-connected Sasakian manifolds. If they have the same index and the same constant ϕ-sectional curvature $c \neq -3$, then they are equivalent; i.e., there exists an isometry $\Phi : M_1 \to M_2$ such that $\Phi_*(\xi_1) = \xi_2$, $\Phi^*\eta_2 = \eta_1$ and $\Phi_* \circ \phi_1 = \phi_2 \circ \Phi_*$.*

Let $\pi : M \to B$ be a Riemannian submersion whose total space admits a Sasakian structure (ϕ, ξ, η, g) such that ξ spans the vertical distribution. For $X, Y \in \mathcal{X}(B)$ with horizontal lifts $\bar{X}, \bar{Y}$, if we put $g_B(X, Y) \circ \pi = g(\bar{X}, \bar{Y})$ and $JX = \pi_*(\phi X)$, then (J, g_B) is a Kählerian structure on B [Morimoto (1964)].

Definition 12.5. A Riemannian submersion $\pi : M \to B$ is called a *canonical fibration* of a Sasakian manifold if B is Kählerian and M has a Sasakian structure (ϕ, ξ, η, g) such that ξ spans the vertical distribution.

Lemma 12.1. *Let* $\pi : (M, \phi, \xi, \eta, g) \to (B, J, g)$ *be a canonical fibration of Sasakian manifold. Then*

(a) *the vertical distribution* $\mathcal{V}$ *is spanned by* ξ*;*
(b) $g = \eta \otimes \eta + \pi^* g'$*;*
(c) $\phi X = (J\pi_*(X))^-$ *for* $X \in TM$*, where* $^-$ *denotes the horizontal lift.*

Proof. Follows from definitions. □

12.3 Sasakian space forms with definite metric

A Sasakian manifold with constant ϕ-sectional curvature is called a *Sasakian space form.*

The curvature tensor of Sasakian space forms was determined by [Ogiue (1964); Takahasi (1969)].

Proposition 12.4. *If a Sasakian manifold* $(M, \phi, \xi, \eta, g, \epsilon)$ *has constant* ϕ*-sectional curvature* c*, then its Riemann curvature tensor satisfies*

$$\begin{aligned}\tilde{R}(X,Y)Z = {} & \frac{c+3\epsilon}{4}(\langle Y,Z\rangle X - \langle X,Z\rangle Y) + \frac{\epsilon c - 1}{4}\{\eta(X)\eta(Z)Y \\ & - \eta(Y)\eta(Z)X + \langle X,Z\rangle \eta(Y)\xi - \langle Y,Z\rangle \eta(X)\xi \\ & + \langle \phi Y, Z\rangle \phi X - \langle \phi X, Z\rangle \phi Y - 2\langle \phi X, Y\rangle \phi Z\}.\end{aligned}$$

The following are the standard models of Sasakian space forms with definite metric.

Example 12.1. Let $\mathbb{R}^{2n+1} = \{(x_1, \ldots, x_n, y_1, \ldots, y_n, z) : x_i, y_i, z \in \mathbf{R}\}$. Consider $\mathbb{R}^{2n+1}$ with its standard contact structure given by

$$\begin{aligned}&\eta = \frac{1}{2}\Big(dz - \sum\nolimits_{i=1}^{n} y_i dx_i\Big), \ \xi = 2\frac{\partial}{\partial z}, \\ &g = \eta \otimes \eta + \frac{1}{4}\sum\nolimits_{i=1}^{n}(dx_i \otimes dx_i + dy_i \otimes dy_i), \qquad (12.5)\\ &\phi\left(\sum_{i=1}^{n}\left(X_i\frac{\partial}{\partial x_i} + Y_i\frac{\partial}{\partial y_i}\right) + Z\frac{\partial}{\partial z}\right) = \sum_{i=1}^{n}\left(Y_i\frac{\partial}{\partial x_i} - X_i\frac{\partial}{\partial y_i}\right) + \sum_{i=1}^{n} Y_i y_i \frac{\partial}{\partial z}.\end{aligned}$$

Then $\mathbb{R}^{2n+1}$ is a Sasakian space form of constant ϕ-sectional curvature -3.

Example 12.2. Consider the unit hypersphere $S^{2n+1}(1) \subset \mathbf{C}^{n+1}$. Denote by $\mathbf{x}$ the position vector field of $S^{2n+1}(1)$ in $\mathbf{C}^{n+1}$ and by g the induced metric. Let $\xi = -J\mathbf{x}$, where J is the complex structure of $\mathbf{C}^{n+1}$. Let η be the dual 1-form given by $\eta(X) = g(\xi, X)$ and ϕ the tensor field of type $(1,1)$

defined by $\phi = \pi \circ J$, where π is the orthogonal projection from $T_p\mathbf{C}^{n+1}$ onto $T_pS^{2n+1}(1)$, $p \in S^{2n+1}(1)$. Then $(S^{2n+1}(1), \phi, \xi, \eta, g)$ is a Riemannian Sasakian space form of constant ϕ-sectional curvature one.

Example 12.3. Let B be a simply-connected bounded domain in $\mathbf{C}^n$ and (J, g') a Kähler structure on B with constant holomorphic sectional curvature $c < 0$. Since the fundamental 2-form Ω of the Kähler structure is exact, $\Omega = d\omega$ for some real analytic 1-form ω.

Let $\pi : B \times \mathbf{R} \to B$ denote the canonical projection and t the coordinate on $\mathbf{R}$. Then $B \times \mathbf{R}$ with $\eta = \pi^*\omega + dt$ and $g = \pi^*g' + \eta \otimes \eta$ is a Sasakian manifold. Regarding η as a connection form on $B \times \mathbf{R}$ and let $\bar{X}$ be the horizontal lift of a vector field X on B. Then by applying the Riemannian submersion technique we have, for orthonormal pair $X, Y \in TB$, that

$$K(\bar{X}, \bar{Y}) = K_B(X, Y) - 3(\eta(\nabla_{\bar{X}}\bar{Y}))^2$$

where K_B is the sectional curvature of B. Since

$$g(\nabla_{\bar{X}}\bar{Y}, \xi) = -g(\bar{Y}, \nabla_{\bar{X}}\xi) = g(\bar{Y}, \phi\bar{X}) = g'(X, JX),$$

$B \times \mathbf{R}$ is a *Sasakian manifold of constant ϕ-sectional curvature* $c - 3$.

A typical example of canonical fibration of Sasakian manifolds is the Hopf fibration $\pi : S^{2n+1}(1) \to CP^n(4)$. Another example is the following.

Example 12.4. Consider the mapping $\pi : \mathbb{R}^{2n+1} \to \mathbf{C}^n$, where $\mathbb{R}^{2n+1}$ is the Sasakian space form of ϕ-sectional curvature -3, $\mathbf{C}^n$ is the usual complex Euclidean n-plane $\mathbf{C}^n$, and π is defined by

$$\pi(x_1, \ldots, x_n, y_1, \ldots, y_n, z) = \frac{1}{2}(y_1, \ldots, y_n, x_1, \ldots, x_n). \tag{12.6}$$

Then $\pi : \mathbb{R}^{2n+1} \to \mathbf{C}^n$ is a Riemannian submersion with totally geodesic fiber. This gives rise to a canonical fibration of $\mathbb{R}^{2n+1}$.

12.4 Sasakian space forms with indefinite metric

The following are Sasakian space forms with indefinite metrics.

Example 12.5. For integers n, s satisfying $n \geq 1$ and $0 \leq s \leq n$, let $S^{2n+1}_{2s}(1)$ be the unit pseudo-sphere defined by (1.16). Denote by $\mathbf{x}$ the position vector field of $S^{2n+1}_{2s}(1)$ in $\mathbf{C}^{n+1}_s$ and let $\xi = -J\mathbf{x}$, where J is the complex structure of $\mathbf{C}^{n+1}_s$. Then ξ is a spacelike unit vector field of $S^{2n+1}_{2s}(1)$. Let η be the dual 1-form given by $\eta(X) = g(\xi, X)$ and ϕ the tensor field defined by $\phi = \pi \circ J$, where π is the orthogonal projection from $T_p(\mathbf{C}^{n+1}_s)$ onto $T_p(S^{2n+1}_{2s}(1))$, $p \in S^{2n+1}_{2s}(1)$.

The following results are known [Takahasi (1969)]:

(a) $S^{2n+1}_{2s}(1)$ is diffeomorphic to $\mathbf{R}^{2s} \times S^{2n+1-2s}$. Thus $S^{2n+1}_{2s}(1)$ is simply-connected for $s \neq n$.
(b) $S^{2n+1}_{2n}(1)$ is connected with infinite cyclic fundamental group.
(c) Put $\tilde{S}^{2n+1}_{2s} = S^{2n+1}_{2s}(1)$ for $s \neq n$ and let $\tilde{S}^{2n+1}_{2n}(1)$ be the universal pseudo-Riemannian covering manifold of $S^{2n+1}_{2n}(1)$. Then the Sasakian structure on $S^{2n+1}_{2n}(1)$ induces a Sasakian structure on $\tilde{S}^{2n+1}_{2n}$.

Consider the pseudo-Riemannian metric g on $S^{2n+1}_{2s}(1)$ induced from the standard metric of $\mathbf{C}^{n+1}_s$. Then $(S^{2n+1}_{2s}(1), \phi, \xi, \eta, g)$ is a *Sasakian manifold with pseudo-Riemannian metric of constant curvature one.*

The following proposition can be found in [Takahasi (1969)]:

Proposition 12.5. *If a Sasakian $(2n+1)$-manifold (M, ϕ, ξ, η, g) is complete, simply connected and of constant ϕ-sectional curvature $c \neq -3$, then it is D-homothetic to $\tilde{S}^{2n+1}_{2s}$, where $2s =$ the index of g if $c > -3$ and $2s = 2n-$the index of g if $c < -3$.*

Example 12.6. Let

$$H^{2n+1}_{2s-1}(-1) = \{z \in \mathbf{C}^{2n+1}_s : b_{s,n+1}(z,z) = -1\}, s \geq 1.$$

Consider the induced metric $\bar{g}$ on $H^{2n+1}_{2s-1}(-1)$. Let $\mathbf{x}$ be the position vector of $H^{2n+1}_{2s-1}(-1)$ in $\mathbf{C}^{n+1}_s$, $\bar{\xi} = J\mathbf{x}$ and $\bar{\eta}$ the dual 1-form of $-\bar{\xi}$, i.e.,

$$\bar{\eta}(X) = -\bar{g}(\bar{\xi}, X), \quad X \in TH^{2n+1}_{2s-1}(-1).$$

Put $\bar{\phi} = \bar{\pi} \circ J$, where J is the complex structure of $\mathbf{C}^{n+1}_s$ and $\bar{\pi}$ the orthogonal projection of $T_p\mathbf{C}^{n+1}_s$ into $T_pH^{2n+1}_{2s-1}(-1)$, $p \in H^{2n+1}_{2s-1}(-1)$. Then $\left(H^{2n-1}_{2s-1}(-1), \bar{\phi}, \bar{\xi}, \bar{\eta}, \bar{g}\right)$ is *a Sasakian manifold with pseudo-Riemannian metric of constant curvature* -1. In particular, the anti-de Sitter space-time $H^{2n+1}_1(-1)$ is a Sasakian manifold of constant curvature -1.

It known that $\left(H^{2n+1}_{2s-1}(-1), -\bar{\phi}, -\bar{\xi}, -\bar{\eta}, -\bar{g}\right)$ is $\left(S^{2n+1}_{2(n-s+1)}(1), \phi, \xi, \eta, g\right)$.

Remark 12.2. It was shown in [Bejancu and Duggal (1993)] that every connected Sasakian real hypersurface of $\mathbb{E}^{2n+2}_{2s}$ is an open part of S^{n+1}_{2s} or of $H^{2n+1}_{2s-1}(-1)$. Also, it is proved in [Calvaruso and Perrone (2010)] that, for $n > 2$, the universal coverings of S^{2n+1}_{2s} and $H^{2n+1}_{2s-1}(-1)$ are the only simply-connected pseudo-Riemannian Sasakian manifolds of constant curvature. Furthermore, the 3-dimensional Lie group $E(1,1)$ admits a flat homogeneous (non-Sasakian) contact Lorentzian metric.

12.5 Legendre submanifolds via canonical fibration

For a contact manifold M^{2n+1}, the condition $\eta \wedge (d\eta)^n \neq 0$ implies that the contact distribution D of M is non-integrable, even locally. However, it is in general possible to find n-dimensional submanifolds whose tangent spaces lie inside the contact distribution.

A submanifold N of a contact manifold M^{2n+1} with contact form η is called an *integral submanifold* if $\eta(X) = 0$ for every $X \in \mathcal{X}(N)$. For any $X, Y \in \mathcal{X}(N)$, we have

$$d\eta(X,Y) = \frac{1}{2}(X\eta(Y) - Y\eta(X) - \eta([X,Y]) = 0.$$

Thus, in terms of associated metrics, $g(X, \phi Y) = 0$ and we see that ϕ maps tangent vectors to normal vectors. For this reason, integral submanifolds are often called *C-totally real submanifolds.* Since ξ is normal to every integral submanifold of M^{2n+1}, the dimension of an integral submanifold of a $(2n+1)$-dimensional contact metric manifold M^{2n+1} is *at most* n.[2]

On the other hand, by Darboux's theorem, we have local coordinates $(x_1, \ldots, x_n, y_1, \ldots, y_n, z)$ with respect to which $\eta = dz - \sum_{i=1}^{n} y_i dx_i$. Thus, $x_i = const, z = const$ define an n-dimensional integral submanifold.

Definition 12.6. An n-dimensional integral submanifold N of a contact metric $(2n+1)$-manifold M^{2n+1} is called a *Legendre submanifold.*

The following link between Legendre submanifolds and Lagrangian submanifolds via canonical fibrations is due to [Reckziegel (1985)].

Proposition 12.6. *Let $\pi : M \to B$ be a canonical fibration of a Sasakian manifold. Then*

(1) *an isometric immersion $f : N \to M$ is horizontal if and only if the characteristic vector field ξ is normal to N and f is C-totally real;*
(2) *an isometric immersion $i : N \to B$ has local horizontal lifts if and only if i is totally real.*

We may apply this proposition to obtain the following.

Case (i): $\mathbb{R}^{2n+1}$. Consider the canonical fibration $\pi : \mathbb{R}^{2n+1} \to \mathbf{C}^n$

$$\pi(x_1, \ldots, x_n, y_1, \ldots, y_n, z) = \frac{1}{2}(y_1, \ldots, y_n, x_1, \ldots, x_n). \tag{12.7}$$

[2]This result is false for almost contact metric manifolds, for instance, $\mathbf{C}^n \times \mathbf{R}$ carries an almost contact metric structure with the usual complex structure on $\mathbf{C}^n$ and with the product metric. But $\mathbf{C}^n$ is an integral submanifold of dimension $2n$.

Let $i : N \to \mathbf{C}^n$ be a Lagrangian isometric immersion of a Riemannian n-manifold N into $\mathbf{C}^n$. Then there is a covering map $\tau : \hat{N} \to N$ and a horizontal immersion $\hat{i} : \hat{N} \to \mathbb{R}^{2n+1}$ such that $i \circ \tau = \pi \circ \hat{i}$. Thus, each Lagrangian immersion can be lifted (locally) (or globally if N is simply-connected) to a Legendre immersion of the same manifold.

Conversely, let $f : \hat{N} \to \mathbb{R}^{2n+1}$ be a Legendre isometric immersion. Then $i = \pi \circ f : N \to \mathbf{C}^n$ is a Lagrangian isometric immersion. Under this correspondence, we have a one-to-one correspondence between Lagrangian submanifold $\hat{N}$ of $\mathbf{C}^n$ and Legendre submanifolds N of $\mathbb{R}^{2n+1}$.

Case (ii): $CP^n(4)$. Consider Hopf's fibration $\pi : S^{2n+1} \to CP^n(4)$. Then π is a Riemannian submersion. Given $z \in S^{2n+1}$, the horizontal space at z is the orthogonal complement of iz with respect to the metric on S^{2n+1} induced from the metric on $\mathbf{C}^{n+1}$. Let $i : N \to CP^n(4)$ be a Lagrangian isometric immersion. Then there is a covering map $\tau : \hat{N} \to N$ and a horizontal immersion $\hat{i} : \hat{N} \to S^{2n+1}$ such that $i \circ \tau = \pi \circ \hat{i}$. Thus each Lagrangian immersion can be lifted locally (or globally if N is simply-connected) to a Legendre immersion of the same Riemannian manifold.

Conversely, if $f : \hat{N} \to S^{2n+1}$ is a Legendre isometric immersion, then $i = \pi \circ f : N \to CP^n(4)$ is again a Lagrangian isometric immersion. Under this correspondence the second fundamental forms h^f and h^i of f and i satisfy $\pi_* h^f = h^i$ by Theorem 11.16. Moreover, h^f is horizontal with respect to π.

Case (iii): $CH^n(-4)$. Consider the anti-de Sitter spacetime

$$H_1^{2n+1}(-1) = \{z \in \mathbf{C}_1^{2n+1} : b_{1,n+1}(z,z) = -1\}$$

with the canonical Sasakian structure given in Example 12.3.

Put $T'_z = \{u \in \mathbf{C}^{n+1} : \langle u, z\rangle = 0\}$, $H_1^1 = \{\lambda \in \mathbf{C} : \lambda\bar{\lambda} = 1\}$, where $\langle\ ,\ \rangle$ is the Hermitian inner product on $\mathbf{C}_1^{n+1}$ whose real part is g_0. Then there is an H_1^1-action on $H_1^{2n+1}(-1)$, $z \mapsto \lambda z$ and at each point $z \in H_1^{2n+1}(-1)$, the vector $\xi = -\mathrm{i}z$ is tangent to the flow of the action. Since the metric g_0 is Hermitian, we have $\langle \xi, \xi\rangle = -1$. The quotient space $H_1^{2n+1}(-1)/\sim$, under the identification induced from the action, is the complex hyperbolic space $CH^n(-4)$ with constant holomorphic sectional curvature -4, with the complex structure J induced from the complex structure J on $\mathbf{C}_1^{n+1}$ via $\pi : H_1^{2n+1}(-1) \to CH^n(4c)$.

Just like case (ii), if $i : N \to CH^n(-4)$ is a Lagrangian immersion, then there is an isometric covering map $\tau : \hat{N} \to N$ and a Legendre immersion $f : \hat{N} \to H_1^{2n+1}(-1)$ such that $i \circ \tau = \pi \circ f$. Thus every Lagrangian immersion can be lifted locally (or globally if N is simply-connected) to a Legendre immersion.

Conversely, if $f : \hat{N} \to H_1^{2n+1}(-1)$ is a Legendre immersion, then $i = \pi \circ f : N \to CH^n(-4)$ is a Lagrangian immersion. Similarly, under this correspondence the second fundamental forms h^f and h^i are related by $\pi_* h^f = h^i$. Also, h^f is horizontal with respect to π.

12.6 Contact slant submanifolds via canonical fibration

Let $(M^{2m+1}, \phi, \xi, \eta, g)$ be an almost contact metric manifold with a Riemannian metric g, the almost contact $(1,1)$-tensor ϕ, and the characteristic vector field ξ. An immersion $i : N \to M^{2m+1}$ is called *contact θ-slant* if

(i) the characteristic vector field ξ of $\tilde{M}^{2m+1}$ is tangent to $i_*(TN)$ and
(ii) for each $X \neq 0$ tangent to $i_*(T_pN)$ and perpendicular to ξ, the angle $\theta(X)$ between $\phi(X)$ and $i_*(T_pN)$ is independent of the choice of X.

The following presents a method to construct slant submanifolds in CP^n and CH^n via Hopf's fibration [Chen and Tazawa (2000)] (see also [Cabrerizo et al. (2002)]). We use the same notations as section 12.2.

Case (1): *Slant submanifolds in $CP^m(4)$.* Consider the Hopf fibration $\pi : S^{2n+1}(1) \to CP^n(4)$. Let $i : N \to CP^m(4)$ be an isometric immersion. Then $\hat{N} = \pi^{-1}(N)$ is a principal circle bundle over N with totally geodesic fibers. The lift $\hat{\imath} : \hat{N} \to S^{2m+1}$ of i is an isometric immersion such that $\pi \circ \hat{\imath} = i \circ \pi$. Conversely, if $\psi : \hat{N} \to S^{2m+1}$ is an isometric immersion, invariant under the action of $\mathbf{C}^* = \mathbf{C} - \{0\}$, then there is a unique isometric immersion $\psi_\pi : \pi(\hat{N}) \to CP^m(4)$ such that $\pi \circ \psi = \psi_\pi \circ \pi$.

We call $\psi_\pi : \pi(\hat{N}) \to CP^m(4)$ *the projection* of $\psi : \hat{N} \to S^{2m+1}$.

Proposition 12.7. *An isometric immersion $i : N \to CP^m(4)$ is θ-slant if and only if the lift $\hat{\imath} : \hat{N} \to S^{2m+1}$ is contact θ-slant.*

Let $h, \hat{h}$ be the second fundamental forms of $i, \hat{\imath}$, respectively. Then

$$\hat{h}(\bar{X}, \bar{Y}) = (h(X,Y))^{\bar{}}, \quad \hat{h}(\bar{X}, V) = (FX)^{\bar{}}, \quad \hat{h}(V,V) = 0 \tag{12.8}$$

for $X, Y \in \mathcal{X}(N)$, where FX is the normal component of JX in $CP^m(4)$. It follows from Proposition 12.7 that in order to obtain the explicit expression of a desired θ-slant submanifold of $CP^m(4)$ with second fundamental form h, it is sufficient to construct a contact θ-slant submanifold of S^{2m+1} whose second fundamental form satisfies $\pi_* \hat{h} = h$, and vice versa.

Case (2): *Slant submanifolds in* $CH^m(-4)$. Consider the generalized Hopf fibration $\pi : H_1^{2n+1}(-1) \to CH^n(-4)$. Let $i : N \to CH^m(-4)$ be an isometric immersion. Then $\hat{N} = \pi^{-1}(N)$ is a principal circle bundle over N with totally geodesic fibers and the lift $\hat{i} : \hat{N} \to H_1^{2m+1}$ is an isometric immersion with $\pi \circ \hat{i} = i \circ \pi$. Conversely, if $\psi : \hat{N} \to H_1^{2m+1}$ is an isometric immersion invariant under the action of $\mathbf{C}^*$, then there exists a unique isometric immersion $\psi_\pi : \pi(\hat{N}) \to CH^m(-4)$ such that $\pi \circ \psi = \psi_\pi \circ \pi$.

Similar to Proposition 12.7 we have

Proposition 12.8. *An isometric immersion* $i : N \to CH^m(-4)$ *is* θ*-slant if and only if the lift* $\hat{i} : \hat{N} \to H_1^{2m+1}$ *is contact* θ*-slant.*

Denote by $\hat{\nabla}$ and $\tilde{\nabla}$ the Levi-Civita connections of $S^{2m+1}(1)$ and of $CP^m(4)$ (resp., of $H_1^{2m+1}(-1)$ and of $CH^m(-4)$), $\bar{X}$ the horizontal lift of X, and by h and $\hat{h}$ the second fundamental forms of i and $\hat{i}$, respectively. Denote by $\breve{\nabla}$ the Levi-Civita connection of $\mathbf{C}^{m+1}$ (resp. of $\mathbf{C}_1^{m+1}$). If N is a θ-slant submanifold of $CP^m(4)$ (resp. of $CH^m(-4)$) and if $z : S^{2m+1} \to \mathbf{C}^{m+1}$ (resp. $z : H_1^{2m+1} \to \mathbf{C}_1^{m+1}$) denotes the standard inclusion, then

$$\breve{\nabla}_{\bar{X}}\bar{Y}* = (\nabla_X Y)\bar{} + (h(X,Y))\bar{} + \langle JX, Y\rangle \mathrm{i}z - \varepsilon \langle X, Y\rangle z \tag{12.9}$$

$$\breve{\nabla}_{\bar{X}} V = \breve{\nabla}_V \bar{X} = (JX)\bar{}, \quad \breve{\nabla}_V V = \varepsilon z, \tag{12.10}$$

for $X, Y \in \mathcal{X}(N)$, where $\varepsilon = 1$ for $CP^m(4)$ and $\varepsilon = -1$ for $CH^m(-4)$.

One may obtain the desired θ-slant submanifold by solving the system (12.9)-(12.10). The general construction procedure goes as follows: First one determines both the intrinsic and extrinsic structures of the θ-slant submanifold in order to find the precise form of the differential system (12.9)-(12.10). Next, one constructs a coordinate system on the associated contact θ-slant submanifold $\pi^{-1}(N)$. After that one may solve the differential system via the coordinate system on $\hat{N}$ to obtain a solution of the system. The solution of the system gives rise to the explicit expression of the associated contact θ-slant submanifold of S^{2m+1} or H_1^{2m+1} which in turn provides the representation of the desired θ-slant submanifold via π.

Case (3): *Slant submanifolds in* $\mathbf{C}^n$. Similarly, one may also apply the canonical fibration $\pi : \mathbb{R}^{2n+1} \to \mathbf{C}^n$ in (12.7) to obtain the one-to-one correspondence between contact slant submanifolds of $\mathbb{R}^{2n+1}$ and slant submanifolds of $\mathbf{C}^n$.

Remark 12.3. For precise constructions of slant submanifolds via the construction method described in this section, see [Chen and Tazawa (2000)].

Chapter 13

δ-invariants, Inequalities and Ideal Immersions

13.1 Motivation

Curvature invariants are the $N^o\,1$ Riemannian invariants and the most natural ones. Curvatures invariants also play key roles in physics. For instance, the magnitude of a force required to move an object at constant speed, according to Newton's law, a constant multiple of the curvature of the trajectory. The motion of a body in a gravitational field is determined, according to Einstein, by the curvatures of spacetime. All sorts of shapes, from soap bubbles to red blood cells, seem to be determined by various curvatures [Osserman (1990)]. Borrow a term from biology, Riemannian invariants are DNA of Riemannian manifolds. Classically, among the Riemannian curvature invariants, people have been studying sectional, scalar and Ricci curvatures in great details (see, for instance, [Berger (2003)]).

One of the most fundamental problems in the theory of submanifolds is the immersibility (or non-immersibility) of a Riemannian manifold in a Euclidean space (or, more generally, in a space form). According to the 1956 celebrated embedding theorem of Nash, every Riemannian manifold can be isometrically embedded in some Euclidean spaces with sufficiently high codimension [Nash (1956)].

The Nash theorem was aimed at the hope that if Riemannian manifolds could be regarded as Riemannian submanifolds, this would then yield the opportunity to use extrinsic help in the study of (intrinsic) Riemannian geometry. However, this hope had not been materialized yet according to [Gromov (1985)]. The main reason for this was the lack of controls of the extrinsic properties of the submanifolds by the known intrinsic invariants.

In view of Nash's theorem, to study embedding problems it is natural to impose some suitable conditions on the immersions. For example, if one imposes the minimality condition, it leads to

Problem 13.1. *Given a Riemannian manifold M, what are necessary conditions for M to admit a minimal isometric immersion in a Euclidean m-space $\mathbb{E}^m$?*

It follows from Corollary 2.1 that a necessary condition for a Riemannian manifold to admit a minimal immersion in any Euclidean space is $Ric \leq 0$. For many years this was the only known necessary Riemannian condition for a general Riemannian manifold to admit a minimal isometric immersion into a Euclidean space. That is why Chern asked in 1968 to search for further Riemannian obstructions for a Riemannian manifold to admit an isometric minimal immersion into a Euclidean space. Also, no solutions to Chern's problem were known for many years.

In order to overcome those difficulties, one needs to introduce new types of Riemannian invariants, different in nature from "classical" invariants. Moreover, one also needs to establish general optimal relationships between the main extrinsic invariants with the new intrinsic invariants on the submanifolds. These are exactly the author's original motivation in early 1990s to introduce his so-called δ-curvature invariants on Riemannian manifolds.

The δ-curvatures are very different in nature from the "classical" scalar and Ricci curvatures; simply due to the fact that both scalar and Ricci curvatures are "total sum" of sectional curvatures on a Riemannian manifold; in contrast, author's δ-curvature invariants are obtained from the scalar curvature by throwing away a certain amount of sectional curvatures.

13.2 Definition of δ-invariants

Let M be a Riemannian n-manifold. Let $K(\pi)$ be the sectional curvature associated with a plane section $\pi \subset T_pM$, $p \in M$. For an orthonormal basis $e_1, \ldots, e_n$ of T_pM, the scalar curvature τ at p is defined to be

$$\tau(p) = \sum_{i<j} K(e_i \wedge e_j).$$

Let L be a subspace of T_pM of dimension $r \geq 2$ and $\{e_1, \ldots, e_r\}$ an orthonormal basis of L. We define the scalar curvature $\tau(L)$ of L by

$$\tau(L) = \sum_{\alpha<\beta} K(e_\alpha \wedge e_\beta), \quad 1 \leq \alpha, \beta \leq r.$$

We simply denote by $\tau_{1\cdots r}$ the scalar curvature of the r-plane section spanned by $e_1, \ldots, e_r$. The scalar curvature $\tau(p)$ of M at p is nothing but the scalar curvature of the tangent space of M at p; if L is a 2-plane section, $\tau(L)$ is the sectional curvature $K(L)$ of L. Geometrically, $\tau(L)$ is nothing but the scalar curvature of the image $\exp_p(L)$ at p.

For an integer $k \geq 0$ let $\mathcal{S}(n,k)$ denote the set consisting of unordered k-tuples $(n_1, \ldots, n_k)$ of integers ≥ 2 such that $n > n_1$ and $n_1 + \cdots + n_k \leq n$. Denote by $\mathcal{S}(n)$ the set of unordered k-tuples with $k \geq 0$ for a fixed n.

For each k-tuple $(n_1, \ldots, n_k) \in \mathcal{S}(n)$, $\delta(n_1, \ldots, n_k)(p)$ is defined by

$$\delta(n_1, \ldots, n_k)(p) = \tau(p) - \inf\{\tau(L_1) + \cdots + \tau(L_k)\}, \tag{13.1}$$

where $L_1, \ldots, L_k$ run over all k mutually orthogonal subspaces of T_pM such that $\dim L_j = n_j$, $j = 1, \ldots, k$. Similarly, $\hat{\delta}(n_1, \ldots, n_k)(p)$ is defined by

$$\hat{\delta}(n_1, \ldots, n_k)(p) = \tau(p) - \sup\{\tau(L_1) + \cdots + \tau(L_k)\}. \tag{13.2}$$

Obviously, $\delta(\emptyset) = \hat{\delta}(\emptyset) = \tau$ for $k = 0$. It is also clear that

$$\delta(n_1, \ldots, n_k) \geq \hat{\delta}(n_1, \ldots, n_k)$$

for any k-tuple $(n_1, n_2, \ldots, n_k) \in \mathcal{S}(n)$.

Definition 13.1. A Riemannian n-manifold M is called an $S(n_1, \ldots, n_k)$-*space* for a given k-tuple $(n_1, n_2, \ldots, n_k) \in \mathcal{S}(n)$ if it satisfies

$$\delta(n_1, \ldots, n_k) = \hat{\delta}(n_1, \ldots, n_k)$$

identically.

Let $\#\mathcal{S}(n)$ denote the cardinal number of $\mathcal{S}(n)$. Then $\#\mathcal{S}(n)$ increases quite rapidly with n. For instance, for

$$n = 2, 3, 4, 5, 6, 7, 8, 9, 10, 11, 12, \cdots, 20, \cdots, \\ 50, \cdots, 100, \cdots, 200, \cdots,$$

$\#\mathcal{S}(n)$ are given respectively by

$$1, 2, 4, 6, 10, 14, 21, 29, 41, 54, 76, \cdots, 626, \cdots, \\ 204\,225, \cdots, 190\,569\,291, \cdots, 3\,972\,999\,029\,387, \cdots.$$

In general, the cardinal number $\#\mathcal{S}(n)$ is equal to $p(n) - 1$, where $p(n)$ is the partition function. The asymptotic behavior of $\#\mathcal{S}(n)$ is given by

$$\#\mathcal{S}(n) \approx \frac{1}{4n\sqrt{3}} \exp\left[\sqrt{\tfrac{2n}{3}}\,\pi\right] \quad \text{as } n \to \infty.$$

Some other invariants of similar nature, i.e., invariants obtained from the scalar curvature by deleting certain amount of sectional curvatures, are also called δ-*invariants*. For instance, we also have the so-called affine δ-invariants, Kählerian δ-invariants, submersion δ-invariant, etc.

13.3 δ-invariants and Einstein and conformally flat manifolds

Let M be a Riemannian n-manifold with $n > 2$ and let $\{e_1, \ldots, e_n\}$ be any orthonormal basis of T_pM, $p \in M$. We denote the scalar curvature of the j-space $\text{Span}\{e_{i_1}, \ldots, e_{i_j}\}$ simply by $\tau_{i_1 \cdots i_j}$.

The following characterization of Einstein spaces from [Chen et al. (2000)] generalizes the well-known characterization of Einstein 4-manifolds of I. M. Singer and J. A. Thorpe [Singer and Thorpe (1969)].

Proposition 13.1. *Let M be a Riemannian 2ℓ-manifold. Then M is an Einstein space if and only if $\tau(L) = \tau(L^\perp)$ for any ℓ-plane section $L \subset T_pM$, $p \in M$.*

Proof. Let L be an arbitrary ℓ-plane section at a point $p \in M$. Choose an orthonormal basis $\{e_1, \ldots, e_{2\ell}\}$ at p such that L is spanned by $e_1, \ldots, e_\ell$.

If M is an Einstein space, the Ricci curvature of M satisfies

$$Ric(e_1) + \cdots + Ric(e_\ell) = Ric(e_{\ell+1}) + \cdots + Ric(e_{2\ell}),$$

which implies $\tau(L) = \tau(L^\perp)$.

Conversely, suppose that $\tau(L) = \tau(L^\perp)$ for any ℓ-plane section $L \subset T_pM$. Then

$$\tau_{12\cdots\ell} - \tau_{2\cdots\ell+1} = \tau_{\ell+1\cdots 2\ell} - \tau_{1\ell+2\cdots 2\ell},$$

which shows that

$$\begin{aligned} K_{12} + \cdots + K_{1\ell} - (K_{(\ell+1)2} + \cdots + K_{(\ell+1)\ell}) \\ = K_{(\ell+1)(\ell+2)} + \cdots + K_{(\ell+1)(2\ell)} - (K_{1(\ell+2)} + \cdots + K_{1(2\ell)}). \end{aligned}$$

Thus $Ric(e_1) = Ric(e_{\ell+1})$. Since $\ell > 1$, this implies that M is an Einstein manifold. □

The following characterization of conformally flat spaces from [Chen et al. (2000)] extends a well-known result of R. S. Kulkarni [Kulkarni (1969)].

Proposition 13.2. *Let M be a Riemannian n-manifold with $n \geq 4$, and let s be an integer satisfying $2 < 2s \leq n$. Then M is a conformally flat manifold if and only if, for any orthonormal set $\{e_1, \ldots, e_{2s}\}$ of vectors, one has*

$$\tau_{1\cdots s} + \tau_{s+1\cdots 2s} = \tau_{1\cdots s-1\, s+1} + \tau_{s\, s+2\cdots 2s}. \tag{13.3}$$

Proof. For $s = 2$ this is Kulkarni's result. However, for the completeness, we include a proof of this special case.

If M is conformally flat, Weyl's conformal curvature tensor vanishes identically. Thus

$$K_{ij} = \frac{1}{n-2}(Ric(e_i) + Ric(e_j)) - \frac{\tau}{(n-1)(n-2)}, \quad i \neq j, \tag{13.4}$$

from which we conclude that

$$K_{ij} + K_{k\ell} = K_{ik} + K_{j\ell}, \quad \text{for distinct } i, j, k, \ell. \tag{13.5}$$

Conversely, if (13.5) holds for distinct i, j, k, ℓ, then, by fixing i, j, k in (13.5) and summing up over all remaining ℓ, one obtains

$$(n-2)K_{ij} + Ric(e_k) = (n-2)K_{ik} + Ric(e_j). \tag{13.6}$$

Fixing i, j in (13.6) and summing up over all remaining k yield (13.4), which implies the vanishing of Weyl's conformal curvature tensor. Therefore, M is conformally flat.

Now, we prove that the theorem for $s > 2$. First we prove that a conformally flat space must satisfies (13.3). So, let us assume that M is conformally flat and let s be any integer satisfying $2 < 2s \leq n$.

From Kulkarni's result we know that

$$K_{is} + K_{k\,s+1} = K_{i\,s+1} + K_{ks}$$

for any $i < s$ and $k > s + 1$. Therefore

$$\begin{aligned}
\tau_{1\cdots s} + \tau_{s+1\cdots 2s} &= \tau_{1\cdots s-1} + \sum\nolimits_{i=1}^{s-1}(K_{is} + K_{s+1\,s+1+i}) + \tau_{s+2\cdots 2s} \\
&= \tau_{1\cdots s-1} + \sum\nolimits_{i=1}^{s-1}(K_{i\,s+1} + K_{s\,s+1+i}) + \tau_{s+2\cdots 2s} \\
&= \tau_{1\cdots s-1\,s+1} + \tau_{s\,j+2\cdots 2s}.
\end{aligned}$$

Next, we prove that (13.3) implies conformal flatness. For this we use (13.3) twice to obtain

$$\begin{aligned}
0 =& (\tau_{1\cdots s} + \tau_{s+1\cdots 2s}) - (\tau_{1\cdots s-1\,s+1} + \tau_{s\,s+2\cdots 2s}) \\
&- ((\tau_{1\cdots s-2\,s+2\,s} + \tau_{s+1\,s-1\,s+3\cdots 2s}) \\
&- (\tau_{1\cdots s-2\,s+2\,s+1} + \tau_{s\,s-1\,s+3\cdots 2s}))
\end{aligned} \tag{13.7}$$

It is clear that K_{ik} does not occur in (13.7) unless both i and k belong to the set $\{s-1, s, s+1, s+2\}$. Taking this into consideration, (13.4) becomes

$$0 = 2(K_{s-1\,s} + K_{s+1\,s+2}) - 2(K_{s-1\,s+1} + K_{s\,s+2}),$$

and Kulkarni's result implies that M is conformally flat. □

$S(n_1, \ldots, n_k)$-spaces are completely determined by the following two theorems [Chen et al. (2000)].

Theorem 13.1. *Let M be a Riemannian n-manifold with $n > 2$. Then*

(1) *for any integer j with $2 \le j \le n-2$, M is an $S(j)$-space if and only if M is a real space form;*
(2) *M is an $S(n-1)$-space if and only if M is an Einstein space.*

Proof. First we prove the following claim: *For a given integer j with $2 \le j \le n-2$, if M is an $S(j)$-space, then it is an $S(j+1)$-space.*

If M is an $S(3)$-space, then τ_{123} is a number, say c, which is independent of the choice of the 3-plane. In particular, from the definition of scalar curvature of a j-plane, we have

$$\tau_{234} = \tau_{1234} - K_{12} - K_{13} - K_{14} = c, \tag{13.8}$$

where K_{ij} denotes the sectional curvature of the 2-plane spanned by e_i, e_j.

On the other hand, since M is an $S(3)$-space, we have

$$\begin{aligned} K_{12} + K_{13} + K_{23} &= c, \\ K_{13} + K_{14} + K_{34} &= c, \\ K_{12} + K_{14} + K_{24} &= c. \end{aligned} \tag{13.9}$$

By summing up the three equations in (13.9) we obtain

$$\tau_{1234} + K_{12} + K_{13} + K_{14} = 3c. \tag{13.10}$$

Combining (13.8) and (13.10) yields $\tau_{1234} = 2c$. Since the orthonormal basis can be chosen arbitrarily, this implies that M is an $S(4)$-space.

In general, if M is an $S(j)$-space, then

$$\tau_{2\cdots j+1} = \tau_{1\cdots j+1} - K_{12} - \cdots - K_{1j+1} = c. \tag{13.11}$$

On the other hand, similar to (13.10), we also have

$$(j-2)\tau_{1\cdots j+1} + K_{12} + \cdots + K_{1j+1} = jc. \tag{13.12}$$

Combining (13.10) and (13.12) yields $(j-1)\tau_{1\cdots j+1} = (j+1)c$. This implies M is an $S(j+1)$-space. Consequently, we have the claim.

Next, we claim that M is an $S(n-1)$-space if and only if M is an Einstein space. This follows obviously from the following identity:

$$\tau(L^{n-1}) = \tau - Ric(e_n),$$

where e_n is a unit vector perpendicular to a hyperplane $L^{n-1} \subset T_pM$.

Finally, we prove that an $S(n-2)$-space is a real space form. This can be done as follows: Let M is an $S(n-2)$-space. Then M is an Einstein space according to the two claims. On the other hand, let e_{n-1}, e_n be any orthonormal vectors at a point p and let L^{n-2} be the $(n-2)$-subspace of the tangent space T_pM perpendicular to e_{n-1} and e_n. We have

$$\tau - Ric(e_{n-1}) - Ric(e_n) + K_{n-1n} = c, \tag{13.13}$$

where c is independent of the choice of L^{n-2}.

Because M is Einstein, (13.13) implies that $K_{n-1\,n}$ is independent of the choice of the plane $e_{n-1} \wedge e_n$. Thus M is a real space form. □

Theorem 13.2. *Let M be a Riemannian n-manifold such that n is not a prime and k an integer ≥ 2. Then*

(1) *if M is an $S(n_1, \ldots, n_k)$-space, it is a real space form unless $n_1 = \cdots = n_k$ and $n_1 + \cdots + n_k = n$.*
(2) *when $n_1 = \cdots = n_k$ and $n_1 + \cdots + n_k = n$ hold, M is an $S(n_1, \ldots, n_k)$-space if and only if M is a conformally flat space.*

Proof. Without loss of generality, we may assume $n_1 \leq n_2 \leq \cdots \leq n_k$. Let $e_1, \ldots, e_n$ be an orthonormal basis of T_pM. If M is an $S(n_1, \ldots, n_k)$-space, then

$$\tau_{12\cdots n_1} + \cdots + \tau_{n_1+\cdots+n_{k-1}+1\cdots n_1+\cdots+n_k} = c, \tag{13.14}$$

where c is independent of the orthonormal basis.

Consider a permutation σ of $\{1, \ldots, n_1 + \cdots + n_k\}$. For each such permutation, there exists a corresponding equation like (13.14) given by

$$\tau_{\sigma(1)\sigma(2)\cdots\sigma(n_1)} + \cdots + \tau_{\sigma(n_1+\cdots+n_{k-1}+1)\cdots\sigma(n_1+\cdots+n_k)} = c. \tag{13.15}$$

By summing up all such equations we obtain

$$c_1\tau_{12\cdots n_1+\cdots+n_k} = c_2 c, \tag{13.16}$$

where c_1, c_2 are positive constants. Thus M is an $S(n_1 + \cdots + n_k)$-space. Hence, according to Theorem 13.1, if $n_1 + \cdots + n_k \leq n-2$, then M is a real space form; if $n_1 + \cdots + n_k = n-1$, then M is an Einstein space; and if $n_1 + \cdots + n_k = n$, we obtain no new information.

In the first case, we are done, so we only have to consider the last two cases. First we assume that M is not Einstein, and prove that M is conformally flat. Next we will show that if M is Einstein, then M is of constant curvature.

Case (1): *M is not Einstein.* Clearly, we have $n_1 + \cdots + n_k = n$. Thus (13.14) becomes

$$\tau_{12\cdots n_1} + \cdots + \tau_{n_1+\cdots+n_{k-1}+1\cdots n} = c. \tag{13.17}$$

Consider a permutation σ of $\{2, \ldots, n\}$. For each such permutation, there is a corresponding equation like (13.17) given by

$$\tau_{1\sigma(2)\cdots\sigma(n_1)} + \cdots + \tau_{\sigma(n_1+\cdots+n_{k-1}+1)\cdots\sigma(n)} = c. \tag{13.18}$$

By summing up all such equations we find

$$c_3 Ric(e_1) + c_4 \tau_{2\cdots n} = c_5 c, \tag{13.19}$$

where

$$\begin{aligned} c_3 &= \binom{n-2}{n_1-2}\binom{n-n_1}{n_2}\cdots\binom{n-n_1-\cdots-n_{k-1}}{n_k}, \\ c_4 &= \binom{n-3}{n_1-3}\binom{n-n_1}{n_2}\cdots\binom{n-n_1-\cdots-n_{k-1}}{n_k} \\ &\quad + \binom{n-3}{n_2-2}\binom{n-n_2-1}{n_1-1}\binom{n-n_1-n_2}{n_3}\cdots\binom{n-n_1-\cdots-n_{k-1}}{n_k} \\ &\quad + \cdots \\ &\quad + \binom{n-3}{n_k-2}\binom{n-n_k-1}{n_1-1}\binom{n-n_1-n_k}{n_2}\cdots\binom{n-n_1-\cdots-n_{k-2}-n_k}{n_{k-1}}, \\ c_5 &= \binom{n-1}{n_1-1}\binom{n-n_1}{n_2}\cdots\binom{n-n_1-\cdots-n_{k-1}}{n_k}, \end{aligned}$$

and $\binom{n-3}{n_1-3}$ is defined to be zero if $n_1 = 2$. These formulas imply that

$$\begin{aligned} &(n_1-2)!\,n_2!\,n_3!\cdots n_k!\,c_3 = (n-2)!\,, \\ &(n_1-1)!\,n_2!\,n_3!\cdots n_k!\,c_4 \\ &\qquad = (n-3)!\left(\sum_{j=2}^{k} n_j(n_j-1) + (n_1-1)(n_1-2)\right). \end{aligned} \tag{13.20}$$

Clearly, (13.19) is equivalent to

$$(c_3 - c_4)Ric(e_1) + c_4\tau = c_5 c.$$

Since M is not Einstein, we must have that $c_3 = c_4$. Now (13.20) implies that $c_3 - c_4 = 0$ if and only if

$$(n_2 + \cdots + n_k)(n_1 - 1) = \sum_{j=2}^{k} n_j(n_j - 1). \tag{13.21}$$

Using the assumption $n_1 \leq n_j$, this implies that $n_1 = \cdots = n_k$.

So M is an $S(n_1, \ldots, n_k)$-space with $n_1 = \ldots = n_k$ and $n_1 + \cdots + n_k = n$. Put $n_1 = j$. Let $\{e_1, \ldots, e_{2j}\}$ be an orthonormal set of vectors and let $L_3, \ldots, L_k$ be mutually orthogonal j-dimensional spaces, which are also orthogonal to $\{e_1, \ldots, e_{2j}\}$. Then

$$\begin{aligned}&\tau_{1\cdots j} + \tau_{j+1\cdots 2j} + \tau(L_3) + \ldots + \tau(L_k)\\ &\quad = \tau_{1\cdots j-1\, j+1} + \tau_{j\, j+2\cdots 2j} + \tau(L_3) + \ldots + \tau(L_k),\end{aligned}$$

which obviously implies (13.3), so M is conformally flat. Before going into the second case, we first prove the converse. So we suppose that M is conformally flat, and suppose $n = kj$ for some natural numbers $k > 1$ and $j > 1$. Let $\{e_1, \ldots, e_{kj}\}$ be any orthonormal basis of M, then (13.3) easily implies that

$$\tau_{1\cdots j} + \tau_{j+1\cdots 2j} + \ldots + \tau_{(k-1)j+1\cdots kj} \tag{13.22}$$

does not depend on the order of the indices. Moreover, if we perform a rotation in the (e_1, e_2)-plane, we obviously do not change the value of (13.22). Hence, since the ordering of the indices is not important, we obtain that, if we perform a rotation in any (e_i, e_m)-plane, then the value of (13.22) remains the same. Since any two orthogonal bases can be mapped into each other by performing consecutive rotations in coordinate planes, we conclude that (13.22) is independent of the choice of orthonormal basis. Hence M is an $S(j, \ldots, j)$-space.

Case (2): *M is Einstein.* Since M is an $S(n_1, \ldots, n_k)$-space, we have (13.14). Now, for any permutation σ of $\{3, \ldots, n\}$, there is a corresponding equation like (13.14). By summing up all such equations, we find

$$(d_1 - 2d_2 + d_3)K_{12} + (d_2 - d_3)(Ric(e_1) + Ric(e_2)) + d_3\tau = d_1 c, \tag{13.23}$$

where $d_1 = c_3$, and d_2, d_3 are constants such that

$$\begin{aligned}d_2 =& \binom{n-3}{n_1-3}\binom{n-n_1}{n_2}\cdots\binom{n-n_1-\cdots-n_{k-1}}{n_k},\\ d_3 =& \binom{n-4}{n_1-4}\binom{n-n_1}{n_2}\cdots\binom{n-n_1-\cdots-n_{k-1}}{n_k}\\ &+\binom{n-4}{n_2-2}\binom{n-n_2-2}{n_1-2}\binom{n-n_1-n_2}{n_3}\cdots\binom{n-n_1-\cdots-n_{k-1}}{n_k}\\ &+\cdots\\ &+\binom{n-4}{n_k-2}\binom{n-n_k-2}{n_1-2}\binom{n-n_1-n_k}{n_2}\cdots\binom{n-n_1-\cdots-n_{k-2}-n_k}{n_{k-1}}.\end{aligned}$$

These formulas easily imply that

$$(n_1-3)!\,n_2!\,n_3!\cdots n_k!\,d_2$$
$$=(n-3)! \quad \textit{unless when } n_1=2, \textit{ then } d_2=0,$$
$$(n_1-2)!\,n_2!\,n_3!\cdots n_k!\,d_3$$
$$=(n-4)!\left(\sum_{j=2}^{k} n_j(n_j-1)+(n_1-2)(n_1-3)\right).$$

Combining these with (13.20) shows that $d_1-2d_2+d_3=0$ if and only if

$$(n-2)(n-3)+(n_1-2)(n_1-3)+\sum_{j=2}^{k} n_j(n_j-1)$$
$$=2(n-3)(n_1-2),$$

which is equivalent to

$$(n-n_1)(n-n_1-1)+\sum_{j=2}^{k} n_j(n_j-1)=0.$$

But this is impossible. Hence, from (13.23) and the hypothesis, we conclude that M is an Einstein manifold. Therefore, M is a real space form. □

13.4 Fundamental inequalities involving δ-invariants

Definition 13.2. Let n be a natural number ≥ 2 and let $n_1,\ldots,n_k$ be k natural numbers. Then $(n_1,\ldots,n_k)$ is called a *partition* of n if

$$n_1+\cdots+n_k=n.$$

Lemma 13.1. *Let $a_1,\ldots,a_n$ be real numbers and k be an integer satisfying $2\leq k\leq n-1$. Then, for any partition $(n_1,\ldots,n_k)$ of n, we have*

$$\sum_{1\leq i_1<j_1\leq n_1} a_{i_1}a_{j_1}+\sum_{n_1+1\leq i_2<j_2\leq n_1+n_2} a_{i_2}a_{j_2}+\cdots$$
$$+\sum_{n_1\cdots+n_{k-1}+1\leq i_1<j_1\leq n} a_{i_k}a_{j_k} \tag{13.24}$$
$$\geq \frac{1}{2k}\{(a_1+\cdots+a_n)^2-k(a_1^2+\cdots+a_n^2)\},$$

with the equality holding if and only if

$$a_1+\cdots+a_{n_1}=\cdots=a_{n_1+\cdots+n_{k-1}+1}+\cdots+a_n. \tag{13.25}$$

Proof. Under the hypothesis we find

$$2k\Bigg\{\sum_{1\le i_1<j_1\le n_1} a_{i_1}a_{j_1} + \sum_{n_1+1\le i_1<j_1\le n_1+n_2} a_{i_2}a_{j_2} + \cdots$$
$$+ \sum_{n_1+\cdots+n_{k-1}+1\le i_1<j_1\le n} a_{i_k}a_{j_k}\Bigg\} - \Bigg(\sum_{\alpha=1}^n a_\alpha\Bigg)^2 + k\sum_{\alpha=1}^n a_\alpha^2$$
$$= 2k\Bigg\{\sum_{1\le i_1<j_1\le n_1} a_{i_1}a_{j_1} + \sum_{n_1+1\le i_1<j_1\le n_1+n_2} a_{i_2}a_{j_2} + \cdots$$
$$+ \sum_{n_1+\cdots+n_{k-1}+1\le i_1<j_1\le n} a_{i_k}a_{j_k}\Bigg\} + (k-1)\sum_{\alpha=1}^n a_\alpha^2 - 2\sum_{1\le\alpha<\beta\le n} a_\alpha a_\beta$$
$$= \Bigg\{\sum_{1\le a_{i_1}\le n_1} a_{i_1} - \sum_{n_1+1\le a_{i_2}\le n_1+n_2} a_{i_2}\Bigg\}^2$$
$$+ \Bigg\{\sum_{1\le a_{i_1}\le n_1} a_{i_1} - \sum_{n_1+n_2+1\le a_{i_3}\le n_1+n_2+n_3} a_{i_3}\Bigg\}^2 + \cdots$$
$$+ \Bigg\{\sum_{n_1+\cdots+n_{k-2}+1\le a_{i_1}\le n_1+\cdots+n_{k-1}} a_{i_{k-1}} - \sum_{n_1+\cdots+n_{k-1}+1\le a_{i_k}\le n} a_{i_k}\Bigg\}^2 \ge 0,$$

with equality holding if and only if (13.24) holds. □

An immediate consequence of Lemma 13.1 is the following.

Corollary 13.1. *Let $a_1, \ldots, a_n, \eta$ be $n+1$ real numbers such that*

$$\Bigg(\sum_{i=1}^n a_i\Bigg)^2 = (n-1)\Bigg(\eta + \sum_{i=1}^n a_i^2\Bigg). \tag{13.26}$$

Then $2a_1a_2 \ge \eta$, with equality holding if and only if

$$a_1 + a_2 = a_3 = \cdots = a_n.$$

Proof. Let $a_1, \ldots, a_n$ be real numbers. Then (13.26) is equivalent to

$$\eta = \frac{(a_1+\cdots+a_n)^2}{n-1} - (a_1^2+\cdots+a_n^2).$$

By choosing $k = n-1$, $n_1 = 2$ and $n_2 = \cdots = n_k = 1$, (13.24) becomes $2a_1a_2 \ge \eta$, with the equality holding if and only if $a_1 + a_2 = a_3 = \ldots = a_n$ according to Lemma 13.1. □

For each $(n_1, \ldots, n_k) \in \mathcal{S}(n)$, let $c(n_1, \ldots, n_k)$ and $b(n_1 \ldots, n_k)$ denote the constants given by

$$
\begin{aligned}
c(n_1,\ldots,n_k) &= \frac{n^2(n+k-1-\sum_{j=1}^k n_j)}{2(n+k-\sum_{j=1}^k n_j)},\\
b(n_1,\ldots,n_k) &= \frac{1}{2}n(n-1)-\frac{1}{2}\sum_{j=1}^k n_j(n_j-1).
\end{aligned}
\tag{13.27}
$$

We have the following general optimal inequality [Chen (2005d)].

Theorem 13.3. *Let $\phi : N \to M$ be an isometric immersion of a Riemannian n-manifold into a Riemannian m-manifold. Then for each k-tuple $(n_1,\ldots,n_k) \in \mathcal{S}(n)$ we have*

$$
\delta(n_1,\ldots,n_k) \le c(n_1,\ldots,n_k)||H||^2 + b(n_1,\ldots,n_k)\max\tilde{K}, \tag{13.28}
$$

where $\max\tilde{K}(p)$ *is the maximum of the sectional curvature of M restricted to 2-plane sections of the tangent space T_pN.*

The equality case of inequality (13.28) *holds at $p \in N$ if and only if the following two conditions hold:*

(a) *there exists an orthonormal basis $e_1,\ldots,e_m$ at p such that the shape operator A at p takes the form*

$$
A_{e_r} = \begin{pmatrix} A_1^r & \ldots & 0 & \\ \vdots & \ddots & \vdots & 0 \\ 0 & \ldots & A_k^r & \\ & 0 & & \mu_r I \end{pmatrix}, \quad r = n+1,\ldots,m, \tag{13.29}
$$

where I is an identity matrix and A_j^r is a symmetric $n_j \times n_j$ submatrix such that

$$
\mathrm{trace}\,(A_1^r) = \cdots = \mathrm{trace}\,(A_k^r) = \mu_r; \tag{13.30}
$$

(b) *for any k mutual orthogonal subspaces $L_1,\ldots,L_k$ of T_pN satisfying*

$$
\delta(n_1,\ldots,n_k) = \tau - \sum_{j=1}^k \tau(L_j) \ \text{ at } p,
$$

we have $\tilde{K}(e_{\alpha_i},e_{\alpha_j}) = \tilde{K}(e_{\alpha_{k+1}},e_{\beta_{k+1}}) = \max\tilde{K}(p)$ for $\alpha_i \in \Delta_i, \alpha_j \in \Delta_j$; $1 \le i \ne j \le k+1$ and for distinct $\alpha_{k+1},\beta_{k+1} \in \Delta_{k+1}$, where

$$
\begin{aligned}
\Delta_1 &= \{1,\ldots,n_1\},\\
&\ \vdots\\
\Delta_k &= \{n_1+\cdots+n_{k-1}+1,\ldots,n_1+\cdots+n_k\},\\
\Delta_{k+1} &= \{n_1+\cdots+n_k+1,\ldots,n\}.
\end{aligned}
$$

Proof. Let N be an n-dimensional submanifold of a Riemannian m-manifold M and p is a point in N. Then Gauss's equation implies that

$$2\tau(p) = n^2||H||^2 - ||h||^2 + 2\tilde{\tau}(T_pN) \tag{13.31}$$

at p, where $||h||^2$ is the squared norm of the second fundamental form h and $\tilde{\tau}(T_pN)$ is the scalar curvature of n-plane section $T_pN \subset T_pM$ of the ambient space M, i.e., $\tilde{\tau}(T_pN) = \sum_{i<j} \tilde{K}(e_i, e_j)$ for an orthonormal basis $e_1, \ldots, e_n$ of T_pN.

Let us put

$$\eta = 2\tau(p) - \frac{n^2(n+k-1-\sum_{j=1}^k n_j)}{n+k-\sum_{j=1}^k n_j}||H||^2 - 2\tilde{\tau}(T_pN). \tag{13.32}$$

Then we find from (13.31) and (13.32) that

$$n^2||H||^2 = \gamma\left(\eta + ||h||^2\right), \quad \gamma = n+k-\sum_{j=1}^n n_j. \tag{13.33}$$

At $p \in N$, let us choose an orthonormal basis $e_1, \ldots, e_m$ of T_pM such that $e_{\alpha_i} \in L_i$ for each $\alpha_i \in \Delta_i$. Moreover, we choose the normal vector e_{n+1} to be in the direction of the mean curvature vector at p. Then (13.33) yields

$$\begin{aligned}\left(\sum_{A=1}^n a_A\right)^2 = \gamma\Bigg[\eta + \sum_{A\neq B}(h_{AB}^{n+1})^2 + \sum_{A=1}^n (a_A)^2 \\ + \sum_{r=n+2}^m \sum_{A,B=1}^n (h_{AB}^r)^2\Bigg],\end{aligned} \tag{13.34}$$

where $a_A = h_{AA}^{n+1}$ with $1 \leq A, B \leq n$. Equation (13.34) is equivalent to

$$\begin{aligned}\left(\sum_{i=1}^{\gamma+1} \bar{a}_i\right)^2 = \gamma\Bigg[\eta + \sum_{i=1}^{\gamma+1}(\bar{a}_i)^2 + \sum_{i\neq j}(h_{ij}^{n+1})^2 + \sum_{r=n+2}^m \sum_{i,j=1}^n (h_{ij}^r)^2 \\ - \sum_{2\leq\alpha_1\neq\beta_1\leq n_1} a_{\alpha_1}a_{\beta_1} - \sum_{\alpha_2\neq\beta_2} a_{\alpha_2}a_{\beta_2} - \cdots - \sum_{\alpha_k\neq\beta_k} a_{\alpha_k}a_{\beta_k}\Bigg],\end{aligned} \tag{13.35}$$

where $\alpha_2, \beta_2 \in \Delta_2, \ \ldots, \ \alpha_k, \beta_k \in \Delta_k$ and

$$\begin{aligned}&\bar{a}_1 = a_1, \ \bar{a}_2 = a_2 + \cdots + a_{n_1}, \\ &\bar{a}_3 = a_{n_1+1} + \cdots + a_{n_1+n_2}, \\ &\quad\vdots \\ &\bar{a}_{k+1} = a_{n_1+\cdots+n_{k-1}+1} + \cdots + a_{n_1+\cdots+n_k}, \\ &\bar{a}_{k+2} = a_{n_1+\cdots+n_k+1}, \ldots, \bar{a}_{\gamma+1} = a_n.\end{aligned}$$

By applying Corollary 13.1 to (13.35) we obtain

$$\begin{aligned}&\sum_{\alpha_1<\beta_1} a_{\alpha_1}a_{\beta_1}+\sum_{\alpha_2<\beta_2} a_{\alpha_2}a_{\beta_2}+\cdots+\sum_{\alpha_k<\beta_k} a_{\alpha_k}a_{\beta_k}\\ &\geq \frac{\eta}{2}+\sum_{A<B}(h^{n+1}_{AB})^2+\frac{1}{2}\sum_{r=n+2}^{m}\sum_{A,B=1}^{n}(h^r_{AB})^2,\end{aligned}\tag{13.36}$$

where $\alpha_i, \beta_i \in \Delta_i$, $i = 1, \ldots, k$.

On the other hand, the equation of Gauss imply that, for each $j \in \{1, \ldots, k\}$, we have

$$\begin{gathered}\tau(L_j)=\sum_{r=n+1}^{m}\sum_{\alpha_j<\beta_j}\left(h^r_{\alpha_j\alpha_j}h^r_{\beta_j\beta_j}-(h^r_{\alpha_j\beta_j})^2\right)+\tilde{\tau}(L_j),\\ \alpha_j,\beta_j\in\Delta_j,\end{gathered}\tag{13.37}$$

where $\tilde{\tau}(L_j)$ is the scalar curvature $\tilde{M}^m$ associated with $L_j \subset T_p$.

Let us put

$$\Delta=\Delta_1\cup\cdots\cup\Delta_k,\ \ \Delta^2=(\Delta_1\times\Delta_1)\cup\cdots\cup(\Delta_k\times\Delta_k).$$

Then we find by combining (13.32), (13.36) and (13.37) that

$$\begin{aligned}\tau(L_1)+\cdots+\tau(L_k)&\geq\frac{\eta}{2}+\frac{1}{2}\sum_{r=n+1}^{m}\sum_{(\alpha,\beta)\notin\Delta^2}(h^r_{\alpha\beta})^2\\ &\quad+\frac{1}{2}\sum_{r=n+2}^{m}\sum_{j=1}^{k}\Bigg(\sum_{\alpha_j\in\Delta_j}h^r_{\alpha_j\alpha_j}\Bigg)^2+\sum_{j=1}^{k}\tilde{\tau}(L_j)\\ &\geq\frac{\eta}{2}+\sum_{j=1}^{k}\tilde{\tau}(L_j)\\ &=\tau-\frac{n^2(n+k-1-\sum n_j)}{2(n+k-\sum n_j)}||H||^2-\tilde{\tau}(T_pM)\\ &\quad+\sum_{j=1}^{k}\tilde{\tau}(L_j).\end{aligned}\tag{13.38}$$

Therefore, by (13.38), we have

$$\begin{aligned}\tau-\sum_{j=1}^{k}\tau(L_j)&\leq\frac{n^2(n+k-1-\sum n_j)}{2(n+k-\sum n_j)}||H||^2\\ &\quad+\tilde{\tau}(T_pM)-\sum_{j=1}^{k}\tilde{\tau}(L_j),\end{aligned}\tag{13.39}$$

which implies that

$$\delta(n_1,\ldots,n_k)\leq\frac{n^2(n+k-1-\sum n_j)}{2(n+k-\sum n_j)}||H||^2+\tilde{\delta}^M(n_1,\ldots,n_k),\tag{13.40}$$

where

$$\tilde{\delta}^M(n_1,\ldots,n_k) := \tilde{\tau}(T_pM) - \inf\{\tilde{\tau}(\tilde{L}_1) + \cdots + \tilde{\tau}(\tilde{L}_k)\} \tag{13.41}$$

with $\tilde{L}_1,\ldots,\tilde{L}_k$ running over all k mutually orthogonal subspaces of T_pN such that $\dim \tilde{L}_j = n_j$. Clearly, (13.40) implies inequality (13.28).

It is easy to see that the equality case of (13.28) holds at the point p if and only if the following two conditions hold:

(i) the inequalities in (13.36) and (13.38) are actually equalities;

(ii) for any k mutual orthogonal subspaces $L_1,\ldots,L_k$ of T_pN satisfying

$$\delta(n_1,\ldots,n_k) = \tau - \sum_{j=1}^{k} \tau(L_j) \tag{13.42}$$

at p, we have $\tilde{K}(e_{\alpha_i}, e_{\alpha_j}) = \tilde{K}(e_{\alpha_{k+1}}, e_{\beta_{k+1}}) = \max \tilde{K}(p)$ for $\alpha_i \in \Delta_i, \alpha_j \in \Delta_j$; $1 \le i \ne j \le k+1$ and for distinct $\alpha_{k+1}, \beta_{k+1} \in \Delta_{k+1}$.

It follows from Corollary 13.1, (13.36) and (13.38) that (i) holds if and only if there exists an orthonormal basis $e_1,\ldots,e_m$ at p such that the shape operator A of N in M at p satisfying conditions (13.29) and (13.30).

The converse can be easily verified. □

The next result is an important special case of Theorem 13.3.

Theorem 13.4. *Let N be a Riemannian n-manifold isometrically immersed in a Riemannian manifold M. If the sectional curvature of M is bounded above by c, then*

$$\delta(n_1,\ldots,n_k) \le c(n_1,\ldots,n_k)||H||^2 + b(n_1,\ldots,n_k)c$$

for any k-tuple $(n_1,\ldots,n_k) \in \mathcal{S}(n)$.

When M is $CP^m(4c)$, Theorem 13.4 gives the following.

Corollary 13.2. *Let N be an n-dimensional submanifold of the complex projective m-space $CP^m(4c)$. Then*

$$\delta(n_1,\ldots,n_k) \le c(n_1,\ldots,n_k)||H||^2 + 4b(n_1,\ldots,n_k)c$$

for any k-tuple $(n_1,\ldots,n_k) \in \mathcal{S}(n)$.

Similarly, if M is $CH^m(4c)$, Theorem 13.4 gives the following.

Corollary 13.3. *Let N be an n-dimensional submanifold of the complex hyperbolic m-space $CH^m(4c)$. Then*

$$\delta(n_1,\ldots,n_k) \le c(n_1,\ldots,n_k)||H||^2 + b(n_1,\ldots,n_k)c$$

for any k-tuple $(n_1,\ldots,n_k) \in \mathcal{S}(n)$.

Another important special case of Theorem 13.3 is the following.

Theorem 13.5. *Let N be an n-dimensional submanifold of a real space form $R^m(c)$. Then for each k-tuple $(n_1, \ldots, n_k) \in \mathcal{S}(n)$ we have*

$$\delta(n_1, \ldots, n_k) \leq c(n_1, \ldots, n_k)||H||^2 + b(n_1, \ldots, n_k)c. \tag{13.43}$$

The equality case of (13.43) *holds at a point $p \in N$ if and only if there exists an orthonormal basis $e_1, \ldots, e_m$ of T_pN such that the shape operator at p satisfies* (13.29) *and* (13.30).

Proof. Follows immediately from (1.14) and Theorem 13.3. □

If N is a totally real submanifold of a Kähler manifold $M^m(4c)$ of constant holomorphic sectional curvature $4c$, we have the following.

Theorem 13.6. *Let N be an n-dimensional totally real submanifold of a Kähler manifold $M^m(4c)$ of constant holomorphic sectional curvature $4c$. Then for any k-tuple $(n_1, \ldots, n_k) \in \mathcal{S}(n)$ we have*

$$\delta(n_1, \ldots, n_k) \leq c(n_1, \ldots, n_k)||H||^2 + b(n_1, \ldots, n_k)c. \tag{13.44}$$

The equality case of (13.44) *holds at a point $p \in N$ if and only if* (13.29) *and* (13.30) *hold at p with respect to a suitable orthonormal basis $e_1, \ldots, e_m$ at p.*

Proof. This is due to the facts that the proof of Theorem 13.3 bases only on the equation of Gauss and that Gauss's equation for a totally real submanifold in $CP^m(4c)$ is the same as the equation of Gauss for submanifolds in a real space form $R^m(c)$. □

Remark 13.1. It was proved in [Chen (2000d)] that every Lagrangian submanifold of a complex space form $M^n(4c)$ that satisfies the equality case of inequality (13.44) identically for some k-tuple $(n_1, \ldots, n_k) \in \mathcal{S}(n)$ is a minimal submanifold. This result extends a similar result for $\delta(2)$ obtained in [Chen et al. (1994, 1996)].

When the ambient space is Euclidean, Theorem 13.5 reduces to the following.

Theorem 13.7. *For a k-tuple $(n_1, \ldots, n_k) \in \mathcal{S}(n)$ and an n-dimensional submanifold N of a Euclidean space $\mathbb{E}^m$ with arbitrary codimension, we have*

$$\delta(n_1, \ldots, n_k) \leq c(n_1, \ldots, n_k)||H||^2. \tag{13.45}$$

The equality case holds at $p \in N$ if and only if there is an orthonormal basis $e_1, \ldots, e_m$ such that the shape operator at p satisfies (13.29) *and* (13.30).

Inequalities (13.43) and (13.45) give prima controls on the most important extrinsic curvature, the squared mean curvature $||H||^2$ by the δ-invariants $\delta(n_1, \ldots, n_k)$ for Riemannian submanifolds in real space forms.

When $k = 0$, Theorem 13.5 reduces to

Corollary 13.4. *Let N be an n-dimensional submanifold of a real space form $R^m(c)$. Then*

$$\tau \leq \frac{n(n-1)}{2}(||H||^2 + c),$$

with the equality holding at a point $p \in N$ if and only if p is a totally umbilical point.

The invariant $\delta(2) = \tau - \inf K$ was extended in [Oprea (2008)] to

$$\delta_\ell := \tau - \theta_\ell, \quad \ell \in [2, n], \tag{13.46}$$

where θ_ℓ is defined by (2.67). Oprea also proved the following two results.

Theorem 13.8. *Let N be an n-dimensional submanifold of a real space form $R^m(c)$ with $n \geq 3$. Then for any $\ell \in [3, n]$ we have*

$$\delta_\ell \leq \frac{n-2}{2}\left\{\frac{n^2}{n-1}||H||^2 + (n+1)c\right\}, \tag{13.47}$$

the equality holding at a point $p \in N$ if and only if there is an orthonormal bases $\{e_1, \ldots, e_n\}$ of T_pN and $\{e_{n+1}, \ldots, e_m\}$ of $T_p^\perp N$ such that the shape operator satisfies

$$A_r = a_r \begin{pmatrix} 0 & 0 \\ 0 & I_{n-1} \end{pmatrix}, \quad a_r \in \mathbf{R},$$

for $r = n+1, \ldots, m$.

Theorem 13.9. *Let N be a Lagrangian submanifold of a complex space form $M^m(4c)$ with $\dim N \geq 3$. Then*

$$\delta_n \leq \frac{(n+1)(n-2)}{2}c + \frac{(3n-1)(n-2)n^2}{2(3n+5)(n-1)}||H||^2.$$

Remark 13.2. Inequalities analogous to (13.42) for submanifolds in various space forms (in particular, for the special case with $k = 1, n_1 = 2$) have been studied by many authors, see for instance, [Aktan et al. (2008); Alegre et al. (2007); Arslan et al. (2010); Carriazo (1999, 2007); Carriazo et al. (2003); Cioroboiu (2003a,b); Cioroboiu and Oiagă (2003); Costache (2008, 2010); Defever et al. (2001); Gupta et al. (2004); Kim et al. (2003a); Kim and Kim (1999); Kim (2003); Lupu (2001); Mihai (2002, 2008); Oiagă and Mihai (1999); Özgür (2010); Özgür and Murathan (2010); Shukla et al. (2010); Vilcu (2010); Yoon (2003)].

Remark 13.3. Inequality analogous to (13.42) with $k = 1, n_1 = 2$ for submanifolds in real and complex space forms endowed with semi-symmetric metric connections are studied in [Mihai and Ozgur (2010a,c)].

13.5 Ideal immersions via δ-invariants

In general, there do not exist direct relationship between δ-invariants. On the other hand, we have the following maximal principle which follows immediately from Theorem 13.7.

Maximum Principle. *Let N be an arbitrary n-dimensional submanifold of a Euclidean m-space $\mathbb{E}^m$. If it satisfies the equality case of* (13.45), *i.e., it satisfies*

$$H^2 = \frac{\delta(n_1, \ldots, n_k)}{c(n_1, \ldots, n_k)} \tag{13.48}$$

for some k-tuple $(n_1, \ldots, n_k) \in \mathcal{S}(n)$, then

$$\delta(n_1, \ldots, n_k) \geq \delta(m_1, \ldots, m_j) \tag{13.49}$$

for any $(m_1, \ldots, m_k) \in \mathcal{S}(n)$.

For a given k-tuple $(n_1, \ldots, n_k) \in \mathcal{S}(n)$, let us put

$$\Delta(n_1, \ldots, n_k) = \frac{\delta(n_1, \ldots, n_k)}{c(n_1, \ldots, n_k)}. \tag{13.50}$$

For any isometric immersion $\phi : N \to \mathbb{E}^m$ of a Riemannian n-manifold N into $\mathbb{E}^m$, Theorem 13.7 gives

$$||H||^2(p) \geq \hat{\Delta}_0(p), \tag{13.51}$$

where $\hat{\Delta}_0$ is the Riemannian invariant on M defined by

$$\hat{\Delta}_0 = \max\left\{\Delta(n_1, \ldots, n_k) : (n_1, \ldots, n_k) \in \mathcal{S}(n)\right\}. \tag{13.52}$$

Inequality (13.51) enables us to introduce the following notion of ideal immersions.

Definition 13.3. An isometric immersion $\phi : N \to \mathbb{E}^m$ of a Riemannian n-manifold N into $\mathbb{E}^m$ is called an *ideal immersion* if it satisfies the equality case of (13.51) identically.

The Maximum Principle yields the following important fact.

Theorem 13.10. *If an isometric immersion $\phi : N \to \mathbb{E}^m$ satisfies the equality sign of* (13.45) *for a given k-tuple $(n_1, \ldots, n_k) \in \mathcal{S}(n)$, then it is an ideal immersion automatically.*

Similarly, we make the following.

Definition 13.4. An isometric immersion $\phi : N \to R^m(c)$ of a Riemannian manifold into a real space form $R^m(c)$ is called an *ideal immersion* if it satisfies (13.43) for some k-tuple $(n_1, \ldots, n_k) \in \mathcal{S}(n)$.

Physical Interpretation of Ideal Immersions. That an isometric immersion $\phi : N \to R^m(c)$ of a Riemannian manifold into a real space form $R^m(c)$ is an *ideal immersion* means that N **receives the least possible amount of tension** (given by $\hat{\Delta}_0(p)$) **from the surrounding space at each point** in N. This is due to inequality (13.43) and the well-known fact that the mean curvature vector field is exactly the tension field for isometric immersions. For an isometric immersion, the squared mean curvature at each point on the submanifold simply measures the amount of tension the submanifold receiving from the surrounding space at that point. For this reason, sometime an ideal immersion is called a *best way of living.*

Ideal immersions as stable critical points of total tension functional. Ideal immersions are closely related with critical point problem in the theory of total mean curvature. This can be seen as follows:

Let N be a compact Riemannian manifold. Denote by $\mathcal{I}(N, R^m(c))$ the family of isometric immersions of N into a real space form $R^m(c)$. For each $\phi \in \mathcal{I}(N, R^m(c))$, the *total tension functional* is defined by

$$T(\phi) = \int_N ||H_\phi||^2 \mathrm{d}V, \tag{13.53}$$

where H_ϕ denotes the mean curvature vector of $\phi : M \to R^m(c)$.

It follows from Theorem 13.5 that an ideal immersion of N into $R^m(c)$ is a critical point of the total tension functional within the class of $\mathcal{I}(N, R^m(c))$ automatically. Moreover, every ideal immersion of N in $R^m(c)$ is stable, i.e., the second variation of $T(\phi)$ is nonnegative for each variation of ϕ in the class of $\mathcal{I}(N, R^m(c))$.

Size of the smallest ball containing an ideal submanifold. According to Nash's embedding theorem, every compact Riemannian n-manifold can be isometrically embedded in any small portion of Euclidean space if the codimension is large enough.

In contrast, the following result states that every ideal immersion of a compact Riemannian manifold into a Euclidean space cannot lie in a very small ball of the Euclidean space. In fact, by applying Theorem 13.7, we can estimate the radius of the smallest ball containing a compact ideal submanifold in the Euclidean space in term of δ-invariants [Chen (1998b)].

Theorem 13.11. *Let $\phi : N \to \mathbb{E}^m$ be an ideal immersion of a compact Riemannian n-manifold N into a Euclidean m-space. Then the radius R of the smallest ball B(R) containing $\phi(N)$ satisfies*

$$R^2 \geq \frac{\mathrm{vol}(N)}{\int_N \hat{\Delta}_0 dV}, \tag{13.54}$$

with the equality sign holding if and only if ϕ is a 1-type ideal immersion.

Proof. Let $\phi : N \to \mathbb{E}^m$ be an ideal immersion of a compact Riemannian n-manifold N into a Euclidean m-space. If we choose the center of mass of N as the origin of $\mathbb{E}^m$, then it follows from Theorem 8.9(a) that

$$\int_N ||H||^2 * 1 \geq \frac{\mathrm{vol}(N)}{R^2}.$$

Combining this with (13.45) gives

$$\int_N \Delta(n_1, \ldots, n_k) *1 \geq \frac{\mathrm{vol}(N)}{R^2}, \tag{13.55}$$

Hence, by combining (13.52) and (13.55), we obtain (13.54). □

13.6 Examples of ideal immersions

Now, we provide some simple examples of ideal submanifolds.

Example 13.1. For an integer $n \geq 3$ and any integer $k \in \{0, 1, \ldots, [\frac{n}{2}]\}$, a spherical cylinder $\mathbb{E}^k \times S^{n-k} \subset \mathbb{E}^k \times \mathbb{E}^{n-k+1}$ is a hypersurface of $\mathbb{E}^{n+1}$ satisfying the equality sign of (13.45) with $(n_1, \ldots, n_k) = (2, \ldots, 2)$.

Example 13.2. For each even number $n = 2k$, a totally umbilical submanifold in a real space form $R^m(c)$ satisfies the equality in (13.43) with $(n_1, \ldots, n_k) = (2, \ldots, 2)$. A totally umbilical submanifold of $R^m(c)$ also satisfies the equality in (13.43) with $k = 0$. In particular, a horosphere in a hyperbolic space H^{2k+1} is an ideal hypersurface.

Example 13.3. Let $\iota_a : S^k(c) \to \mathbb{E}^{k+1}$ be a standard embedding. For two given positive numbers c_1, c_2, the product embedding

$$\iota_1 \times \iota_2 : S^k(c_1) \times S^k(c_2) \to \mathbb{E}^{2k+2}$$

is an ideal embedding in $\mathbb{E}^{2k+2}$ which satisfies the equality in (13.43) with $(n_1, \ldots, n_k) = (2, \ldots, 2)$.

More generally, if $\ell, n-\ell$ are positive integers with $\ell = kt, n-\ell = ks$ for some integers $k > 1, t, s$, the standard embedding of $S^{\ell}(c_1) \times S^{n-\ell}(c_2)$ in $\mathbb{E}^{n+2}$ is an ideal embedding associated with a k-tuple $(n_1, \ldots, n_k)$ satisfying $n_1 = \cdots = n_k = t + s$.

Example 13.4. Let B^{2k} be an austere submanifold of $S^{n+1}(1)$ with zero relative nullity. For each point $p \in B^{2k}$, if we put

$$N_pB^{2k} = \{\xi \in T_pS^{n+1} | \langle \xi, \xi \rangle = 1 \text{ and } \langle \xi, T_pB^{2k} \rangle = 0\},$$

then NB^{2k} lies in $S^{n+1}(1)$ and the identity map from NB^{2k} into $S^{n+1}(1)$ is an austere immersion on an open dense subset U of NB^{2k}. Moreover, U is an austere hypersurface of $S^{n+1}(1)$ with relative nullity $n - 2k$ and NB^{2k} satisfies the equality in (13.43) with $(n_1, \ldots, n_k) = (2, \ldots, 2)$.

Example 13.5. Let B be a minimal surface of $S^{m-n+2}(1)$ of $\mathbb{E}^{m-n+3}$ centered at the origin o and let $C(B)$ be the 3-dimensional minimal cone in $\mathbb{E}^{m-n+3}$ over B with vertex at o. Then the direct product $N = C(B) \times \mathbb{E}^{n-3}$ is a minimal $\delta(2)$-ideal submanifold of $\mathbb{E}^m$.

13.7 δ-invariants of curvature-like tensor

Definition 13.5. Let (N, g) be any Riemannian n-manifold and let T be a $(0, 4)$-tensor field on N. Then T is called a *curvature-like tensor* if it satisfies the following symmetric properties:

$$\begin{aligned} &T(X, Y, Z, W) = -T(Y, X, Z, W), \\ &T(X, Y, Z, W) = -T(X, Y, W, Z), \\ &T(X, Y, Z, W) + T(X, Z, W, Y) + T(X, W, Y, Z) = 0. \end{aligned}$$

If T is a curvature-like $(0, 4)$-tensor field, then we can talk about the sectional curvature $T(\pi)$ associated with a 2-plane section $\pi \subset T_pN$, $p \in N$ in the same way as for Riemann curvature tensor. Further, let L be a linear subspace of T_pN of dimension $r \geq 2$ and let $\{e_1, \ldots, e_r\}$ be an orthonormal basis of L. We define the scalar curvature $\tau_T(L)$ of L by

$$\tau_T(L) = \sum_{\alpha<\beta} T(e_\alpha \wedge e_\beta), \quad 1 \leq \alpha, \beta \leq r. \tag{13.56}$$

For each $(n_1, \ldots, n_k) \in \mathcal{S}(n)$ we define the δ-invariant $\delta_T(n_1, \ldots, n_k)$ in the same way as $\delta(n_1, \ldots, n)$, i.e.,

$$\delta_T(n_1, \ldots, n_k)(p) = \tau_T(p) - \inf\{\tau_T(L_1) + \cdots + \tau_T(L_k)\}, \tag{13.57}$$

where $\tau_T(p) = \tau_T(T_pN)$ and $L_1, \ldots, L_k$ run over all k mutually orthogonal subspaces of T_pN such that $\dim L_j = n_j$, $j = 1, \ldots, k$.

Lemma 13.2. *If a curvature-like* $(0,4)$*-tensor field* T *can be written as*

$$T(X,Y,Z,W) = T_1(X,Y,Z,W) + c\{g(Y,Z)g(X,W) - g(X,Z)g(Y,W)\},$$

then T_1 *is curvature-like and for each* k*-tuple* $(n_1, \ldots, n_k) \in \mathcal{S}(n)$ *we have*

$$\delta_T(n_1, \ldots, n_k) = \delta_{T_1}(n_1, \ldots, n_k) + b(n_1, \ldots, n_k)c, \tag{13.58}$$

where δ_{T_1} *is the* δ*-invariant of* T_1 *and* $b(n_1, \ldots, n_k)$ *is defined by* (13.27).

Proof. Follows easily from the definitions. □

Let (N,g) be any n-dimensional Riemannian manifold and (B,g) any Riemannian vector bundle over N. Let μ be a B-valued symmetric $(1,2)$-tensor field. If T is a $(0,4)$-tensor field on N such that

$$\begin{aligned} T(X,Y,Z,W) &= g(\mu(Y,Z), \mu(X,W)) - g(\mu(X,Z), \mu(Y,W)) \\ &\quad + c\{g(Y,Z)g(X,W) - g(X,Z)g(Y,W)\} \end{aligned} \tag{13.59}$$

for $X,Y,Z,W \in \mathcal{X}(N)$, then clearly T is curvature-like. Equation (13.59) is called an *algebraic Gauss equation.*

Because the proof of Theorem 13.5 bases only on Gauss's equation, the same proof gives the following (cf. [Chen et al. (2005a)]).

Theorem 13.12. *Let* (N,g) *be a Riemannian* n*-manifold and* T *be a* $(0,4)$*-tensor field satisfying* (13.59). *Then for each* k*-tuple* $(n_1, \ldots, n_k) \in \mathcal{S}(n)$ *we have*

$$\begin{aligned} \delta_T(n_1, \ldots, n_k) \le \frac{c(n_1, \ldots, n_k)}{n^2} & g(\operatorname{trace}\mu, \operatorname{trace}\mu) \\ & + b(n_1, \ldots, b_n)c, \end{aligned} \tag{13.60}$$

where $\operatorname{trace}\mu = \sum_{i=1}^n \mu(e_i, e_i)$.

The equality case of inequality (13.60) *holds at a point* $p \in N$ *if and only if, there exists an orthonormal basis* $\{e_1, \ldots, e_n\}$ *at* p *such that with respect this basis every linear map* μ_ξ, $\xi \in B_p$, *of the tangent space* T_pN, *defined by* $g(\mu_\xi X, Y) = g(\mu(X,Y), \xi)$ *for all* $X, Y \in T_pN$ *takes the following form:*

$$\mu_\xi = \begin{pmatrix} A_1^\xi & & & 0 \\ & \ddots & & \\ & & A_k^\xi & \\ 0 & & & \eta_\xi I \end{pmatrix}, \tag{13.61}$$

where I is an identity submatrix and $\{A_j^\xi\}_{j=1}^k$ are symmetric $n_j \times n_j$ submatrices satisfying

$$\text{trace}(A_1^\xi) = \cdots = \text{trace}\,(A_k^\xi) = \eta_\xi \tag{13.62}$$

for some η_ξ.

Some situations in which such T and α satisfying an algebraic Gauss equation exist are

- submanifolds of a real space form $R^m(c)$; in this case
$$\begin{aligned} T(X,Y,Z,W) &= R(X,Y,Z,W) \\ &\quad - c\{g(Y,Z)g(X,W) - g(X,Z)g(Y,W)\} \end{aligned}$$
and μ is the second fundamental form,
- totally real submanifolds of a complex space form $M^n(4c)$; T and μ are as in the previous case,
- positive definite centroaffine hypersurfaces; μ being the difference tensor K defined by (17.5), and
$$\begin{aligned} T(X,Y,Z,W) &= \varepsilon\{h(Y,Z)h(X,W) - h(X,Z)h(Y,W)\}, \\ &\quad - \hat{R}(X,Y,Z,W). \end{aligned}$$

For any k-tuple $(n_1,\dots,n_k) \in \mathcal{S}(n)$, we put $\sigma_k = n_1 + \cdots + n_k$ and

$$\Delta_1 = \{1,\dots,n_1\},\dots,\Delta_k = \{n_1{+}\cdots{+}n_{k-1}{+}1,\dots,\sigma_k\}.$$

If the equality sign of (13.60) holds for some k-tuple $(n_1,\dots,n_k) \in \mathcal{S}(n)$, then there exists an orthonormal basis $\{e_1,\dots,e_n\}$ of T_pN such that, with respect to this basis, each linear map μ_ξ takes the form of (13.61)-(13.62). With respect to the orthonormal basis $\{e_1,\dots,e_n\}$ so chosen, we put

$$L_j = \text{Span}\,\{e_\alpha \,:\, \alpha \in \Delta_j\}. \tag{13.63}$$

Also, let L_{k+1} be the linear subspace of T_pN spanned by $e_{\sigma_k+1},\dots,e_n$.

Now suppose that the vector bundle B is the tangent bundle TN. If $g(\mu(X,Y),Z)$ is totally symmetric, we have $\mu_X Y = \mu(X,Y)$; so every μ_X is a symmetric endomorphism of TN. Then (13.59) becomes

$$T(X,Y) = [\mu_X,\mu_Y]\,, \tag{13.64}$$

where $T(X,Y)$ is the endomorphism defined by

$$g(T(X,Y)Z,W) = T(X,Y,Z,W).$$

We can say more about the case when equality in (13.60) holds.

Lemma 13.3. *Let (N, g) be a Riemannian n-manifold, T a curvature-like $(0,4)$-tensor field and μ a symmetric $(1,2)$-tensor field satisfying (13.59). Suppose that $g(\mu(X,Y),Z)$ is totally symmetric. If equality holds in (13.60) for some $(n_1,\ldots,n_k)\in\mathcal{S}(n)$ at $p\in N$, then there is an orthonormal basis $\{e_1,\ldots,e_n\}$ of T_pN such that μ satisfies*

$$\mu_{L_i}L_i\subset L_i,\quad \mu_{L_i}L_j=\{0\}, \tag{13.65}$$

$$\mu_X e_{\sigma_k+1}=\cdots=\mu_X e_n=0,\quad X\in T_pN, \tag{13.66}$$

$$\sum_{\alpha\in\Delta_j}\mu_{e_\alpha}e_\alpha=0, \tag{13.67}$$

for $1\le i\neq j\le k$, where $\sigma_k=n_1+\cdots+n_k$.

Proof. From (13.61) follows that $\mu_X L_j\subset L_j$ for all X. Hence the first statement of (13.65) is obvious. Next we take $1\le i\neq j\le k+1$ and observe that $\mu_{L_i}L_j\subset L_j$ and $\mu_{L_j}L_i\subset L_i$, therefore $\mu_{L_i}L_j=0$. Equations (13.66) and (13.67) now follow from (13.62). □

Lemma 13.3 immediately implies the following (cf. [Chen (2000d); Chen et al. (2005a)]).

Theorem 13.13. *Let (N,g) be a Riemannian n-manifold, T a curvature-like $(0,4)$-tensor field and μ a symmetric $(1,2)$-tensor field satisfying (13.59). Suppose that $g(\mu(X,Y),Z)$ is totally symmetric. If the equality case of (13.60) is satisfied at a point p for some k-tuple $(n_1,\ldots,n_k)\in\mathcal{S}(n)$, then* trace $\mu=0$ *at p.*

From Theorems 13.12 and 13.13 we immediately have the following corollary, which we need for later use.

Corollary 13.5. *Let (N,g) be a Riemannian n-manifold, T a curvature-like $(0,4)$-tensor field and μ a symmetric $(1,2)$-tensor field satisfying (13.59). Suppose that $g(\mu(X,Y),Z)$ is totally symmetric. If the equality case of (13.60) is satisfied at a point p for some k-tuple $(n_1,\ldots,n_k)\in\mathcal{S}(n)$, then also the equality case of (13.60) associated with the ℓ-tuple $(m_1,\ldots,m_\ell)$ is satisfied, where*

$$m_1=\sum_{j=1}^{k_1}n_j,\ \ m_2=\sum_{j=k_1+1}^{k_1+k_2}n_j,\ \ldots,\ m_\ell=\sum_{j=k_1+\cdots+k_{\ell-1}+1}^{k}n_j.$$

13.8 A dimension and decomposition theorem

In this section, we always assume that (N, g) is a Riemannian n-manifold, T is a curvature-like $(0,4)$-tensor field and μ a symmetric $(1,2)$-tensor field satisfying (13.59).

Suppose that

$$g(\mu(X,Y),Z) = g(Y,\mu(X,Z)) \tag{13.68}$$

$$(\nabla'_X\mu)(Y,Z) = (\nabla'_Y\mu)(X,Z) \tag{13.69}$$

for $X, Y, Z \in TN$, where ∇' is the Levi-Civita connection of N, i.e., μ is a totally symmetric Codazzi tensor.

First we put

$$\operatorname{Im}\mu(p) = \operatorname{Span}\{\mu(X,Y) : X, Y \in T_pN\}, \tag{13.70}$$

$$\mathcal{N}_p = \{Z \in T_pN : \mu(X,Z) = 0, \forall X \in T_pN\}. \tag{13.71}$$

Lemma 13.4. *We have* $(\operatorname{Im}\mu(p))^\perp = \mathcal{N}_p$, *where* $(\operatorname{Im}\mu(p))^\perp$ *is the orthogonal complementary subspace of* $\operatorname{Im}\mu(p)$ *in* T_pN.

Proof. Follows from

$$\begin{aligned} X \in (\operatorname{Im}\mu(p))^\perp &\iff g(X, \mu_Y Z) = 0, \quad \forall Y, Z \in T_pN \\ &\iff g(\mu_Y X, Z) = 0, \quad \forall Y, Z \in T_pN \\ &\iff \mu(X,Y) = 0, \quad \forall Y \in T_pN \\ &\iff X \in \mathcal{N}_p. \end{aligned}$$

□

Now, we assume that the equality is achieved in (13.60) identically for some k-tuple $(n_1, \ldots, n_k) \in \mathcal{S}(n)$. So, we may define $\mathcal{D}_j, j = 1, \ldots, k$, to be the distributions given by $L_j = \operatorname{Span}\{e_\alpha : \alpha \in \Delta_j\}$.

The purpose of this section is to investigate what happens if $\dim(\operatorname{Im}\mu) = n$, which implies that $\sigma_k = n$, by Lemma 13.3 and Lemma 13.4. Thus, let us assume that $\dim(\operatorname{Im}\mu) = n$ at every point.

Lemma 13.5. *We have*

$$\mu_X(\mathcal{D}_i) \subset \mathcal{D}_i, \quad \forall X \in TN, \tag{13.72}$$

$$\sum_{\alpha \in \Delta_j} \mu_{e_\alpha} e_\alpha = 0 \tag{13.73}$$

for $1 \leq i \neq j \leq k$. *Moreover, the distributions* $\mathcal{D}_1, \ldots, \mathcal{D}_k$ *are completely integrable and their leaves are minimal submanifolds of* (N, g).

Proof. By Lemma 13.3, we have (13.72) and (13.73). Now, we claim that $\nabla_{Y_j}X_i \in \mathcal{D}_i \oplus \mathcal{D}_j$ for any vector fields X_i in $\mathcal{D}_i$ and Y_j in $\mathcal{D}_j$, $1 \leq i \neq j \leq k$. This can be seen as follows: Since $n_1 + \cdots + n_k = n$, we get $k \geq 2$ from the definition of $\mathcal{S}(n)$. If $k = 2$, there is nothing to prove. So, we assume $k \geq 3$. For any distinct integers $i, j, \ell \in \{1, \ldots, k\}$, (13.69) and (13.72) imply that

$$\mu(X, [Y, Z]) = \mu(Y, \nabla'_Z X) - \mu(Z, \nabla'_Y X) \tag{13.74}$$

for X, Y and Z belong to $\mathcal{D}_i, \mathcal{D}_j$ and $\mathcal{D}_\ell$, respectively.

It follows from (13.70) that $\mu(Z, \nabla'_Y X), \mu(Y, \nabla'_Z X)$ and $\mu([Y, Z], X)$ belong to $\mathcal{D}_\ell, \mathcal{D}_j$ and $\mathcal{D}_i$, respectively. Hence $\mu(Z, \nabla'_Y X) = 0$. Thus, by applying $\dim(\operatorname{Im} K) = n$, (13.72), and Lemma 13.4, we obtain the claim.

By applying (13.69) and (13.72), we also have

$$\mu([X, Y], Z) = \mu(X, \nabla'_Y Z) - \mu(Y, \nabla'_X Z) \tag{13.75}$$

for vector fields X, Y in $\mathcal{D}_i$ and Z in $\mathcal{D}_j$ with $j \neq i$,

On the other hand, it follows from (13.72) that $\mu(X, \nabla'_Y Z) - \mu(Y, \nabla'_X Z)$ belongs to $\mathcal{D}_i$ and $\mu([X, Y], Z)$ belongs to $\mathcal{D}_j$. Hence, $\mu([\mathcal{D}_i, \mathcal{D}_i], \mathcal{D}_j) = \{0\}$ for $1 \leq i \neq j \leq k$. Thus, by using $\dim(\operatorname{Im} K) = n$ and (13.72), we obtain $[\mathcal{D}_i, \mathcal{D}_i] \subset \mathcal{D}_i$ which implies that $\mathcal{D}_i$ is an integrable distribution.

Finally, we prove that leaves of each $\mathcal{D}_i$ are minimal submanifolds of (N, g). This can be seen as follows. Assume that $\{e_1, \ldots, e_n\}$ is a local h-orthonormal frame of N such that e_{α_i} belongs to $\mathcal{D}_i$ for each $\alpha_i \in \Delta_i$.

We put

$$\mu(X, Y) = \sum_{t=1}^{n} \mu^t(X, Y) e_t.$$

For any vector fields X, Y in $\mathcal{D}_i$ and Z in $\mathcal{D}_j$ with $j \neq i$, we find from (13.68) and (13.72) that

$$\begin{aligned} &\sum_{t=1}^{n} (Z\mu^t(X, Y)) e_t + \sum_{t=1}^{n} \mu^t(X, Y) \nabla'_Z e_t \\ &\qquad - \mu(\nabla'_Z X, Y) - \mu(X, \nabla'_Z Y) \\ &\quad = -\mu(\nabla'_X Z, Y) - \mu(Z, \nabla'_X Y). \end{aligned} \tag{13.76}$$

From (13.72) we know that the first sum in the above formula belongs to $\mathcal{D}_i$. Thus, using (13.68) and (13.72), we obtain

$$\begin{aligned} -\sum_{t=1}^{n} \mu^t(X, Y) g(\nabla'_Z e_t, e_\alpha) &= g(\mu(Z, \nabla'_X Y), e_\alpha) \\ &= g(\nabla'_X Y, \mu(Z, e_\alpha)) \end{aligned} \tag{13.77}$$

for any $\alpha \in \Delta_j$ with $j \neq i$. On the other hand, we have

$$-\sum_{t=1}^{n} \mu^t(X,Y)g(\nabla'_Z e_t, e_\alpha) = g(\mu(X,Y), \nabla'_Z e_\alpha). \tag{13.78}$$

Thus, by combining (13.77) and (13.78), we get

$$g(\nabla'_X Y, \mu(Z, e_\alpha)) = g(\nabla'_Z e_\alpha, \mu(X,Y)). \tag{13.79}$$

Hence, by (13.73), we obtain

$$g(N_j, \mu(X,Y)) = \sum_{\alpha\in\Delta_j} g(\nabla'_X Y, \mu(e_\alpha, e_\alpha)) = 0 \tag{13.80}$$

for any X, Y in $\mathcal{D}_i$ with $i \neq j$, where $N_j = \sum_{\alpha\in\Delta_j} \nabla'_{e_\alpha} e_\alpha$.

Equations (13.72), (13.80) and the condition $\dim(\operatorname{Im} K) = n$ imply that N_j is in $\mathcal{D}_j$ which shows that leaves of $\mathcal{D}_j$ are minimal submanifolds of (N, g). This completes the proof of Lemma 13.5.

Let us put $\mu_{\alpha\beta}^{\gamma} = g(\mu(e_\alpha, e_\beta), e_\gamma)$. Then (13.68) yields $\mu_{\alpha\beta}^{\gamma} = \mu_{\alpha\gamma}^{\beta} = \mu_{\beta\gamma}^{\alpha}$. By Lemma 13.5, we get $\mu(\mathcal{D}_1, \mathcal{D}_1) = \mathcal{D}_1$. Thus, for any $\alpha, \beta, \delta \in \Delta_1$ and any $Z \in \mathcal{D}_1^{\perp}$, we find

$$\begin{aligned} g((\nabla'\mu)(e_\alpha, e_\beta, e_\delta), Z) &= g(\nabla'_{e_\alpha}\mu(e_\beta, e_\delta), Z) = -g(\mu(e_\beta, e_\delta), \nabla'_{e_\alpha} Z) \\ &= -\sum_{\gamma\in\Delta_1} \mu_{\beta\delta}^{\gamma} g(e_\gamma, \nabla'_{e_\alpha} Z) = \sum_{\gamma\in\Delta_1} \mu_{\beta\delta}^{\gamma} g(\nabla'_{e_\alpha} e_\gamma, Z). \end{aligned}$$

Combining this with (17.16) yields

$$\sum_{\gamma\in\Delta_1} \mu_{\alpha\delta}^{\gamma} g(\nabla'_{e_\beta} e_\gamma, Z) = \sum_{\gamma\in\Delta_1} \mu_{\beta\delta}^{\gamma} g(\nabla'_{e_\alpha} e_\gamma, Z) \tag{13.81}$$

for $\alpha, \beta, \delta \in \Delta_1$ and Z in $\mathcal{D}_1^{\perp}$. □

Theorem 13.14. *Let (N, g) be a Riemannian n-manifold, T is a curvature-like $(0,4)$-tensor field and μ a symmetric $(1,2)$-tensor field satisfying the algebraic Gauss equation (13.59) and the total symmetry conditions (13.68) and (13.69). If $\dim(\operatorname{Im}\mu) = n$, then the leaves of the integrable distributions $\mathcal{D}_i$ are totally geodesic submanifolds of N.*

Proof. We prove the lemma for $\mathcal{D}_1$. For a vector field $Z \in \mathcal{D}_1^{\perp}$ we put $\sigma_Z(X,Y) = g(\nabla'_X Y, Z)$ for vector fields X, Y in $\mathcal{D}_1$. Then σ_Z defines a symmetric $(0,2)$-tensor which is essentially the second fundamental form (in the normal direction Z) of the leaves of $\mathcal{D}_1$ in N. Suppose that $\mathcal{D}_1$ is not autoparallel, then there exists a vector field $Z \in \mathcal{D}_1^{\perp}$ such that $\sigma_Z \neq 0$.

Case (1): *σ_Z has only one eigenvalue.* We know from Lemma 13.5 that every leaf of $\mathcal{D}_1$ is minimal in N. Thus the unique eigenvalue of σ_Z vanishes, which contradicts $\sigma_Z \neq 0$.

Case (2): *σ_Z has at least two eigenvalues.* In this case, we may choose an orthonormal basis $\{e_1, \ldots, e_{n_1}\}$ of $\mathcal{D}_1$ which diagonalizes σ_Z so that

$$\sigma_Z(e_\alpha, e_\beta) = \mu_\alpha \delta_{\alpha\beta}, \quad \alpha, \beta \in \Delta_1 \tag{13.82}$$

for some $\mu_1, \ldots, \mu_{n_1}$. From (13.81) and (13.82) we find

$$\mu_{\alpha\beta}^{\delta}(\mu_\alpha - \mu_\beta) = 0, \;\; \alpha, \beta, \delta \in \Delta_1,$$

which implies that

$$\mu_{\alpha\beta}^{\delta} = 0, \quad \forall \delta \in \Delta_1, \;\; \text{whenever} \;\; \mu_\alpha \neq \mu_\beta. \tag{13.83}$$

If the multiplicity of an eigenvalue, say μ_1, of σ_Z is one, (13.83) gives

$$\mu_{\beta\delta}^{1} = 0, \quad \beta, \delta \in \Delta_1; \; \beta \neq 1. \tag{13.84}$$

On the other hand, from Theorem 13.13, we also have $\sum_{j=1}^{n_1} \mu_{jj}^1 = 0$. Thus, by combining this with (13.84) and Lemma 13.5, we obtain $\mu_{e_1} = 0$ which contradicts to the assumption: $\dim(\operatorname{Im}\mu) = n$. Hence, we know that each eigenvalue of σ_Z has multiplicity at least two. Let us assume that $\zeta_1, \ldots, \zeta_\ell$ are the distinct eigenvalues of σ_Z with multiplicities $m_1, \ldots, m_\ell$. Let $\mathcal{E}_1, \ldots, \mathcal{E}_\ell$ denote the corresponding eigenspace distributions.

From (13.83), we discover that $\mu_X(\mathcal{E}_j) \subset \mathcal{E}_j$ for $j = 1, \ldots, \ell$, and X in $\mathcal{D}_1$. Therefore, it follows from Theorem 13.12 that N satisfies the equality case of (13.60) associated with $(m_1, \ldots, m_\ell, n_2, \ldots, n_k) \in \mathcal{S}(n)$ as well. We can now apply the same reasoning on each of $\mathcal{E}_1, \ldots, \mathcal{E}_\ell$. But, for each $\mathcal{E}_i$, by assumption, σ_Z has only one eigenvalue, which has to vanish as in Case 1. Consequently, σ_Z vanishes on $\mathcal{D}_1$, which is again a contradiction. □

By applying Theorem 13.14, we obtain the following decomposition theorem [Chen et al. (2005a)].

Theorem 13.15. *Under the assumptions of Theorem 13.14, N is locally isometric to the Riemannian product $N_1 \times \cdots \times N_k$, where N_i denotes the leaf of the integrable distribution $\mathcal{D}_i$ for each i.*

Proof. From Theorem 13.14 and Corollary 13.5, we also know that $\oplus_{j \in F} \mathcal{D}_j$ is also autoparallel for each proper subset F of $\{1, \ldots, k\}$. Hence, we may apply the de Rham decomposition theorem to conclude that N is locally the Riemannian product $N_1 \times \cdots \times N_k$. □

Chapter 14

Some Applications of δ-invariants

14.1 Applications of δ-invariants to minimal immersions

Because inequality (13.28) and (13.43) provide optimal relation between δ-invariants and the squared mean curvature, one important application of δ-invariants is to provide many solutions to Problem 13.1.

Theorem 14.1. *Let N be a Riemannian n-manifold. If there exists a point $p \in N$ and a k-tuple $(n_1, \ldots, n_k) \in \mathcal{S}(n)$ such that*

$$\delta(n_1, \ldots, n_k)(p) > \frac{1}{2}\left(n(n-1) - \sum n_j(n_j - 1)\right) c, \tag{14.1}$$

then N does not admits a minimal isometric immersion into any Riemannian manifold M whose sectional curvature is bounded above by c.

Proof. If N is an n-dimensional submanifold of a Riemannian manifolds M whose sectional curvature is bounded above by c, then it follows from (13.27) and (13.28) that

$$\delta(n_1, \ldots, n_k) \leq c(n_1, \ldots, n_k)||H||^2 + \frac{1}{2}\left(n(n-1) - \sum n_j(n_j - 1)\right) c$$

for every k-tuple $(n_1, \ldots, n_k) \in \mathcal{S}(n)$ at every point $p \in N$. Consequently, if (14.1) holds at some point p, then N cannot be minimal in M. $\square$

In particular, Theorem 14.1 gives the following answer to Problem 13.1.

Corollary 14.1. *Let N be a Riemannian n-manifold. If there exists a k-tuple $(n_1, \ldots, n_k) \in \mathcal{S}(n)$ such that $\delta(n_1, \ldots, n_k)(p) > 0$ at some point $p \in N$, then N admits no minimal isometric immersion into any non-positively curved Riemannian m-manifold. In particular, N admits no minimal immersion into a Euclidean space with arbitrary codimension.*

Remark 14.1. For each integer $n \geq 3$ and each k-tuple $(n_1, \ldots, n_k) \in \mathcal{S}(n)$, the condition on $\delta(n_1, \ldots, n_k)$ in Corollary 14.1 is sharp. This can be seen as follows: Let $\phi_j : N_j^{n_j} \to \mathbb{E}^{m_j}$, $j = 1, \ldots, k$, be k minimal submanifolds and let ι be a totally geodesic embedding of a Euclidean $(n - \sum_j n_j)$-space into a Euclidean space $\mathbb{E}^m$. Then the invariant $\delta(n_1, \ldots, n_k)$ of the direct product $N_1^{n_1} \times \cdots \times N_k^{n_k} \times \mathbb{E}^{n - \sum n_j}$ vanishes identically. Clearly, the product immersion

$$\phi_1 \times \cdots \times \phi_k \times \iota : N_1^{n_1} \times \cdots \times N_k^{n_j} \times \mathbb{E}^{n - \sum n_j} \longrightarrow \mathbb{E}^{m_1} \times \cdots \times \mathbb{E}^{m_k} \times \mathbb{E}^m$$

is a minimal immersion. Consequently, the condition given in Corollary 14.1 is sharp for each k-tuple $(n_1, \ldots, n_k)$.

Another application of Theorem 13.3 is the following.

Proposition 14.1. *Let $N_1^{n_1}, \ldots, N_k^{n_k}$ be Riemannian manifolds of dimension ≥ 2 with scalar curvature ≤ 0. Then a minimal isometric immersion*

$$\phi : N = N_1^{n_1} \times \cdots \times N_k^{n_k} \times \mathbb{E}^{n - \sum n_j} \to \mathbb{E}^m$$

of $N_1^{n_1} \times \cdots \times N_k^{n_k} \times \mathbb{E}^{n - \sum n_j}$ into a Euclidean m-space is a product immersion $\phi_1 \times \cdots \times \phi_k \times \iota$ of minimal immersions $\phi_j : N_j^{n_j} \to \mathbb{E}^{m_j}$, $j = 1, \ldots, k$, and a totally geodesic immersion ι.

Proof. Under the hypothesis, the invariant $\delta(n_1, \ldots, n_k)$ of N vanishes identically. Thus, for any minimal immersion ϕ of N into a Euclidean space, the equality sign of inequality (13.45) holds identically. Hence, according to Theorem 13.7, the second fundamental form h of ϕ satisfies

$$h(X_i, X_j) = 0, \quad 1 \leq i \neq j \leq k + 1$$

for X_j tangent to the j-th component of the Riemannian product. Hence, after applying Moore's lemma, we obtain the proposition. □

Remark 14.2. There exist many Riemannian manifolds with Ricci tensor $Ric \leq 0$, but with some positive δ-invariants. For instance, the following example was constructed in [Suceavă (2001)].

Let $\varepsilon > 0$ and put $U = \{(x, y, u, v) \in \mathbf{R}^4 : y > \frac{\varepsilon}{2} \text{ and } v > 0\}$. Denote by N^4 the Riemannian 4-manifold (U, g) with the metric

$$g = \frac{1}{y^2}(dx^2 + dy^2) + \frac{2\varepsilon \tan^{-1} y}{v^2}(du^2 + dv^2).$$

Then, for sufficiently small $\varepsilon > 0$, N^4 satisfies $Ric < 0$ and $\delta(2, 2) > 0$.

14.2 Applications of δ-invariants to spectral geometry

Another important application of δ-invariants is to derive new intrinsic results via the theory of finite type submanifolds.

For any isometric immersion of a compact Riemannian n-manifold N into a Euclidean space, Theorem 7.5 gives

$$\lambda_q \operatorname{vol}(N) \geq n \int_M ||H||^2 *1 \geq \lambda_p \operatorname{vol}(N), \tag{14.2}$$

where p and q are the lower and upper order of N, respectively. Either equality sign of (14.2) holds if and only if the immersion is of 1-type with order $\{q\}$ or order $\{p\}$, respectively.

Theorem 14.2. *Let $\phi : N \to \mathbb{E}^m$ be an isometric immersion of a compact Riemannian n-manifold N into $\mathbb{E}^m$. Then, for any k-tuple $(n_1, \ldots, n_k) \in \mathcal{S}(n)$, we have*

$$\lambda_q \geq \frac{n}{\operatorname{vol}(N)} \int_N \Delta(n_1, \ldots, n_k) *1, \tag{14.3}$$

where q is the upper order of the immersion.

The equality case of (14.3) holds if and only if ϕ is a 1-type ideal immersion associated with $(n_1, \ldots, n_k)$, i.e., ϕ is a 1-type immersion of order $\{q\}$ satisfying $H^2 = \Delta(n_1, \ldots, n_k)$ identically.

Proof. Follows immediately from Theorem 13.7 and (14.2). □

Remark 14.3. Inequality (14.3) provides a way to estimate the upper order of an isometric immersion of a compact Riemannian manifold into a Euclidean space in terms of δ-invariants.

For example, let $N = S^2\left(\frac{1}{a^2}\right) \times S^2\left(\frac{1}{b^2}\right)$ be the Riemannian product of two 2-spheres with curvature $\frac{1}{a^2}, \frac{1}{b^2}$, respectively. Assume that

$$a^2 + b^2 = 1, \quad 0 < b < (s(s+1))^{-1/2},$$

for some natural number $s \in \mathbf{N}_0$. Then the first s nonzero eigenvalues $\lambda_1, \lambda_2, \ldots, \lambda_s$ of the Laplacian Δ of N are given by

$$\frac{2}{a^2}, \frac{6}{a^2}, \ldots, \frac{s(s+1)}{a^2},$$

respectively. Since $\hat{\Delta}_0 = \frac{1}{4a^2b^2}$ and $\hat{\Delta}_0 = \frac{1}{4a^2b^2} > \lambda_s$, inequality (14.3) implies that the upper order of any isometric immersion of $S^2\left(\frac{1}{a^2}\right) \times S^2\left(\frac{1}{b^2}\right)$ into any Euclidean space is at least $s+1$. Consequently, regardless of the codimension, if b is small, the upper order q of any isometric immersion of $S^2\left(\frac{1}{a^2}\right) \times S^2\left(\frac{1}{b^2}\right)$ into any Euclidean space must be large.

Theorem 13.13 and the Maximum Principle imply the following.

Theorem 14.3. *If a compact Riemannian n-manifold N admits a 1-type isometric immersion of order* $\{p\}$ *into a Euclidean space, then*

$$\lambda_p \geq \frac{n}{\text{vol}(N)} \int_N \Delta(n_1, \dots, n_k) *1, \tag{14.4}$$

for any k-tuple $(n_1, \dots, n_k) \in \mathcal{S}(n)$.

Proof. If N is a Riemannian n-manifold which admits a 1-type isometric immersion of order $\{p\}$ into a Euclidean space, then $p = q$. Thus, (14.3) becomes (14.4). □

An immediate consequence of Theorem 14.3 is the following.

Corollary 14.2. *If a compact Riemannian n-manifold N admits an ideal 1-type isometric immersion of order* $\{p\}$ *into a Euclidean space associated with a k-tuple* $(n_1, \dots, n_k)$, *then*

$$\lambda_p \geq \frac{n}{\text{vol}(N)} \int_N ||H||^2 *1. \tag{14.5}$$

In particular, when $\delta(n_1, \dots, n_k)$ *is constant, we have* $\lambda_p \geq n||H||^2$.

Proof. Under the hypothesis, we have $\Delta(n_1, \dots, n_k) = ||H||^2$. Thus, (14.4) implies (14.5). In particular, if $\delta(n_1, \dots, n_k)$ is constant, we find from (14.4) and the equality case of (13.45) that

$$\lambda_p \geq \frac{n}{\text{vol}(N)} \int_N \Delta(n_1, \dots, n_k) *1 = n\Delta(n_1, \dots, n_k) = n||H||^2.$$
□

Theorem 13.7 also yields immediately the following necessary intrinsic condition for a compact Riemannian manifold to admit an ideal immersion.

Theorem 14.4. *Let N be a compact Riemannian n-manifold. Then*

(1) *if N admits an ideal immersion into a Euclidean space associated with a k-tuple* $(n_1, \dots, n_k)$, *then*

$$\lambda_1 \leq \frac{n}{\text{vol}(N)} \int_N \Delta(n_1, \dots, n_k) *1; \tag{14.6}$$

(2) *if N satisfies*

$$\lambda_p \leq \frac{n}{\text{vol}(N)} \int_N \hat{\Delta}_0 *1, \tag{14.7}$$

then every 1-type isometric immersion of order $\{p\}$ *from N into an arbitrary Euclidean space is an ideal immersion;*

(3) *an ideal immersion of N satisfies the equality case of* (14.6) *if and only if the immersion is a 1-type ideal immersion of order* $\{1\}$.

Proof. This follows from Theorem 13.7, Theorem 13.13, (8.48), (14.2) and the definition of ideal immersions. □

Theorem 14.4 gives the following obstruction to ideal immersions.

Corollary 14.3. *If a compact Riemannian n-manifold N satisfies*

$$\lambda_1 > \frac{n}{\mathrm{vol}(N)} \int_N \hat{\Delta}_0 *1, \tag{14.8}$$

then N does not admit any ideal immersion into any Euclidean space.

The next result provides a sharp relationship between the δ-invariant $\hat{\Delta}_0$ and the first two nonzero eigenvalues λ_1, λ_2 of Δ for ideal immersions.

Theorem 14.5. *Let $\phi : N \to \mathbb{E}^m$ be an ideal immersion of a compact Riemannian n-manifold into $\mathbb{E}^m$. Then*

$$\int_N \hat{\Delta}_0 *1 \geq \frac{1}{n^2} \left\{ n(\lambda_1 + \lambda_2) - R^2 \lambda_1 \lambda_2 \right\} \mathrm{vol}(N), \tag{14.9}$$

where R is the radius of the smallest ball $B(R)$ containing $\phi(N)$.

The equality sign of (14.9) *holds if and only if $\phi(N)$ lies in the boundary $\partial(B(R))$ of $B(R)$ and ϕ is* 1*-type ideal immersion of order* $\{1\}$, *or* 1*-type ideal immersion of order* $\{2\}$, *or* 2*-type ideal immersion of order* $\{1,2\}$. *Moreover, if the equality case of* (14.9) *holds, then N is mass-symmetric in $\partial(B(R))$ and $\hat{\Delta}_0$ is a constant on N.*

Proof. Assume that $\phi : N \to \mathbb{E}^m$ is an ideal immersion of a compact Riemannian n-manifold into $\mathbb{E}^m$. Let us choose the center of mass as the origin of $\mathbb{E}^m$. Then we find from Theorem 8.10(1) that

$$\int_M ||H||^2 *1 \geq \frac{1}{n^2} \{n(\lambda_1 + \lambda_2) - R^2 \lambda_1 \lambda_2\} \mathrm{vol}(N). \tag{14.10}$$

Since ϕ is an ideal immersion, by combining (14.10) with the equality case of (13.51) we obtain (14.9).

The remaining part follows from Theorem 8.10(1). □

14.3 Applications of δ-invariants to homogeneous spaces

For compact homogeneous spaces we have the following sharp estimate of λ_1 in terms of δ-invariants.

Theorem 14.6. *If N is a compact homogeneous Riemannian n-manifold with irreducible isotropy action, then the first nonzero eigenvalue λ_1 of the Laplacian on N satisfies*

$$\lambda_1 \geq n\,\Delta(n_1, \ldots, n_k) \tag{14.11}$$

for any k-tuple $(n_1, \ldots, n_k) \in \mathcal{S}(n)$. Consequently, we have

$$\lambda_1 \geq n\,\hat{\Delta}_0. \tag{14.12}$$

The equality sign of (14.12) *holds if and only if N admits a 1-type ideal immersion into a Euclidean space.*

Proof. This follows immediately from Theorem 14.3 and the fact that every compact homogeneous Riemannian manifold with irreducible isotropy action admits a 1-type isometric immersion of order $\{1\}$ and that every δ-invariant $\Delta(n_1, \ldots, n_k)$ of a compact homogeneous Riemannian manifold is constant. □

Remark 14.4. When $k = 0$, inequality (14.11) reduces to the following well-known result for compact homogeneous Riemannian n-manifolds with irreducible isotropy action [Nagano (1961)]:

$$\lambda_1 \geq n\rho, \tag{14.13}$$

where $\rho = 2\tau/(n(n-1))$ is the normalized scalar curvature. In general, we have $\hat{\Delta}_0 \geq \rho$, and $\hat{\Delta}_o > \rho$ for most Riemannian manifolds.

Remark 14.5. If a compact homogeneous Riemannian manifold N admits an ideal immersion of order $\{1\}$, then Corollary 14.2 implies immediately the following

$$\lambda_1 \geq n||H||^2. \tag{14.14}$$

This is simply due to the fact that each δ-invariant $\delta(n_1, \ldots, n_k)$ is constant on N.

For homogeneous Riemannian manifolds, we also have the following.

Theorem 14.7. *A compact homogeneous Riemannian n-manifold with irreducible isotropy action admits an ideal immersion into some Euclidean space if and only if it satisfies $\lambda_1 = n\,\hat{\Delta}_0$.*

Proof. This follows from Theorems 14.4 and 14.6 and the fact that each $\delta(n_1, \ldots, n_k)$ on a homogeneous Riemannian manifold is constant. □

Remark 14.6. Clearly, a standard hypersphere in a Euclidean space is an ideal immersion. Besides spheres, there exist many compact homogeneous Riemannian manifolds which admit ideal immersions in the Euclidean space. For instance, the following three compact homogeneous spaces:

$$SU(3)/T^2, \quad Sp(3)/Sp(1)^3, \quad \text{and } F_4/\text{Spin}(8)$$

admit ideal immersions in $\mathbb{E}^8, \mathbb{E}^{14}$ and $\mathbb{E}^{26}$ associated with $(3,3) \in \mathcal{S}(6)$, $(3,3,3,3) \in \mathcal{S}(12)$ and $(12,12) \in \mathcal{S}(24)$, respectively. Their corresponding ideal immersions in $\mathbb{E}^8$, $\mathbb{E}^{14}$ and $\mathbb{E}^{26}$ are given by the corresponding standard isoparametric hypersurfaces with 3 distinct principal curvatures in S^7, S^{13} and S^{25}. É. Cartan found that these three hypersurfaces are precisely the tubes of constant radius over the standard embeddings of CP^2, HP^2 and $\mathcal{O}P^2$ in S^7, S^{13}, and S^{25}, respectively [Cartan (1939)].

Remark 14.7. For compact irreducible homogeneous spaces N, one may apply Theorem 14.6 to determine whether N admits an ideal immersion. In principle, both λ_1 (using Freudenthal's formula [Freudenthal (1954)] for Casimir's operator) and the δ-invariant $\hat{\Delta}_0$ are "computable" for compact irreducible homogeneous spaces. For many compact irreducible symmetric spaces G/H with classical group G, λ_1 has been computed in [Nagano (1961)].

Let $RP^n(1), CP^n(4), QP^n(4)$ and $\mathcal{O}P^2(4)$ denote the real, complex and quaternion projective n-spaces and the Cayley plane with maximal sectional curvature $1, 4, 4, 4$, respectively. Then

$$\begin{aligned}
&\lambda_1 = 2(n+1), \ \hat{\Delta}_0 = 1 \ \text{ for } RP^n(1);\\
&\lambda_1 = 4(n+1), \ \hat{\Delta}_0 = \Delta(n,n) = \frac{n+3}{n} \ \text{ for } CP^n(4) \ (n>1);\\
&\lambda_1 = 8(n+1), \ \hat{\Delta}_0 = \Delta(n,n,n,n) = \frac{n+3}{n} \ \text{ for } QP^n(4) \ (n>1);\\
&\lambda_1 = 48, \ \hat{\Delta}_0 = \Delta(2,2,2,2,2,2,2,2) = \frac{145}{56} \ \text{ for } \mathcal{O}P^2(4).
\end{aligned} \tag{14.15}$$

Theorem 14.7 and (14.15) imply the following.

Corollary 14.4. *RP^n, CP^n, QP^n with $n > 1$ and the Cayley plane $\mathcal{O}P^2$ equipped with a standard Riemannian metric do not admit ideal immersions into any Euclidean space.*

Next, we prove the following.

Proposition 14.2. *CP^1 is the only irreducible Hermitian symmetric space which admits an ideal immersion into a Euclidean space for arbitrary codimension.*

Proof. For an n-dimensional compact homogeneous Einstein-Kähler manifold M with positive scalar curvature, we have $n\lambda_1 = 4\tau$ [Obata (1966)]. Thus, $\lambda_1 > n\Delta(n_1, \ldots, n_k)$ if and only if

$$\left(1 + \frac{1}{n + k - 1 - \sum_{j=1}^{k} n_j}\right) \delta(n_1, \ldots, n_k) < 2\tau. \tag{14.16}$$

The coefficient of $\delta(n_1, \ldots, n_k)$ in (14.16) is equal to two if and only if one of the three cases occurs:
(1) $k = 0, n = 2$;
(2) $k = 1, n = n_1 + 1$; and
(3) $k = 2, n_1 + n_2 = n$.

When N is an irreducible Hermitian symmetric space of compact type, this observation together with geometric properties of irreducible symmetric spaces of compact type imply that (14.16) holds unless $n = 2, k = 0$; namely, unless the ideal submanifold is a complex projective line. □

14.4 Applications of δ-invariants to rigidity problems

Although ideal immersions of a given Riemannian manifold in a Euclidean space is not necessarily unique, very often Theorem 13.5 can be applied to obtain rigidity theorems for isometric immersions of Riemannian manifolds into Euclidean spaces with arbitrary codimension; in particular, for open portions of a homogeneous spaces.

The philosophy of such rigidity comes from the fact that, for a given Riemannian manifold N isometrically immersed in a Euclidean space, inequality (13.45) provides us a lower bound of the squared mean curvature. When the inequality is actually an equality, N is ideal according to the Maximum Principle. In such case Theorem 13.7 implies that the shape operator takes a very special simple form. Often, such information are sufficient to conclude the rigidity of the immersions *without imposing any global assumption.*

Here we provide merely two of such applications.

Theorem 14.8. *Let N be an open portion of a unit n-sphere $S^n(1)$. Then, for any isometric immersion of N into a Euclidean m-space $\mathbb{E}^m$, we have*

$$||H||^2 \geq 1 \tag{14.17}$$

regardless of codimension.

The equality case of (14.17) holds identically if and only if N is immersed as an open portion of a hypersphere in a totally geodesic $\mathbb{E}^{n+1} \subset \mathbb{E}^m$.

Proof. Let N be an open portion of a unit n-sphere $S^n(1)$ isometrically immersed in $\mathbb{E}^m$. Then $\tau = n(n-1)/2$. If we choose $k = 0$, then we have $\delta(n_1, \dots, n_k) = \tau$. Hence, by applying Theorem 13.7, we obtain (14.17).

If the equality case of (14.17) holds identically, N is totally umbilical in $\mathbb{E}^m$ according to Theorem 13.7. Hence, N is immersed as an open part of a unit hypersphere in a totally geodesic $\mathbb{E}^{n+1} \subset \mathbb{E}^m$.

The converse is clear. □

Theorem 14.9. *Let N be an open portion of $\mathbb{E}^{n_1-1} \times S^{n-n_1+1}(1)$ with $n > n_1$. Then for any isometric immersion of N into $\mathbb{E}^m$ we have*

$$||H||^2 \geq \left(\frac{n-n_1+1}{n}\right)^2 \tag{14.18}$$

regardless of codimension.

The equality case of (14.18) *holds identically if and only if N is immersed as an open portion of a spherical hypercylinder*

$$\mathbb{E}^{n_1-1} \times S^{n-n_1+1}(1) \subset \mathbb{E}^{n_1-1} \times \mathbb{E}^{n-n_1} \subset \mathbb{E}^m,$$

where $\mathbb{E}^{n+1} = \mathbb{E}^{n_1+1} \times \mathbb{E}^{n-n_1}$ is totally geodesic in $\mathbb{E}^m$.

Proof. Let N be an open portion of $\mathbb{E}^{n_1-1} \times S^{n-n_1+1}(1)$ isometrically immersed in $\mathbb{E}^m$. Then

$$\delta(n_1) = \frac{(n-n_1+1)(n-n_1)}{2}, \quad c(n_1) = \frac{n^2(n-n_1)}{2(n-n_1+1)}. \tag{14.19}$$

Thus, after applying Theorem 13.7, we obtain inequality (14.18).

Suppose that the equality sign of (14.18) holds identically. Then

$$||H||^2 = \left(\frac{n-n_1+1}{n}\right)^2 > 0. \tag{14.20}$$

Moreover, according to Theorem 13.7, there exists an orthonormal frame $\{e_1, \dots, e_n, e_{n+1}, \dots, e_m\}$ such that the shape operator satisfies

$$A_{e_r} = \begin{pmatrix} B_r & 0 \\ 0 & \mu_r I \end{pmatrix}, \quad r = n+1, \dots, m, \tag{14.21}$$

where B_r is a symmetric $n_1 \times n_1$ submatrix such that trace$(B_r) = \mu_r$ and I is the $(n-n_1) \times (n-n_1)$ matrix. For simplicity, we may choose e_{n+1} to be a unit vector field in the direction of the mean curvature vector H. Then it follows from (14.21) that

$$A_{e_{n+1}} = \begin{pmatrix} B_{n+1} & 0 \\ 0 & \mu I \end{pmatrix}, \quad A_{e_r} = \begin{pmatrix} B_r & 0 \\ 0 & 0 \end{pmatrix}, \quad r = n+2, \dots .m, \tag{14.22}$$

$$\text{trace}\,(B_{n+1}) = \mu \neq 0, \quad \text{trace}\,(B_{n+2}) = \cdots = \text{trace}\,(B_m) = 0. \tag{14.23}$$

It follows from (14.20) and (14.23) that $\mu^2 = 1$. We may assume that $\mu = 1$, after replacing e_{n+1} by $-e_{n+1}$ if necessary.

It follows from (14.22) and Gauss's equation that

$$K_{ij} = 1, \;\; n_1 + 1 \leq i \neq j \leq n. \tag{14.24}$$

Thus, $e_{n_1+1}, \ldots, e_n$ are tangent to the second factor of $\mathbb{E}^{n_1-1} \times S^{n-n_1+1}(1)$. Hence, there exists a unit vector $u \in \text{Span}\{e_1, \ldots, e_{n_1}\}$ tangent to the second factor. Without loss of generality, we may put $u = e_{n_1}$.

Since $K_{n_1 j} = 1$ for $n_1 + 1 \leq j \leq n$, Gauss's equation and (14.22) imply that $h^{n+1}_{n_1 n_1} = 1$. Similarly, since $K_{ij} = 0$ for $1 \leq i \leq n_1 - 1; n_1 \leq j \leq n$, Gauss's equation and (14.22) also imply that $h^{n+1}_{ii} = 0$, $i = 1, \ldots, n_1 - 1$. Hence, we may choose $e_1, \ldots, e_{n_1-1}$ such that $A_{e_{n+1}}$ takes the form:

$$A_{e_{n+1}} = \begin{pmatrix} 0 & \cdots & 0 & k_1 & \\ \vdots & \vdots & \vdots & \vdots & 0 \\ 0 & \cdots & 0 & k_{n_1-1} & \\ k_1 & \cdots & k_{n_1-1} & 1 & \\ & 0 & & & I \end{pmatrix}. \tag{14.25}$$

Since $K_{in_1-1} = 0, 1 \leq i \leq n_1 - 1$, Gauss's equation and (14.22), (14.23), (14.25) imply that $k_1 = \cdots = k_{n_1-1} = B_r = 0$. Hence, we find

$$A_{e_{n+1}} = \begin{pmatrix} 0 & 0 \\ 0 & I \end{pmatrix}, \;\; A_{e_{n+2}} = \cdots = A_n = 0, \tag{14.26}$$

where I is an $(n-n_1+1) \times (n-n_1+1)$-identity matrix. Thus, after applying Moore's lemma we conclude that N is a direct product of a totally geodesic submanifold and a totally umbilical submanifold. Consequently, N is immersed as an open portion of a spherical hypercylinder in a totally geodesic $\mathbb{E}^{n+1}$ as described in the theorem. □

14.5 Applications to warped products

Next, we apply the method in section 13.3 to obtain the following sharp relationship between warping functions and the squared mean curvature [Chen and Dillen (2008)] (see, also [Chen and Wei (2009)]).

Theorem 14.10. *Let $\phi : N_1 \times_{f_2} N_2 \times \cdots \times_{f_\ell} N_\ell \to M$ be an isometric immersion of a multiply warped product $N_1 \times_{f_2} N_2 \times \cdots \times_{f_\ell} N_\ell$ into an arbitrary Riemannian manifold M. Then*

$$\sum_{j=2}^{\ell} n_j \frac{\Delta f_j}{f_j} \leq \frac{n^2(\ell - 1)}{2\ell} ||H||^2 + n_1(n - n_1) \max \tilde{K}, \;\; n = \sum_{j=1}^{\ell} n_j, \tag{14.27}$$

where $\max \tilde{K}(p)$ *denotes the maximum of the sectional curvature* $\tilde{K}$ *of* M *restricted to plane sections in* T_pN *at* $p \in N$.

The equality sign of (14.27) *holds identically if and only if the following two conditions hold:*

(1) ϕ *is mixed totally geodesic such that* $\operatorname{trace} h_1 = \cdots = \operatorname{trace} h_\ell$;
(2) *At each point* $p \in N$, $\tilde{K}$ *satisfies* $\tilde{K}(u,v) = \max \tilde{K}(p), \forall u \in T^1_{p_1}N_1$ *and* $\forall v \in T^1_{(p_2,\cdots,p_k)}(N_2 \times \cdots \times N_\ell)$.

Proof. Let $\hat{\nabla}, \hat{R}$, *etc.*, denote the Levi-Civita connection, the Riemann curvature tensor, etc., of the Riemannian product $N_1 \times N_2 \times \cdots \times N_\ell$; and by ∇, R, etc. the corresponding quantities of the multiply warped product $N = N_1 \times_{f_2} N_2 \times \cdots \times_{f_\ell} N_\ell$.

If we put $H_i = -\nabla((\ln f_i) \circ \pi_1)$, then Lemma 4.2 implies that the sectional curvature K of $N_1 \times_{f_2} N_2 \times \cdots \times_{f_\ell} N_\ell$ satisfies

$$\begin{aligned} K(X_1, X_i) &= \frac{1}{f_i}\left((\nabla_{X_1}X_1)f_i - X_1^2 f_i\right), \\ K(X_i, X_j) &= -\frac{\langle \nabla f_i, \nabla f_j\rangle}{f_i f_j}, \quad i,j = 2,\ldots,\ell, \end{aligned} \tag{14.28}$$

for each unit vector $X_i \in \mathcal{D}_i$. In particular, (14.28) yields

$$\Delta f_i = f_i \sum_{j=1}^{n_1} K(e_j, X_i), \quad i = 2,\ldots,\ell, \tag{14.29}$$

for a unit vector $X_i \in \mathcal{D}_i$, where $\{e_1,\ldots,e_{n_1}\}$ is an orthonormal frame of N_1. If $\phi : N \to M$ is an isometric immersion of a multiply warped product N into an Riemannian m-manifold M, Gauss's equation implies that, at each point $p \in N$, we have

$$2\tau = n^2||H||^2 - S_h + 2\tilde{\tau}(T_pN), \tag{14.30}$$

where $n_i = \dim N_i$, $n = \sum_{i=1}^{\ell} n_i$, S_h is defined by (2.31), and $\tilde{\tau}(T_pN)$ is the scalar curvature of M, restricted to T_pN.

Let us put

$$\eta = 2\tau - n^2\Big(1 - \frac{1}{\ell}\Big)||H||^2 - 2\tilde{\tau}(T_pN). \tag{14.31}$$

Then we find from (14.30) and (14.31) that

$$n^2||H||^2 = \ell\,(\eta + S_h)\,. \tag{14.32}$$

Put

$$\Delta_1 = \{1,\ldots,n_1\}, \ \ldots, \ \Delta_\ell = \{n_1 + \cdots + n_{\ell-1} + 1, \ldots, n\}.$$

Let us choose an orthonormal basis $\{e_1, \ldots, e_m\}$ at p such that, for each $j \in \Delta_i$, $e_j \in \mathcal{D}_j$, $j = 1, \ldots, \ell$. Moreover, we choose the normal vector e_{n+1} in $H(p)$. Then it follows from (14.32) that

$$\begin{aligned}\left(\sum_{A=1}^{n} a_A\right)^2 &- \ell \sum_{A=1}^{n} (a_A)^2 \\ &= \ell\left[\eta + \sum_{A \neq B} (h_{AB}^{n+1})^2 + \sum_{r=n+2}^{m} \sum_{A,B=1}^{n} (h_{AB}^r)^2\right],\end{aligned} \tag{14.33}$$

where

$$a_A = h_{AA}^{n+1},\ h_{AB}^r = \langle h(e_A, e_B), e_r\rangle,\ 1 \leq A, B \leq n;\ n+1 \leq r \leq m.$$

After apply Lemma 13.1 to (14.33) we obtain

$$\begin{aligned}&\sum_{\alpha_1 < \beta_1} a_{\alpha_1} a_{\beta_1} + \sum_{\alpha_2 < \beta_2} a_{\alpha_2} a_{\beta_2} + \cdots + \sum_{\alpha_\ell < \beta_\ell} a_{\alpha_\ell} a_{\beta_\ell} \\ &\quad \geq \frac{\eta}{2} + \sum_{A<B} (h_{AB}^{n+1})^2 + \frac{1}{2} \sum_{r=n+2}^{m} \sum_{A,B=1}^{n} (h_{AB}^r)^2,\end{aligned} \tag{14.34}$$

where $\alpha_i, \beta_i \in \Delta_i$, $i = 1, \ldots, \ell$.

On the other hand, from the equation of Gauss and (14.29) we find

$$\begin{aligned}\sum_{i=2}^{\ell} n_i \frac{\Delta f_i}{f_i} &= \sum_{j \in \Delta_1} \sum_{\beta \in \Delta_2 \cup \cdots \cup \Delta_\ell} K(e_j, e_\beta) \\ &= \tau - \sum_{1 \leq j_1 < j_2 \leq n_1} K(e_{j_1}, e_{j_2}) - \sum_{n_1+1 \leq \alpha < \beta \leq n} K(e_\alpha, e_\beta) \\ &= \tau - \tilde{\tau}(\mathcal{D}_1) - \sum_{r=n+1}^{m} \sum_{1 \leq j_1 < j_2 \leq n_1} \left(h_{j_1 j_1}^r h_{j_2 j_2}^r - (h_{j_1 j_2}^r)^2\right) \\ &\quad - \tilde{\tau}(\mathcal{D}_2 \oplus \cdots \oplus \mathcal{D}_k) - \sum_{r=n+1}^{m} \sum_{n_1+1 \leq \alpha < \beta < n} \left(h_{\alpha\alpha}^r h_{\beta\beta}^r - (h_{\alpha\beta}^r)^2\right).\end{aligned} \tag{14.35}$$

Thus, by combining (14.31), (14.34) and (14.35), we obtain

$$\begin{aligned}\sum_{i=2}^{\ell} n_i \frac{\Delta f_i}{f_i} &\leq \tau - \tilde{\tau}(T(N)) + \sum_{j \in \Delta_1} \sum_{\beta \in \Delta_2 \cup \cdots \cup \Delta_\ell} \tilde{K}(e_j, e_\beta) \\ &\quad - \frac{1}{2} \sum_{r=n+2}^{m} \sum_{A,B=1}^{n} (h_{AB}^r)^2 - \sum_{1 \leq j \leq n_1} \sum_{n_1+1 \leq \alpha \leq n} (h_{j\alpha}^{n+1})^2 \\ &\quad + \sum_{r=n+2}^{m} \sum_{1 \leq j_1 < j_2 \leq n_1} \left((h_{j_1 j_2}^r)^2 - h_{j_1 j_1}^r h_{j_1 j_2}^r\right)\end{aligned} \tag{14.36}$$

$$
\begin{aligned}
&+\sum_{r=n+2}^{m}\sum_{n_1+1\le\alpha<\beta<n}\left((h^r_{\alpha\beta})^2-h^r_{\alpha\alpha}h^r_{\beta\beta}\right)-\frac{\eta}{2}\\
&=\tau-\tilde{\tau}(T(N))+\sum_{j\in\Delta_1}\sum_{\beta\in\Delta_2\cup\cdots\cup\Delta_\ell}\tilde{K}(e_j,e_\beta)-\frac{\eta}{2}\\
&-\sum_{r=n+1}^{m}\sum_{1\le j\le n_1}\sum_{n_1+1\le\alpha\le n}(h^r_{j\alpha})^2\\
&-\frac{1}{2}\sum_{r=n+2}^{m}\Bigg(\sum_{1\le j\le n_1}h^r_{jj}\Bigg)^2-\frac{1}{2}\sum_{r=n+2}^{m}\Bigg(\sum_{n_1+1\le\alpha\le n}h^r_{\alpha\alpha}\Bigg)^2\\
&\le\tau-\tilde{\tau}(T(N))+n_1(n-n_1)\max\tilde{K}-\frac{\eta}{2}\\
&=\frac{n^2(\ell-1)}{2\ell}H^2+n_1(n-n_1)\max\tilde{K}.
\end{aligned}
$$

This proves inequality (14.32).

If the equality sign of (14.32) holds, then all of inequalities in (14.34) and (14.36) become equalities. Hence, by applying Lemma 13.1, we conclude that if the equality sign of (14.32) holds, then ϕ is a mixed totally geodesic submanifold which satisfies trace $h_1 = \cdots =$ trace h_ℓ holds identically. This gives condition (1). From the equality case of the last inequality in (14.36) we get condition (2).

The converse can be easily verified. □

Theorem 14.10 is an extension of the following result [Chen (2002a)].

Theorem 14.11. *Let $\phi : N_1 \times_f N_2 \to R^m(c)$ be an isometric immersion of a warped product $N_1 \times_f N_2$ into a real space form $R^m(c)$. Then*

$$\frac{\Delta f}{f} \le \frac{(n_1+n_2)^2}{4n_2}||H||^2 + n_1 c, \tag{14.37}$$

where $n_i = \dim N_i$, $i = 1, 2$, and Δ is the Laplacian of N_1. The equality sign of (14.37) *holds identically if and only if ϕ is mixed totally geodesic and* trace $h_1 =$ trace h_2.

Moreover, if the the equality sign of (14.37) *holds identically, then either*

(a) *ϕ is a minimal immersion and the warping function f is an eigenfunction of the Laplacian Δ with eigenvalue $n_1 c$, or*
(b) *$\Delta f \ne n_1 c f$ and locally ϕ is a non-minimal warped product immersion*

$$(\phi_1, \phi_2) : N_1 \times_f N_2 \to M_1 \times_\rho M_2,$$

where $M_1 \times_\rho M_2$ is a warped product representation of $R^m(c)$.

Proof. Under the hypothesis, it follows from Theorem 14.10 that ϕ satisfies (14.37), with the equality sign holding identically if and only if ϕ satisfies the following two properties:

(1) ϕ is mixed totally geodesic and
(2) $\operatorname{trace} h_1 = \operatorname{trace} h_2$.

Now, assume that the equality case of (14.37) holds identically. If $\Delta f = n_1 cf$ holds, then the equality case of (14.37) implies that ϕ is minimal. Thus we have (a).

Next, let us assume that $\Delta f \neq n_1 cf$. Then it follows from the equality of (14.37) that ϕ is non-minimal. Since ϕ is mixed totally geodesic by property (1), Nölker's theorem implies that there exists a warped product representation $M_1 \times_\rho M_2$ of $R^m(c)$ such that ϕ is locally a warped product immersion $\phi = (\phi_1, \phi_2) : N_1 \times_f N_2 \to M_1 \times_\rho M_2$, where $M_1 \times_\rho M_2$ is a warped product representation of $R^m(c)$. This gives (b). □

Some consequences of Theorems 14.10 and 14.11 are the following.

Corollary 14.5. *Let $N_1 \times_f N_2$ be a warped product of two Riemannian manifolds whose warping function f is a harmonic function. Then*

(1) *$N_1 \times_f N_2$ admits no isometric minimal immersion into any Riemannian manifold of negative curvature;*
(2) *every isometric minimal immersion from $N_1 \times_f N_2$ into a Euclidean space is a warped product immersion.*

Proof. Let $\phi : N_1 \times_f N_2 \to M$ be an isometric minimal immersion of $N_1 \times_f N_2$ into a Riemannian manifold M. If f is a harmonic function on N_1, then inequality (14.27) of Theorem 14.10 implies $\max \tilde{K} \geq 0$ on $N = N_1 \times_f N_2$. This shows that $N_1 \times_f N_2$ does not admit any isometric minimal immersion into any Riemannian manifold of negative curvature.

When M is a Euclidean space, the minimality of $N_1 \times_f N_2$ and the harmonicity of f imply that the equality sign of (14.37) holds identically. Hence, ϕ is a mixed totally geodesic submanifold according to Theorem 14.10. Therefore, by Nölker's theorem, ϕ is locally a warped product immersion. □

Corollary 14.6. *Let f be an eigenfunction of Laplacian Δ on N_1 with eigenvalue $\lambda > 0$. Then every Riemannian warped product $N_1 \times_f N_2$ does not admits any isometric minimal immersion into any Riemannian manifold of non-positive curvature.*

Proof. Let f be an eigenfunction of Δ on N_1 with eigenvalue $\lambda > 0$. Suppose that $N_1 \times_f N_2$ admits an isometric minimal immersion into M. Then (14.27) implies that $n_1 \max \tilde{K} \geq \lambda > 0$. Therefore, M cannot be non-positively curved. □

Corollary 14.7. *Let N_1 be a compact manifold. Then*

(1) *every Riemannian warped product $N_1 \times_f N_2$ does not admit an isometric minimal immersion into any Riemannian manifold of negative curvature;*
(2) *every Riemannian warped product $N_1 \times_f N_2$ does not admit an isometric minimal immersion into a Euclidean space.*

Proof. Assume N_1 is compact. Let $\phi : N_1 \times_f N_2 \to M$ be an isometric minimal immersion of $N_1 \times_f N_2$ into a non-positively curved Riemannian manifold M. Then inequality (14.27) implies that

$$\frac{\Delta f}{f} \leq n_1 \max \tilde{K} \leq 0.$$

Since the warping function f is positive, we find $\Delta f \leq 0$. Hence, it follows from Hopf's lemma and the compactness of N_1 that f is a positive constant. Consequently, we have $\max \tilde{K} = 0$, which implies (1).

If $M = \mathbb{E}^m$, the equality case of (14.37) holds. Hence ϕ is mixed totally geodesic. Thus, by Moore's lemma, ϕ is a product immersion, say

$$\phi = (\phi_1, \phi_2) : (N_1, g_1) \times (N_2, f^2 g_2) \to \mathbb{E}^{m_1} \times \mathbb{E}^{m_2} = \mathbb{E}^m.$$

Since ϕ is minimal, $\phi_1 : N_1 \to \mathbb{E}^{m_1}$ is also minimal. But is impossible since N_1 is compact. □

Example 14.1. There exist many minimal immersions of a warped product $N_1 \times_f N_2$ with harmonic warping function f into a Euclidean space. For instance, if N_2 is a minimal submanifold of the unit $(m-1)$-hypersphere $S^{m-1}(1) \subset \mathbb{E}^m$ centered at the origin, then the minimal cone $C(N_2)$ over N_2 with vertex at the origin of $\mathbb{E}^m$ is a warped product $\mathbb{R}_+ \times_s N_2$ whose warping function $f = s$ is a harmonic function. Here s is the coordinate function of the positive real line $\mathbb{R}_+$. This provides many examples of minimal warped products in $\mathbb{E}^m$ which satisfy the equality case of (14.37).

Remark 14.8. Let N_1 is a complete, non-compact Riemannian manifold. A function f on N_1 is called an *L^q-function* if

$$||f||_{L^q} := \left(\int_{N_1} |f|^q *1 \right)^{\frac{1}{q}}$$

converges. By computing growth estimate of the warping function, it was proved in [Chen and Wei (2009)] that if f is an L^q-function on N_1 for some $q > 1$, then for any Riemannian manifold N_2 the warped product $N_1 \times_f N_2$ does not admit any isometric minimal immersion into any non-positively curved Riemannian manifold.

Example 14.1 illustrates that Theorems 14.10 and 14.11 and Corollary 14.5 are optimal. The following examples show that Corollaries 14.6 and 14.7 are also optimal.

Example 14.2. There are many isometric minimal immersions of warped products $N_1 \times_f N_2$ into a hyperbolic space such that the warping function f is an eigenfunction with negative eigenvalue. For example, $\mathbf{R} \times_{e^x} \mathbb{E}^{n-1}$ admits an isometric minimal immersion into $H^{n+1}(-1)$.

Example 14.3. There are many isometric minimal immersions of $N_1 \times_f N_2$ into Euclidean space with compact N_2. For examples, a hypercaternoid in $\mathbb{E}^{n+1}$ is a minimal hypersurfaces which is isometric to a warped product $\mathbf{R} \times_f S^{n-1}$. Also, for a compact minimal submanifold N_2 of $S^{m-1} \subset \mathbb{E}^m$, the minimal cone $C(N_2)$ is a warped product $R_+ \times_s N_2$.

Example 14.4. Contrast to Euclidean and hyperbolic spaces, the unit m-sphere S^m admits warped product minimal submanifolds $N_1 \times_f N_2$ such that N_1, N_2 are both compact. The simplest examples are minimal Clifford tori $M_{k,n-k}$, $k = 2, \ldots, n-1$, in S^{n+1} defined by

$$M_{k,n-k} = S^k\left(\sqrt{\frac{k}{n}}\right) \times S^{n-k}\left(\sqrt{\frac{n-k}{n}}\right).$$

For warped products in complex space forms we have the following results from [Chen (2002d, 2003b); Mihai (2004a)].

Theorem 14.12. *Let $\phi : N_1 \times_f N_2 \to CH^m(4c)$, $c < 0$, be an isometric immersion of a warped product $N_1 \times_f N_2$ into $CH^m(4c)$. Then*

$$\frac{\Delta f}{f} \leq \frac{(n_1 + n_2)^2}{4n_2} ||H||^2 + n_1 c, \tag{14.38}$$

where $n_i = \dim N_i$, $i = 1, 2$, and Δ is the Laplacian operator of N_1.

The equality sign of (14.38) holds identically if and only if the following three conditions are satisfied:

(1) *ϕ is mixed totally geodesic;*
(2) $\operatorname{trace} h_1 = \operatorname{trace} h_2$*;*

(3) $J\mathcal{H} \perp \mathcal{V}$, *where* J *is the almost complex structure of* $CH^m(4c)$ *and* $\mathcal{V}$ *is the vertical distribution of the warped product.*

Theorem 14.13. *Let* $\phi : N_1 \times_f N_2 \to CP^m(4c)$, $c > 0$, *be an isometric immersion of a warped product into* $CP^m(4c)$. *Then*

$$\frac{\Delta f}{f} \leq \frac{(n_1 + n_2)^2}{4n_2}||H||^2 + (3 + n_1)c. \tag{14.39}$$

The equality sign of (14.39) *holds identically if and only if we have*

(1) $n_1 = n_2 = 1$;
(2) f *is an eigenfunction of the Laplacian of* N_1 *with eigenvalue* 4;
(3) ϕ *is totally geodesic and holomorphic.*

Some applications of Theorems 14.12 and 14.13 are the following.

Corollary 14.8. *If* f *is a positive function on a Riemannian* n_1*-manifold* N_1 *such that* $(\Delta f)/f > 3 + n_1$ *at some point* $p \in N_1$, *then, for any Riemannian manifold* N_2, *the warped product* $N_1 \times_f N_2$ *does not admit any isometric minimal immersion into* $CP^m(4)$ *for any* m.

Corollary 14.9. *Let* $N_1 \times_f N_2$ *be a warped product whose warping function* f *is a harmonic function. Then* $N_1 \times_f N_2$ *does not admit an isometric minimal immersion into any complex hyperbolic space.*

Corollary 14.10. *If* f *is an eigenfunction of Laplacian on* N_1 *with eigenvalue* $\lambda > 0$, *then* $N_1 \times_f N_2$ *does not admits an isometric minimal immersion into any complex hyperbolic space.*

Corollary 14.11. *If* N_1 *is a compact Riemannian manifold, then every warped product* $N_1 \times_f N_2$ *does not admit an isometric minimal immersion into any complex hyperbolic space.*

Another interesting application of Theorem 14.12 is the following [Chen (2002d)]

Corollary 14.12. *If* $\dim N_1 = \dim N_2$, *then the warping function* f *of every warped product decomposition* $N_1 \times_f N_2$ *of a real space form is an eigenfunction of the Laplacian.*

Proof. Assume that $N_1 \times_f N_2$ is a warped product decomposition of a real space form $R^{2n_1}(\varepsilon)$ of constant curvature ε with $\dim N_1 = \dim N_2 = n_1$. Let c be a negative number $< \varepsilon$. Then locally there is a totally umbilical isometric immersion j of $N_1 \times_f N_2$ into the real hyperbolic space $H^{2n_1+1}(c)$ of constant curvature c. Denote by $\iota : H^{2n_1+1}(c) \to CH^{2n_1+1}(4c)$ the totally geodesic Lagrangian embedding of $H^{2n_1+1}(c)$ into $CH^{2n_1+1}(4c)$. Then the composition $\phi = \iota \circ j$:

$$N_1 \times_f N_2 \xrightarrow{\text{totally umbilical}} H^{2n_1+1}(c) \xrightarrow[\text{Lagrangian}]{\text{totally geodesic}} CH^{2n_1+1}(4c)$$

is an immersion satisfying (1), (2) and (3) of Theorem 14.12. Hence, ϕ satisfies the equality case of (14.38) by Theorem 14.12. Thus

$$\frac{\Delta f}{f} = n_1 H^2 + n_1 c.$$

Since the composition ϕ is a totally umbilical totally real immersion, it has constant squared mean curvature. Thus, the warping function f is an eigenfunction of the Laplacian. □

14.6 Applications to Einstein manifolds

Let p be a point in a Riemannian n-manifold N and q a natural number $\leq \frac{n}{2}$. For a given point $p \in N$, let $\pi_1, \ldots, \pi_q$ be q mutually orthogonal plane sections in T_pN. Define the invariant $K_q^{\inf}(p)$ to be the infimum of the average of the sectional curvatures $K(\pi_1), \ldots, K(\pi_q)$, i.e.,

$$K_q^{\inf}(p) = \inf_{\pi_1 \perp \cdots \perp \pi_q} \frac{1}{q} \left(K(\pi_1) + \cdots + K(\pi_q) \right), \tag{14.40}$$

where $\pi_1, \ldots, \pi_q$ run over all mutually orthogonal q plane sections in T_pN. For each natural number $q \leq \frac{n}{2}$, define the invariant δ_q^{Ric} on N by

$$\delta_q^{Ric}(p) = \sup_{u \in T_p^1 M} Ric(u,u) - \frac{2q}{n} K_q^{\inf}(p), \tag{14.41}$$

where u runs over all unit vectors in $T_p^1 N$.

Remark 14.9. The invariants δ_q^{Ric} and $\delta(2)$ are different in nature; e.g. for Einstein 4-manifolds, we have

$$\delta_2^{Ric} = Ric - K_2^{\inf}(\pi), \quad \delta(2) = \tau - \inf K(\pi). \tag{14.42}$$

where π runs over all plane sections at each given point.

The next two results on Einstein manifolds were obtained in [Chen (2004b)].

Theorem 14.14. *For any integer $k \geq 2$ and any isometric immersion of an Einstein $2k$-manifold N into a real space form $R^m(c)$, we have*

$$\delta_k^{Ric} \leq 2(k-1)\left(||H||^2 + c\right). \tag{14.43}$$

The equality sign of (14.43) holds identically if and only if one of the following two cases occurs:

(1) *N is a minimal Einstein submanifold such that, with respect to some suitable orthonormal frame $\{e_1, \ldots, e_{2k}, e_{2k+1}, \ldots, e_m\}$, the shape operator of N takes the following form:*

$$A_r = \begin{pmatrix} A_1^r & & 0 \\ & \ddots & \\ 0 & & A_k^r \end{pmatrix}, \quad r = 2k+1, \ldots, m;$$

(2) *N is a pseudo-umbilical Einstein submanifold such that, with respect to some suitable orthonormal frame $\{e_1, \ldots, e_{2k}, e_{2k+1}, \ldots, e_m\}$, the shape operator satisfies*

$$A_r = \begin{pmatrix} A_1^r & & 0 \\ & \ddots & \\ 0 & & A_k^r \end{pmatrix}, \quad r = 2k+2, \ldots, m,$$

where A_j^r, $j = 1, \ldots, k$, are symmetric 2×2 submatrices satisfy

$$\mathrm{trace}\,(A_1^r) = \cdots = \mathrm{trace}\,(A_k^r) = 0.$$

Theorem 14.15. *Let $\phi : N \to R^m(c)$ be an isometric immersion of an Einstein n-manifold N into a real space form $R^m(c)$. Then, for every natural number $q < \frac{n}{2}$, we have*

$$\delta_q^{Ric} \leq \frac{n(n-q-1)}{n-q}||H||^2 + \left(n - 1 - \frac{2q}{n}\right)c, \tag{14.44}$$

with the equality sign of (14.44) holding identically if and only if N is a totally geodesic submanifold.

From Theorems 14.14 and 14.15 we obtain easily the following.

Corollary 14.13. *If a Riemannian n-manifold N admits an isometric immersion into a Euclidean space which satisfies*

$$\delta_q^{Ric} > \frac{n(n-q-1)}{n-q}||H||^2, \tag{14.45}$$

for some natural number $q \leq \frac{n}{2}$ at some point, then N is not Einsteinian.

Corollary 14.14. *If an Einstein n-manifold satisfies*

$$\delta_q^{Ric} > \left(n - 1 - \frac{2q}{n}\right) c \tag{14.46}$$

for some natural number $q \leq \frac{n}{2}$ at some point, then it admits no minimal isometric immersion into a real space form $R^m(c)$ for any $m > n$.

Corollary 14.15. *Let N be a compact Einstein n-manifold with finite fundamental group π_1 or with null first betti number. If there is a natural number $q \leq \frac{n}{2}$ such that $\delta_q^{Ric} > 0$, then N admits no Lagrangian isometric immersion into any complex Euclidean n-space.*

Remark 14.10. Inequality (14.43) is sharp, e.g. each hypersphere S^{2k} of $\mathbb{E}^{2k+1}$ and the standard embedding of $S^2(1) \times S^2(1)$ into $\mathbb{E}^6$ are umbilical and pseudo-umbilical Einstein submanifolds of Euclidean spaces satisfying the equality case of (14.43).

Remark 14.11. The minimal Clifford torus: $S^2(1) \times S^2(1) \subset S^5(1)$ is a minimal hypersurface of $S^5(1)$ satisfying the equality case of (14.43) with $c = 1$. This example and the example in Remark 14.10 show that both cases (1) and (2) of Theorem 14.14 do occur.

Remark 14.12. The following example shows that inequality (14.43) does not hold for arbitrary submanifolds in general. Consider the spherical hypercylinder: $S^2(1) \times \mathbb{E}^{2k-2} \subset \mathbb{E}^{2k+1}$. Then $\delta_k^{Ric} = 1$ and $||H||^2 = k^{-2}$; and thus

$$\delta_k^{Ric} = 1 > \frac{2(k-1)}{k^2} = 2(k-1)||H||^2, \quad k \geq 2.$$

14.7 Applications to conformally flat manifolds

Let N be a Riemannian n-manifold. For a given ℓ-subspace $L \subset T_pN$, $p \in N$, define the *Ricci curvature* of L, $Ric(L)$, as

$$Ric(L) = Ric(e_1, e_1) + \ldots + Ric(e_\ell, e_\ell), \tag{14.47}$$

where $\{e_1, \ldots, e_\ell\}$ is an orthonormal basis of L and Ric is the Ricci tensor.

For a k-tuple $(n_1, \ldots, n_k)$ in $\mathcal{S}(n)$, we define a δ-invariant $\sigma(n_1, \ldots, n_k)$ on the Riemannian manifold N by

$$\sigma(n_1, \ldots, n_k) = \tau - \inf\left\{\frac{(n-1)\sum_{j=1}^k (n_j - 1)Ric(L_j)}{(n-1)(n-2) + \sum_{j=1}^k n_j(n_j - 1)}\right\}, \tag{14.48}$$

where $L_1, \ldots, L_k$ run over all k mutually orthogonal subspaces of T_pN such that $\dim L_j = n_j$, $j = 1, \ldots, k$. Put

$$\alpha(n_1, \ldots, n_k) = \frac{n^2(n-1)(n-2)\big(n+k-1-\sum_{j=1}^{k} n_j\big)}{2\big((n-1)(n-2)+\sum_{j=1}^{k} n_j(n_j-1)\big)\big(n+k-\sum_{j=1}^{k} n_j\big)},$$

$$\beta(n_1, \ldots, n_k) = \frac{(n-1)(n-2)\big(n(n-1)-\sum_{j=1}^{k} n_j(n_j-1)\big)}{2(n-1)(n-2)+2\sum_{j=1}^{k} n_j(n_j-1)}.$$

We have the following general optimal inequality for conformally flat submanifolds [Chen (2005a)].

Theorem 14.16. *If N is a conformally flat n-manifold isometrically immersed in a real space form $R^m(c)$, then for each k-tuple $(n_1, \ldots, n_k) \in \mathcal{S}(n)$ we have*

$$\sigma(n_1, \ldots, n_k) \leq \alpha(n_1, \ldots, n_k)||H||^2 + \beta(n_1, \ldots, n_k)c. \tag{14.49}$$

The equality case of inequality (14.49) *holds at a point $p \in N$ if and only if, there exists an orthonormal basis $e_1, \ldots, e_m$ of T_pN, such that the shape operators of N in $R^m(c)$ at p takes the form:*

$$A_r = \begin{pmatrix} \begin{matrix} A_1^r & \ldots & 0 \\ \vdots & \ddots & \vdots \\ 0 & \ldots & A_k^r \end{matrix} & 0 \\ 0 & \mu_r I \end{pmatrix}, \quad r = n+1, \ldots, m, \tag{14.50}$$

where I is an identity matrix and A_j^r is a symmetric $n_j \times n_j$ submatrix satisfying $\operatorname{trace}(A_1^r) = \cdots = \operatorname{trace}(A_k^r) = \mu_r$.

Remark 14.13. Inequality (14.49) does not hold for arbitrary Riemannian submanifolds of real space forms in general. This can be seen from the following example.

Example 14.5. Let $N = S^{n-2}(1) \times \mathbb{E}^2 \subset \mathbb{E}^{n+1} = \mathbb{E}^{n-1} \times \mathbb{E}^2$ be the standard isometrically embedding of $S^{n-2}(1) \times \mathbb{E}^2$ in $\mathbb{E}^{n+1}$ as a spherical hypercylinder over the unit $(n-2)$-sphere. If we choose $k = 1$ and $n_1 = 2$, then $\inf Ric(\pi) = 0$, where π runs over all 2-planes of T_qN at a given point $q \in N$. Hence, find from (14.48) that

$$\sigma(2) = \tau - \frac{n-1}{n^2-3n+4} \inf_{\pi} Ric(\pi) = \frac{(n-2)(n-3)}{2}.$$

On the other hand, we have

$$\alpha(2) = \frac{n^2(n-2)^2}{2(2n^2-3n+4)}, \quad ||H||^2 = \frac{(n-2)^2}{n^2}.$$

Hence

$$\frac{(n-2)(n-3)}{2} = \sigma(2) > \alpha(2)||H||^2 = \frac{(n-2)^4}{2(2n^2-3n+4)}$$

for $n > 4$, which shows that the inequality (14.49) does not hold for an arbitrary submanifold in a Euclidean space in general.

The following corollaries are immediate consequences of Theorem 14.16.

Corollary 14.16. *If a Riemannian n-manifold N admits an isometric immersion into a Euclidean space such that*

$$\sigma(n_1,\ldots,n_k) > \alpha(n_1,\ldots,n_k)||H||^2$$

at a point $p \in N$ for some k-tuple $(n_1,\ldots,n_k) \in \mathcal{S}(n)$, then N is not conformally flat.

Corollary 14.17. *Let N be a conformally flat n-manifold. If there exists a k-tuple $(n_1,\ldots,n_k)$ in $\mathcal{S}(n)$ such that $\sigma(n_1,\ldots,n_k) > 0$ at some points in N, then N does not admit any minimal isometric immersion into a Euclidean space.*

Definition 14.1. A Riemannian submanifold is called *coordinate-minimal* with respect to a coordinate system $\{x_1,\ldots,x_n\}$ if the second fundamental form h satisfies $\sum_{i=1}^n h(\partial_{x_i},\partial_{x_i}) = 0$, where $\partial_{x_i} = \partial/\partial x_i$, $i = 1,\ldots,n$, are coordinate vector fields.

We have the following classification theorem for conformally flat manifolds [Chen and Garay (2006)].

Theorem 14.17. *Let $\phi : N \to \mathbb{E}^m$ be an isometric immersion of a conformally flat n-manifold N with $n \geq 4$ into $\mathbb{E}^m$. Then*

$$\sigma(2) \leq \frac{n^2(n-2)^2}{2(n^2-3n+4)}||H||^2. \tag{14.51}$$

The equality sign of (14.51) holds identically if and only if one of the following five cases occurs:

(a) *N is an open portion of a totally geodesic n-plane;*
(b) *N is an open portion of a spherical hypercylinder $S^{n-1} \times \mathbf{R}$ lying in a totally geodesic $\mathbb{E}^{n+1} \subset \mathbb{E}^m$;*
(c) *N is an open portion of a round hypercone lying in a totally geodesic $\mathbb{E}^{n+1} \subset \mathbb{E}^m$;*

(d) *$m \geq n+3$ and N is the loci of $(n-2)$-spheres defined by*

$$\phi = \left(\Psi(s,t), \Big(\frac{1}{4c^2} - c^2(s^2+t^2)\Big)F\right), \quad c > 0,$$

where $F : S^{n-2} \to \mathbb{E}^{n-1}$ is a unit hypersphere in $\mathbb{E}^{n-1}$, and $\Psi : N_1 \to \mathbb{E}^{m-n+1}$ is a coordinate-minimal immersion with respect to

$$g_{N_1} = (1 - 4c^4s^2)ds^2 - 8c^4 st ds dt + (1 - 4c^4t^2)dt^2;$$

(e) *$m \geq n+3$ and N is the loci of $(n-2)$-spheres defined by*

$$\phi = (P(s,t), f(s,t)F),$$

where F is a unit hypersphere in $\mathbb{E}^{n-1}$ and $P : N_2 \to \mathbb{E}^{m-n+1}$ is a coordinate-minimal surface with respect to

$$g_P = (f\Delta f + f_t^2)ds^2 - 2f_s f_t ds dt + (f\Delta f + f_s^2)dt^2,$$

where $\Delta = -(\partial_s^2 + \partial_t^2)$ and f is a positive solution of the system:

$$(\Delta f)K_s = (\Delta f)_s - f_s, \ (\Delta f)K_t = (\Delta f)_t + f_t, \ \Delta f > 0$$

with $K = \ln(f^2 \Delta \ln f)$.

14.8 Applications of δ-invariants to general relativity

Consider an isometric embedding $\phi : N \to M$ of a pseudo-Riemannian n-manifold N into a pseudo-Riemannian m-manifold M. Let L be an r-plane section in T_pN spanned by orthonormal vectors $e_1, \ldots, e_r$ is defined as

$$\tau(L) = \sum_{1 \leq \alpha < \beta \leq r} K(e_\alpha \wedge e_\beta).$$

We put $\sigma(L) = \tau(L) - \tilde{\tau}(L)$, where $\tau(L)$ and $\tilde{\tau}(L)$ denote the scalar curvature of L in N and M, respectively.

Let us define the δ-invariants $\Lambda(n_1, \ldots, n_k)$ and $\hat{\Lambda}(n_1, \ldots, n_k)$ of a pseudo-Riemannian manifold N as follows [Haesen and Verstraelen (2004)]:

$$\begin{aligned} \Lambda(n_1, \ldots, n_k) &= \tau - \inf\{\sigma(L_1) + \cdots + \sigma(L_k)\}, \\ \hat{\Lambda}(n_1, \ldots, n_k) &= \tau - \sup\{\sigma(L_1) + \cdots + \sigma(L_k)\}, \end{aligned} \tag{14.52}$$

where $L_1, \ldots, L_k$ run over all mutually orthogonal *non-null* plane sections with $\dim L_i = n_i, i = 1, \ldots, k$. Notice that in this definition the plane sections $L_1, \ldots, L_k$ can be timelike or spacelike.

The following result was proved in [Haesen and Verstraelen (2004); Haesen (2005)].

Theorem 14.18. *Let N be a Riemannian or Lorentzian n-manifold. If N is isometrically embedded in a pseudo-Riemannian $(n+1)$-manifold M with diagonalizable Ricci tensor $\widetilde{Ric}$, (i.e., there exists an orthonormal basis $\{e_1, \dots, e_{n+1}\}$ of M such that $\widetilde{Ric} = \sum_{a=1}^{n+1} \lambda_a e_a \otimes e_a$). Then, for any k-tuple $(n_1, \dots, n_k) \in \mathcal{S}(n)$, we have*

$$c(n_1, \dots, n_k)||H||^2 \geq \Lambda(n_1, \dots, n_k) - \frac{1}{2}\left(\sum_{\alpha=1}^{n} \varepsilon_\alpha \lambda_\alpha - \lambda_{n+1}\right) \tag{14.53}$$

if $index(M) = index(N) + 1$, and

$$c(n_1, \dots, n_k)||H||^2 \leq \Lambda(n_1, \dots, n_k) - \frac{1}{2}\left(\sum_{\alpha=1}^{n} \varepsilon_\alpha \lambda_\alpha - \lambda_{n+1}\right) \tag{14.54}$$

if $index(M) = index(N)$.

Equality holds if and only if the shape operator, with respect to the eigenframe of the Ricci tensor $\widetilde{Ric}$, satisfies

$$A_{e_r} = \begin{pmatrix} \begin{matrix} A_1^r & \dots & 0 \\ \vdots & \ddots & \vdots \\ 0 & \dots & A_k^r \end{matrix} & 0 \\ 0 & \mu_r I_s \end{pmatrix},$$

where A_j^r is a symmetric $n_j \times n_j$-matrix with $\operatorname{trace}(A_j^r) = \mu_r$.

For an isometric embedding of a pseudo-Riemannian manifold N into another pseudo-Riemannian manifold M, we choose an orthonormal frame $\{e_1, \dots, e_n\}$ of tangent bundle and $\{e_{n+1}, \dots, e_m\}$ of the normal bundle. When N is Lorentzian, we shall assume that N is time-orientable, i.e., there exists a global nowhere-zero timelike vector field, denoted by e_n.

Definition 14.2. An embedding $\phi : N_1^n \to M_1^{n+1}$ is called *causal-type preserving* if $\nabla_{e_i} e_r$ is spacelike, $\forall\, r = n+1, \dots, m$ and $\forall\, i = 1, \dots, n$, where ∇ is the Levi-Civita connection of the ambient space.

Definition 14.3. An embedding $\phi : N_1^n \to M_{m-n+1}^m$ is called causal-type preserving if $\nabla_{e_n} e_r$ is timelike, $\forall\, r = n+1, \dots, m$.

It follows from definitions that causal-type preserving embeddings have $h_{in}^r = 0$, $\forall\, i = 1, \dots, n-1$.

Definition 14.4. An isometric embedding $\phi : N \to M$ is called an *ideal embedding* if and only if there exists a k-tuple $(n_1, \dots, n_k) \in \mathcal{S}(n)$ such that in a neighborhood U of a point $p \in N$ there is equality in (14.53) or (14.54), respectively.

Remark 14.14. Since the standard embedding of de Sitter spacetime and anti-de Sitter spacetimes in pseudo-Euclidean spaces are totally umbilical, both de Sitter and anti-de Sitter spacetimes admit ideal embedding. Further, by direct computation, one can show that the embeddings of Robertson-Walker spacetimes given in section 5.6 are also ideal.

In 1916, Einstein and Hilbert independently constructed Einstein's field equation for the pure gravitational field. The remaining part of his scientific career Einstein searched for an unification between his description of gravity and the other known forces, in particular electromagnetism. Several attempts have been made by using e.g. a non-symmetric metric, a connection with torsion, etc.

T. Kaluza noticed in 1919 that when he solved Einstein's equation for General Relativity using five dimensions, Maxwell's equations for electromagnetism emerged spontaneously. In 1926, O. Klein suggested that this fifth dimension would be compactified and unobservable on experimentally accessible energy scales.

This idea of compactifying the extra dimension has now dominated the search for a unified theory and lead to the 11D supergravity theory and more recently 10D superstring theory. Recently, this strategy of using higher dimensions to unify different forces is also an active area of research in particle physics.

Instead of compactifying the extra dimensions other approaches have also been developed during the last decade. For example, one particular variant of Kaluza-Klein theory is space-time-matter theory or induced matter theory, chiefly promulgated by P. Wesson and other members of the so-called Space-Time-Matter Consortium. In this version of the theory, it is noted that solutions to the equation $Ric = 0$, may be re-expressed so that in four dimensions, these solutions satisfy Einstein's field equation with the precise form of the energy-momentum tensor following from the Ricci-flat condition on the 5D space.

Since the energy-momentum tensor is normally understood to be due to concentrations of matter in 4D space, the above result can be interpreted as saying that 4D matter is induced from geometry in a Ricci-flat 5D space. In particular, the soliton solutions of $Ric = 0$ can be shown to contain the Robertson-Walker metric in both matter-dominated (early universe) and radiation-dominated (present universe) forms. The general equations can be shown to be sufficiently consistent with classical tests of general relativity to be acceptable on physical principles, while still leaving considerable freedom to also provide interesting cosmological models. In this way, we arrive back at an old idea of Einstein, namely geometrizing matter. Matter is induced on the four-dimensional spacetime by the properties of the fifth dimension.

The mathematical justification of this model is given by the Campbell-Magaard theorem which states that every pseudo-Riemannian manifold with analytic metric can be locally and isometrically embedded as a hypersurface in a Ricci-flat space (cf. [Campbell (1926)]). Several problems remain. For example, there are several ways to embed a given 4D spacetime in a 5D Ricci-flat manifold and vice versa, given a 5D Ricci-flat manifold we can extract different 4D spacetimes. The theory gives no criterion how to choose a particular embedding. In classical general relativity there exists a analogous problem. Namely, how to choose a particular solution of the field equations from the infinite solutions.

Using methods of submanifold theory we can give criterions to look for certain special solutions. For example, [Pavsić (1986)] and [Chervon et al. (2004)] consider the spacetime as a surface with $H = 0$ in a higher-dimensional embedding space.

As an extension, the above mentioned inequalities leading to the notion of ideal embeddings, in which the submanifold receives the least amount of tension from the surrounding space, are mathematically meaningful criterions with a physical interpretation.

It was proposed in [Haesen and Verstraelen (2004); Haesen (2005)] to apply Theorem 14.18 to study the following problems:

(1) *Look within a class of solutions of the Einstein field equations in 4D for those in which the 3D space is ideally embedded.*
(2) *Look for those spacetimes which are ideally embedded in a 5D Ricci-flat space.*

Finally, it maybe worth to point out that the explicit realization of Robertson-Walker spacetimes as affine hypersurfaces have been constructed in [Chen (2007a)]. Since many physics quantities are represented by the metric and its curvatures, most physical quantities of the Robertson-Walker spacetimes are preserved via the realizations. These embeddings allow us to study Robertson-Walker spacetimes and their submanifolds using affine differential geometry. Via such realization, if additional physical quantities were induced from the extra dimension via the embeddings it would then be neutral, in the sense that it is neither spacelike nor timelike.

Chapter 15

Applications to Kähler and Para-Kähler Geometry

15.1 A vanishing theorem for Lagrangian immersions

The following is a vanishing theorem for Lagrangian immersions [Chen (1998a)].

Theorem 15.1. *If a compact n-manifold N, $n \geq 2$, has finite fundamental group or null first Betti number, then each Lagrangian immersion of N into any Einstein-Kähler manifold has minimal points, i.e., the mean curvature vector must vanishes at some points.*

Proof. Let $\phi : N \to M$ be a Lagrangian immersion of a compact n-manifold N into a Kähler manifold M. Let us choose a local orthonormal frame $\{e_1, \ldots, e_n, e_{1^*}, \ldots, e_{n^*}\}$ such that $e_1, \ldots, e_n$ are tangent to N; thus $e_{1^*} = Je_1, \ldots, e_{n^*} = Je_n$ are normal to N. Denote by $\{\omega^1, \ldots, \omega^n\}$ the dual frame of $\{e_1, \ldots, e_n\}$.

From (2.89), (2.92) and (2.94) we have

$$\omega_i^{i^*} = \sum_{j=1}^{n} h_{ij}^{i^*} \omega^j, \tag{15.1}$$

$$d\omega_i^{j^*} = -\sum_k \omega_k^{j^*} \wedge \omega_i^k - \sum_k \omega_{k^*}^{j^*} \wedge \omega_i^{k^*} + \frac{1}{2} \sum_{k,\ell} K_{ik\ell}^{j^*} \omega^k \wedge \omega^\ell. \tag{15.2}$$

Also, by applying Lemma 9.2(1), we find $h_{jk}^{i^*} = h_{ik}^{j^*} = h_{ij}^{k^*}$.

Let Θ be the 1-form defined by

$$\Theta = \sum\nolimits_i \omega_i^{i^*}. \tag{15.3}$$

Then

$$\Theta = \sum_{i,j} h_{ij}^{i^*} \omega^j = \sum_{i,j} h_{ii}^{j^*} \omega^j = \sum_j (\operatorname{trace} h^{j^*}) \omega^j,$$

which gives

$$\Theta(X) = -n\,\langle JH, X\rangle\,, \quad \forall X \in TN. \tag{15.4}$$

By taking exterior differentiation of (15.3) and by applying (15.2), we find

$$d\Theta = \frac{1}{2}\sum\nolimits_{i,j,k} K^{i^*}_{ijk}\omega^j \wedge \omega^k. \tag{15.5}$$

By applying the first Bianchi identity and the curvature identities of Kähler manifolds from Proposition 1.6, we know that the curvature tensor of M satisfies

$$\begin{aligned}\sum_i K^{i^*}_{ijk} &= \sum_i R(e_{i^*}, e_i; e_j, e_k)\\ &= -\sum_i R(e_i, e_j; e_{i^*}, e_k) - \sum_i R(e_j, e_{i^*}; e_i, e_k)\\ &= -\sum_i R(e_j, e_i; e_i, e_{k^*}) + \sum_i R(e_{j^*}, e_i; e_i, e_k)\\ &= \sum_i R(e_{j^*}, e_{i^*}; e_{i^*}, e_k) - \sum_i R(e_j, e_{i^*}; e_{i^*}, e_{k^*}).\end{aligned}$$

Thus

$$\sum_i K^{i^*}_{ijk} = \frac{1}{2}\widetilde{Ric}(e_{j^*}, e_k) - \frac{1}{2}\widetilde{Ric}(e_j, e_{k^*}) = \widetilde{Ric}(e_{j^*}, e_k), \tag{15.6}$$

where $\widetilde{Ric}$ is the Ricci tensor of M. Now, assume that M is Einstein-Kähler. Then (15.6) yields $d\Theta = 0$. Consequently, Θ defines a cohomology class $[\Theta] \in H^1(N; \mathbf{R})$. Next, let us assume that either N has finite fundamental group $\pi_1(N)$ or it has null first Betti number $b_1(N)$.

Case (1): $b_1(N) = 0$. If the mean curvature vector H of N does not vanish at every point, then (15.4) implies that Θ is nowhere zero. Thus, Θ cannot be an exact 1-form, since M is compact. Therefore, the first cohomology group $H^1(N; \mathbf{R})$ of N is non-trivial. Thus, Poincaré's duality theorem implies that the first Betti number of N is non-trivial, which is a contradiction. Consequently, the mean curvature vector of N must vanishes at some points.

Case (2): $\pi_1(N) = 0$. If the mean curvature vector H does not vanish at every point, then the first cohomology group $H^1(N; \mathbf{R})$ of N is non-trivial just as in case (1). Hence, N is not simply-connected, which is a contradiction.

Case (3): $\pi_1(N) \neq 0$ *and* $\pi_1(N)$ *is finite.* Let $\pi : \hat{N} \to N$ be the universal covering map of N. Then $\hat{N}$ is also a compact n-manifold, since $\pi_1(N)$ is finite. Consider the composition $\hat{\phi} = \phi \circ \pi : \hat{N} \to N \to M$. Then $\hat{\phi}$ is a Lagrangian immersion of $\hat{N}$ into the Einstein-Kähler manifold M, since ϕ does. Because $\hat{N}$ is simply-connected, case (2) implies that $\hat{\phi}$ has at least one minimal point, say $\hat{p}$. Clearly, $p = \pi(\hat{p})$ is also a minimal point of ϕ. Thus, N must have some minimal points in N. □

Remark 15.1. Theorem 15.1 is sharp. This can be seen from the following example. Consider the standard imbedding

$$\phi : T^n = S^1 \times \cdots \times S^1 \to \mathbf{C}^n = \mathbf{C} \times \cdots \times \mathbf{C}.$$

With respect to the Kählerian structure of $\mathbf{C}^n$, T^n is a compact Lagrangian submanifold which have nonzero constant mean curvature. Clearly, T^n has non-trivial first Betti number and infinite fundamental group $\pi_1(T^n)$.

The following are some easy consequences of Theorem 15.1.

Corollary 15.1. *There do not exist Lagrangian immersions of nonzero constant mean curvature from a topological n-sphere into any Einstein-Kähler manifold.*

Proof. Follows immediately from Theorem 15.1. □

Corollary 15.2. *Every Lagrangian immersion of constant mean curvature from a symmetric space of compact type into an Einstein-Kähler manifold is a minimal Lagrangian immersion.*

Proof. Follows immediately from Theorem 15.1. □

Corollary 15.3. *If N is a complete Riemannian manifold whose Ricci curvature satisfies $Ric(u) \geq c^2 > 0$ for any unit vector $u \in T^1N$, where c is a positive number, then every Lagrangian immersion of constant mean curvature from N into an Einstein-Kähler manifold is a minimal immersion.*

Proof. This follows from Theorem 15.1 and a result of [Myers (1941)] which states that every complete Riemannian manifold whose Ricci curvature satisfies $Ric(u) \geq c^2 > 0, u \in T^1N$, is a compact manifold with finite fundamental group. □

A compact Riemannian manifold of positive constant sectional curvature is called a *spherical space form.*

Applying Theorem 15.1 and Gauss's equation, we have the following.

Corollary 15.4. *Every Lagrangian isometric immersion of constant mean curvature from a spherical space form into a complex projective space of constant holomorphic sectional curvature is totally geodesic.*

Proof. If N is a spherical space form, it has finite fundamental group. Thus, every Lagrangian immersion of constant mean curvature from N into a complex projective n-space CP^n is a minimal immersion by Theorem 15.1. Consequently, the corollary follows from a result of [Chen and Ogiue (1974a)] which states that the only Lagrangian isometric minimal immersion of a Riemannian n-manifold of positive constant sectional curvature into CP^n is the totally geodesic one. □

Corollary 15.5. *There do not exist Lagrangian isometric immersions from a compact Riemannian manifold with positive Ricci curvature into any flat Kähler manifold or into any complex hyperbolic space.*

Proof. Suppose that N is a compact Riemannian manifold with positive Ricci curvature. If $\phi : N \to M$ is a Lagrangian isometric immersion of N into a Kähler manifold M of constant holomorphic sectional curvature $c \leq 0$, then, by Theorem 15.1, ϕ has at least one minimal point, say $p \in N$. Thus, after applying the equation of Gauss we conclude that the Ricci tensor of N is negative semi-definite at p which is a contradiction. □

15.2 Obstructions to Lagrangian isometric immersions

A result of [Gromov (1970)] states that a compact n-manifold N admits a Lagrangian immersion into the complex Euclidean n-space $\mathbf{C}^n$ if and only if the complexification $TN \otimes \mathbf{C}$ of the tangent bundle of N is trivial. This result implies that there are no topological obstructions to Lagrangian immersions for compact 3-manifolds in $\mathbf{C}^3$, because the tangent bundle of every 3-manifold is always trivial.

On the other hand, from the Riemannian geometric point of view, one may ask the following basic question.

Problem 15.1. *What are the necessary conditions for a compact Riemannian manifold to admit a Lagrangian isometric immersion into* $\mathbf{C}^n$*?*

Corollary 15.5 provides a solution to this problem in term of Ricci curvature. Another application of δ-invariants is to provide further solutions to this problem.

Theorem 15.2. *Let N be a compact Riemannian n-manifold with null first Betti number or with finite fundamental group. If there exists a k-*

tuple $(n_1,\ldots,n_k)\in\mathcal{S}(n)$ *such that* $\delta(n_1,\ldots,n_k)>0$, *then* N *does not admit any Lagrangian isometric immersion into* $\mathbf{C}^n$.

Proof. Let N be a compact Riemannian n-manifold with null first Betti number or with finite fundamental group. Let us assume that N admits a Lagrangian isometric immersion into $\mathbf{C}^n$. If N satisfies $\delta(n_1,\ldots,n_k)>0$ for some k-tuple $(n_1,\ldots,n_k)\in\mathcal{S}(n)$, then Theorem 13.7 implies that the mean curvature vector H of N is nowhere zero.

Let Φ be the 1-form associated with JH, i.e., $\Phi(X)=\langle JH,X\rangle$ for each $X\in TN$. Then Φ is nowhere zero. It follows from the proof of Theorem 15.1 that Φ is a closed 1-form. Thus, Φ represents a cohomology class $[\Phi]\in H^1(N;\mathbf{R})$. Because N is compact and Φ is nowhere zero, Φ cannot be exact. Therefore, $[\Phi]$ is a non-trivial cohomology class which implies that the first cohomology group $H^1(N;\mathbf{R})$ is non-trivial. Hence, $b_1(N)\neq 0$; thus N is not simply-connected.

Since N is a compact Riemannian n-manifold with $\delta(n_1,\ldots,n_k)>0$ and with finite fundamental group $\pi_1(N)\neq 0$, the universal Riemannian covering map lifts the Lagrangian immersion of N to a Lagrangian immersion for the universal Riemannian covering space. Since N is compact and $\pi_1(N)$ is finite, the universal covering space is also compact. Because the covering map preserves the invariant $\delta(n_1,\ldots,n_k)$, the same argument as previous case applied to the universal covering yields a contradiction. □

Remark 15.2. The condition on $\delta(n_1,\ldots,n_k)$ given in Theorem 15.2 is sharp. This can be seen as follows: Consider the *Whitney sphere* W^n defined by the Whitney immersion $w:S^n\to\mathbf{C}^n$ given by

$$w(y_0,y_1,\ldots,y_n)=\frac{1+\mathrm{i}\,y_0}{1+y_0^2}(y_1,\ldots,y_n), \tag{15.7}$$

with $y_0^2+y_1^2+\cdots+y_n^2=1$. The Whitney immersion is a Lagrangian immersion with a unique self-intersection point at $w(-1,0,\ldots,0)=w(1,0,\ldots,0)$. For each n-tuple $(n_1,\ldots,n_k)\in\mathcal{S}(n)$, we have $\delta(n_1,\ldots,n_k)\geq 0$; and $\delta(n_1,\ldots,n_k)=0$ occurs only at the unique point of self-intersection.

Remark 15.3. The assumptions on the finiteness of $\pi_1(N)$ and vanishing of $b_1(N)$ given in Theorem 15.2 are both necessary for $n\geq 3$. This can be seen from the following example: Let $F:S^1\to\mathbf{C}$ be the unit circle in the complex plane given by $F(s)=e^{\mathrm{i}s}$ and let $\iota:S^{n-1}\to\mathbb{E}^n$ $(n\geq 3)$ be the unit hypersphere in $\mathbb{E}^n$ centered at the origin. Denote by $\phi:S^1\times S^{n-1}\to\mathbf{C}^n$ the *complex extensor* defined by [Chen (1997a)]

$$\phi(s,p)=F(s)\otimes\iota(p),\ \ p\in S^{n-1}.$$

Then ϕ is a Lagrangian isometric immersion of $N = S^1 \times S^{n-1}$ into $\mathbf{C}^n$ which carries each pair $\{(u,p), (-u,-p)\}$ of points in $S^1 \times S^{n-1}$ to a point in $\mathbf{C}^n$. Clearly, $\pi_1(N) = \mathbb{Z}$ and $b_1(N) = 1$. Moreover, for each k-tuple $(n_1, \ldots, n_k) \in \mathcal{S}(n)$, the δ-invariant $\delta(n_1, \ldots, n_k)$ is a positive constant. This example shows that both conditions on $\pi_1(N)$ and $b_1(N)$ cannot be removed.

Similarly, by applying Corollaries 13.3 and 13.6, we have the following two results by using the same arguments as Theorem 15.2.

Corollary 15.6. *Let N be a compact Riemannian n-manifold with finite fundamental group or with null first Betti number. If*

$$\delta(n_1, \ldots, n_k) > \frac{1}{2}\sum_{j=1}^{k} n_j(n_j - 1) - \frac{1}{2}n(n-1)$$

for some k-tuple $(n_1, \ldots, n_k) \in \mathcal{S}(n)$, then N does not admit a Lagrangian isometric immersion into the complex hyperbolic n-space $CH^n(-4)$.

Corollary 15.7. *Let N be a compact Riemannian n-manifold with finite fundamental group or with null first Betti number. If*

$$\delta(n_1, \ldots, n_k) > \frac{1}{2}n(n-1) - \frac{1}{2}\sum_{j=1}^{k} n_j(n_j - 1)$$

for some k-tuple $(n_1, \ldots, n_k) \in \mathcal{S}(n)$, then N does not admit a Lagrangian isometric immersion in the complex projective n-space $CP^n(4)$.

15.3 Improved inequalities for Lagrangian submanifolds

Let N be a Lagrangian submanifold of a complex space form $M^n(4c)$. Then inequality (13.44) holds. If we let T be the curvature tensor of N and let μ be the second fundamental form in Theorem 13.13, then we obtain the following result from Theorems 13.6 and 13.13.

Theorem 15.3. *Let N be a Lagrangian submanifold of a complex space from $M^n(4c)$ of constant holomorphic sectional curvature $4c$. Then, for any k-tuple $(n_1, \ldots, n_k) \in \mathcal{S}(n)$, we have*

$$\delta(n_1, \ldots, n_k) \leq c(n_1, \ldots, n_k)||H||^2 + b(n_1, \ldots, n_k)c. \tag{15.8}$$

If the equality of (15.8) *holds identically for some k-tuple $(n_1, \ldots, n_k) \in \mathcal{S}(n)$, then N is a minimal submanifold.*

Recently, inequality (15.8) was improved in [Chen and Dillen (2010)] to the following.

Theorem 15.4. *Let N be a Lagrangian submanifold of a complex space form $M^n(4c)$. Then, for any k-tuple $(n_1, \ldots, n_k) \in \mathcal{S}(n)$, we have*

$$\begin{aligned}\delta(n_1,\ldots,n_k) \leq &\frac{n^2\Big\{n - \sum_{i=1}^k n_i + 3k - 1 - 6\sum_{i=1}^k (2+n_i)^{-1}\Big\}}{2\Big\{n - \sum_{i=1}^k n_i + 3k + 2 - 6\sum_{i=1}^k (2+n_i)^{-1}\Big\}}||H||^2 \\ &+ \frac{1}{2}\left\{n(n-1) - \sum_{i=1}^k n_i(n_i-1)\right\}c.\end{aligned} \tag{15.9}$$

The equality sign of (15.9) holds at a point $p \in N$ if and only if there is an orthonormal basis $\{e_1, \ldots, e_n\}$ at p such that with respect to this basis the second fundamental form h takes the following form

$$\begin{aligned} &h(e_{\alpha_i}, e_{\beta_i}) = \sum_{\gamma_i} h^{\gamma_i}_{\alpha_i\beta_i} Je_{\gamma_i} + \frac{3\delta_{\alpha_i\beta_i}}{2+n_i}\lambda Je_{\mu+1}, \quad \sum_{\alpha_i=1}^{n_i} h^{\gamma_i}_{\alpha_i\alpha_i} = 0, \\ &h(e_{\alpha_i}, e_{\alpha_j}) = 0, \quad i \neq j, \\ &h(e_{\alpha_i}, e_{\mu+1}) = \frac{3\lambda}{2+n_i} Je_{\alpha_i}, \quad h(e_{\alpha_i}, e_u) = 0, \\ &h(e_{\mu+1}, e_{\mu+1}) = 3\lambda Je_{\mu+1}, \quad h(e_{\mu+1}, e_u) = \lambda Je_u, \\ &h(e_u, e_v) = \lambda\delta_{uv} Je_{\mu+1}, \end{aligned} \tag{15.10}$$

for $1 \leq i, j \leq k$; $\mu+2 \leq u, v \leq n$ and $\lambda = \frac{1}{3}h^{\mu+1}_{\mu+1\mu+1}$, where $\mu = n_1 + \cdots + n_k$.

Proof. Let $(n_1, \ldots, n_k) \in \mathcal{S}(n)$, $p \in N$, and let $L_1, \ldots, L_k$ be k mutually orthogonal subspaces of T_pN with $\dim L_j = n_j$, $j = 1, \ldots, k$. We choose an orthonormal basis $\{e_1, \ldots, e_n\}$ at a point p such that

$$L_j = \text{Span}\,\{e_{n_1+\cdots+n_{j-1}+1}, \ldots, e_{n_1+\cdots+n_j}\}, \quad j = 1, \ldots, k. \tag{15.11}$$

For simplicity let us put $\mu = n_1 + \cdots + n_k$. We also put

$$\begin{aligned} \Delta_1 &= \{1, \ldots, n_1\}, \\ &\;\;\vdots \\ \Delta_k &= \{n_1 + \cdots + n_{k-1} + 1, \ldots, \mu\}, \\ \Delta_{k+1} &= \{\mu+1, \ldots, n\}. \end{aligned} \tag{15.12}$$

Now, we use of the following convention on the ranges of indices unless mentioned otherwise:

$$\begin{aligned} &\alpha_i, \beta_i, \gamma_i \in \Delta_i, \quad i, j \in \{1, \ldots, k\}; \quad A, B, C \in \{1, \ldots, n\}; \\ &r, s, t \in \Delta_{k+1}; \quad u, v \in \{\mu+2, \ldots, n\}. \end{aligned} \tag{15.13}$$

Since

$$\tau=\sum_{A=1}^{n}\sum_{B<C}(h^A_{BB}h^A_{CC}-(h^A_{BC})^2), \tag{15.14}$$

$$\tau(L_i)=\sum_{A}\sum_{\alpha_i<\beta_i}(h^A_{\alpha_i\alpha_i}h^A_{\beta_i\beta_i}-(h^A_{\alpha_i\beta_i})^2), \tag{15.15}$$

we have

$$\begin{aligned}\tau-\sum_{i=1}^{k}\tau(L_i)&=\sum_{A}\sum_{i<j}\sum_{\alpha_i,\alpha_j}(h^A_{\alpha_i\alpha_i}h^A_{\alpha_j\alpha_j}-(h^A_{\alpha_i\alpha_j})^2)\\&+\sum_{A}\sum_{r<s}(h^A_{rr}h^A_{ss}-(h^A_{rs})^2)+\sum_{A,i}\sum_{\alpha_i,r}(h^A_{\alpha_i\alpha_i}h^A_{rr}-(h^A_{\alpha_i r})^2)\\&\leq\sum_{A}\Bigg\{\sum_{r<s}h^A_{rr}h^A_{ss}+\sum_{i}\sum_{\alpha_i,r}h^A_{\alpha_i\alpha_i}h^A_{rr}+\sum_{i<j}\sum_{\alpha_i,\alpha_j}h^A_{\alpha_i\alpha_i}h^A_{\alpha_j\alpha_j}\Bigg\}\\&\qquad-\sum_{i}\sum_{\alpha_i,s}(h^{\alpha_i}_{ss})^2-\sum_{B\neq r}\sum_{r=\mu+1}^{n}(h^r_{BB})^2,\end{aligned} \tag{15.16}$$

with the equality sign holding if and only if

$$h^{\alpha_i}_{\alpha_j\alpha_\ell}=h^{\alpha_j}_{\alpha_i\beta_i}=h^r_{\alpha_i\alpha_j}=h^{\alpha_i}_{st}=h^r_{st}=0 \tag{15.17}$$

for distinct $i,j,\ell\in\{1,\dots,k\}$ and distinct $r,s,t\in\Delta_{k+1}$.

For a given $i\in\{1,\dots,k\}$ and a given $\gamma_i\in\Delta_i$, we have

$$\begin{aligned}0&\leq\sum_{j=1}^{k}\sum_{r\in\Delta_{k+1}}^{n}\Bigg(\sum_{\alpha_j\in\Delta_j}h^{\gamma_i}_{\alpha_j\alpha_j}-3h^{\gamma_i}_{rr}\Bigg)^2+3\sum_{r<s}(h^{\gamma_i}_{rr}-h^{\gamma_i}_{ss})^2\\&\quad+3\sum_{\ell<j}\Bigg(\sum_{\alpha_\ell\in\Delta_\ell}h^{\gamma_i}_{\alpha_\ell\alpha_\ell}-\sum_{\alpha_j\in\Delta_j}h^{\gamma_i}_{\alpha_j\alpha_j}\Bigg)^2\\&=(n-\mu+3k-3)\sum_{j}\Bigg(\sum_{\alpha_j}h^{\gamma_i}_{\alpha_j\alpha_j}\Bigg)^2-6\sum_{j}\sum_{\alpha_j,r}h^{\gamma_i}_{\alpha_j\alpha_j}h^{\gamma_i}_{rr}\\&\quad-6\sum_{r<s}h^{\gamma_i}_{rr}h^{\gamma_i}_{ss}-6\sum_{\ell<j}h^{\gamma_i}_{\alpha_\ell\alpha_\ell}\sum_{\alpha_j\in\Delta_j}h^{\gamma_i}_{\alpha_j\alpha_j}\\&\quad+3(n-\mu+3k-1)\sum_{r}(h^{\gamma_i}_{rr})^2\\&=(n-\mu+3k-3)(h^{\gamma_i}_{11}+\cdots+h^{\gamma_i}_{nn})^2-2(n-\mu+3k)\times\\&\quad\Bigg\{\sum_{r<s}h^{\gamma_i}_{rr}h^{\gamma_i}_{ss}+\sum_{j=1}^{k}\sum_{\alpha_j,r}h^{\gamma_i}_{\alpha_j\alpha_j}h^{\gamma_i}_{rr}+\sum_{\ell<j}h^{\gamma_i}_{\alpha_\ell\alpha_\ell}h^{\gamma_i}_{\alpha_j\alpha_j}-\sum_{s}(h^{\gamma_i}_{ss})^2\Bigg\}.\end{aligned}$$

Thus we find

$$\begin{aligned}\sum_{r<s} h^{\gamma_i}_{rr} h^{\gamma_i}_{ss} + \sum_{j=1}^{k}\sum_{\alpha_j, r} h^{\gamma_i}_{\alpha_j\alpha_j} h^{\gamma_i}_{rr} + \sum_{\ell<j} h^{\gamma_i}_{\alpha_\ell\alpha_\ell} h^{\gamma_i}_{\alpha_j\alpha_j} - \sum_{s} (h^{\gamma_i}_{ss})^2 \\ \le \frac{n-\mu+3k-3}{2(n-\mu+3k)} (h^{\gamma_i}_{11} + \cdots + h^{\gamma_i}_{nn})^2,\end{aligned} \tag{15.18}$$

with the equality holding if and only if

$$\sum_{\alpha_j\in\Delta_j} h^{\gamma_i}_{\alpha_j\alpha_j} = 3h^{\gamma_i}_{ss}, \quad j=1,\ldots,k, \quad s\in\Delta_{k+1}. \tag{15.19}$$

Since

$$\frac{n-\mu+k-1}{n-\mu+k+2} < \frac{n-\mu+3k-1-6\sum_{i=1}^{k}(2+n_i)^{-1}}{n-\mu+3k+2-6\sum_{i=1}^{k}(2+n_i)^{-1}}.$$

We obtain from (15.18) that

$$\begin{aligned}\sum_{r<s} h^{\gamma_i}_{rr} h^{\gamma_i}_{ss} + \sum_{j=1}^{k}\sum_{\alpha_j, r} h^{\gamma_i}_{\alpha_j\alpha_j} h^{\gamma_i}_{rr} + \sum_{\ell<j} h^{\gamma_i}_{\alpha_\ell\alpha_\ell} h^{\gamma_i}_{\alpha_j\alpha_j} - \sum_{s} (h^{\gamma_i}_{ss})^2 \\ \le \frac{n-\mu+3k-1-6\sum_{i=1}^{k}(2+n_i)^{-1}}{2\{n-\mu+3k+2-6\sum_{i=1}^{k}(2+n_i)^{-1}\}} \left(\sum_{A=1}^{n} h^{\gamma_i}_{AA}\right)^2,\end{aligned} \tag{15.20}$$

with the equality sign holding if and only if

$$\sum_{\alpha_j\in\Delta_j} h^{\gamma_i}_{\alpha_j\alpha_j} = 3h^{\gamma_i}_{ss} = 0, \quad j=1,\ldots,k, \quad s\in\Delta_{k+1} \tag{15.21}$$

for $i \in \{1,\ldots,k\}$. Put

$$w = \frac{2}{3}\left\{n-\mu+3k+2-\sum_{j=1}^{k}\frac{6}{2+n_j}\right\}.$$

From

$$\begin{aligned}\sum_{i=1}^{k}\frac{n_i}{2+n_i} &= k - \sum_{i=1}^{k}\frac{2}{2+n_i},\\ \sum_{j\neq i}\frac{n_j}{2+n_j} &= k - \sum_{j}\frac{2}{2+n_j} - \frac{n_i}{2+n_i},\end{aligned}$$

we find, for each $t \in \{\mu+1,\ldots,n\}$, that

$$0 \le \sum_{i}\sum_{r\neq t}\frac{2+n_i}{3n_i}\left(\sum_{\alpha_i} h^t_{\alpha_i\alpha_i} - \frac{3n_i}{2+n_i} h^t_{rr}\right)^2$$

$$+\sum_i \sum_{\alpha_i<\beta_i} \frac{w}{n_i}(h^t_{\alpha_i\alpha_i}-h^t_{\beta_i\beta_i})^2+\sum_{\substack{r<s\\ r,s\neq t}}(h^t_{rr}-h^t_{ss})^2$$

$$+\sum_{i<j}\left(\frac{\sqrt{(2+n_i)n_j}}{\sqrt{(2+n_j)n_i}}\sum_{\alpha_i}h^t_{\alpha_i\alpha_i}-\frac{\sqrt{(2+n_j)n_i}}{\sqrt{(2+n_i)n_j}}\sum_{\alpha_j}h^t_{\alpha_j\alpha_j}\right)^2$$

$$+\frac{1}{3}\sum_{r\neq t}(h^t_{tt}-3h^t_{rr})^2+\sum_i\frac{n_i}{2+n_i}\left(h^t_{tt}-\frac{2+n_i}{n_i}\sum_{\alpha_i}h^t_{\alpha_i\alpha_i}\right)^2$$

$$=\sum_i\left\{(n-\mu+2)\frac{2+n_i}{3n_i}-\frac{w}{n_i}+\sum_{j\neq i}\frac{(2+n_i)n_j}{(2+n_j)n_i}\right\}\left(\sum_{\alpha_i}h^t_{\alpha_i\alpha_i}\right)^2$$

$$-2\sum_i\sum_{\alpha_i}h^t_{\alpha_i\alpha_i}h^t_{rr}+\left\{n-\mu+1+\sum_i\frac{3n_i}{2+n_i}\right\}\sum_{r\neq t}(h^t_{rr})^2$$

$$+w\sum_i\sum_{\alpha_i}(h^t_{\alpha_i\alpha_i})^2-2\sum_{i<j}h^t_{\alpha_i\alpha_i}h^t_{\alpha_j\alpha_j}-2\sum_{\substack{r<s\\ r,s\neq t}}h^t_{rr}h^t_{ss}$$

$$+\left\{\frac{n-\mu-1}{3}+\sum_i\frac{n_i}{2+n_i}\right\}(h^t_{tt})^2-2h^t_{tt}\sum_{r\neq t}h^t_{rr}-2h^t_{tt}\sum_i\left(\sum_{\alpha_i}h^t_{\alpha_i\alpha_i}\right)^2$$

$$=\frac{1}{3}\left\{n-\mu+3k-1-\sum_i\frac{6}{2+n_i}\right\}\left(\sum_{\alpha_i}h^t_{\alpha_i\alpha_i}\right)^2-2\sum_i\sum_{\alpha_i}h^t_{\alpha_i\alpha_i}h^t_{rr}$$

$$+\left\{n-\mu+3k+1-\sum_i\frac{6}{2+n_i}\right\}\sum_{r\neq t}(h^t_{rr})^2-2h^t_{tt}\sum_i\left(\sum_{\alpha_i}h^t_{\alpha_i\alpha_i}\right)^2$$

$$-2\sum_{i<j}h^t_{\alpha_i\alpha_i}h^t_{\alpha_j\alpha_j}-2\sum_{\substack{r<s\\ r,s\neq t}}h^t_{rr}h^t_{ss}-2h^t_{tt}\sum_{r\neq t}h^t_{rr}+w\sum_i\sum_{\alpha_i}(h^t_{\alpha_i\alpha_i})^2$$

$$+\frac{1}{3}\left\{n-\mu+3k-1-\sum_i\frac{6}{2+n_i}\right\}(h^t_{tt})^2$$

$$=w\left\{\frac{n-\mu+3k-1-6\sum_{i=1}^k(2+n_i)^{-1}}{2\{n-\mu+3k+2-6\sum_{i=1}^k(2+n_i)^{-1}\}}\left(\sum_{A=1}^n h^t_{AA}\right)^2\right.$$

$$-\sum_{r<s}h^t_{rr}h^t_{ss}-\sum_{i<j}\sum_{\alpha_i,\alpha_j}h^t_{\alpha_i\alpha_i}h^t_{\alpha_j\alpha_j}-\sum_i\sum_{\alpha_i,r}h^t_{\alpha_i\alpha_i}h^t_{rr}$$

$$\left.+\sum_{s\neq t}(h^t_{ss})^2+\sum_i\sum_{\alpha_i}(h^t_{\alpha_i\alpha_i})^2\right\}.$$

Hence, we find

$$\sum_{r<s} h^t_{rr}h^t_{ss} + \sum_i \sum_{\alpha_i, r} h^t_{\alpha_i\alpha_i}h^t_{rr} + \sum_{i<j}\sum_{\alpha_i,\alpha_j} h^t_{\alpha_i\alpha_i}h^t_{\alpha_j\alpha_j} - \sum_{B\neq t}(h^t_{BB})^2$$
$$\leq \frac{n-\mu+3k-1-6\sum_{i=1}^k (2+n_i)^{-1}}{2\{n-\mu+3k+2-6\sum_{i=1}^k (2+n_i)^{-1}\}}\left(\sum_{A=1}^n h^r_{AA}\right)^2, \tag{15.22}$$

with equality holding if and only if

$$h^t_{tt} = (2+n_i)h^t_{\alpha_i\alpha_i} = 3h^t_{ss}, \;\; i=1,\ldots,k, \;\; \mu+1 \leq s \neq t \leq n. \tag{15.23}$$

By combining (15.16), (15.20) and (15.22), we obtain inequality (15.9).

Equality in (15.9) implies that the inequalities (15.16), (15.20) and (15.22) become equalities. Thus, we have

$$\sum_{\alpha_j\in\Delta_j} h^{\gamma_i}_{\alpha_j\alpha_j} = 3h^{\gamma_i}_{ss} = 0, \;\; j=1,\ldots,k, \;\; s\in\Delta_{k+1}, \tag{15.24}$$

$$h^t_{tt} = (2+n_i)h^t_{\alpha_i\alpha_i} = 3h^t_{ss}, \;\; i=1,\ldots,k, \tag{15.25}$$

$$h^{\alpha_i}_{\alpha_j\alpha_\ell} = h^{\alpha_j}_{\alpha_i\beta_i} = h^r_{\alpha_i\alpha_j} = h^{\alpha_i}_{st} = h^r_{st} = 0, \;\; \mu+1\leq s\neq t\leq n, \tag{15.26}$$

for distinct $i,j,\ell \in \{1,\ldots,k\}$ and distinct $r,s,t\in\Delta_{k+1}$.

It follows from (15.21) that the mean curvature vector H lies in $\mathrm{Span}\{e_{k+1},\ldots,e_n\}$. Therefore, we may choose e_{k+1} in the direction of H. Hence, we conclude that conditions (15.24)-(15.26) are equivalent to (15.10) due to the totally symmetry of h. □

An immediate consequence of Theorem 15.4 is the following.

Corollary 15.8. *Let N be a Lagrangian submanifold of a complex space form $M^n(4c)$. Then the scalar curvature of N satisfies*

$$\tau \leq \frac{n^2(n-1)}{2(n+2)}||H||^2 + \frac{n(n-1)}{2}c. \tag{15.27}$$

The equality sign of (15.27) holds identically if and and only if N is a Lagrangian H-umbilical submanifold with $\lambda = 3\mu$.

Proof. If we choose $k=0$, then (15.10) reduces to (15.27). The remaining part follows immediately from (15.11). □

Remark 15.4. Inequality (15.27) with $c=0$ and $n=2$ was proved in [Castro and Urbano (1993)]. The general inequality for $c=0$ and arbitrary n was established in [Borrelli et al. (1995)]; and for $c\neq 0$ and arbitrary n in [Chen (1996d); Chen and Vrancken (1996)]; independently in [Castro and Urbano (1995)] for $c\neq 0$ with $n=2$. Also, Lagrangian H-submanifolds with $\lambda = 3\mu$ in complex space forms have been completely determined. For instance, it was proved in [Borrelli et al. (1995); Ros and Urbano (1998)] that, up to dilations, a Lagrangian H-submanifold with $\lambda = 3\mu$ in $\mathbf{C}^n$ is an open portion of the Whitney n-sphere.

Another consequence of Theorem 15.4 is the following [Oprea (2005)].

Corollary 15.9. *Let N be a Lagrangian submanifold of a complex space form $M^n(4c)$. Then the Ricci curvature of N satisfies*

$$Ric(u) \leq \frac{n(n-1)}{4}||H||^2 + (n-1)c \tag{15.28}$$

for each $u \in T^1N$.

Proof. If we choose $k = 1$ and $n_1 = n - 1$, then (15.38) becomes

$$\delta(n-1) \leq \frac{n(n-1)}{4}||H||^2 + (n-1)c. \tag{15.29}$$

Since $\delta(n-1) = \max\limits_{u \in T^1N} Ric(u)$, (15.29) implies (15.28). □

The following proposition is due to [Deng (2009)].

Proposition 15.1. *Let N be a Lagrangian submanifold of a complex space form $M^n(4c)$. Then the Ricci curvature of N satisfies*

$$Ric(u) = \frac{n(n-1)}{4}||H||^2 + (n-1)c \tag{15.30}$$

for each $u \in T^1N$ if and only if either N is totally geodesic or N is Lagrangian H-umbilical with $\lambda = 3\mu$.

Remark 15.5. When $k = 1$ and $n_1 = 2$, inequality (15.9) is due to [Oprea (2007)]. Submanifolds in complex space forms satisfying the equality case of (15.9) for $k = 1, n_1 = 2$ have been investigated rather detailed in [Bolton et al. (2009,a); Bolton and Vrancken (2007)].

Consider the direct product $\mathbb{E}^n \times \mathbb{E}^n$ of two Euclidean n-spaces with natural Euclidean metric and the natural coordinates $(x_1, \ldots, x_n, y_1, \ldots, y_n)$. The product $\mathbb{E}^n \times \mathbb{E}^n$ has a natural complex structure J defined by $J\frac{\partial}{\partial x_i} = \frac{\partial}{\partial y_i}$, $J\frac{\partial}{\partial y_i} = -\frac{\partial}{\partial x_i}$, $i = 1, \ldots, n$. We denote the pair $(\mathbb{E}^n \times \mathbb{E}^n, J)$ by $\mathbf{C}^n$, which is the complex Euclidean n-space.

Consider the graph in $\mathbf{C}^n = (\mathbb{E}^n \times \mathbb{E}^n, J)$ defined by a smooth map $f = (f_1, \ldots, f_n) : D \to \mathbb{E}^n$ on an open domain $D \subset \mathbb{E}^n$. Then the graph

$$L(x_1, \ldots, x_n) = (x_1, \ldots, x_n, f_1(x_1, \ldots, x_n), \ldots, f_n(x_1, \ldots, x_n))$$

is a Lagrangian submanifold of $\mathbf{C}^n$ if and only if $\left(\frac{\partial f^i}{\partial x_j}\right)$ is a symmetric matrix. In particular, if D is simply connected, then there exists a function $F : D \to \mathbf{R}$ with $f = \nabla F$. Hence, the Lagrangian graph takes the form

$$L(x_1, \dots, x_n) = \big(x_1, \dots, x_n, F_{x_1}, \dots, F_{x_n}\big) \tag{15.31}$$

with $F_{x_i} = \partial F/\partial x_i$ for $i = 1, \dots, n$.

The next theorem shows that, for each k-tuple $(n_1, \dots, n_k) \in \mathcal{S}(n)$, inequality (15.9) is the best possible [Chen and Dillen (2010)].

Theorem 15.5. *For each k-tuple $(n_1, \dots, n_k) \in \mathcal{S}(n)$, there exists a non-minimal Lagrangian submanifold in $\mathbf{C}^n$ which satisfies the equality case of (15.9) at a point.*

Proof. For a given real number $\lambda > 0$, consider the Lagrangian graph

$$L(x_1, \dots, x_n) = \big(x_1, \dots, x_n, F_{x_1}, \dots, F_{x_n}\big) \tag{15.32}$$

with

$$F = \sum_{i=1}^{k} \frac{3\lambda}{2(2 + n_i)} \sum_{\alpha_i \in \Delta_i} x_{\alpha_i}^2 x_{\mu+1} + \frac{\lambda}{2} \sum_{r=\mu+1}^{n} x_{\mu+1} x_r^2 \tag{15.33}$$

with $\mu = \sum_{i=1}^{k} n_i$. Then $e_1 = (1, 0, \dots, 0), \dots, e_n = (0, \dots, 0, \overbrace{1}^{n-th}, 0, \dots, 0)$ form an orthonormal basis of T_oN at $o = (0, \dots, 0)$. It is easy to verify that the coefficients of the second fundamental form h satisfies

$$h_{AB}^{C} = \frac{\partial^3 F}{\partial x_A \partial x_B \partial x_C}, \quad A, B, C = 1, \dots, n.$$

Thus, it follows from (15.33) that the second fundamental form satisfies (15.10) at o. □

From [Chen and Dillen (2010)] we also have the following.

Theorem 15.6. *Let N be a Lagrangian submanifold of a complex space form $M^n(4c)$. Then for any integer $n_1 \in [2, n-1]$ we have*

$$\begin{aligned}\delta(n_1) \leq\ & \frac{n^2\{n_1(n - n_1) + 2n - 2\}}{2\{n_1(n - n_1) + 2n + 3n_1 + 4\}} ||H||^2 \\ & + \frac{1}{2}\{n(n-1) - n_1(n_1 - 1)\}c,\end{aligned} \tag{15.34}$$

Moreover, if the equality sign of (15.34) holds identically for some $n_1 \leq n - 2$, then N is a minimal submanifold of $M^n(4c)$.

Remark 15.6. Let $F(s) = r(s)e^{i\theta(s)} : I \to \mathbf{C}^*$ be a unit speed curve whose curvature κ of F satisfies $\kappa(s) = (n+1)\theta'(s) \neq 0$ and let $\iota : S^{n-1}(1) \to \mathbb{E}^n$ be the unit hypersphere centered at the origin. It was proved in [Chen and Dillen (2010)] that the complex extensor $F \otimes \iota : I \times S^{n-1}(1) \to \mathbf{C}^n$ is a non-minimal Lagrangian submanifold satisfying the equality case of (15.34) with $n_1 = n-1$. (Further non-minimal example for $n = 3$ can be found in [Bolton and Vrancken (2007)].)

15.4 Totally real δ-invariants δ_k^r and their applications

Let N be a Kähler manifold of complex dimension $n \geq 2$. A plane section $\pi \subset T_pN$, $p \in N$, is called *totally real* if $J\pi$ is orthogonal to π. Denote by $K(\pi^r)$ the sectional curvature of a totally real plane section π^r. Let

$$(\inf K^r)(p) = \inf_{\pi^r \subset T_pN} K(\pi^r),$$

where $K(\pi^r)$ runs over all totally real plane sections at $p \in N$.

Proposition 15.2. *Let N be a Kähler submanifold with* $\dim_{\mathbf{c}} N \geq 2$ *in a complex space form $M^m(4c)$. Then* $\inf K^r \leq c$*, with the equality holding identically if and only if N is a totally geodesic Kähler submanifold.*

Proof. Under the hypothesis, it follows from Proposition 9.2 that

$$\begin{aligned} K(X,Y) + K(X,JY) = &\frac{1}{4}\{H(X+JY) + H(X-JY) \\ &+ H(X+Y) + H(X-Y) - H(X) - H(Y)\} \end{aligned} \tag{15.35}$$

for orthonormal vectors X and Y with $g(X,JY) = 0$. For each $p \in N$, put

$$W_p = \big\{(X,Y) : g(X,Y) = g(X,JY) = 0 \ \forall X,Y \in T_p^1N\big\}. \tag{15.36}$$

Then W_p is a closed subset of $T_p^1N \times T_p^1N$. It is easy to verify that if $\{X,Y\}$ spans a totally real plane section, then $\{X+JY, X-JY\}$ and $\{X+Y, X-Y\}$ also span totally real plane sections.

Define a function $\hat{H} : W_p \to \mathbf{R}$ by

$$\hat{H}(X,Y) = H(X) + H(Y), \ \ (X,Y) \in W_p.$$

Suppose that $(\bar{X},\bar{Y}) \in W_p$ such that $\hat{H}$ attains an absolute maximum value at $(\bar{X},\bar{Y})$. Then (15.35) implies that

$$4\{K(\bar{X},\bar{Y}) + K(\bar{X},J\bar{Y})\} \leq \hat{H}(\bar{X},\bar{Y}). \tag{15.37}$$

On the other hand, it follows from Proposition 9.2 that a holomorphic sectional curvature $H(X)$ of a Kähler submanifold N in complex space form $M^m(4c)$ satisfies $H(X) \leq 4c$. Thus (15.37) gives

$$K(\bar{X}, \bar{Y}) + K(\bar{X}, J\bar{Y}) \leq 2c,$$

which implies inequality $\inf K^r \leq c$.

Now, suppose that $\inf K^r = c$ holds identically. Then (15.35) yields

$$\begin{aligned} H(X + JY)+&H(X - JY) + H(X + Y) + H(X - Y) \\ &- H(X) - H(Y) \geq 8c, \end{aligned} \tag{15.38}$$

for orthonormal vectors X, Y with $g(X, JY) = 0$. Put

$$\begin{aligned} H_1 &= H(X+JY) + H(X-JY), \\ H_2 &= H(X+Y) + H(X-Y), \\ H_3 &= H(X) + H(Y). \end{aligned} \tag{15.39}$$

Case (a): $H_3 \geq H_1, H_2$. From (15.38) we find

$$H(X + Y) + H(X - Y) \geq 8c.$$

Combining this with $H \leq 4c$ yields

$$H(X + Y) = H(X - Y) = 4c \tag{15.40}$$

for orthonormal vectors X, Y with $g(X, JY) = 0$. Since $X \in N$ with $n \geq 2$ lies in a totally real 2-plane, every nonzero vector in T_pN can be expressed as the sum of two orthonormal vectors X, Y with $g(X, JY) = 0$. Thus, (15.40) implies that N has constant holomorphic sectional curvature $4c$. Therefore, N is a totally geodesic Kähler submanifold in $M^{n+p}(4c)$.

Case (b): $H_2 \geq H_1, H_3$. In this case, after replacing X and Y by $(X+Y)/2$ and $(X - Y)/2$, respectively, we find from (15.38) that

$$H(X + Y + JX + JY) + H(X + Y - JX - JY) + H_3 - H_2 \geq 8c.$$

Since $H_2 \geq H_3$ and $H \leq 4c$, we obtain

$$8c \geq H(X + Y + JX + JY) + H(X + Y - JX - JY) \geq 8c.$$

Consequently,

$$H(X + Y + JX + JY) = H(X + Y - JX - JY) = 4c$$

for any orthonormal vectors X, Y with $g(X, JY) = 0$. Since every nonzero tangent vector can be expressed as $X + Y + J(X + Y)$ for some orthonormal vectors X, Y with $g(X, JY) = 0$, we conclude that N is of constant holomorphic sectional curvature $4c$. Therefore, N is totally geodesic.

Case (c): $H_1 \geq H_2, H_3$. Can be done in the same way as case (b). □

For each $k \in \mathbf{R}$, define a *totally real δ-invariant* δ_k^r on N by

$$\delta_k^r(p) = \tau(p) - k \inf K^r(p), \quad p \in N, \tag{15.41}$$

where $\inf K^r(p) = \inf_{\pi^r}\{K(\pi^r)\}$ and π^r runs over all totally real plane sections in T_pN. For a Kähler submanifold N of a Kähler m-manifold M^m, we choose a local field of orthonormal frames $e_1, \ldots, e_n, e_{1^*} = Je_1, \ldots, e_{n^*} = Je_n$, $\xi_1, \ldots, \xi_{m-n}$, $\xi_{1^*} = J\xi_1, \ldots, \xi_{(m-n)^*} = J\xi_{m-n}$ in M^m in such a way that, restrict to N, $e_1, \ldots, e_n, e_{1^*}, \ldots, e_{n^*}$ are tangent to N. With respect to such frames, the complex structure J on M^m is given by

$$J = \begin{pmatrix} 0 & -I_n & 0 & 0 \\ I_n & 0 & 0 & 0 \\ 0 & 0 & 0 & -I_{m-n} \\ 0 & 0 & I_{m-n} & 0 \end{pmatrix}, \tag{15.42}$$

where I_p is the $p \times p$ identity matrix. From Proposition 9.3, we have

$$A_{J\xi_r} = JA_r, \quad JA_r = -A_rJ, \quad r = 1, \ldots, n, 1^*, \ldots, (m-n)^*, \tag{15.43}$$

where $A_r = A_{\xi_r}$. It follows from (15.42) and (15.43) that

$$A_\alpha = \begin{pmatrix} A'_\alpha & A''_\alpha \\ A''_\alpha & -A'_\alpha \end{pmatrix}, \quad A_{\alpha^*} = \begin{pmatrix} -A''_\alpha & A'_\alpha \\ A'_\alpha & A''_\alpha \end{pmatrix}, \quad \alpha = 1, \ldots, m-n, \tag{15.44}$$

where A'_α and A''_α are $n \times n$ matrices.

Theorem 15.7. *If N is a Kähler submanifold of complex dimension $n \geq 2$ in a complex space form $M^m(4c)$, then*

(a) *for each real number $k \in (-\infty, 4]$, δ_k^r satisfies*

$$\delta_k^r \leq (2n^2 + 2n - k)c; \tag{15.45}$$

(b) *inequality* (15.45) *fails for every $k > 4$;*

(c) *$\delta_k^r = (2n^2 + 2n - k)c$ holds identically for some $k \in (-\infty, 4)$ if and only if N is a totally geodesic;*

(d) *$\delta_4^r = (2n^2 + 2n - 4)c$ holds at $p \in N$ if and only if there is an orthonormal basis $e_1, \ldots, e_n, e_{1^*} = Je_1, \ldots, e_{n^*} = Je_n, \xi_1, \ldots, \xi_{m-n}, \xi_{1^*} = J\xi, \ldots, \xi_{(m-n)^*} = J\xi_{m-n}$ of $T_pM^m(4c)$ such that*

$$\begin{gathered} A_\alpha = \begin{pmatrix} A'_\alpha & A''_\alpha \\ A''_\alpha & -A'_\alpha \end{pmatrix}, \quad A_{\alpha^*} = \begin{pmatrix} -A''_\alpha & A'_\alpha \\ A'_\alpha & A''_\alpha \end{pmatrix}, \\ A'_\alpha = \begin{pmatrix} a_\alpha & b_\alpha & 0 \\ b_\alpha & -a_\alpha & \\ 0 & & 0 \end{pmatrix}, \quad A''_\alpha = \begin{pmatrix} a^*_\alpha & b^*_\alpha & 0 \\ b^*_\alpha & -a^*_\alpha & \\ 0 & & 0 \end{pmatrix} \end{gathered} \tag{15.46}$$

for some $n \times n$ matrices A'_α, A''_α, with respect to this basis.

Proof. Under the hypothesis, we find from (9.16) that

$$2\tau = 4n(n+1)c - ||h||^2. \tag{15.47}$$

Thus

$$\tau \leq 2n(2n+1)c, \tag{15.48}$$

with the equality holding if and only if N is totally geodesic. Let π be a given totally real plane section $\pi \subset T_pN$, $p \in N$. We choose an orthonormal basis $e_1, \ldots, e_n, e_{1^*}, \ldots, e_{n^*}, \xi_1, \ldots, \xi_{m-n}, \xi_{1^*}, \ldots, \xi_{(m-n)^*}$ such that $\pi = \text{Span}\,\{e_1, e_2\}$. With respect to such a basis, we have

$$A_\alpha = \begin{pmatrix} A'_\alpha & A''_\alpha \\ A''_\alpha & -A'_\alpha \end{pmatrix}, \quad A_{\alpha^*} = \begin{pmatrix} -A''_\alpha & A'_\alpha \\ A'_\alpha & A''_\alpha \end{pmatrix}, \quad \alpha = 1, \ldots, m-n. \tag{15.49}$$

By applying (15.47)-(15.49), we find

$$\begin{aligned} 4n(n+1)c - 2\tau &= 4\sum_{\alpha=1}^{m-n} \{||A'_\alpha||^2 + ||A''_\alpha||^2\} \\ &\geq 4\sum_{\alpha=1}^{m-n} \{(h_{11}^\alpha)^2 + (h_{22}^\alpha)^2 + 2(h_{12}^\alpha)^2 + (h_{11}^{\alpha^*})^2 \\ &\quad + (h_{22}^{\alpha^*})^2 + 2(h_{12}^{\alpha^*})^2\} \\ &\geq -8\sum_{\alpha=1}^{m-n} \{h_{11}^\alpha h_{22}^\alpha - (h_{12}^\alpha)^2 + h_{11}^{\alpha^*} h_{22}^{\alpha^*} - (h_{12}^{\alpha^*})^2\} \\ &= -8K(\pi) + 8c. \end{aligned} \tag{15.50}$$

Thus

$$\tau - 4\,K(\pi) \leq (2n^2 + 2n - 4)c. \tag{15.51}$$

Since (15.51) holds for any totally real plane sections, we get

$$\tau - 4\,\inf K^r \leq (2n^2 + 2n - 4)c. \tag{15.52}$$

From (15.47) and (15.52) we derive, for any positive number μ, that

$$(\mu+1)\tau - 4\,\inf\,K^r \leq \big\{(\mu+1)(2n^2+2n) - 4\big\}\,c. \tag{15.53}$$

Thus $\delta_k^r \leq (2n^2 + 2n - k)c$, for any $k \in (0, 4)$. Combining this with (15.48) and (15.52) gives (15.45) for $k \in [0, 4]$. Inequality (15.45) with $k < 0$ follows from (15.48) and Proposition 15.2.

For (b) let us consider the complex quadric defined by

$$Q_2 = \big\{(z_0, z_1, z_2, z_3) \in CP^3(4c) : z_0^2 + z_1^2 + z_2^2 + z_3^2 = 0\big\}, \tag{15.54}$$

where $[z_0, z_1, z_2]$ is a homogeneous coordinate system. Since the scalar curvature τ and inf K^r of Q_2 are given by $\tau = 8c$ and inf $K^r = 0$. Thus

$$\delta_k^r = 8c \tag{15.55}$$

for any k. Since $(2n^2 + 2n - k)c = (12 - k)c$ holds for $n = 2$, (15.55) yields $\delta_k^r > (2n^2 + 2n - k)c$ for $k > 4$. Hence (15.45) fails for every $k > 4$.

In order to prove (c), let us assume that N is a Kähler submanifold of $M^{n+p}(4c)$ satisfying $\delta_k^r = (2n^2 + 2n - k)c$ for some $k \in (-\infty, 4)$. We divide the proof into three cases:

If $\delta_0^r = (2n^2 + 2n)c$, then (15.47) implies that N is totally geodesic.

If $\delta_k^r = (2n^2 + 2n - k)c$ holds for some $k \in (0, 4)$, then (15.45) and the definition of δ_k^r yield

$$(2n^2 + 2n - k)c = \left(1 - \frac{k}{4}\right)\delta_0^r + \frac{k}{4}\delta_4^r \leq (2n^2 + 2n - k)c, \tag{15.56}$$

which implies that $\delta_0^r = (2n^2 + 2n)c$. Thus N is totally geodesic.

If $\delta_k^r = (2n^2 + 2n - k)c$ holds for some $k \in (-\infty, 0)$, then (15.48), the definition of δ_k^r, and Proposition 15.2 imply that

$$(2n^2 + 2n - k)c = \tau - k \inf K^r \leq (2n^2 + 2n - k)c. \tag{15.57}$$

In particular, this gives $\delta_0^r = (2n^2 + 2n)c$. Hence N is totally geodesic.

Conversely, it is easy to verify that every totally geodesic Kähler submanifold of $M^m(4c)$ satisfies $\delta_k^r = (2n^2 + 2n - k)c$ identically for any k.

For (d) let us assume that N satisfies $\delta_4^r = (2n^2 + 2n - 4)c$. It follows from the proof of (a) that all inequalities in (15.50) become equalities. Thus

$$\begin{aligned} &h_{11}^r + h_{22}^r = 0, \quad h_{1j}^r = h_{2j}^r = h_{jk}^r = 0, \\ &r = 1, \ldots, m-n, 1^*, \ldots, (m-n)^*, \quad j, k = 3, \ldots, n. \end{aligned} \tag{15.58}$$

Hence the shape operator satisfies (15.46) with respect to some orthonormal basis $e_1, \ldots, e_n, Je_1, \ldots, Je_n, \xi_1, \ldots, \xi_{m-n}, J\xi_1, \ldots, J\xi_{m-n}$.

Conversely, if the shape operator at p satisfies (15.46) with respect to an orthonormal basis $e_1, \ldots, e_m, Je_1, \ldots, Je_n, \xi_1, \ldots, \xi_{m-n}, J\xi_1, \ldots, J\xi_{m-n}$, then Gauss's equation gives $\inf K^r = K(e_1, e_2)$. Moreover, from (15.46) and (15.47), we also have

$$\begin{aligned} 4n(n+1)c - 2\tau &= 8 \sum_{\alpha=1}^{p} \left\{a_\alpha^2 + b_\alpha^2 + {a_\alpha^*}^2 + {b_\alpha^*}^2\right\} \\ &= -8K(e_1, e_2) + 8c. \end{aligned}$$

Hence $\delta_4^r = (2n^2 + 2n - 4)c$. □

Definition 15.1. A Kähler submanifold N of a Kähler manifold M^m is called *strongly minimal* if A'_α and A''_α in (15.44) satisfy

$$\text{trace}\, A'_\alpha = \text{trace}\, A''_\alpha = 0, \quad \text{for } \alpha = 1, \ldots, m-n.$$

Theorem 15.8. *A complete Kähler submanifold N of complex dimension $n \geq 2$ in $CP^m(4)$ satisfies $\delta_4^r = 2(n^2+n-2)$ identically if and only if*

(1) *N is a totally geodesic Kähler submanifold, or*
(2) *$n = 2$ and N is a strongly minimal Kähler surface in $CP^m(4)$.*

Proof. Let N be a Kähler submanifold with $\dim_{\mathbf{C}} = n \geq 2$ in $CP^m(4)$ satisfying $\delta_4^r = 2(n^2+n-2)$ identically. Then it follows from Theorem 15.7 that the shape operator satisfies (15.46) with respect to some orthonormal frame $e_1, \ldots, e_n, Je_1, \ldots, Je_n, \xi_1, \ldots, \xi_{m-n}, J\xi_1, \ldots, J\xi_{m-n}$.

Recall that the dimension $\mu(p)$ of the relative null space at p is called the nullity at p. The subset G of N on which $\mu(p)$ assumes the minimum, say μ, is open in N. The scalar μ is called the index of relative nullity of N. From (15.46) we find $\mu \geq 2n-4$. Hence, if N is complete, then Theorem 4.2.1 of [Abe (1971)] implies that N is totally geodesic unless $n = 2$.

When $n = 2$, the shape operator of N in $CP^m(4)$ satisfies

$$\begin{gathered} A_\alpha = \begin{pmatrix} A'_\alpha & A''_\alpha \\ A''_\alpha & -A'_\alpha \end{pmatrix}, \quad A_{\alpha^*} = \begin{pmatrix} -A''_\alpha & A'_\alpha \\ A'_\alpha & A''_\alpha \end{pmatrix}, \\ A'_\alpha = \begin{pmatrix} a_\alpha & b_\alpha \\ b_\alpha & -a_\alpha \end{pmatrix}, \quad A''_\alpha = \begin{pmatrix} a^*_\alpha & b^*_\alpha \\ b^*_\alpha & -a^*_\alpha \end{pmatrix} \end{gathered} \tag{15.59}$$

for some functions $a_\alpha, b_\alpha, a^*_\alpha, b^*_\alpha$, $\alpha = 1, \ldots, m-n$. This shows that N is a strongly minimal Kähler surface.

Conversely, assume that N^n is either totally geodesic in $CP^m(4)$ or a strongly minimal Kähler surface in $CP^m(4)$. If N is totally geodesic, then N is a $CP^n(4)$. In this case, we have $\tau = 2n(n+1)$ and $\inf K^r = 1$. Thus $\delta_4^r = 2(n^+n-2)$ holds. If N is a strongly minimal Kähler surface with $n = 2$, then statement (d) of Theorem 15.7 implies that N satisfies $\delta_4^r = 2(n^2+n-2)$ identically. □

Theorem 15.9. *A complete Kähler submanifold N with* $\dim_{\mathbf{C}} N = n \geq 2$ *of $\mathbf{C}^m$ satisfies $\delta_4^r = 0$ identically if and only if*

(1) *N is a complex n-plane of $\mathbf{C}^m$, or*
(2) *N is a complex cylinder over a strongly minimal Kähler surface.*

Proof. Assume that N is a complete Kähler submanifold of $\mathbf{C}^m$ satisfying $\delta_4^r = 0$ identically. Then Theorem 15.7 implies that the shape operator satisfies (15.46). By applying (15.46) we obtain $\mu \geq 2n-4\,(n \geq 2)$ where μ is defined in the proof of Theorem 15.8. Thus, by the main theorem of [Abe (1972)], N is a complex cylinder over a strongly minimal Kähler surface, unless N is totally geodesic.

The converse is easy to verify. □

Definition 15.2. A pseudo-Riemannian n-manifold N is called *framed-Einstein* if there exists a pseudo-orthonormal frame $\{e_1, \ldots, e_n\}$ on N such that the Ricci tensor Ric of N satisfies

$$Ric(e_i, e_i) = \gamma g(e_i, e_i), \quad i = 1, \ldots, n,$$

for some function γ on N.

Clearly, every Einstein manifold is automatically framed-Einsteinian, but not the converse.

The following result provides a simple relationship between strongly minimal surfaces and framed-Einsteinian.

Proposition 15.3. *Let N be a strongly minimal Kähler surface in a complex space form. Then N is a framed-Einstein Kähler surface.*

Proof. If N is a strongly minimal Kähler surface in a complex space form $M^{2+p}(c)$ whose shape operator satisfies (15.59) with respect to an orthonormal frame $e_1, e_2, Je_1, Je_2, \xi_1, \ldots, \xi_p, J\xi_1, \ldots, J\xi_p$, then

$$A_\alpha^2 = \begin{pmatrix} \lambda_\alpha & 0 & 0 & \mu_\alpha \\ 0 & \lambda_\alpha & -\mu_\alpha & 0 \\ 0 & -\mu_\alpha & \lambda_\alpha & 0 \\ \mu_\alpha & 0 & 0 & \lambda_\alpha \end{pmatrix}, \tag{15.60}$$

where

$$\lambda_a = a_\alpha^2 + b_\alpha^2 + a_\alpha^{*\,2} + b_\alpha^{*\,2}, \quad \mu_\alpha = 2a_\alpha^2 b_\alpha^{*\,2} - 2b_\alpha^2 a_\alpha^{*\,2}.$$

On the other hand, it follows from the equation of Gauss that

$$S(X,Y) = 6cg(X,Y) - 2\sum_{\alpha=1}^{p} g(A_\alpha^2 X, Y). \tag{15.61}$$

Now, we find rom (15.60) and (15.61) that

$$S(e_1,e_1) = S(e_2,e_2) = S(Je_1,Je_1) = S(Je_2,Je_2) = 6\,c - 2\sum_{\alpha=1}^{p} \lambda_\alpha.$$

Thus, N is framed-Einsteinian. □

Remark 15.7. For a Kähler n-manifold N, invariant δ_k^r was extended in [Suceavă (2003)] to

$$\delta_{\ell,k}^r(p) = \tau(p) - \frac{k}{\ell-1} \inf_{L_\ell^r} \tau(L_\ell^r), \quad p \in N,$$

where L_ℓ^r runs over all totally real ℓ-subspaces of T_pN. For each integer $\ell \in [2, n]$, B. Suceavă extended inequality (15.45) to

$$\delta_{\ell,k}^r(p) \le \left\{2n^2 + 2n - \frac{k}{4}\binom{\ell}{2}\right\}c \tag{15.62}$$

for Kähler submanifolds in a complex space form $M^m(4c)$. He also proved that, for $\ell \ge 3$, the equality sign holds only for totally geodesic N.

15.5 Examples of strongly minimal Kähler submanifolds

Clearly, every totally geodesic Kähler submanifold of a complex space form is strongly minimal. The following provides some non-trivial examples of strongly minimal Kähler surfaces.

Example 15.1. Consider the *complex quadric* Q_n in $CP^{n+1}(4)$ defined by

$$Q_n = \{(z_0, z_1, \ldots, z_{n+1}) \in CP^{n+1}(4) : z_0^2 + z_1^2 + \cdots + z_{n+1}^2 = 0\},$$

where $\{z_0, z_1, \ldots, z_{n+1}\}$ is a homogeneous coordinate system of $CP^{n+1}(4)$.

For Q_2 we have $\tau = 8$, $\inf K^r = 0$ and $\delta_4^r = 8$. Thus Q_2 is a non-totally geodesic Kähler submanifold satisfying (15.50) with $n = 2$. Thus, by Theorem 15.8, Q_2 is a strongly minimal Kähler surface in $CP^3(4)$.

Since Q_2 is an Einstein-Kähler surface whose Ricci tensor satisfies $Ric = 4g$, Gauss' equation yields $g(A_1^2X, Y) = g(X, Y)$, $X, Y \in TQ_2$. Thus, with respect to a suitable choice of $e_1, e_2, Je_1, Je_2, \xi_1, J\xi_1$, we have

$$A_1 = \begin{pmatrix} A_1' & A_1'' \\ A_1'' & -A_1' \end{pmatrix}, \quad A_{1^*} = \begin{pmatrix} -A_1'' & A_1' \\ A_1' & A_1'' \end{pmatrix},$$

where $A_1' = \begin{pmatrix} 1 & 0 \\ 0 & -1 \end{pmatrix}$, $A_1'' = \begin{pmatrix} 0 & 0 \\ 0 & 0 \end{pmatrix}$. Thus Q_2 is strongly minimal in $CP^3(4)$.

Example 15.2. The Kähler surface

$$\{z \in \mathbf{C}^3 : z_1^2 + z_2^2 + z_3^2 = 1\}$$

is a strongly minimal complex surface in $\mathbf{C}^3$.

Example 15.3. [Suceavă (2003)] The Kähler surfaces

$$\{z \in \mathbf{C}^3 : z_1 + z_2 + z_3^2 = k\}, \quad k \in \mathbf{C},$$

are strongly minimal Kähler surfaces in $\mathbf{C}^3$.

15.6 Kählerian δ-invariants δ^c and their applications to Kähler submanifolds

Let N be a submanifold of a complex space form $M^m(4c)$. For each $X \in TN$ we put

$$JX = PX + FX,$$

where PX and FX are the tangential and the normal components of JX, respectively. For a linear r-subspace $L \subset T_pN$, we put

$$\Psi(L) = \sum_{1 \le i < j \le r} \langle Pe_i, e_j \rangle^2, \tag{15.63}$$

where $\{e_1, \ldots, e_r\}$ is an orthonormal basis of L. Then $\Psi(L)$ is independent of the choice of the orthonormal basis $\{e_1, \ldots, e_r\}$ of L.

Lemma 15.1. *Let N be an n-dimensional submanifold of a complex space form $M^m(4c)$. Then, for any k mutually orthogonal subspaces $L_1, \ldots, L_k$ of T_pN with $\dim L_j \ge 2$, $j = 1, \ldots, k$, we have*

$$\begin{aligned} \tau - \sum_{j=1}^{k} \tau(L_j) &\le c(n_1, \ldots, n_k)||H||^2 \\ &+ \frac{c}{2}\left\{ n(n-1) - \sum_{j=1}^{k} n_j(n_j - 1) + 3||P||^2 - 6\sum_{j=1}^{k} \Psi(L_j) \right\}, \end{aligned} \tag{15.64}$$

where $||P||^2$ is the squared norm of P.

Equality sign in (15.64) holds if and only if there is an orthonormal basis $e_1, \ldots, e_n, e_{n+1}, \ldots, e_{2m}$ at p such that

(a) *$S(n_1, \ldots, n_k) = \tau(L_1) + \cdots + \tau(L_k)$, where $e_{n_1+\cdots+n_{j-1}+1}, \ldots, e_{n_1+\cdots+n_j}$ span L_j for $j = 1. \ldots, k$ and*
(b) *the shape operator of N satisfies (13.29) and (13.30).*

Proof. Under the hypothesis, Gauss's equation and (9.5) yield

$$2\tau = n^2||H||^2 - ||h||^2 + n(n-1)c + 3c||P||^2. \tag{15.65}$$

Put

$$\eta = 2\tau - \frac{n^2(n + k - 1 - \sum_j n_j)}{n + k - \sum_j n_j}||H||^2 - n(n-1)c - 3c\,||P||^2. \tag{15.66}$$

From (15.65) and (15.66) we find

$$n^2||H||^2 = \left(n + k - \sum_{j=1}^{k} n_j\right)\left(||h||^2 + \eta\right). \tag{15.67}$$

Let $L_1, \ldots, L_k$ be mutually orthogonal subspaces of T_pN. By applying an argument similar to the proof of Theorem 13.3, we have

$$\tau(L_1) + \cdots + \tau(L_k) \geq \frac{\eta}{2} + \sum_{j=1}^{k} \frac{n_j(n_j-1)}{2} c + 3c \sum_{j=1}^{k} \Psi(L_j).$$

Combining this with (15.66) yields (15.64). The equality case can be verified in a way similar as that of Theorem 13.3. □

Let N be a Kähler manifold with complex dimension n. For each k-tuple $(2n_1, \ldots, 2n_k) \in \mathcal{S}(2n)$, the *Kählerian δ-invariant* $\delta^c(2n_1, \ldots, 2n_k)$ of N is defined by

$$\delta^c(2n_1, \ldots, 2n_k) = \tau - \inf\{\tau(L_1^c) + \cdots + \tau(L_k^c)\}, \tag{15.68}$$

where $L_1^c, \ldots, L_k^c$ run over all k mutually orthogonal complex subspaces of T_pN, $p \in N$, with real dimensions $2n_1, \ldots, 2n_k$, respectively.

For a Kähler submanifold we have the following [Chen (1998c)].

Theorem 15.10. *Let N be a complex n-dimensional Kähler submanifold of a complex space form $M^m(4c)$. Then for each k-tuple $(2n_1, \ldots, 2n_k) \in \mathcal{S}(2n)$ we have*

$$\delta^c(2n_1, \ldots, 2n_k) \leq 2\left\{n(n+1) - \sum_{j=1}^{k} n_j(n_j+1)\right\}c. \tag{15.69}$$

The equality case of inequality (15.69) holds at a point $p \in N$ if and only if, there exists an orthonormal basis

$$e_1, \ldots, e_{n_1}, Je_1, \ldots, Je_{n_1}, \ldots, e_{2(n_1+\cdots+n_{k-1})+1}, \ldots, e_{2(n_1+\cdots+n_{k_1})+n_k},$$
$$Je_{2(n_1+\cdots+n_{k-1})+1}, \ldots, Je_{2(n_1+\cdots+n_{k-1})+n_k}, e_{2n+1}, \ldots, e_{2m}$$

of $T_pM^m(4c)$ such that the shape operators at p satisfies

$$A_{e_r} = \begin{pmatrix} A_1^r & \ldots & 0 & \\ \vdots & \ddots & \vdots & 0 \\ 0 & \ldots & A_k^r & \\ & 0 & & 0 \end{pmatrix}, \quad r = 2n+1, \ldots, 2m,$$

where each A_j^r is a symmetric $2n_j \times 2n_j$ submatrix with zero trace.

Proof. Using the fact that Kähler submanifolds are minimal, this can be proved in a similar way as Theorem 13.3. □

Remark 15.8. Contrast to inequality (15.48), there exist non-totally geodesic Kähler submanifolds of $M^m(4c)$ satisfying the equality case of (15.69) identically. For instance, let $N = Q_n$ be the complex quadric in

$CP^{n+1}(4)$. Then $\tau = 2n^2$ and $\delta^c(2,\ldots,2) = 2n(n-1)$, in which 2 in $\delta^c(2,\ldots,2)$ repeats n times. Q_n satisfies the equality case of (15.69) identically for $(2,\ldots,2) \in \mathcal{S}(2n)$.

Also, the direct product of k Kähler submanifolds of complex Euclidean spaces is a Kähler submanifold of a complex Euclidean space satisfying the equality case of (15.69) for some suitable k-tuples.

15.7 Applications of δ-invariants to real hypersurfaces

Definition 15.3. A real hypersurface N of almost Hermitian manifold (M, g, J) is called a *Hopf hypersurface* if $U = -J\xi$ is a principal curvature vector, i.e., an eigenvector of the shape operator A_ξ, where ξ is a unit normal vector of N. The vector field U is called the *Hopf vector field.*

Choose a point $p \in CH^m(-4)$ and any direction $\xi \in T_p(CH^m(-4))$. For each $r > 0$, let $q(r) = \exp_p(r\xi)$ and γ_ξ be the geodesic with initial direction ξ that joins p to $q(r)$. Then p is on each geodesic hypersphere $G_r(q(r))$ centered at $q(r)$ with radius r. It is known that, as $q(r)$ recedes from p as $r \to \infty$, the $G_r(q(r))$ approach a limiting hypersurface, called a *horosphere* of $CH^m(-4)$.

Theorem 15.11. *Let N be a real hypersurface of a complex hyperbolic m-space $CH^m(-4)$. Then, for $\bar{\delta}_2^k := \delta(2,\ldots,2)$ (2 appears k times), we have*

$$\bar{\delta}_2^k \leq \frac{(2m-1)^2(2m-k-2)}{2(2m-k-1)}||H||^2 - (2m^2 - 4k - 2) \tag{15.70}$$

for any natural number $k \leq m-1$.

Equality sign of (15.70) *holds identically for some k if and only if one of the following two cases occurs:*

(a) *m is odd, $k = m-1$ and N is an open portion of a tubular hypersurface of radius $r \in \mathbb{R}_+$ over a totally geodesic $CH^{(m-1)/2}(-4)$;*
(b) *$k = m-1$ and N is an open portion of a horosphere of $CH^m(-4)$.*

Proof. Let N be a real hypersurface in $CH^m(-4)$. Then for k mutually orthogonal plane sections $\pi_1,\ldots,\pi_k$ we have

$$||P||^2 - 2\sum_{j=1}^{k}\Psi(\pi_j) \geq 2m - 2k - 2, \tag{15.71}$$

with equality holding if and only if $\pi_1,\ldots,\pi_k$ are mutually orthogonal complex planes. Combining (15.64) of Lemma 15.1 and (15.71) gives (15.70).

Assume that the equality case of (15.70) holds identically. Then there is an orthonormal frame $\{e_1, \ldots, e_{2m-1}\}$ satisfying conditions (a) and (b) of Lemma 15.1; moreover, $\pi_1, \ldots, \pi_k$ are complex planes.

Let ξ be a unit normal vector at p and $a_1, \ldots, a_{2m-1}$ are the principal curvatures. Since $\pi_1, \ldots, \pi_k$ are complex plane, $J\xi$ must lies in the subspace $\mathcal{D}(p)$ spanned by $e_{2k+1}, \ldots, e_{2m-1}$. By condition (b), $J\xi$ is a principal curvature vector with principal curvature, say a_{2m-1} satisfying

$$a_{2m-1} = a_1 + a_2 = \cdots = a_{2k-1} + a_{2k}.$$

Thus, N is a Hopf hypersurface. Hence, the principal curvature corresponding to $J\xi$ is constant [Berndt (1989)].

On the other hand, it is known that the shape operator $A = A_\xi$ of a Hopf hypersurface in a complex space form $M^m(4c)$ satisfies

$$2c \langle X, PY \rangle = \alpha \langle PX, AY \rangle - \alpha \langle AX, PY \rangle - 2 \langle PAX, AY \rangle, \tag{15.72}$$

where $A(J\xi) = \alpha J\xi$. By choosing $X = Pe_{2j-1}, Y = e_{2j-1}$, we obtain

$$(2a_{2j-1} - a_{2m-1})a_{2j} = 2c + a_{2j-1}a_{2m-1}, \quad j = 1, \ldots, k. \tag{15.73}$$

Combining this with $a_{2m-1} = a_{2j-1} + a_{2j}$ showing that N has constant principal curvatures. So, by applying the classification of Hopf hypersurfaces in $CH^m(-4)$ with constant principal curvatures [Berndt (1989)], we conclude that N is orientable and it is an open part of one of the following:

(1) a tubular hypersurface with radius $r \in \mathbf{R}_+$ over a totally geodesic $CH^\ell(-4)$ for an integer $\ell \in \{0, \ldots, m-1\}$;
(2) a tubular hypersurface with radius $r \in \mathbf{R}_+$ over a totally geodesic $RH^m(-1)$;
(3) a horosphere of $CH^m(-4)$.

Thus, N has principal curvatures $\{2\coth(2r), \tanh(r), \coth(r)\}$ of multiplicities $\{1, 2\ell, 2(m-\ell-1)\}$ in case (1); principal curvatures $\{2\tanh(2r), \tanh(r), \coth(r)\}$ of multiplicities $\{1, m-1, m-1\}$ in case (2); and principal curvatures $\{2, 1\}$ of multiplicities $\{1, 2m-2\}$ in case (3). Also, it is known that the multiplicity of the principal curvature with respect to the Hopf vector field $J\xi$ is one.

If case (1) occurs, $a_n = 2\coth(2r)$. Thus, either $\coth(2r) = \tanh(r)$, or $\coth(2r) = \coth(r)$, or $2\coth(2r) = \tanh(r) + \coth(r)$ holds. But the first two are impossible and the third one is an identity. Thus, we conclude from Lemma 15.1 that m must be odd and $k = 2\ell = m-1$. It is direct to verify that case (2) cannot occurs and that a horosphere in $CH^m(-4)$ satisfies the equality case of (15.67) with $k = m-1$. $\square$

Theorem 15.12. *Let N be a real hypersurface of $CP^m(4)$ $(m \geq 2)$. Then*

$$\bar{\delta}_2^k \leq \frac{(2m-1)^2(2m-k-2)}{2(2m-k-1)}||H||^2 + 2m^2 - k - 2 \tag{15.74}$$

for any natural number $k \leq m-1$.

If N is a Hopf hypersurface, the equality sign of (15.74) *holds for some k if and only if one of the following three cases occurs:*

(1) *$k = m-1$ and N is an open part of a geodesic sphere with radius $\frac{\pi}{4}$;*
(2) *m is odd, $k = m-1$, and N is an open part of a tubular hypersurface with radius $r \in (0, \frac{\pi}{2})$ over a totally geodesic $CP^{(m-1)/2}(4c)$;*
(3) *$m = 2$, $k = 1$, and N is an open part of a tubular hypersurface over a complex quadric Q_1 with radius $r = \tan^{-1}\left(\frac{1}{2}(1+\sqrt{5}-\sqrt{2+2\sqrt{5}}\,)\right)$.*

Proof. Inequality (15.74) follows from Lemma 15.1 and $||P||^2 = 2m-2$.

If N is a Hopf hypersurface, $J\xi$ is a principal vector. Put $A(J\xi) = \alpha J\xi$, where $A = A_{2m} = A_\xi$. Then α is locally a constant [Takagi (1975)].

Assume that N satisfies the equality in (15.74). Then by Lemma 15.1 there exists an orthonormal basis $\{e_1, \ldots, e_{2m-1}\}$ satisfying (a) and (b) of Lemma 15.1. Also, we know that $\pi_1, \ldots, \pi_k$ are totally real planes. Since $J\xi$ is an eigenvector of A, α is one of the eigenvalues $a_1, \ldots, a_{2m-1}$.

Because $J\xi$ is an eigenvector of A, the subspace L is orthogonal ξ and $J\xi$ is a complex subspace containing other eigenvectors of A. Let $X \in L$ be a unit eigenvector of A with eigenvalue a. From (15.72) we find

$$(2a - \alpha)A(JX) = (2 + \alpha a)JX. \tag{15.75}$$

Thus, if $2a \neq \alpha$, JX is an eigenvector of A with eigenvalue $\frac{2+\alpha a}{2a-\alpha}$ which is not equal to $\frac{\alpha}{2}$. Consequently, the eigenvalues of A are given by

$$\{a_1, \ldots, a_{2m-1}\} = \left\{b_1, \ldots, b_\ell, \frac{2+\alpha b_1}{2b_1-\alpha}, \ldots, \frac{2+\alpha b_\ell}{2b_\ell-\alpha}, \frac{\alpha}{2}, \ldots, \frac{\alpha}{2}, \alpha\right\}, \tag{15.76}$$

where $\alpha/2$ repeats $2m - 2\ell - 2$ times.

From (13.29), (13.30), (15.76) and α being constant, we conclude that N has constant principal curvatures. Thus, by a result of [Kimura (1986)], N is an open part of a homogeneous real hypersurface. Hence, a result of [Takagi (1975)] implies that N is an open part of one of the following:

(a_1) a geodesic sphere with radius $r \in (0\frac{\pi}{2})$;
(a_2) a tubular hypersurface with radius $r \in (0, \frac{\pi}{4})$ over a complex totally geodesic $CP^\ell(4)$, where $1 \leq \ell \leq m-2$;
(b) a tubular hypersurface with radius $r \in (0, \frac{\pi}{4})$ over a complex quadric Q_{m-1};

(c) a tubular hypersurface with radius $r \in (0, \frac{\pi}{4})$ over the Segre embedding of $CP^1 \times CP^{(m-1)/2}$;
(d) a tubular hypersurface with radius $r \in (0, \frac{\pi}{4})$ over a Plücker embedding of the complex Grassmann manifold $G^{\mathbb{C}}_{2,3}$ in CP^9;
(e) a tubular hypersurface with radius $r \in (0, \frac{\pi}{4})$ over a canonical embedding of the Hermitian symmetric space $SO(10)/U(5)$ in CP^{15}.

Beside the principal curvature $\alpha = 2\cot(2r)$ with multiplicity one, the other principal curvatures are given by $\{\cot(r)\}$ with multiplicity $2m-2$ for case (a_1); $\{\cot(r), -\tan(r)\}$ with even multiplicities for case (a_2); $\{-\tan(r-\frac{\pi}{4}), \cot(r-\frac{\pi}{4}\}$ of multiplicities $\{m-1, m-1\}$ for case (b); $\{\cot(r), \cot(r-\frac{\pi}{4}), \cot(r-\frac{\pi}{2}), \cot(r-\frac{3\pi}{4})\}$ with appropriate even multiplicities for cases (c), (d) and (e).

For case (a_1) the principal curvatures satisfy (13.29) and (13.30) if and only if $\cot(r) = 1$. This gives rise to a geodesic sphere with radius $\frac{\pi}{4}$. For case (a_2), (13.29) and (13.30) hold if and only if m is odd and $k = m-1$. For case (b), (13.29) and (13.30) hold if and only if $m = 2$ and $u = \tan(r)$ satisfies one of the following two polynomial equations:

$$\begin{aligned} &u^4 - 2u^3 - 2u^2 - 2u + 1 = 0, \\ &u^4 + 2u^3 - 2u^2 + 2u + 1 = 0. \end{aligned} \tag{15.77}$$

The first equation in (15.77) has only two real roots given by

$$u_1 = \frac{1}{2}\left(1+\sqrt{5}-\sqrt{2+2\sqrt{5}}\,\right), \quad u_2 = \frac{1}{u_1} = \frac{1}{2}\left(1+\sqrt{5}+\sqrt{2+2\sqrt{5}}\,\right).$$

The only real roots of the second equation in (15.77) are $-u_1$ and $-u_2$. Also, u_1 is the only real root of the first equation in (15.77) lying between 0 and 1 and the second equation has no positive roots. Therefore, N is a hypersurface of the third kind given in the theorem. It is direct to verify that the principal curvatures for hypersurfaces of types (c), (d) and (e) never satisfy (13.29) and (13.30).

The converse is easy to verify. □

15.8 Applications of δ-invariants to para-Kähler manifolds

Let N be a spacelike Lagrangian submanifold of a para-Kähler $2n$-manifold $(M^{2n}_n, g, \mathcal{P})$ and $e_1, \ldots, e_n$ an orthonormal frame of N. We put $e_{i^*} = \mathcal{P}e_i$, $i = 1, \ldots, n$. Then $e_{1^*}, \ldots, e_{n^*}$ form a timelike orthonormal frame of the normal bundle.

Let $\omega_B^A, A, B = 1, \ldots, n, 1^*, \ldots, n^*$ be the connection forms defined by (2.86). Then Lemma 10.3(a) implies that

$$\omega_{j^*}^{i^*} = -\omega_j^i, \quad i, j = 1, \ldots, n. \tag{15.78}$$

Lemma 15.2. *Let N be a spacelike Lagrangian submanifold of a para-Kähler manifold $(M_n^{2n}, g, \mathcal{P})$. If M_n^{2n} is Einstein, then $\Theta = \sum_{i=1}^n \omega_i^{i^*}$ is a closed form.*

Proof. It follows from (2.92) that

$$d\Theta = -\sum_{i,j,k} h_{jk}^{i^*} \omega_k \wedge \omega_i^j - \sum_{i,j,k} h_{ij}^{k^*} \omega^j \wedge \omega_{k^*}^{i^*} + \frac{1}{2} \sum_{i,k,\ell} K_{ik\ell}^{i^*} \omega^k \wedge \omega^\ell.$$

Combining this with (15.78) gives

$$d\Theta = \frac{1}{2} \sum\nolimits_{i,k,\ell} K_{ik\ell}^{i^*} \omega^k \wedge \omega^\ell. \tag{15.79}$$

By applying the first Bianchi identity, Proposition 1.7 and curvature identities of para-Kähler manifolds from Lemma 10.1, we find

$$\begin{aligned} \sum_{i=1}^n K_{ijk}^{i^*} &= \sum_{i=1}^n R(e_{i^*}, e_i; e_j, e_k) \\ &= -\sum_{i=1}^n R(e_i, e_j; e_{i^*}, e_k) - \sum_{i=1}^n R(e_j, e_{i^*}; e_i, e_k) \\ &= -\sum_{i=1}^n R(e_j, e_i; e_i, e_{k^*}) + \sum_{i=1}^n R(e_{j^*}, e_i; e_i, e_k) \\ &= -\sum_{i=1}^n R(e_{j^*}, e_{i^*}; e_{i^*}, e_k) + \sum_{i=1}^n R(e_i, e_{j^*}, e_k, e_i) \\ &= -\sum_{i=1}^n R(e_{i^*}, e_{j^*}; e_k, e_{i^*}) + \sum_{i=1}^n R(e_i, e_{j^*}, e_k, e_i) \\ &= \widetilde{Ric}(e_{j^*}, e_k). \end{aligned} \tag{15.80}$$

(15.79) and (15.80) imply that $d\Theta = 0$ whenever M_n^{2n} is Einstein. □

It follows from (2.88), (2.89) and Lemma 10.3(c) that

$$\Theta = -\sum\nolimits_{i,j} h_{ij}^{i^*} \omega^j = -\sum\nolimits_{i,j} h_{ii}^{j^*} \omega^j.$$

Thus $\Theta(e_k) = n \langle H, \mathcal{P}e_k \rangle$ for $k = 1, \ldots, n$. Hence, we may apply the same arguments of the proof of Theorem 15.1 to obtain the following vanishing theorem.

Theorem 15.13. *If a compact n-manifold N has finite fundamental group or null first Betti number, then each Lagrangian immersion of N into any Einstein para-Kähler manifold must have a minimal point.*

The following two corollaries are easy consequences of Theorem 15.12.

Corollary 15.10. *There do not exist immersions of nonzero constant mean curvature from a topological n-sphere into any Einstein para-Kähler manifold as a spacelike Lagrangian submanifold.*

Corollary 15.11. *If N is a complete Riemannian manifold satisfying Ricci curvature $Ric(u) \geq c^2 > 0, u \in T^1N$, for some positive real number c, then every Lagrangian immersion of constant mean curvature from N into any Einstein para-Kähler manifold is a minimal immersion.*

For spacelike Lagrangian submanifolds in para-Kähler space forms, we have the following.

Theorem 15.14. *If N is a spacelike Lagrangian submanifold of a para-Kähler space form $M^{2n}_n(4c)$, then for each k-tuple $(n_1, \ldots, n_k) \in \mathcal{S}(n)$ we have*

$$\begin{aligned}\delta(n_1,\ldots,n_k) \geq{}& \frac{n^2\{n-\sum_i n_i+3k-1-6\sum_i(2+n_i)^{-1}\}}{2\{n-\sum_i n_i+3k+2-6\sum_i(2+n_i)^{-1}\}}\langle H,H\rangle \\ &+\frac{1}{2}\Big\{n(n-1)-\sum_{i=1}^k n_i(n_i-1)\Big\}c.\end{aligned} \tag{15.81}$$

The equality sign of (15.81) holds at a point $p \in N$ if and only if there is an orthonormal basis $\{e_1, \ldots, e_n\}$ of T_pN such that with respect to this basis the second fundamental form h satisfies

$$\begin{aligned}
&h(e_{\alpha_i}, e_{\beta_i}) = \sum_{\gamma_i} h^{\gamma_i}_{\alpha_i\beta_i}\mathcal{P}e_{\gamma_i} + \frac{3\delta_{\alpha_i\beta_i}}{2+n_i}\lambda\mathcal{P}e_{\mu+1}, \ \sum_{\alpha_i=1}^{n_i} h^{\gamma_i}_{\alpha_i\alpha_i} = 0,\\
&h(e_{\alpha_i}, e_{\alpha_j}) = 0, \ \ i \neq j,\\
&h(e_{\alpha_i}, e_{\mu+1}) = \frac{3\lambda}{2+n_i}\mathcal{P}e_{\alpha_i}, \ \ h(e_{\alpha_i}, e_u) = 0,\\
&h(e_{\mu+1}, e_{\mu+1}) = 3\lambda\mathcal{P}e_{\mu+1}, \ \ h(e_{\mu+1}, e_u) = \lambda\mathcal{P}e_u,\\
&h(e_u, e_v) = \lambda\delta_{uv}\mathcal{P}e_{\mu+1},
\end{aligned} \tag{15.82}$$

for $1 \leq i, j \leq k$; $\mu+2 \leq u, v \leq n$ and $\lambda = \frac{1}{3}h^{\mu+1}_{\mu+1\mu+1}$, where $\mu = n_1+\cdots+n_k$.

Proof. This can be proved in the same way as Theorem 15.4. □

The following result shows that inequality (15.81) is sharp.

Proposition 15.4. *For each k-tuple $(n_1,\ldots,n_k)\in\mathcal{S}(n)$, $n\geq 3$,, there exists a non-minimal Lagrangian spacelike submanifold in $(\mathbb{E}^{2n}_n,g_0,\mathcal{P})$ which satisfies the equality case of* (15.81) *at a point.*

Proof. Given a k-tuple $(n_1,\ldots,n_k)\in\mathcal{S}(n)$, let us put $\mu=\sum_{i=1}^k n_i$. Let $\Delta_1,\ldots,\Delta_{k+1}$ be defined by (15.12). Consider the Lagrangian graph in the para-Kähler n-plane $\mathcal{P}K^n$ defined by

$$L(y_1,\ldots,y_n)=\big(F_{y_1},\ldots,F_{y_n},y_1,\ldots,y_n\big),$$

where $F_{y_1},\ldots,F_{y_n}$ are the first partial derivatives of

$$F=\sum_{i=1}^{k}\frac{3\lambda y_{\mu+1}}{2(2+n_i)}\sum_{\alpha_i\in\Delta_i}y_{\alpha_i}^2+\frac{\lambda y_{\mu+1}}{2}\sum_{r=\mu+1}^{n}y_r^2,$$

for a nonzero real number λ. Then

$$e_1=(1,0,\ldots,0),\ldots,e_n=(0,\ldots,0,\overbrace{1}^{n-th},0,\ldots,0)$$

form an orthonormal basis at $o=(0,\ldots,0)$. After applying Lemma 10.4, it is direct to verify that the second fundamental of this Lagrangian graph satisfies (15.82) at o. Consequently, this non-minimal Lagrangian graph satisfies the equality case of (15.66) at o. □

An immediate consequence of Theorem 15.14 is the following.

Corollary 15.12. *Let N be a Riemannian n-manifold. If there is a k-tuple $(n_1,\ldots,n_k)\in\mathcal{S}(n)$ such that*

$$\delta(n_1,\ldots,n_k)(p)<\frac{1}{2}\Big\{n(n-1)-\sum_{i=1}^{k}n_i(n_i-1)\Big\}c$$

at some point $p\in N$. Then N cannot be isometrically immersed as a Lagrangian minimal submanifold in a para-Kähler space form $M^{2n}_n(4c)$.

Further, by apply Lemma 15.2, Theorem 15.14 and the same arguments as in the proof of Theorem 15.2, we may obtain the following.

Corollary 15.13. *Let N be a compact Riemannian n-manifold with null first Betti number or with finite fundamental group. If there is a k-tuple $(n_1,\ldots,n_k)\in\mathcal{S}(n)$ such that*

$$\delta(n_1,\ldots,n_k)<0,$$

then N does not admit any Lagrangian isometric immersions into the para-Kähler n-plane $(\mathbb{E}^{2n}_n,g_0,\mathcal{P})$.

Chapter 16

Applications to Contact Geometry

16.1 δ-invariants and submanifolds of Sasakian space forms

Let $(M^{2m+1}(c), \phi, \xi, \eta, g)$ be a Sasakian space form of constant ϕ-sectional curvature c. Proposition 12.4 implies that the sectional curvature satisfies

$$\frac{3+c}{4} \leq \tilde{K} \leq c \text{ if } c \geq 1 \text{ and } c \leq \tilde{K} \leq \frac{3+c}{4} \text{ if } c < 1. \tag{16.1}$$

By combining (16.1) and Theorem 13.3 we obtain immediately the following results for arbitrary submanifolds in Sasakian space forms.

Proposition 16.1. *If N is an n-dimensional submanifold of a Sasakian space form $M^{2m+1}(c)$ of constant ϕ-sectional curvature $c \geq 1$, then, for any k-tuple $(n_1, \ldots, n_k) \in \mathcal{S}(n)$, we have*

$$\delta(n_1, \ldots, n_k) \leq c(n_1, \ldots, n_k)||H||^2 + b(n_1, \ldots, n_k)c. \tag{16.2}$$

Proposition 16.2. *If N is an n-dimensional Legendre submanifold of a Sasakian space form $M^{2m+1}(c)$ of constant ϕ-sectional curvature $c < 1$, then, for any k-tuple $(n_1, \ldots, n_k) \in \mathcal{S}(n)$, we have*

$$\delta(n_1, \ldots, n_k) \leq c(n_1, \ldots, n_k)||H||^2 + \left(\frac{3+c}{4}\right)b(n_1, \ldots, n_k). \tag{16.3}$$

Remark 16.1. When N is a Legendre submanifold, inequality (16.3) is due to [Defever et al. (1997, 2001)].

Two immediate consequences of Propositions 16.1 and 16.2 are the following.

Corollary 16.1. *Let N be a Riemannian n-manifold. If there exists a point $p \in N$ such that $\delta(n_1, \ldots, n_k)(p) > b(n_1, \ldots, n_k)c$ for some k-tuple $(n_1, \ldots, n_k)$, then N cannot isometric immersed as a minimal submanifold in any Sasakian space form $M^{2m+1}(c)$ for $c \geq 1$, regardless of codimension.*

Corollary 16.2. *Let N be a Riemannian n-manifold. If there is a point $p \in N$ such that $\delta(n_1, \ldots, n_k)(p) > b(n_1, \ldots, n_k)(3+c)/4$ for some k-tuple $(n_1, \ldots, n_k)$, then N cannot isometric immersed as a minimal submanifold in any Sasakian space form $M^{2m+1}(c)$ for $c < 1$, regardless of codimension.*

16.2 δ-invariants and Legendre submanifolds

For the next theorem we need the following two lemmas from [Blair (2010)].

Lemma 16.1. *Let N be a Legendre submanifold of a K-contact manifold M. Then $A_\xi = 0$.*

Proof. Under the hypothesis, this follows from the following identities:

$$\begin{aligned}\langle A_\xi X, Y\rangle &= \langle h(X,Y), \xi\rangle = \langle \nabla_X Y, \xi\rangle \\ &= -\langle Y, \nabla_X \xi\rangle = -\langle Y, \phi(X)\rangle = 0\end{aligned}$$

for X, Y tangent to N, where ∇ is the Levi-Civita connection of M. □

Lemma 16.2. *Let N be a Legendre submanifold of a Sasakian manifold M^{2n+1}. Then $A_{\phi X}Y = A_{\phi Y}X$ for $X, Y \in TN$.*

Proof. This follows from the following identities:

$$\begin{aligned}\langle A_{\phi X} Z, Y\rangle &= \langle h(Y,Z), \phi X\rangle = \langle \nabla_Y Z, \phi X\rangle \\ &= -\langle Z, \nabla_Y(\phi X)\rangle = -\langle Z, (\nabla_Y \phi)X + \phi(\nabla_Y X)\rangle \\ &= \langle \phi Z, \nabla_X Y\rangle = \langle A_{\phi Z} X, Y\rangle\end{aligned}$$

for X, Y, Z tangent to N, where we have applied (12.4). □

Inequalities (16.2) and (16.3) are not sharp for Legendre submanifolds. In fact, we can improve them as follows.

Theorem 16.1. *Let N be a Legendre submanifold of a Sasakian space form $M^{2n+1}(c)$. Then for any k-tuple $(n_1, \ldots, n_k) \in \mathcal{S}(n)$ we have*

$$\begin{aligned}\delta(n_1, \ldots, n_k) \le\, & \frac{n^2\{n - \sum_i n_i + 3k - 1 - 6\sum_i (2+n_i)^{-1}\}}{2\{n - \sum_i n_i + 3k + 2 - 6\sum_i (2+n_i)^{-1}\}} ||H||^2 \\ & + \left(\frac{3+c}{4}\right) b(n_1, \ldots, n_k).\end{aligned} \tag{16.4}$$

The equality sign of (16.4) holds at a point $p \in N$ if and only if there is an orthonormal basis $\{e_1, \ldots, e_n\}$ of T_pN such that with respect to this *basis the second fundamental form h satisfies*

$$\begin{aligned}
&h(e_{\alpha_i}, e_{\beta_i}) = \sum_{\gamma_i} h^{\gamma_i}_{\alpha_i\beta_i}\phi e_{\gamma_i} + \frac{3\delta_{\alpha_i\beta_i}}{2+n_i}\lambda\phi e_{N+1}, \ \sum_{\alpha_i=1}^{n_i} h^{\gamma_i}_{\alpha_i\alpha_i} = 0, \\
&h(e_{\alpha_i}, e_{\alpha_j}) = 0, \ \ i \neq j, \\
&h(e_{\alpha_i}, e_{\mu+1}) = \frac{3\lambda}{2+n_i}\phi e_{\alpha_i}, \ \ h(e_{\alpha_i}, e_u) = 0, \\
&h(e_{\mu+1}, e_{\mu+1}) = 3\lambda\phi e_{\mu+1}, \ \ h(e_{\mu+1}, e_u) = \lambda\phi e_u, \\
&h(e_u, e_v) = \lambda\delta_{uv}\phi e_{\mu+1},
\end{aligned} \tag{16.5}$$

for $1 \leq i, j \leq k; \mu+2 \leq u, v \leq n$ *and* $\lambda = \frac{1}{3}h^{\mu+1}_{\mu+1\mu+1}$, *where* $\mu = n_1 + \cdots + n_k$.

Proof. Let N be a Legendre submanifold of a Sasakian space form $M^{2n+1}(c)$. Then it follows from Proposition 12.4 that the curvature tensor $\tilde{R}$ of $M^{2n+1}(c)$ satisfies

$$\tilde{R}(X, Y; Z, W) = \frac{c+3}{4}\{\langle X, W\rangle \langle Y, Z\rangle - \langle X, Z\rangle \langle Y, W\rangle\}$$

for X, Y, X, W tangent to N. Thus, after applying Gauss's equation, we find that the curvature tensor R of N satisfies

$$\begin{aligned}
R(X, Y; Z, W) = \ &\frac{c+3}{4}\{\langle X, W\rangle \langle Y, Z\rangle - \langle X, Z\rangle \langle Y, W\rangle\} \\
&+ \langle h(X, W), h(Y, Z)\rangle - \langle h(X, Z), h)Y, W)\rangle .
\end{aligned}$$

It follows from Lemmas 16.1 and 16.2 that $A_\xi = 0$ and $\sigma(X, Y, Z) := \langle h(X, Y), \phi Z\rangle$ is totally symmetric. Hence, we may apply the same argument as the proof of Theorem 15.4 to derive (16.4).

The remaining part can be proved in the same way as the corresponding part of Theorem 15.4. □

For a real number $\lambda > 0$, an integer $n \geq 3$, and a k-tuple $(n_1, \ldots, n_k) \in \mathcal{S}(n)$, we put

$$F = \sum_{i=1}^{k} \frac{3\lambda x_{\mu+1}}{2(2+n_i)} \sum_{\alpha_i \in \Delta_i} x^2_{\alpha_i} + \frac{\lambda x_{\mu+1}}{2} \sum_{r=\mu+1}^{n} x^2_r, \tag{16.6}$$

with $\mu = \sum_{i=1}^{k} n_i$ and $\Delta_1 = \{1, \ldots, n_1\}, \ldots, \Delta_k = \{\mu - n_k + 1, \ldots, \mu\}$.

Consider the mapping $\psi : \mathbf{R}^n \to \mathbb{R}^{2n+1}$ given by

$$\psi(x_1, \ldots, x_n) = \big(x_1, \ldots, x_n, F_{x_1}, \ldots, F_{x_n}, F\big). \tag{16.7}$$

Then

$$\frac{\partial\psi}{\partial x_i} = \Big(0, \ldots, 0, \overbrace{1}^{i-th}, 0, \ldots, 0, F_{x_1x_i}, \ldots, F_{x_nx_i}, F_{x_i}\Big) \tag{16.8}$$

for $i = 1, \ldots, n$.

Since

$$\eta = \frac{1}{2}dz - \frac{1}{2}\sum_{i=1}^{n} F_{x_i} dx_i,$$

it follows from (16.8) that $\eta(\frac{\partial}{\partial x_i}) = 0$ for $i = 1, \ldots, n$. Thus, (16.7) defines a Legendre immersion of $\mathbf{R}^n$ into $\mathbb{R}^{2n+1}$. Let $\pi : \mathbb{R}^{2n+1} \to \mathbf{C}^n$ denote the Riemannian submersion defined by

$$\pi(x_1, \ldots, x_n, y_1, \ldots, y_n, z) = \frac{1}{2}(y_1, \ldots, y_n, x_1, \ldots, x_n). \tag{16.9}$$

Then $\pi \circ \psi : \mathbf{R}^n \to \mathbf{C}^n$ is exactly the Lagrangian immersion $L : \mathbf{R}^n \to \mathbf{C}^n$ defined by (15.32) and (15.33). Thus, the second fundamental form of L satisfies (15.10) at the point o as we already seen in the proof of Theorem 15.4. Hence, by applying the Legendre-Lagrangian correspondence described in section 12.4 we conclude that (16.7) defines a non-minimal Legendre submanifold of $\mathbb{R}^{2n+1}$ which satisfies the equality case of (16.4) at o (for $c = -3$). Consequently, we obtain the following.

Proposition 16.3. *For each k-tuple $(n_1, \ldots, n_k) \in \mathcal{S}(n)$, there exists a non-minimal Legendre submanifold of $\mathbb{R}^{2n+1}$ which satisfies the equality case of (16.4) at a point.*

Remark 16.2. Proposition 16.3 shows that (16.4) cannot be improved.

16.3 Scalar and Ricci curvatures of Legendre submanifolds

For scalar curvature of Legendre submanifolds in Sasakian space forms, we have the following optimal inequality.

Corollary 16.3. *Let N be a Legendre submanifold of a Sasakian space form $M^{2n+1}(c)$. Then the scalar curvature τ of N*

$$\tau \leq \frac{n^2(n-1)}{2(n+2)}||H||^2 + \frac{(3+c)n(n-1)}{8}, \tag{16.10}$$

with the equality sign holding if and only if there is an orthonormal frame $\{e_1, \ldots, e_n\}$ such that the second fundamental form h satisfying

$$\begin{aligned} &h(e_1, e_1) = 3\mu\phi e_1, \ h(e_1, e_j) = \mu\phi e_j, \\ &h(e_i, e_j) = \mu\delta_{ij}\phi e_1, \ \ 2 \leq i, j \leq n, \end{aligned} \tag{16.11}$$

for some function μ.

Proof. When $k = 0$, Theorem 16.1 reduces to the corollary. □

Remark 16.3. For Legendre submanifolds in $\mathbb{R}^{2n+1}$, this corollary is due to [Blair and Carriazo (2001)].

For Ricci curvature of Legendre submanifolds in Sasakian space forms, we have the following optimal inequality.

Theorem 16.2. *Let N be a Legendre submanifold of a Sasakian space form $M^{2n+1}(c)$. Then the Ricci curvature of N satisfies*

$$Ric(u) \leq \frac{n(n-1)}{4}||H||^2 + \frac{(n-1)(3+c)}{4}, \quad \forall u \in T^1N, \tag{16.12}$$

with the equality sign holding identically if and only if either

(1) *N is totally geodesic, or*
(2) *$n = 2$ and the second fundamental form of N satisfies*

$$h(e_1, e_1) = 3\mu e_{1^*}, \ h(e_1, e_2) = \mu e_{2^*}, \ h(e_2, e_2) = \mu e_{1^*}$$

for some function μ with respect some orthonormal frame e_1, e_2 of N, where $e_{i^*} = \phi e_i$.

Proof. Let N be a Legendre submanifold of a Sasakian space form $M^{2n+1}(c)$. If we choose $k = 1$ and $n_1 = n - 1$, then (16.4) becomes (16.11). Now, by applying Lemma 12.1, Lemma 16.1 and (16.12), we may prove the remaining part of the theorem exactly in the same way as that of Theorem 10.4. □

Remark 16.4. It follows from section 12.4 that there exists a one-to-one correspondence between Legendre submanifolds in the Sasakian $S^{2n+1}(1)$ (resp. $\mathbb{R}^{2n+1}$ or $H_1^{2n+1}(-1)$) whose second fundamental form satisfy (16.5) and Lagrangian submanifolds in $CP^n(4)$ (resp. $\mathbf{C}^n$ or $CH^n(-4)$) whose second fundamental form satisfy (15.10). Thus, in view of Remark 15.4, Corollaries 15.8 and 16.3, Proposition 15.2, and Theorem 16.2, we know that there exist Legendre submanifolds in Sasakian space forms whose scalar curvature satisfy the equality case of (16.10) or whose Ricci curvature satisfy the equality case of (16.12).

16.4 Contact δ-invariants $\tilde{\delta}^c(n_1, \ldots, n_k)$ and applications

Let (N, ϕ, ξ, η, g) be a $(2n+1)$-dimensional almost contact metric manifold with $n \geq 1$ and let $(n_1, \ldots, n_k) \in \mathcal{S}(2n+1)$. Then the *contact δ-invariant* $\tilde{\delta}^c(n_1, \ldots, n_k)$ is defined by

$$\tilde{\delta}^c(n_1, \ldots, n_k)(p) = \tau(p) - \inf\{\tau(L_1) + \cdots + \tau(L_k)\},$$

where $L_1, L_2, \ldots, L_k$ run over all mutually orthogonal subspaces of T_pN such that L_1 contains the characteristic vector field ξ and $\dim L_j = n_j$ for $j = 1, \ldots, k$. Clearly, $\tilde{\delta}^c(n_1, \ldots, n_k)$ depends on the almost contact metric structure of N.

Definition 16.1. Let N be an almost contact metric manifold and let $L_1, \ldots, L_k$ be mutually orthogonal subspaces of T_pN with $\dim L_j \geq 2$, $j = 1, \ldots, k$. A plane section $\pi \subset T_pN$ is said to be *orthogonal to* $L_1, \ldots, L_k$ if there exists orthonormal vectors $\{\bar{e}_1, \bar{e}_2\}$ such that $\pi = \mathrm{Span}\{\bar{e}_1, \bar{e}_2\}$ and, moreover, one of the following three cases occurs:

(1) $\bar{e}_1 \in L_i$ and $\bar{e}_2 \in L_j$ with $1 \leq i \neq j \leq k$;

(2) $\bar{e}_1 \in L_i$ for some $i \in \{1, \ldots, k\}$ and $\bar{e}_2 \perp L_1, \ldots, L_k$;

(3) $\bar{e}_1, \bar{e}_2 \perp L_1, \ldots, L_k$.

We call an orthonormal frame $\{e_1, \ldots, e_{2n+1}\}$ of TN an *orthonormal ξ-frame* if e_1 is parallel to ξ.

Theorem 16.3. *Let N be an almost contact metric $(2n+1)$-manifold isometrically immersed in a Riemannian m-manifold M. Then, for each point $p \in N$ and each k-tuple $(n_1, \ldots, n_k) \in \mathcal{S}(2n+1)$, we have*

$$\tilde{\delta}^c(n_1, \ldots, n_k) \leq c(n_1, \ldots, n_k)||H||^2 + b(n_1, \ldots, n_k) \max \tilde{K}, \tag{16.13}$$

where $\max \tilde{K}(p)$ is the maximum of the sectional curvature of M restricted to 2-plane sections of the tangent space T_pN.

Moreover, the equality case of (16.13) holds at p if and only if the following two conditions hold:

(a) *there exist an orthonormal ξ-basis $\{e_1, \ldots, e_{2n+1}\}$ of T_pN and an orthonormal basis $\{e_{2n+2}, \ldots, e_m\}$ of $T_p^{\perp}N$ such that the shape operator with respect to $\{e_1, \ldots, e_m\}$ satisfies*

$$A_{e_r} = \begin{pmatrix} A_1^r & \ldots & 0 & \\ \vdots & \ddots & \vdots & 0 \\ 0 & \ldots & A_k^r & \\ & 0 & & \mu_r I \end{pmatrix}, \quad r = 2n+2, \ldots, m, \tag{16.14}$$

where I is an identity matrix and A_j^r are symmetric $n_j \times n_j$ submatrices satisfying

$$\mathrm{trace}\,(A_1^r) = \cdots = \mathrm{trace}\,(A_k^r) = \mu_r; \tag{16.15}$$

(b) *there exist k mutually orthogonal subspaces $L_1, \ldots, L_k$ of T_pN with $\xi \in L_1$ and $\tilde{\delta}^c(n_1, \ldots, n_k)(p) = \tau(p) - \sum_{j=1}^k \tau(L_j)$ such that any plane section $\pi \subset T_pN$ orthogonal to $L_1, \ldots, L_k$ satisfies $\tilde{K}(\pi) = \max \tilde{K}(p)$.*

Proof. This can be proved in the same way as Theorem 13.3 □

As an immediate consequence of Theorem 16.3, we have the following.

Theorem 16.4. *Let N be an almost contact metric $(2n+1)$-manifold isometrically immersed in a real space form $R^m(c)$ of constant curvature c. Then for any k-tuple $(n_1,\dots,n_k)\in\mathcal{S}(2n+1)$ we have*

$$\tilde{\delta}^c(n_1,\dots,n_k)\le c(n_1,\dots,n_k)||H||^2+b(n_1,\dots,n_k)c. \tag{16.16}$$

The equality case of inequality (16.16) *holds at a point $p\in N$ if and only if there exists an orthonormal ξ-basis $\{e_1,\dots,e_{2n+1}\}$ of T_pN and an orthonormal basis $\{e_{2n+2},\dots,e_m\}$ of $T_p^{\perp}N$ such that the shape operator with respect to $\{e_1,\dots,e_m\}$ satisfies*

$$A_{e_r}=\begin{pmatrix} \begin{matrix} A_1^r & \dots & 0\\ \vdots & \ddots & \vdots\\ 0 & \dots & A_k^r\end{matrix} & 0\\ 0 & \mu_r I\end{pmatrix},\quad r=2n+2,\dots,m, \tag{16.17}$$

where A_j^r are symmetric $n_j\times n_j$ submatrices satisfying

$$\operatorname{trace}(A_1^r)=\cdots=\operatorname{trace}(A_k^r)=\mu_r. \tag{16.18}$$

Remark 16.5. Theorems 16.3 and 16.4 are due to [Chen and Martín-Molina (2010)]. A special case was obtained in [Chen and Mihai (2005)]. Notice that Theorem 16.4 is quite different from Propositions 16.1 and 16.2, since the target space in Theorem 16.4 is a real space form instead of Sasakian space forms in the later case.

Theorem 16.4 implies immediately the following.

Corollary 16.4. *Let N be an almost contact metric $(2n+1)$-manifold. If there is a k-tuple $(n_1,\dots,n_k)\in\mathcal{S}(2n+1)$ such that $\tilde{\delta}^c(n_1,\dots,c_k)(p)>0$ at some point p, then N cannot be isometrically immersion in any Euclidean space as a minimal submanifold.*

Definition 16.2. Let N be an almost contact metric $(2n+1)$-manifold immersed in a Riemannian m-manifold M^m. If N satisfies the equality case of (16.13) for some k-tuple $(n_1,\dots,n_k)$, then an orthonormal ξ-frame $\{e_1,\dots,e_{2n+1}\}$ satisfying (16.14) and (16.15) is called an *adapted ξ-frame.*

Let N be an almost contact metric $(2n+1)$-manifold isometrically immersed in a Riemannian m-manifold M. If N satisfies the equality case of

(16.13) for some k-tuple $(n_1, \ldots, n_k) \in \mathcal{S}(2n+1)$. Then, with respect to an adapted ξ-frame $\{e_1, \ldots, e_{2n+1}\}$, we have

$$h(e_\alpha, Y) = \frac{(2n+1)\langle e_\alpha, Y\rangle}{2n+k+1-\sum_{j=1}^k n_j} H, \ \ \forall Y \in TN, \tag{16.19}$$

for $\alpha \in \{1 + \sum_{j=1}^k n_j, \ldots, 2n+1\}$.

The following result was given in [Chen and Martín-Molina (2010)].

Proposition 16.4. *Let N be a contact metric $(2n+1)$-manifold isometrically immersed in a real space form $R^m(c)$. If N satisfies the equality case of (16.16) for a k-tuple $(n_1, \ldots, n_k) \in \mathcal{S}(2n+1)$ with $\sum_{j=1}^k n_j \le n$, then N is a minimal submanifold of $R^m(c)$.*

Proof. Let N be a contact metric $(2n+1)$-manifold isometrically immersed in $R^m(c)$ which satisfies the equality case of (16.16) for some k-tuple $(n_1, \ldots, n_k)$ with $\sum_{j=1}^k n_j \le n$. Then Theorem 16.4 implies that there exists an adapted ξ-basis $\{e_1, \ldots, e_{2n+1}\}$ of T_pN and an orthonormal basis $\{e_{2n+2}, \ldots, e_m\}$ of $T_p^\perp N$ such that (16.17) and (16.18) hold.

Now, assume that N is non-minimal in $R^m(c)$. Put

$$\mathcal{D}(p) = \left\{X \in T_pN : h(X,Y) = \frac{(2n+1)\langle X, Y\rangle}{2n+k+1-\sum_{j=1}^k n_j} H, \forall Y \in T_pN\right\}.$$

It follows from (16.19) that $\dim \mathcal{D}(p) \ge 2n+1-\sum_{j=1}^k n_j$. Clearly, $\dim \mathcal{D}(p)$ is constant on some nonempty open submanifold, say U. Since $H \ne 0$, (16.17) and (16.18) imply that, $\eta(X) = 0$ for $X \in \mathcal{D}$. Thus, $\mathcal{D}$ is a contact distribution on $U \subset N$. Let $\mathcal{D}^\perp$ be the orthogonal complementary distribution of $\mathcal{D}$. Then $h(\mathcal{D}, \mathcal{D}^\perp) = \{0\}$. Thus

$$(\bar\nabla_X h)(Y,Z) = -h(\nabla'_X Y, Z) - h(Y, \nabla'_X Z)$$

$X, Y \in \mathcal{D}$ and $Z \in \mathcal{D}^\perp$. After applying the equation of Codazzi, we find

$$\begin{aligned} h([X,Y],Z) &= -(\bar\nabla_X h)(Y,Z) - h(\nabla'_Y X, Z) - h(Y, \nabla'_X Z) \\ &= h(X, \nabla'_Y Z) - h(Y, \nabla'_X Z) \\ &= \frac{2n+1}{2n+k+1-\sum_{j=1}^k n_j} (\langle X, \nabla'_Y Z\rangle - \langle Y, \nabla'_X Z\rangle) H \\ &= \frac{2n+1}{2n+k+1-\sum_{j=1}^k n_j} \langle [X,Y], Z\rangle H, \end{aligned}$$

which gives $[X, Y] \in \mathcal{D}$. Hence $\mathcal{D}$ is an integrable distribution on U whose leaves are integral submanifolds of N.

On the other hand, since the maximal dimension of integral submanifolds is n, we get $2n + 1 - \sum_{j=1}^k n_j \le n$. This contradicts the assumption $\sum_{j=1}^k n_j \le n$. □

Remark 16.6. The condition $\sum_{j=1}^{k} n_i \leq n$ in Proposition 16.4 is necessary. This can be seen as follows: Consider $\psi : \mathbb{R} \times S^2(1) \to \mathbb{E}^4$ defined by

$$\psi(t, \theta, \varphi) = \big(t, \cos\theta\cos\varphi, \sin\theta\cos\varphi, \sin\varphi\big),$$

where $\mathbb{E}^4$ is the Euclidean 4-space endowed with the standard flat metric.

Define a contact metric structure (ϕ, ξ, η, g) on $\mathbb{R} \times S^2(1)$ by

$$\begin{aligned}
&\eta = \cos\theta dt + \sin\theta d\varphi,\\
&\xi = \cos\theta\frac{\partial}{\partial t} + \sin\theta\frac{\partial}{\partial\varphi},\\
&\phi\Big(\frac{\partial}{\partial t}\Big) = -\tan\theta\frac{\partial}{\partial\theta}, \quad \phi\Big(\frac{\partial}{\partial\varphi}\Big) = \frac{\partial}{\partial\theta},\\
&\phi\Big(\frac{\partial}{\partial\theta}\Big) = \cos\varphi\left(\sin\theta\frac{\partial}{\partial t} - \cos\theta\frac{\partial}{\partial\varphi}\right),\\
&g = dt^2 + d\varphi^2 + \cos^2\varphi d\theta^2.
\end{aligned}$$

Then

$$\begin{aligned}
&\eta \wedge d\eta = dt \wedge d\theta \wedge d\varphi \neq 0,\\
&g(\phi X, \phi Y) = g(X, Y) - \eta(X)\eta(Y).
\end{aligned}$$

Thus, this non-minimal contact metric hypersurface in $\mathbb{E}^4$ satisfies the equality case of (16.16) with $k = n = 1, n_1 = 2$ and $\dim N = 3$.

16.5 K-contact submanifold satisfying the basic equality

Theorem 16.5. *Let N be a K-contact $(2n+1)$-manifold isometrically immersed in a real space form $R^m(c)$. If there exists a k-tuple $(n_1, \ldots, n_k) \in \mathcal{S}(2n+1)$ with $\sum_{j=1}^{k} n_j \leq 2n$ such that the equality case of (16.16) holds, then $c \geq 1$. Moreover, if $c = 1$, then N is a Sasakian manifold whose characteristic vector field ξ lies in relative nullity subspace, i.e. $\xi \in \mathcal{N}$.*

Proof. Let N be a K-contact manifold isometrically immersed in $R^m(c)$. Assume that the equality case of (16.16) holds for a k-tuple $(n_1, \ldots, n_k) \in \mathcal{S}(2n+1)$ with $\sum_{j=1}^{k} n_j \leq 2n$. Then Theorem 16.4 implies that there exist an orthonormal ξ-basis $\{e_1, \ldots, e_{2n+1}\}$ of T_pN and an orthonormal basis $\{e_{2n+2}, \ldots, e_m\}$ of $T_p^{\perp}N$ such that (16.17) and (16.18) hold.

Since $\{e_1, \ldots, e_{2n+1}\}$ is an orthonormal ξ-frame, e_1 is parallel to ξ. Thus Gauss's equation and (16.17) imply that, for any unit tangent vector

e_j perpendicular to ξ, we have

$$K(\xi, e_j) = c + \langle h(e_1, e_1), h(e_j, e_j)\rangle - \langle h(e_1, e_j), h(e_1, e_j)\rangle$$
$$= \begin{cases} c + \langle h(e_1, e_1), h(e_j, e_j)\rangle - |h(e_1, e_j)|^2, & \text{if } j = 2, \ldots, n_1, \\ c + \langle h(e_1, e_1), h(e_j, e_j)\rangle, & \text{if } n_1 + 1 \leq j \leq 2n + 1. \end{cases}$$

On the other hand, since N is K-contact, $K(\xi, X) = 1$ for any $X \perp \xi$. Hence

$$\begin{aligned} n_1 - 2 &= \sum_{j=2}^{n_1} K(\xi, e_j) - K(\xi, e_{2n+1}) \\ &= (n_1 - 2)c + \Big\langle h(e_1, e_1), \sum_{j=1}^{n_1} h(e_j, e_j) \Big\rangle \\ &\quad - \sum_{j=1}^{n_1} |h(e_1, e_j)|^2 - \langle h(e_1, e_1), h(e_{2n+1}, e_{2n+1})\rangle . \end{aligned}$$

Thus, after applying (16.17) and (16.18), we deduce that

$$(n_1 - 2)(c - 1) = \sum_{j=1}^{n_1} |h(e_1, e_j)|^2 \geq 0. \tag{16.20}$$

If $n_1 > 2$, then (16.20) yields $c \geq 1$ immediately. If $n_1 = 2$, then (16.20) implies that $h(e_1, e_j) = 0$, $j = 1, \ldots, 2n + 1$. Combining this with (16.17) shows that $h(\xi, X) = 0$ for $X \in TN$, due to the fact that e_1 is parallel to ξ. Thus, from (16.19) and $K(\xi, X) = 1$ we get $c = 1$. Therefore, $c \geq 1$ holds for both cases.

Now, suppose that $c = 1$. Then (16.20) implies that $h(\xi, X) = 0$ for $X \in TN$. Hence, ξ lies in the relative nullity space $\mathcal{N}$.

Finally, after applying $h(\xi, X) = 0$ and Gauss's equation, we find

$$R(X, Y)\xi = \eta(Y)X - \eta(X)Y,$$

which implies that N is Sasakian (cf. [Blair (2010)]). □

For K-contact hypersurfaces satisfying the equality case of (16.16), we have the following [Chen and Martín-Molina (2010)].

Theorem 16.6. *Let N be a K-contact hypersurface of a real space form $R^{2n+2}(c)$. If N satisfies the equality case of inequality (16.16) for some k-tuple $(n_1, \ldots, n_k) \in S(2n+1)$ with $\sum_{j=1}^{k} n_j \leq 2n$, then $c = 1$. Moreover, N is a Sasakian manifold of constant curvature one immersed in $R^{2n+2}(1)$ as a totally geodesic hypersurface.*

Chapter 17

Applications to Affine Geometry

In 1872, F. Klein (1849-1925) stated his famous *Erlangen program*: geometry is the study of invariants with respect to a given transformation group. Affine differential geometry is an area of differential geometry in which the differential invariants are invariant under volume-preserving affine transformations. The name affine differential geometry follows from Klein's Erlangen program.

The basic difference between affine and Riemannian differential geometry is that in the affine differential geometry, one introduces volume elements over a manifold instead of metrics.

G. Tzitzéica proved in 1908 that for a surface in Euclidean 3-space $\mathbb{E}^3$ the ratio of the Gaussian curvature to the fourth power of the support function from the origin o is invariant under an affine transformation fixed o [Tzitzéica (1908)]. He defined an S-surface to be any surface for which this ratio is constant. Later, these S-surfaces are called proper affine spheres centered at o. W. Blaschke in collaboration with Berwald, Liebmann, Pick, Radon, Reidemeister and Thomsen among others, developed a systematic study of affine differential geometry during the period 1916–1923.

Blaschke and his school tried to develop the affine differential geometry as analogously as possible to Euclidean geometry. Later contributions of Cartan, Norden, Brathel, Calabi, Nomizu and others emphasized more structure aspects.

During the last three decades the study of affine differential geometry was influenced among others by the solution of problems like the classification of all affine complete locally strongly convex affine spheres and the affine Bernstein problem.

For general references on affine differential geometry, we refer to [Nomizu and Sasaki (1994); Simon (2000); Simon et al. (1991); Li et al. (1993)].

17.1 Affine hypersurfaces

Let $\mathbf{R}^m$ denote the standard real vector m-space. For an ordered pair (p,q) of points in $\mathbf{R}^m$, define $x = \overrightarrow{pq}$ to be the vector $q - p \in \mathbf{R}^m$. The vector space $\mathbf{R}^m$ together with the mapping

$$\mathbf{R}^m \times \mathbf{R}^m \to \mathbf{R}^m : (p,q) \mapsto \overrightarrow{pq}$$

is called an *affine m-space.*

Let Y be a vector field on $\mathbf{R}^m$ and let $x(t), a \leq t \leq b$, be a smooth curve in $\mathbf{R}^m$. If we take an affine coordinate system $\{x_1, \ldots, x_m\}$ and write

$$Y = \sum_{i=1}^{m} Y^i \frac{\partial}{\partial x_i} \quad \text{and} \quad x(t) = (x_1(t), \ldots, x_m(t)),$$

then the covariant derivative D_tY of Y along the curve $x(t)$ is defined by

$$D_tY = \sum_{i=1}^{m} \frac{dY^i(x_i)}{dt}\frac{\partial}{\partial x_i} = \sum_{i,j=1}^{m} \frac{\partial Y^i}{\partial x_j}\frac{dx_j}{dt}\frac{\partial}{\partial x_i}.$$

Thus, D_tY is a generalization of the directional derivative of functions to vector fields. If X is a tangent vector at a point x_o, then $\nabla_X Y$ is defined by $\nabla_X Y = (D_tY)_t$, where $x(t)$ is a curve with initial point x_o and initial tangent vector X. From this definition, ∇ is an affine connection.

Consider an affine m-space $\mathbf{R}^m$ with the usual affine connection ∇ as above. Then the curvature tensor associated with ∇ vanishes. Thus ∇ is a flat affine connection.

Definition 17.1. An immersion $\phi : N \to \mathbf{R}^{n+k}$ of an n-manifold N into $\mathbf{R}^{n+k}$ is called an *affine immersion* if there exists a rank-k distribution E on N (i.e., each E_p is k-dimensional) with $E_p \subset T_{\phi(p)}\mathbf{R}^{n+k}$ for each $p \in N$ such that

$$T_{\phi(p)}\mathbf{R}^{n+k} = \phi_*(T_p(M)) \oplus E_p \quad \text{(direct sum)}, \tag{17.1}$$

$$(\nabla_X(\phi_*(Y)))_p = \phi_{*_p}(\nabla'_X Y) + \alpha_p(X,Y), \tag{17.2}$$

for $X, Y \in \mathcal{X}(M)$, where $\alpha_p(X,Y)$ is the E_p-component of $(\nabla_X \phi_*(Y))_p$ according to the direct sum (17.1). And (17.2) decomposes $\nabla_X \phi_*(Y)$ into the tangential component $\phi_{*_p}(\nabla'_X Y)$ and the component $\alpha_p(X,Y)$ in E_p according to (17.2).

By *an affine hypersurface* N in $\mathbf{R}^{n+1}$ we mean a codimension-one affine immersion $\phi : N \to \mathbf{R}^{n+1}$ with a transversal vector field ξ, i.e., ξ is a vector field which is not tangent to N at each point on N.

The *formulas of Gauss and Weingarten* for an affine hypersurface $\phi : N \to \mathbf{R}^{n+1}$ are given respectively by

$$\nabla_X \phi_*(Y) = \phi_*(\nabla'_X Y) + h(X,Y)\xi, \tag{17.3}$$

$$\nabla_X \xi = -\phi_*(SX) + \tau(X)\xi, \tag{17.4}$$

for $X, Y \in \mathcal{X}(N)$, where $\phi_*(\nabla'_X Y)$ and $-\phi_*(SX)$ are the tangential components of $\nabla_X \phi_*(Y)$ and $\nabla_X \xi$, respectively.

The induced affine connection ∇', the $(1,1)$-tensor S, the 1-form τ, and the symmetric $(0,2)$-tensor h are called the *induced (affine) connection*, the *affine shape operator*, the *torsion form*, and the *affine fundamental form* (relative to the transversal vector field ξ), respectively.

The transversal vector field ξ of $\phi : N \to \mathbf{R}^{n+1}$ is called a *relative normalization* (or *equiaffine*) if $\tau = 0$ holds identically, i.e., $\nabla_X \xi$ is tangent to N for each $X \in TN$.

Notice that ∇' and h depend upon the choice of transverse vector field ξ. We consider only those hypersurfaces for which h is non-degenerate. Interestingly, this is a property of the hypersurface N and not depend upon the choice of transverse vector field ξ. When h is nondegenerate, it defines a pseudo-Riemannian metric on N, called the *relative metric*. Throughout this book, we assume that the affine fundamental form h is definite.

For tangent vectors $X_1, \ldots, X_n$ of N let (h_{ij}) be the $n \times n$ matrix given by $h_{ij} = h(X_i, X_j)$. Define an (affine) *volume element* v on N given by

$$v(X_1, \ldots, X_n) = \det(h_{ij})^{1/2}.$$

Again, this is a natural definition to make. If $N = \mathbb{E}^n$ and h is the Euclidean inner product, then $v(X_1, \ldots, X_n)$ is the standard Euclidean volume spanned by $X_1, \ldots, X_n$. Since h depends on the choice of transverse vector field ξ, it follows that v does too. Let $\hat{\nabla}$, $\hat{K}$, $\hat{R}$ and $\hat{\kappa}$ denote the Levi-Civita connection, the sectional curvature, the curvature tensor and the normalized scalar curvature of h, respectively.

The *difference tensor* K is the symmetric $(1,2)$-tensor field defined by

$$K_X Y = K(X,Y) = \nabla_X Y - \hat{\nabla}_X Y. \tag{17.5}$$

The *Tchebychev form* T, the *Tchebychev vector field* $T^\#$, the *Pick invariant* J, and the *relative theorema egregium* are given by

$$T(X) = \frac{1}{n}\,\mathrm{Trace}\,\{Y \to K(X,Y)\}, \tag{17.6}$$

$$h(T^\#, X) = T(X), \tag{17.7}$$

$$h(C,C) = 4h(K,K) = 4n(n-1)J, \tag{17.8}$$

$$\hat{\kappa} = J + H - \frac{n}{n-1} h(T^\#, T^\#), \tag{17.9}$$

where $H = (1/n)\,\mathrm{trace}\, S$ is the *relative mean curvature*.

Let us consider a more general situation. Suppose an n-manifold N is provided with a nondegenerate metric, say h, and an affine connection ∇^1. Define a new affine connection ∇^2 by requiring that

$$Xh(Y,Z) = h(\nabla^1_X Y, Z) + h(Y, \nabla^2_X Z) \tag{17.10}$$

holds for all $X, Y, Z \in \mathcal{X}(N)$. Equation (17.10) determines a unique connection ∇^2, which is called the *conjugate connection*[1] of ∇^1 relative to h.

Proposition 17.1. *Let ∇^2 be the conjugate connection of ∇^1 relative to h. Then the torsion tensors T_1 and T_2 of ∇^1 and ∇^2 satisfy*

$$(\nabla^1_X h)(Y,Z) + h(Y, T_1(X,Z)) = (\nabla^1_Z h)(Y,X) + h(Y, T_2(X,Z)). \tag{17.11}$$

Proof. It follows from (17.10) and the definition of $(\nabla^1_X h)(Y,Z)$ that

$$(\nabla^1_X h)(Y,Z) = h(Y, \nabla^2_X Z) - h(Y, \nabla^1_X Z),$$
$$(\nabla^1_Z h)(Y,X) = h(Y, \nabla^2_Z X) - h(Y, \nabla^1_Z X).$$

Thus, by applying the definition of torsion tensor fields, we have

$$\begin{aligned}
&(\nabla^1_X h)(Y,Z) - (\nabla^1_Z h)(Y,X) \\
&\quad = h(Y, \nabla^2_X Z - \nabla^2_Z X) - h(Y, \nabla^1_X Z - \nabla^1_Z X) \qquad (17.12)\\
&\quad = h(Y, [X,Z] + T_2(X,Z)) - h(Y, [X,Z] + T_1(X,Z)) \\
&\quad = h(Y, T_2(X,Z) - T_1(X,Z)).
\end{aligned}$$

□

An immediate consequence of Proposition 17.1 is the following.

Corollary 17.1. *Let ∇^1 be a torsion-free affine connection on N. Then the conjugate connection ∇^2 of ∇^1 is torsion-free if and only if the pair (∇^1, h) satisfies Codazzi's equation $(\nabla^1_X h)(Y,Z) = (\nabla^1_Z h)(Y,X)$ for $X, Y, Z \in TN$.*

If (∇^1, h) satisfies Codazzi's equation, the totally symmetric $(0,3)$-tensor $C(X,Y,Z) = (\nabla^1_X h)(Y,Z)$ is called the *cubic form*.

17.2 Centroaffine hypersurfaces

Let N be an n-manifold and let $\phi : N \to \mathbf{R}^{n+1}$ be a non-degenerate hypersurface whose position vector field is nowhere tangent to N. Then ϕ can be regarded as a transversal field along itself. We call $\xi = -\phi$ the *centroaffine*

[1] A pseudo-Riemannian manifold (N, h) with a pair of conjugate connections is called a *statistical manifold*. It is shown in [Amari (1985)] that statistical manifolds play some important role in statistics.

normal. The ϕ together with this normalization is called a *centroaffine hypersurface.*

The *centroaffine structure equations* are given by

$$\nabla_X \phi_*(Y) = \phi_*(\nabla'_X Y) + h(X,Y)\xi, \tag{17.13}$$

$$\nabla_X \xi = -\phi_*(X), \tag{17.14}$$

where ∇ is the canonical flat connection of $\mathbf{R}^{n+1}$, the induced centroaffine connection ∇' is a torsion-free connection, and h is a non-degenerate symmetric $(0,2)$-tensor field, called the *centroaffine metric.*

The corresponding equations of Gauss and Codazzi are given by

$$R(X,Y)Z = h(Y,Z)X - h(X,Z)Y, \tag{17.15}$$

$$(\nabla'_X h)(Y,Z) = (\nabla'_Y h)(X,Z). \tag{17.16}$$

Moreover, we have

$$(\nabla'_X h)(Y,Z) = -2h(K_X Y, Z). \tag{17.17}$$

From now on we assume that the centroaffine hypersurface is definite, i.e., h is definite, so h defines a pseudo-Riemannian metric on N. In order to consider only a positive-definite metric, we now make the following changes: if h is negative definite, we introduce a transversal vector field $\xi = -\phi$ and a $(0,2)$-tensor h is replaced by $-h$. In both cases, (17.13) and (17.16) hold, whereas (17.14) and (17.15) change sign. In case $\xi = -\phi$, we call N positive definite; in case $\xi = \phi$, we call N negative definite.

It is known that the centroaffine metric h is definite if and only if the hypersurface is *locally strongly convex.* For this the following terminology is used:

(i) The centroaffine hypersurface N is said to be of *elliptic type* if, for any point $\phi(p) \in \mathbf{R}^{n+1}$ with $p \in N$, the origin of $\mathbf{R}^{n+1}$ and the hypersurface are on the same side of the tangent hyperplane $\phi_*(T_pN)$; in this case the centroaffine normal vector field is given by $\xi = -\phi$.

(ii) The centroaffine hypersurface N is said to be of *hyperbolic type* if, for any point $\phi(p) \in \mathbf{R}^{n+1}$, the origin of $\mathbf{R}^{n+1}$ and the hypersurface are on the different side of the tangent hyperplane $\phi_*(T_pN)$; in this case the centroaffine normal vector field is given by $\xi = \phi$.

Consider a hypersurface N of the equiaffine space. If the Tchebychev form T vanishes on N, then N is called a *proper affine hypersphere* centered at the origin. If the difference tensor K vanishes, then N is a quadric, centered at the origin. In particular, an ellipsoid if N is positive definite; and a two-sheeted hyperboloid if N is negative definite.

For centroaffine geometry one has

$$h(K_XY, Z) = h(Y, K_XZ), \tag{17.18}$$

$$\hat{R}(X,Y)Z = K_YK_XZ - K_XK_YZ + \varepsilon(h(Y,Z)X - h(X,Z)Y), \tag{17.19}$$

$$(\hat{\nabla}K)(X,Y,Z) = (\hat{\nabla}K)(Y,Z,X) = (\hat{\nabla}K)(Z,X,Y), \tag{17.20}$$

$$\hat{\kappa} = J + \varepsilon - \frac{n}{n-1}h(T^{\#}, T^{\#}), \tag{17.21}$$

where $\varepsilon = 1$ or -1 according to N is of elliptic or of hyperbolic type. Equation (17.18) implies that K_X is self-adjoint with respect to h.

17.3 Graph hypersurfaces

An affine hypersurface $\phi : N \to \mathbf{R^{n+1}}$ is called a *graph hypersurface* if the transversal vector field ξ is a constant vector field. A result of [Nomizu and Pinkall (1987)] states that N locally is affine equivalent to the graph immersion of some function F. Again in case that h is nondegenerate, it defines a semi-Riemannian metric, called the *Calabi metric* of the graph hypersurface.

For graph hypersurfaces, the following equations hold as well:

$$\nabla_X\phi_*(Y) = \phi_*(\nabla'_XY) + h(X,Y)\xi, \tag{17.22}$$

$$(\nabla'_Xh)(Y,Z) = (\nabla'_Yh)(X,Z), \tag{17.23}$$

$$h(K_XY, Z) = h(Y, K_XZ), \tag{17.24}$$

$$\hat{R}(X,Y)Z = K_YK_XZ - K_XK_YZ, \tag{17.25}$$

$$(\nabla'_Xh)(Y,Z) = -2h(K_XY, Z). \tag{17.26}$$

However, equations (17.14), (17.15), (17.19) and (17.21) shall be replaced respectively by

$$\nabla_X\xi = 0, \tag{17.27}$$

$$R(X,Y)Z = 0, \tag{17.28}$$

$$\hat{R}(X,Y)Z = K_YK_XZ - K_XK_YZ, \tag{17.29}$$

$$\hat{\kappa} = J - \frac{n}{n-1}h(T^{\#}, T^{\#}), \tag{17.30}$$

respectively. If $T = 0$, the graph hypersurface is called an *improper affine hypersphere*. If we choose $\xi = (0, \ldots, 0, 1)$, then we can assume that locally

N is given by $x_{n+1} = F(x_1, \ldots, x_n)$. It turns out that $(x_1, \ldots, x_n)$ are ∇'-flat coordinates on N and that the Calabi metric is given by

$$h\left(\frac{\partial}{\partial x_i}, \frac{\partial}{\partial x_j}\right) = \frac{\partial^2 F}{\partial x_i \partial x_j}. \tag{17.31}$$

Moreover, N is an improper affine sphere if and only if the Hessian determinant $\det\left(\frac{\partial^2 F}{\partial x_i \partial x_j}\right)$ is constant.

Let N_1 and N_2 be two improper affine hyperspheres in $\mathbf{R}^{p+1}$ and $\mathbf{R}^{q+1}$ defined respectively by the equations $x_{p+1} = F_1(x_1, \ldots, x_p)$ and $y_{q+1} = F_2(y_1, \ldots, y_q)$. Then one can define a new improper affine hypersphere N in $\mathbf{R}^{p+q+1}$ by

$$z = F_1(x_1, \ldots, x_p) + F_2(y_1, \ldots, y_q), \tag{17.32}$$

where $(x_1, \ldots, x_p, y_1, \ldots, y_q, z)$ are the coordinates on $\mathbf{R}^{p+q+1}$. The affine normal of N is $(0, \ldots, 0, 1)$. Obviously, the Calabi metric on N is the product metric. Following [Dillen and Vrancken (1994)] we call this composition the *Calabi composition* of N_1 and N_2.

17.4 A general optimal inequality for affine hypersurfaces

Let N be a definite centroaffine hypersurface in $\mathbf{R}^{n+1}$. Analogous to invariant $\delta(n_1, \ldots, n_k)$, one can define *affine δ-invariant* $\delta^{\#}(n_1, \ldots, n_k)$ by

$$\delta^{\#}(n_1, \ldots, n_k)(p) = \hat{\tau}(p) - \sup\{\hat{\tau}(L_1) + \cdots + \hat{\tau}(L_k)\}, \tag{17.33}$$

where $L_1, \ldots, L_k$ run over all k mutually h-orthogonal subspaces of T_pN such that $\dim L_j = n_j$, $j = 1, \ldots, k$ and $\hat{\tau}(L)$ is the scalar curvature of L *with respect to induced affine metric h.*

We have the following optimal inequality.

Theorem 17.1. *Let N be a definite centroaffine hypersurface in $\mathbf{R}^{n+1}$. Then, for each k-tuple $(n_1, \ldots, n_k) \in \mathcal{S}(n)$, we have*

$$\begin{aligned} \delta^{\#}(n_1, \ldots, n_k) &\geq \frac{1}{2}\varepsilon b(n_1, \ldots, n_k) \\ &+ \frac{n^2\{(n - \sum_{i=1}^k n_i + 3k - 1) - 6\sum_{i=1}^k (2+n_i)^{-1}\}}{2\{(n - \sum_{i=1}^k n_i + 3k + 2) - 6\sum_{i=1}^k (2+n_i)^{-1}\}} ||T^{\#}||_h^2, \end{aligned} \tag{17.34}$$

where $||T^{\#}||_h^2 = h(T^{\#}, T^{\#})$ and $\varepsilon = 1$ or -1, according to N is positive definite or negative definite.

The equality sign holds at a point $p \in N$ if and only if there exists an h-orthonormal basis $\{e_1, \ldots, e_n\}$ at p such that, with respect to this basis, K takes the following form

$$\begin{aligned}
&K(e_{\alpha_i}, e_{\beta_i}) = \sum\nolimits_{\gamma_i} K^{\gamma_i}_{\alpha_i\beta_i} e_{\gamma_i} + \frac{3\delta_{\alpha_i\beta_i}}{2+n_i}\lambda e_{\mu+1},\\
&\sum\nolimits_{\alpha_i=1}^{n_i} K^{\gamma_i}_{\alpha_i\alpha_i} = 0, \qquad\qquad (17.35)\\
&K(e_{\alpha_i}, e_{\alpha_j}) = 0, \ \ i \neq j,\\
&K(e_{\alpha_i}, e_{\mu+1}) = \frac{3\lambda}{2+n_i} e_{\alpha_i}, \ \ K(e_{\alpha_i}, e_u) = 0,\\
&K(e_{\mu+1}, e_{\mu+1}) = 3\lambda e_{\mu+1}, \ \ K(e_{\mu+1}, e_u) = \lambda e_u,\\
&K(e_u, e_v) = \lambda\delta_{uv} e_{\mu+1}, \quad \lambda = \frac{1}{3} K^{\mu+1}_{\mu+1\mu+1},
\end{aligned}$$

for $1 \leq i, j \leq k; \mu + 2 \leq u, v \leq n$, where $\mu = n_1 + \cdots + n_k$.

Proof. Let N be a definite centroaffine hypersurface. Then it follows from (17.18) and (17.19) that the curvature tensor $\hat{R}$ of (N, h) satisfies

$$\begin{aligned}
h(\hat{R}(X,Y)Z,W) &= h(K_X Z, K_Y W) - h(K_Y Z, K_X W)\\
&\quad + \varepsilon(h(X,W)h(Y,Z) - h(X,Z)h(Y,W)),
\end{aligned}$$

which is the same as the corresponding formula for Lagrangian submanifolds in a complex space form of constant holomorphic sectional curvature 4ε. Moreover, (17.18) shows that K is totally symmetric. Consequently, we may apply the same arguments as the proof of Theorem 15.3 to obtain the desired result. □

Remark 17.1. Theorem 17.1 improves some results of [Chen et al. (2005a); Bolton et al. (2009); Scharlach et al. (1997)].

Two immediate consequences of Theorem 17.1 are the following [Chen et al. (2005a)].

Corollary 17.2. *Let N be a Riemannian n-manifold. If there is a k-tuple $(n_1, \ldots, n_k)$ in $\mathcal{S}(n)$ such that*

$$\delta^{\#}(n_1, \ldots, n_k) < \frac{b(n_1, \ldots, n_k)}{2} \tag{17.36}$$

at some point, then N cannot be realized as an elliptic proper affine hypersphere in $\mathbf{R}^{n+1}$. In particular, if there is a k-tuple $(n_1, \ldots, n_k)$ such that $\delta^{\#}(n_1, \ldots, n_k) \leq 0$ at some point in N, then N cannot be realized as an elliptic proper affine hypersphere in $\mathbf{R}^{n+1}$.

Corollary 17.3. *Let N be a Riemannian n-manifold. If there is a k-tuple $(n_1, \ldots, n_k)$ in $\mathcal{S}(n)$ such that*

$$\delta^{\#}(n_1, \ldots, n_k) < -\frac{b(n_1, \ldots, n_k)}{2} \tag{17.37}$$

at some point, then N cannot be realized as a hyperbolic proper affine hypersphere in $\mathbf{R}^{n+1}$.

A *hyperovaloid* in $\mathbf{R}^{n+1}$ is a compact strictly convex hypersurface embedded in $\mathbf{R}^{n+1}$. The inequalities in (17.33) give rise to some global centroaffine curvature invariants for hyperovaloids. They also provide simple characterizations of hyperellipsoids in terms of these global invariants.

Corollary 17.4. *Consider a centroaffine hyperovaloid $\phi : N \to \mathbf{R}^{n+1}$ with* $\dim N = n \geq 3$ *and with normalization as in* (17.13) *with $\xi = -\phi$. Then for any $(n_1, \ldots, n_k) \in \mathcal{S}(n)$ we have*

$$\begin{aligned} &\int_N \Bigg(\delta^{\#}(n_1, \ldots, n_k) \\ &\qquad - \frac{n^2\{n - \sum_{i=1}^k n_i + 3k - 1 - 6\sum_{i=1}^k (2+n_i)^{-1}\}}{2\{n - \sum_{i=1}^k n_i + 3k + 2 - 6\sum_{i=1}^k (2+n_i)^{-1}\}} ||T^{\#}||_h^2 \Bigg) \omega(h) \\ &\geq \frac{1}{2} b(n_1, \ldots, n_k) \mathrm{vol}(N, h), \end{aligned} \tag{17.38}$$

where $\omega(h)$ is the volume form and $\mathrm{vol}(N, h)$ *is the volume of (N, h).*

The equality sign of (17.38) *holds if and only if N is a hyperellipsoid centered at the origin.*

Proof. Inequality (17.38) is an easy consequence of (17.33). If the equality case of (17.38) holds, then Theorem 17.1 implies that N is a proper affine hypersphere. Hence, by applying a theorem of Blaschke and Deicke (see page 124 of [Nomizu and Sasaki (1994)]), we conclude that N is a hyperellipsoid centered at the origin.

The converse is easy to see. □

For graph hypersurfaces we have the following.

Theorem 17.2. *Let N be a graph hypersurface in $\mathbf{R}^{n+1}$ with positive definite Calabi metric. Then, for each k-tuple $(n_1, \ldots, n_k) \in \mathcal{S}(n)$, we have*

$$\begin{aligned} &\delta^{\#}(n_1, \ldots, n_k) \\ &\qquad \geq \frac{n^2\{n - \sum_{i=1}^k n_i + 3k - 1 - 6\sum_{i=1}^k (2+n_i)^{-1}\}}{2\{n - \sum_{i=1}^k n_i + 3k + 2 - 6\sum_{i=1}^k (2+n_i)^{-1}\}} ||T^{\#}||_h^2. \end{aligned} \tag{17.39}$$

The equality holds in (17.39) *at a point* $p \in N$ *if and only if there exists an* h*-orthonormal basis* $\{e_1, \dots, e_n\}$ *at* p *such that, with respect to this basis,* K *takes the following form*

$$\begin{aligned}
&K(e_{\alpha_i}, e_{\beta_i}) = \sum\nolimits_{\gamma_i} K^{\gamma_i}_{\alpha_i\beta_i} e_{\gamma_i} + \frac{3\delta_{\alpha_i\beta_i}}{2+n_i}\lambda e_{\mu+1},\\
&\sum\nolimits_{\alpha_i=1}^{n_i} K^{\gamma_i}_{\alpha_i\alpha_i} = 0, \quad K(e_{\alpha_i}, e_{\alpha_j}) = 0, \quad i \neq j,\\
&K(e_{\alpha_i}, e_{\mu+1}) = \frac{3\lambda}{2+n_i} e_{\alpha_i}, \quad K(e_{\alpha_i}, e_u) = 0,\\
&K(e_{\mu+1}, e_{\mu+1}) = 3\lambda e_{\mu+1}, \quad K(e_{\mu+1}, e_u) = \lambda e_u,\\
&K(e_u, e_v) = \lambda \delta_{uv} e_{\mu+1}, \quad \lambda = \frac{1}{3} K^{\mu+1}_{\mu+1\mu+1},
\end{aligned}$$

for $1 \leq i, j \leq k; \mu + 2 \leq u, v \leq n$, *where* $\mu = n_1 + \cdots + n_k$.

Proof. For graph hypersurfaces we have

$$h(\hat{R}(X,Y)Z,W) = h(K_X Z, K_Y W) - h(K_Y Z, K_X W).$$

Hence, this theorem can be proved in the same way as Theorem 17.1. □

From Theorem 17.2 we immediately obtain the following [Dillen and Vrancken (2005)].

Corollary 17.5. *If* (N,h) *is a Riemannian* n*-manifold such that the affine* δ*-invariant of* (N,h) *satisfies* $\delta^{\#}(n_1,\dots,n_k) < 0$ *at some point* $p \in N$ *for some* k*-tuple* $(n_1,\dots,n_k) \in \mathcal{S}(n)$, *then* (N,h) *cannot be realized as improper affine sphere in some affine space.*

In order to show that inequality (17.39) is the best possible, we provide the following.

Example 17.1. Let λ be a positive number. Consider the graph hypersurface F given by

$$F : \mathbb{R}^n \to \mathbb{R}^{n+1} : (x_1,\dots,x_n) \mapsto (x_1,\dots,x_n, f(x_1,\dots,x_n)),$$

where

$$f(x_1,\dots,x_n) = -\sum_{i=1}^{k} \frac{3\lambda}{2+n_i} \sum_{\alpha_i \in \Delta_i} x_{\alpha_i}^2 x_{\mu+1} - \lambda \sum_{r=\mu+1}^{n} x_{\mu+1} x_r^2$$

with $\mu = \sum_{i=1}^{k} n_i$. Then at the origin o we have

$$\left(\nabla'_{\frac{\partial F}{\partial x_i}} h\right)\left(\frac{\partial F}{\partial x_j}, \frac{\partial F}{\partial x_k}\right)(0,\dots,0) = \frac{\partial^3 f}{\partial x_i \partial x_j \partial x_k}(0,\dots,0).$$

Thus, after computing the difference tensor K using (17.26), we conclude from Theorem 17.2 that F attains the equality of (17.39) at the point o. Hence, this hypersurface verifies the equality case of (17.39) at o and the Tchebychev vector field does not vanish. Therefore, the constant in (17.39) cannot be improved.

17.5 A realization problem for affine hypersurfaces

For a Riemannian n-manifold (N, g) equipped with Levi-Civita connection ∇', Cartan and Norden studied nondegenerate affine immersions $\psi : N \to \mathbf{R}^{n+1}$ with a transversal vector field ξ such that the induced affine connection is exactly the Levi-Civita connection ∇'. The Cartan-Norden theorem states that if ψ is such an affine immersion, then either ∇' is flat and ψ is a graph immersion or ∇' is not flat and $\mathbf{R}^{n+1}$ admits a parallel Riemannian metric relative to which ψ is an isometric immersion and ξ is perpendicular to $\psi(N)$ (cf. [Nomizu and Sasaki (1994)]).

In this section, we present another realization problem from a different view point; namely, we consider the following [Chen (2005c)].

Realization Problem: *Which Riemannian manifolds (N, g) can be immersed as affine hypersurfaces in an affine space such that the induced affine metric h on N is exactly the given Riemannian metric g?*

Definition 17.2. We say that a warped product $N_1 \times_f N_2$ *can be realized as an affine hypersurface* if there exists a codimension one affine immersion from $N_1 \times_f N_2$ into some affine space in such a way that the induced fundamental form h is exactly the warped product metric of $N_1 \times_f N_2$.

Let $S^n(a^2)$, $H^n(-a^2)$, and $\mathbb{E}^n$ denote the n-sphere of constant curvature a^2, the hyperbolic n-space of constant curvature $-a^2$, and the Euclidean n-space, respectively. Assume that $S^n(a^2)$, $H^n(-a^2)$, and $\mathbb{E}^n$ are equipped with their canonical metrics.

The following results show that there exist many warped product which can be realized either as graph or as centroaffine hypersurfaces.

Proposition 17.2. *Let f be a positive function defined on an open interval $I \subset \mathbf{R}$. Then*

(a) *every warped product $I \times_f \mathbf{R}$ can be realized as a graph surface in an affine 3-space $\mathbf{R}^3$;*
(b) *for each integer $n > 2$, the warped product $I \times_f H^{n-1}(-a^2)$ can be realized as a graph hypersurface in $\mathbf{R}^{n+1}$;*
(c) *if $f'(s) \neq 0$ on I, then the warped product $I \times_f \mathbb{E}^{n-1}, n > 2$, can be realized as a graph hypersurface in $\mathbf{R}^{n+1}$;*
(d) *if $f'(s)^2 > a^2$ on I for some positive number a, then the warped product $I \times_f S^{n-1}(a^2), n > 2$, can be realized as a graph hypersurface in $\mathbf{R}^{n+1}$.*

Proof. We may assume that the open interval I contains 0. Consider the warped product surface $I\times_f \mathbf{R}$ with warped product metric: $ds^2+f^2(s)dt^2$. If we put

$$\begin{aligned}\psi(s,t) = \Bigg(&f(s)\exp\left\{\int_0^s \frac{\sqrt{f'^2+1}}{f}ds\right\}\sinh t,\\ &f(s)\exp\left\{\int_0^s \frac{\sqrt{f'^2+1}}{f}ds\right\}\cosh t,\ \int_0^s \frac{f\,ds}{f'-\sqrt{f'^2+1}}\Bigg),\end{aligned}\tag{17.40}$$

then ψ satisfies

$$\begin{aligned}\psi_{ss} &= \frac{ff''+f'^2+1}{f\sqrt{f'^2+1}}\psi_s+\xi,\\ \psi_{st} &= \frac{f'+\sqrt{f'^2+1}}{f}\psi_t,\\ \psi_{tt} &= f\big(\sqrt{f'^2+1}-f'\big)\psi_s+f^2\xi,\end{aligned}$$

where $\xi=(0,0,1)$. Thus ψ is a realization of $I\times_f\mathbf{R}$ as a graph surface in $\mathbf{R}^3$ with the warped product metric $ds^2+f^2(s)dt^2$ as the Calabi metric h and ξ as the Calabi normal.

Next, assume that a is a positive real number. Let N^- be the warped product manifold $I\times_f H^{n-1}(-a^2)$ with the warped product metric

$$ds^2+f^2(s)\left(du_2^2+\cosh^2(au_2)du_3^2+\cdots+\prod_{j=2}^{n-1}\cosh^2(au_j)du_n^2\right). \tag{17.41}$$

Consider the immersion ϕ of N^- into $\mathbf{R}^{n+1}$ given by

$$\begin{aligned}\phi(s,u_1,\ldots,u_n) &= f(s)\exp\left\{\int_0^s \frac{\sqrt{f'^2+a^2}}{f}ds\right\}\Bigg(\sinh(au_2),\\ &\sinh(au_3)\cosh(au_2),\ldots,\sinh(au_n)\prod_{j=2}^{n-1}\cosh(au_j),\prod_{j=2}^{n}\cosh(au_j),0\Bigg)\\ &-\frac{1}{a^2}\left(0,\ldots,0,\int_0^s f\big(\sqrt{f'^2+a^2}+f'\big)ds\right).\end{aligned}$$

Then a straightforward computation yields

$$\begin{aligned}\phi_{ss} &= \frac{ff''+f'^2+a^2}{f\sqrt{f'^2+a^2}}\phi_s+\xi,\\ \phi_{su_j} &= \frac{f'+\sqrt{f'^2+a^2}}{f}\phi_{u_j},\quad j=2,\ldots,n,\\ \phi_{u_iu_j} &= a\tanh(au_i)\phi_{u_j},\quad 2\le i<j\le n, \\ \phi_{u_ju_j} &= \left(f\sqrt{f'^2+a^2}\phi_s-ff'\phi_s+f^2\xi\right)\prod_{i=2}^{j-1}\cosh^2(au_i)\end{aligned}\tag{17.42}$$

$$-a\sum_{k=2}^{j-1}\left(\frac{\sinh(2au_k)}{2}\prod_{i=k+1}^{j-1}\cosh^2(au_i)\right)\phi_{u_k},\ j=2,\dots,n,$$

with $\xi=(0,\dots,0,1)$. It follows from (17.42) that ϕ is a graph hypersurface whose Calabi metric is given by (17.41). Hence, ϕ is a realization of M^- as a graph hypersurface with the warped product metric (17.41) as the Calabi metric h and ξ as the Calabi normal.

If $f'(s)\neq 0$, let N^o denote $I\times_f \mathbb{E}^{n-1}$ with the warped product metric:

$$ds^2+f^2(s)\big(du_2^2+du_3^2+\cdots+du_n^2\big). \tag{17.43}$$

Consider the immersion ϕ^o of N^o into $\mathbf{R}^{n+1}$ defined by

$$\phi^o(s,u_2,\dots,u_n)=\left(u_2,\dots,u_n,\int_0^s\frac{ds}{f(s)f'(s)}-\sum_{j=2}^n u_j^2,\int_0^s\frac{f(s)}{f'(s)}ds\right).$$

Then a straightforward computation yields

$$\begin{aligned}\phi^o_{ss}&=-\frac{f'^2+ff''}{ff'}\phi^o_s+\xi,\\ \phi^o_{su_j}&=\phi^o_{u_iu_j}=0,\\ \phi^o_{u_ju_j}&=-2ff'\phi^o_s+f^2\xi,\ \ 2\le i\neq j\le n,\end{aligned} \tag{17.44}$$

where $\xi=(0,\dots,0,2)$ is a transversal vector field. It follows from (17.44) that ϕ^o is a graph hypersurface whose Calabi metric is the warped product metric given by (17.43). Thus, ϕ^o is a realization of the warped product manifold $I\times_f\mathbb{E}^{n-1}$ as a graph hypersurface with (17.43) as the Calabi metric h and ξ as the Calabi normal.

Finally, assume that f satisfies $f'(s)^2>a^2$ with $a>0$. Let N^+ denote $I\times_f S^{n-1}(a^2)$ equipped with the warped product metric

$$ds^2+f^2(s)\left(du_2^2+\cos^2(au_2)du_3^2+\cdots+\prod_{j=2}^{n-1}\cos^2(au_j)du_n^2\right). \tag{17.45}$$

Consider the immersion ϕ^+ of N^+ into $\mathbf{R}^{n+1}$ defined by

$$\begin{aligned}\phi^+(s,u_2,\dots,u_n)&=f(s)\exp\left\{\int_0^s\frac{\sqrt{f'^2-a^2}}{f}ds\right\}\\ &\quad\times\left(\sin(au_2),\dots,\sin(au_n)\prod_{j=2}^{n-1}\cos(au_j),\prod_{j=2}^{n}\cos(au_j),0\right)\\ &\quad-\left(0,\dots,0,\int_0^s\frac{fds}{\sqrt{f'^2-a^2}-f'}\right).\end{aligned}$$

Then a straightforward computation yields

$$\begin{aligned}
\phi^+_{ss} &= \frac{ff''+f'^2-a^2}{f\sqrt{f'^2-a^2}}\phi^+_s + \xi,\\
\phi^+_{su_j} &= \frac{f'+\sqrt{f'^2-a^2}}{f}\phi^+_{u_j}, \quad j=2,\ldots,n,\\
\phi^+_{u_iu_j} &= -a\tan(au_i)\phi^+_{u_j}, \quad 2\le i<j\le n,\\
\phi^+_{u_ju_j} &= \Big(f\sqrt{f'^2-a^2}\phi^+_s - ff'\phi^+_s + f^2\xi\Big)\prod_{i=2}^{j-1}\cos^2(au_i)\\
&\quad + a\sum_{k=2}^{j-1}\frac{\sin(2au_k)}{2}\prod_{i=k+1}^{j-1}\cos^2(au_i)\phi^+_{u_k}, \quad j=2,\ldots,n,
\end{aligned} \tag{17.46}$$

where $\xi=(0,\ldots,0,1)$ is a transversal vector field. It follows from (17.46) that ϕ^+ is a graph hypersurface whose Calabi metric is the warped product metric (17.45). Hence, ϕ^+ is a realization of N^+ as a graph hypersurface with the warped product metric (17.45) as the Calabi metric h and ξ as the Calabi normal. ∎

Remark 17.2. If the warping function f of $I\times_{f(s)}\mathbb{E}^{n-1}$ satisfies $f'(s)=0$ on I, then f is a positive real number, say c. In this case, $I\times_f\mathbb{E}^{n-1}$ can be realized as a graph hypersurface in $\mathbf{R}^{n+1}$ by the following immersion

$$\phi(s,u_2,\ldots,u_n)=\left(s,u_2,\ldots,u_n,\frac{s^2}{2}+\frac{c^2}{2}\sum_{j=2}^n u_j^2\right).$$

Proposition 17.3. *Let $n>2$ and $f=f(s)$ be a positive function defined on an open interval I and a a positive number. Then*

(a) *if $f'(s)^2>f^2(s)-a^2$, then $I\times_f H^{n-1}(-a^2)$ can be realized as a centroaffine hypersurface in $\mathbf{R}^{n+1}$;*
(b) *if $f'(s)^2>f(s)^2$, then $I\times_f\mathbb{E}^{n-1}$ can be realized as a centroaffine hypersurface in $\mathbf{R}^{n+1}$;*
(c) *if $f'(s)^2>f(s)^2+a^2$, then $I\times_f S^{n-1}(a^2)$ can be realized as a graph hypersurface in $\mathbf{R}^{n+1}$.*

Proof. Without loss of generality, we may assume that the open interval I contains 0. Let N^-, N^o and N^+ be the warped products defined as before. Suppose that f satisfies $f'(s)^2>f^2(s)-a^2$. Consider the immersion η^- of N^- into $\mathbf{R}^{n+1}$ given by

$$\eta^-(s,u_2,\ldots,u_n)=f(s)\exp\left\{\int_0^s\frac{\sqrt{f'^2-f^2+a^2}}{f}ds\right\}$$

$$\times \left(\sinh(au_2), \ldots, \sinh(au_n) \prod_{j=2}^{n-1} \cosh(au_j), \prod_{j=2}^{n} \cosh(au_j), 0 \right)$$
$$+ \left(0, \ldots, 0, \exp\left(\int_0^s \frac{f ds}{f' - \sqrt{f'^2 - f^2 + a^2}} \right) \right).$$

Then by a straightforward computation we find

$$\begin{aligned}
\eta^-_{ss} &= \frac{f'^2 + ff'' - 2f^2 + a^2}{f\sqrt{f'^2 - f^2 + a^2}} \eta^-_s + \eta^-, \\
\eta^-_{su_j} &= \frac{f' + \sqrt{f'^2 - f^2 + a^2}}{f} \eta^-_{u_j}, \quad j = 2, \ldots, n, \\
\eta^-_{u_i u_j} &= a \tanh(au_i) \eta^-_{u_j}, \quad 2 \le i < j \le n, \\
\eta^-_{u_j u_j} &= \left(f\sqrt{f'^2 - f^2 + a^2} \eta^-_s - ff'\eta^-_s + f^2 \eta^- \right) \sum_{i=2}^{j-1} \cosh^2(au_i) \\
&\quad - a \sum_{k=2}^{j-1} \left(\frac{\sinh(2au_k)}{2} \prod_{i=k+1}^{j-1} \cosh^2(au_i) \right) \phi^-_{u_k}, \quad j = 2, \ldots, n,
\end{aligned} \tag{17.47}$$

which implies that η^- is a centroaffine hypersurface with (17.41) as the centroaffine metric. Thus, η^- is a realization of N^- as a centroaffine hypersurface with (17.41) as the centroaffine metric.

If f satisfies $f'(s)^2 > f^2(s)$, we consider the immersion

$$\eta(s, u_2, \ldots, u_n) = f(s) \exp\left(\int_0^s \frac{\sqrt{f'^2 - f^2}}{f} ds \right) (1, u_2, \ldots, u_n, 0)$$
$$+ \left(0, \ldots, 0, \exp\left(\int_0^s \frac{f ds}{f' - \sqrt{f'^2 - f^2}} \right) \right).$$

A direct computation shows that η satisfies (17.47) with $a = 0$. Hence, η defines a centroaffine hypersurface with (17.43) as its centroaffine metric. Thus, η is a realization of N^o as a centroaffine hypersurface with the warped product metric as its centroaffine metric. If $f'(s)^2 > f^2(s) + a^2$, we consider

$$\eta^+(s, u_2, \ldots, u_n) = f(s) \exp\left\{ \int_0^s \frac{\sqrt{f'^2 - f^2 - a^2}}{f} ds \right\}$$
$$\times \left(\sin(au_2), \ldots, \sin(au_n) \prod_{j=2}^{n-1} \cos(au_j), \prod_{j=2}^{n} \cos(au_j), 0 \right)$$
$$+ \left(0, \ldots, 0, \exp\left(\int_0^s \frac{f ds}{f' - \sqrt{f'^2 - f^2 - a^2}} \right) \right).$$

Then η^+ is a realization of N^+ as a centroaffine hypersurface with the warped product metric as its centroaffine metric. □

Remark 17.3. Robertson-Walker spacetimes can also be realized explicitly as centroaffine and graph hypersurfaces [Chen (2007a)].

17.6 Applications to affine warped product hypersurfaces

17.6.1 *Centroaffine hypersurfaces*

Theorem 17.3. *Let $N_1 \times_f N_2$ be the warped product of two Riemannian manifolds N_1 and N_2. If $N_1 \times_f N_2$ is realized as a centroaffine hypersurface in $\mathbf{R}^{n+1}$, then the warping function satisfies*

$$\frac{\Delta f}{f} \geq n_1\varepsilon - \frac{(n_1+n_2)^2}{4n_2} h(T^{\#}, T^{\#}), \tag{17.48}$$

where $n = n_1 + n_2$, $n_i = \dim N_i$, $i = 1, 2$, Δ is the Laplacian of N_1, and $\varepsilon = 1$ or -1 according to the centroaffine hypersurface is elliptic or hyperbolic.

Proof. Let $N = N_1 \times_f N_2$ be a warped product manifold equipped with the warped product metric $h = h_1 + f^2 h_2$, where h_1, h_2 are the metrics on N_1, N_2, respectively. Assume that (N, h) is realized via $\phi : N_1 \times_f N_2 \to \mathbf{R}^{n+1}$ as a locally strongly convex centroaffine hypersurface in $\mathbf{R}^{n+1}$ with h as the centroaffine metric. Denote by $\hat{\nabla}, \hat{K}$, *etc.* the Levi-Civita connection, sectional curvature, *etc.* of (N, h) as in section 17.1.

Since $N_1 \times_f N_2$ is a warped product, we have

$$\hat{\nabla}_X Z = \hat{\nabla}_Z X = (X \ln f) Z \tag{17.49}$$

for h-unit vector fields X, Z tangent to N_1, N_2, respectively. Hence

$$\begin{aligned} \hat{K}(X \wedge Z) &= h(\hat{\nabla}_Z \hat{\nabla}_X X - \hat{\nabla}_X \hat{\nabla}_Z X, Z) \\ &= \frac{1}{f}\{(\hat{\nabla}_X X) f - X^2 f\}. \end{aligned} \tag{17.50}$$

For simplicity, we use the following ranges of indices:

$$1 \leq i, j, k \leq n_1; \quad n_1 + 1 \leq s, t \leq n; \quad 1 \leq \alpha, \beta, \gamma \leq n,$$

unless mentioned otherwise. If we choose a local h-orthonormal frame $\{e_1, \ldots, e_n\}$ such that $e_1, \ldots, e_{n_1}$ and $e_{n_1+1}, \ldots, e_n$ are tangent to N_1 and N_2, respectively, then for each $s \in \{n_1 + 1, \ldots, n\}$ we have

$$\frac{\Delta f}{f} = \sum_{j=1}^{n_1} \hat{K}(e_j \wedge e_s), \tag{17.51}$$

where $\hat{K}(e_j \wedge e_s)$ denotes the sectional curvature of the plane section spanned by e_j and e_s. Let us put

$$\delta = \frac{n^2}{2} h(T^{\#}, T^{\#}) + n(n-1)\hat{\kappa} - 2n_1 n_2 \varepsilon. \tag{17.52}$$

Then (17.21) and (17.52) imply that

$$\frac{n^2}{2} h(T^{\#}, T^{\#}) = h(K, K) - \delta + \big(n(n-1) - 2n_1 n_2\big)\varepsilon. \tag{17.53}$$

At a point $p \in N$, we choose a h-orthonormal basis $e_1^*, \ldots, e_n^*$ of T_pN such that e_1^* is in the direction of the Tchebychev vector field $T^{\#}$ at p (when $T^{\#} = 0$ at p, we may choose $e_1^*, \ldots, e_n^*$ to be an arbitrary h-orthonormal basis). If we put $K_{\alpha\beta}^{\gamma^*} = h(K(e_\alpha, e_\beta), e_\gamma^*)$, then (17.53) becomes

$$\begin{aligned} \frac{n^2}{2} h(T^{\#}, T^{\#}) - \sum_{\alpha=1}^{n} \big(K_{\alpha\alpha}^{1^*}\big)^2 = \sum_{\alpha \neq \beta} \big(K_{\alpha\beta}^{1^*}\big)^2 + \sum_{r=2}^{n} \sum_{\alpha,\beta} (K_{\alpha\beta}^{r^*})^2 \\ + \big(n(n-1) - 2n_1 n_2\big)\varepsilon - \delta. \end{aligned} \tag{17.54}$$

Hence we can apply Corollary 13.1 to the left-hand-side of (17.54) to obtain

$$\begin{aligned} \sum_{i<j} K_{ii}^{1^*} K_{jj}^{1^*} + \sum_{s<t} K_{ss}^{1^*} K_{tt}^{1^*} \geq \sum_{\alpha<\beta} \big(K_{\alpha\beta}^{1^*}\big)^2 + \frac{1}{2} \sum_{r=2}^{n} \sum_{\alpha,\beta}^{n} (K_{\alpha\beta}^{r^*})^2 \\ + \frac{n(n-1) - 2n_1 n_2}{2}\varepsilon - \frac{\delta}{2}, \end{aligned} \tag{17.55}$$

with equality holding if and only if

$$K_{11}^{1^*} + \cdots + K_{n_1 n_1}^{1^*} = K_{n_1+1, n_1+1}^{1^*} + \cdots + K_{nn}^{1^*}. \tag{17.56}$$

On the other hand, (17.19) gives

$$\hat{K}(e_\alpha \wedge e_\beta) = \varepsilon - h(K(e_\alpha, e_\alpha), K(e_\beta, e_\beta)) + h(K(e_\alpha, e_\beta), K(e_\alpha, e_\beta)).$$

Combining with (17.51) gives

$$\begin{aligned} \frac{n_2 \Delta f}{f} &= \frac{n(n-1)}{2}\hat{\kappa} - \sum_{j<k} \hat{K}(e_j \wedge e_k) - \sum_{s<t} \hat{K}(e_s \wedge e_t) \\ &= \frac{n(n-1)}{2}\hat{\kappa} - \frac{n_1(n_1-1)}{2}\varepsilon + \sum_{\alpha} \sum_{j<k} \big(K_{jj}^{\alpha^*} K_{kk}^{\alpha^*} - (K_{jk}^{\alpha^*})^2\big) \\ &\quad - \frac{n_2(n_2-1)}{2}\varepsilon + \sum_{\alpha} \sum_{s<t} \big(K_{ss}^{\alpha^*} K_{tt}^{\alpha^*} - (K_{st}^{\alpha^*})^2\big) \\ &= \frac{n(n-1)}{2}(\hat{\kappa} - \varepsilon) + n_1 n_2 \varepsilon + \sum_{r=2}^{n} \sum_{j<k} \big(K_{jj}^{r^*} K_{kk}^{r^*} - (K_{jk}^{r^*})^2\big) \end{aligned} \tag{17.57}$$

$$+\sum_{r=2}^{n}\sum_{s<t}\left(K^{r^*}_{ss}K^{r^*}_{tt}-(K^{r^*}_{st})^2\right)+\sum_{j<k}K^{1^*}_{jj}K^{1^*}_{kk}$$
$$+\sum_{s<t}K^{1^*}_{ss}K^{1^*}_{tt}-\sum_{j<k}(K^{1^*}_{jk})^2-\sum_{s<t}(K^{1^*}_{st})^2.$$

Therefore, (17.55) and (17.57) yield

$$\begin{aligned}
\frac{n_2\Delta f}{f} &\geq \frac{n(n-1)}{2}\hat{\kappa}+\sum_{r=2}^{n}\sum_{j<k}\left(K^{r^*}_{jj}K^{r^*}_{kk}-(K^{r^*}_{jk})^2\right)+\sum_{j,t}(K^{1^*}_{jt})^2\\
&\quad+\frac{1}{2}\sum_{r=2}^{n}\sum_{\alpha,\beta}(K^{r^*}_{\alpha\beta})^2+\sum_{r=2}^{n}\sum_{s<t}\left(K^{r^*}_{ss}K^{r^*}_{tt}-(K^{r^*}_{st})^2\right)-\frac{\delta}{2}\\
&=\frac{n(n-1)}{2}\hat{\kappa}+\sum_{r=1}^{n}\sum_{j,t}(K^{r^*}_{jt})^2+\frac{1}{2}\sum_{r=2}^{n}\Big(\sum_{j}K^{r^*}_{jj}\Big)^2 \qquad (17.58)\\
&\quad+\frac{1}{2}\sum_{r=2}^{n}\Big(\sum_{t}K^{r^*}_{tt}\Big)^2-\frac{\delta}{2}\\
&\geq \frac{n(n-1)}{2}\hat{\kappa}-\frac{\delta}{2}.
\end{aligned}$$

Thus, by combining (17.52) and (17.58), we find

$$\frac{n_2\Delta f}{f}\geq n_1n_2\varepsilon-\frac{n^2}{4}h(T^{\#},T^{\#}),$$

which implies inequality (17.48). □

Two immediate consequences of Theorem 17.3 are the following.

Corollary 17.6. *If the warping function f of a warped product $N_1\times_f N_2$ satisfies $\Delta f\leq 0$ at some point on N_1, then $N_1\times_f N_2$ cannot be realized as an elliptic proper affine hypersphere in $\mathbf{R}^{n+1}$.*

Corollary 17.7. *If the warping function f of a warped product $N_1\times_f N_2$ satisfies $(\Delta f)/f<-\dim N_1$ at some point on N_1, then $N_1\times_f N_2$ cannot be realized as a hyperbolic proper affine hypersphere in $\mathbf{R}^{n+1}$.*

If the warping function is harmonic, Corollary 17.6 yields

Corollary 17.8. *Every warped product manifold $N_1\times_f N_2$ with harmonic warping function cannot be realized as an elliptic proper affine hypersphere in $\mathbf{R}^{n+1}$.*

Another interesting application of Theorem 17.3 is the following.

Corollary 17.9. *If N_1 is a compact Riemannian manifold, then every warped product manifold $N_1 \times_f N_2$ with arbitrary warping function cannot be realized as an elliptic proper affine hypersphere in* $\mathbf{R}^{n+1}$.

Proof. Assume N_1 is compact and that the warped product $N_1 \times_f N_2$ can be realized as an elliptic proper affine hypersphere in $\mathbf{R}^{n+1}$. Then we find from Theorem 17.3 that $(\Delta f)/f \geq n_1 = \dim N_1$. Since the warping function is positive, $\Delta f > 0$ on N_1. But this is impossible since N_1 is a compact Riemannian manifold according to Hopf's lemma. □

Theorem 17.4. *Let $\phi : N_1 \times_f N_2 \to \mathbf{R}^{n+1}$ be a realization of a warped product $N_1 \times_f N_2$ as a centroaffine hypersurface. If the warping function satisfies the equality case of* (17.48) *identically, then*

(a) *the Tchebychev vector field $T^\#$ vanishes identically;*
(b) *the warping function f is an eigenfunction of the Laplacian Δ with eigenvalue $n_1\varepsilon$;*
(c) *$N_1 \times_f N_2$ is realized as a proper affine hypersphere centered at the origin.*

Proof. Assume that $\phi : N_1 \times_f N_2 \to \mathbf{R}^{n+1}$ is a realization of the warped product $N_1 \times_f N_2$ as a centroaffine hypersurface which satisfies the equality case of (17.48) identically. Let us choose local h-orthonormal frames $e_1, \ldots, e_n$ and $e_1^*, \ldots, e_n^*$ in the same way as in the proof of Theorem 17.3.

If we put $L_1 = \text{Span}\{e_1, \ldots, e_{n_1}\}$ and $L_2 = \text{Span}\{e_{n_1+1}, \ldots, e_n\}$, then it follows from the proof of Theorem 17.3 that all of the inequalities in (17.55), (17.57) and (17.58) become equalities. Hence we obtain

$$\sum_{j=1}^{n_1} K_{jj}^{1^*} = \sum_{s=n_1}^{n} K_{ss}^{1^*}, \tag{17.59}$$

$$K_{js}^{1^*} = K_{js}^{r^*} = \sum_{j=1}^{n_1} K_{jj}^{r^*} = \sum_{s=n_1+1}^{n} K_{ss}^{r^*} = 0 \tag{17.60}$$

for $1 \leq j \leq n_1$; $n_1 + 1 \leq s \leq n$; $2 \leq r \leq n$. In particular, (17.59) implies that $K(L_1, L_2) = \{0\}$. Thus we get $K(L_1, L_1) \subset L_1$ and $K(L_2, L_2) \subset L_2$ according to (17.19). Hence, by using (17.59), we find

$$\sum_{j=1}^{n_1} K_{jj}^{1^*} = \sum_{s=n_1}^{n} K_{ss}^{1^*} = 0.$$

Consequently, we obtain $T^{\#} = 0$ which implies (a) and (c).

Statement (b) follows from (a) and the equality case of (17.48). □

The following examples illustrate that the results obtained in this section are best possible.

Example 17.2. Let $M = N_1 \times_{\cos s} N_2$ be the warped product of the open interval $N_1 = (-\pi, \pi)$ and an open portion N_2 of the unit $(n-1)$-sphere $S^{n-1}(1)$ equipped with the warped product metric:

$$h = ds^2 + \cos^2 s \left(du_2^2 + \cos^2 u_2 du_3^2 + \cdots + \prod_{j=2}^{n-1} \cos^2 u_j du_n^2 \right). \tag{17.61}$$

Consider the immersion of N into the affine $(n+1)$-space defined by

$$\left(\sin s, \sin u_2 \cos s, \ldots, \sin u_n \cos s \prod_{j=2}^{n-1} \cos^2 u_j, \cos s \prod_{j=2}^{n} \cos u_j \right).$$

Then N is a centroaffine elliptic hypersurface whose centroaffine metric is (17.61) and it satisfies $T^{\#} = 0$. Moreover, the warping function $f = \cos s$ satisfies $(\Delta f)/f = 1 = \varepsilon n_1$. Hence, this centroaffine hypersurface satisfies the equality case of (17.48) identically. Consequently, the estimate given in Theorem 17.3 is optimal for centroaffine elliptic hypersurfaces.

Example 17.3. Let $N = \mathbf{R} \times_{\cosh s} H^{n-1}(-1)$ be the warped product of the real line and the unit hyperbolic space $H^{n-1}(-1)$ equipped with warped product metric:

$$h = ds^2 + \cosh^2 s \left(du_2^2 + \cosh^2 u_2 du_3^2 + \cdots + \prod_{j=2}^{n-1} \cosh^2 u_j du_n^2 \right). \tag{17.62}$$

Consider the immersion of N into the affine $(n+1)$-space defined by

$$\left(\sinh s, \sinh u_2 \cosh s, \ldots, \sinh u_n \cosh s \prod_{j=2}^{n-1} \cosh^2 u_j, \cosh s \prod_{j=2}^{n} \cosh u_j \right).$$

Then N is a centroaffine hyperbolic hypersurface whose centroaffine metric is (17.62) and it satisfies $T^{\#} = 0$. Moreover, $f = \cosh s$ satisfies $(\Delta f)/f = -1 = \varepsilon n_1$. Thus, this centroaffine hypersurface satisfies the equality case of (17.48) identically. Consequently, the estimate given in Theorem 17.3 is optimal for centroaffine hyperbolic hypersurfaces as well.

17.6.2 Graph hypersurfaces

For warped product graph hypersurfaces we have the following.

Theorem 17.5. *If a warped product $N_1 \times_f N_2$ can be realized as a graph hypersurface in $\mathbf{R}^{n+1}$, then the warping function satisfies*

$$\frac{\Delta f}{f} \geq -\frac{(n_1+n_2)^2}{4n_2} h(T^\#, T^\#), \tag{17.63}$$

where $n = n_1 + n_2$, $n_1 = \dim N_1$ and $n_2 = \dim N_2$.

Proof. For graph hypersurfaces in $\mathbf{R}^{n+1}$ we have (17.25) instead of (17.19). Thus, by an argument similar to the proof of Theorem 17.3, we obtain the theorem. □

Theorem 17.4 implies the following.

Corollary 17.10. *If the warping function f of a warped product $N_1 \times_f N_2$ satisfies $\Delta f < 0$ at some point on N_1, then $N_1 \times_f N_2$ cannot be realized as an improper affine hypersphere in $\mathbf{R}^{n+1}$.*

Another application of Theorem 17.5 is the following.

Corollary 17.11. *If N_1 is a compact Riemannian manifold, then every warped product $N_1 \times_f N_2$ cannot be realized as an improper affine hypersphere in $\mathbf{R}^{n+1}$.*

Proof. Assume N_1 is a compact Riemannian manifold and that the warped product $N_1 \times_f N_2$ can be realized as an improper affine hypersphere in $\mathbf{R}^{n+1}$. Then Theorem 17.4 yields $\Delta f \geq 0$. Because N_1 is assumed to be a compact Riemannian manifold, the Hopf lemma implies that f is a positive constant, say c. Thus, the warped product $N_1 \times_f N_2$ is the Riemannian product $(N_1, h_1) \times (N_2, c^2 h_2)$. Since $TM_1 \oplus \xi$ is parallel along N_1, N_1 is contained in an (n_1+1)-dimensional affine subspace as a graph hypersurface with ξ as the Calabi normal. But this is impossible since N_1 is compact and ξ is a nonzero constant transversal vector field.

Theorem 17.6. *Let $\phi : N_1 \times_f N_2 \to \mathbf{R}^{n+1}$ be a realization of a warped product as a graph hypersurface. If the warping function satisfies the equality case of (17.63) identically, then*

(a) *the Tchebychev vector field $T^\#$ vanishes identically;*
(b) *the warping function f is a harmonic function;*
(c) *$N_1 \times_f N_2$ is realized as an improper affine hypersphere.*

Proof. This can be proved in the same way as Theorem 17.4. □

An application of Theorem 17.6 is the following.

Theorem 17.7. *If the Calabi metric of an improper affine hypersphere in an affine space is the direct product metric of k Riemannian manifolds, then the improper affine hypersphere is locally the Calabi composition of k improper affine spheres.*

Proof. Assume that $\phi : N_1 \times \cdots \times N_k \to \mathbf{R}^{n+1}$ is a graph hypersurface whose Calabi metric is the direct product metric of k Riemannian manifolds $N_1, \ldots, N_k$. Then each $TN_j \oplus \xi$ is parallel along N_j. So, each N_j lies in an affine (n_j+1)-subspace $\mathbf{R}^{n_j+1}$ of $\mathbf{R}^{n+1}$, $n_j = \dim N_j$. Thus, we can regard each N_j as graph hypersurface with ξ as its Calabi normal.

Since each N_j is totally geodesic in $N_1 \times \cdots \times N_k$, the induced connection, the Calabi metric and the difference tensor of N_j are the restrictions of the corresponding quantities of $N_1 \times \cdots \times N_k$ to N_j. Thus, each N_j is an improper affine hypersphere with the Calabi normal in the direction of ξ. Hence, if we choose affine coordinates $(x_1, \ldots, x_n, x_{n+1})$ on $\mathbf{R}^{n+1}$, $n = n_1 + \cdots + n_k$, such that $\xi = (0, \ldots, 0, 1)$ and each $\mathbf{R}^{n_j+1}$ is given by the equations $x_i = 0$ for $i \notin \{n_1 + \cdots + n_{j-1} + 1, \ldots, n_1 + \cdots + n_j\}$, then N_j is locally given as the graph of a function F_j.

Let $N_j^{\perp}$ be the direct product $N_1 \times \cdots \times N_{j-1} \times N_{j+1} \times \cdots \times N_k$. Then $N_1 \times \cdots \times N_k = N_j \times N_j^{\perp}$ for each j. Denote by L_j the distribution spanned by the vectors tangent to the j-th factor N_j and $L_j^{\perp}$ the h-orthogonal distribution of L_j in the tangent bundle of $N_j \times N_j^{\perp}$. From the assumption we find $f = 1$ and $T^{\#} = 0$. Thus the equality case of (17.63) holds for each $N_j \times N_j^{\perp}$. Now, it follows from the proof of Theorem 17.5 that $K(L_j, L_j^{\perp}) = \{0\}$. Combining this with (17.11) and (17.15) gives $D_X \phi_*(Y) = 0$ for $X \in TN_j, Y \in TN_j^{\perp}$. Hence, after applying a suitable translation, the improper affine hypersphere is the Calabi composition of k improper affine spheres (cf. [Dillen and Vrancken (1994, 1996)])

$$\big(x, \ldots, x_n, F_1(x_1, \ldots, x_{n_1}) + \cdots + F_k(x_{n-n_k+1}, \ldots, x_n)\big). \qquad \square$$

Example 17.4. Let $N = \mathbf{R} \times_s N_2$ be the warped product of the real line and an open part N_2 of $S^{n-1}(1)$ with the warped product metric:

$$h = ds^2 + s^2\left(du_2^2 + \cos^2 u_2 du_3^2 + \cdots + \prod_{j=2}^{n-1} \cos^2 u_j du_n^2\right). \tag{17.64}$$

Consider the immersion of N into the affine $(n+1)$-space defined by

$$s\left(\sin u_2, \sin u_3 \cos u_2, \ldots, \sin u_n \prod_{j=2}^{n-1} \cos^2 u_j, \textstyle\prod_{j=2}^{n} \cos u_j, \dfrac{s}{2}\right).$$

Then N is a graph hypersurface whose Calabi normal is $\xi = (0, \ldots, 0, 1)$ and it satisfies $T^{\#} = 0$. Moreover, the Calabi metric is given by the warped product metric (17.64). Clearly, the warping function is a harmonic function. Thus, this hypersurface satisfies the equality case of (17.63) identically. Consequently, the estimate given in Theorem 17.4 is optimal.

Remark 17.4. Example 17.2 shows that conditions "$\Delta f \leq 0$" in Corollary 17.6 and "harmonicity" in Corollary 17.8 are both necessary. Example 17.2 implies that condition "N_1 is a compact Riemannian manifold" in Corollary 17.9 is necessary. Example 17.3 shows that condition "$(\Delta f)/f < -\dim N_1$" in Corollary 17.7 is sharp. Finally, Example 17.4 implies that "$\Delta f < 0$" in Corollary 17.10 is optimal as well.

17.7 Eigenvalues of Tchebychev's operator $K_{T^{\#}}$

Let N be a centroaffine or graph hypersurface with positive definite induced affine metric h. Let K denote the difference tensor defined by (17.5) and $T^{\#}$ the Tchebychev vector field. For simplicity, we call $K_{T^{\#}}$ the *Tchebychev operator.*

Denote by $\hat{K}(\pi)$ the h-sectional curvature of a 2-plane section $\pi \subset T_pN$. For a k-plane section L^k of T_pN and a h-unit vector in $u \in L^k$, we choose a h-orthonormal basis $\{e_1, \ldots, e_k\}$ of L^k with $e_1 = u$. Then the k-Ricci curvature $\hat{S}_{L^k}(u)$ and the scalar curvature $\hat{\tau}(L^k)$ are defined by

$$\hat{S}_{L^k}(u) = \hat{K}_{12} + \cdots + \hat{K}_{1k}, \quad \hat{\tau}(L^k) = \sum_{1 \leq i < j \leq k} \hat{K}_{ij}. \tag{17.65}$$

For each integer $k \in [2, n]$, we define an invariant $\hat{\theta}_k$ on N by

$$\hat{\theta}_k(p) = \left(\frac{1}{k-1}\right) \sup_{L^k, u} \hat{S}_{L^k}(u), \quad u \in T_p^1 N, \tag{17.66}$$

where L^k runs over all linear k-subspaces in the tangent space T_pN at p and u runs over all h-unit vectors in L^k.

The *relative K-null subspace* $\mathcal{N}_p^K$ of N in $\mathbf{R}^{n+1}$ is defined by

$$\mathcal{N}_p^K = \{X \in T_pN : K(X, Y) = 0 \; \forall Y \in T_pN\}.$$

When $\dim \mathcal{N}_p^K$ is constant, $\mathcal{N}^K = U_{p \in N} \mathcal{N}_p^K$ defines a subbundle of the tangent bundle, called the *relative K-null subbundle.*

17.7.1 Centroaffine hypersurfaces

For centroaffine hypersurface in $\mathbf{R}^{n+1}$ we have the following [Chen (2006)].

Theorem 17.8. *Let $\phi : N \to \mathbf{R}^{n+1}$ be a locally strongly convex centroaffine hypersurface in $\mathbf{R}^{n+1}$ and $p \in N$. Then for any integer $k \in [2, n]$ we have*

1. *If $\hat{\theta}_k(p) \neq \varepsilon$, then every eigenvalue of Tchebychev's operator $K_{T^\#}$ at p is greater than $\left(\frac{n-1}{n}\right)(\varepsilon - \hat{\theta}_k(p))$;*
2. *If $\hat{\theta}_k(p) = \varepsilon$, every eigenvalue of $K_{T^\#}$ at p is ≥ 0;*
3. *A nonzero vector $X \in T_pN$ is an eigenvector of the operator $K_{T^\#}$ with eigenvalue $\left(\frac{n-1}{n}\right)(\varepsilon - \hat{\theta}_k(p))$ if and only if $\hat{\theta}_k(p) = \varepsilon$ and X lies in the relative K-null space $\mathcal{N}_p^K$ at p, where $\varepsilon = 1$ or -1 according to N is of elliptic or hyperbolic type.*

Proof. Let $\phi : N \to \mathbf{R}^{n+1}$ be a locally strongly convex centroaffine hypersurface in $\mathbf{R}^{n+1}$ and $\{e_1, \ldots, e_n\}$ an arbitrary h-orthonormal basis of T_pN. From the definition of Tchebychev vector field and (17.15), we have

$$2\hat{\tau} = n(n-1)\varepsilon + h(K, K) - n^2 h(T^\#, T^\#). \tag{17.67}$$

It follows from the Cauchy-Schwarz inequality that every endomorphism A of T_pN satisfies

$$n\, h(A, A) \geq (\operatorname{trace} A)^2, \tag{17.68}$$

with equality holding if and only if A is proportional to the identity map I. By applying (17.67) and (17.68), we find

$$2\hat{\tau} \geq n(n-1)\varepsilon - n(n-1)h(T^\#, T^\#), \tag{17.69}$$

with the equality holding at p if and only if

(a) $K_{T^\#}$ is proportional to the identity map and
(b) $K_Z = 0$ for Z perpendicular to $T^\#$ at p.

Let $L_{i_1 \cdots i_k}$ be the k-plane section spanned by the orthonormal vectors $e_{i_1}, \ldots, e_{i_k}$. It follows from (17.65) that

$$\hat{\tau}(L_{i_1 \cdots i_k}) = \frac{1}{2} \sum_{i \in \{i_1, \cdots, i_k\}} \hat{S}_{L_{i_1 \cdots i_k}}(e_i), \tag{17.70}$$

$$\hat{\tau}(p) = \frac{(k-2)!(n-k)!}{(n-2)!} \sum_{1 \leq i_1 < \cdots < i_k \leq n} \hat{\tau}(L_{i_1 \cdots i_k}). \tag{17.71}$$

By combining (17.66), (17.70) and (17.71) we find

$$\hat{\tau} \leq \frac{n(n-1)}{2} \hat{\theta}_k. \tag{17.72}$$

Thus (17.69) and (17.72) ensure that

$$h(T^{\#}, T^{\#}) \geq \varepsilon - \hat{\theta}_k. \tag{17.73}$$

Hence $T^{\#}$ vanishes at a point p only when $\hat{\theta}_k(p) \geq \epsilon$. Consequently, if $T^{\#}(p) = 0$, statements (1) and (2) hold automatically.

Next, assume $T^{\#}(p) \neq 0$. Since $K_{T^{\#}}$ is self-adjoint with respect to h, we may choose a h-orthonormal basis $e_1, \ldots, e_n$ which diagonalizes $K_{T^{\#}}$. Let e_1^* be the h-unit vector at p in the direction of $T^{\#}$ and let us choose h-orthonormal vectors $e_2^*, \ldots, e_n^*$ at p perpendicular to $T^{\#}$. Then we have

$$K_{e_1^*} = \begin{pmatrix} a_1 & 0 & \ldots & 0 \\ 0 & a_2 & \ldots & 0 \\ \vdots & \vdots & \ddots & \vdots \\ 0 & 0 & \ldots & a_n \end{pmatrix} \tag{17.74}$$

and trace $(K_{e_r^*}) = 0$ for $r = 2, \ldots, n$.

If we put $K_{ij}^{r^*} = h(K(e_i, e_j), e_r^*)$, then (2.9) implies that

$$K_{ij} = \varepsilon - a_i a_j + \sum_{r=2}^{n} \left(K_{ij}^{r^*}\right)^2 - \sum_{r=2}^{n} K_{ii}^{r^*} K_{jj}^{r^*}, \quad 1 \leq i \neq j \leq n. \tag{17.75}$$

Now, by applying the same argument as the proof of Theorem 2.10, we find

$$\begin{aligned} a_1 \sum_{j=1}^{n} a_j &\geq (n-1)\big(\varepsilon - \hat{\theta}_k(p)\big) + a_1^2 \\ &\geq (n-1)\big(\varepsilon - \hat{\theta}_k(p)\big), \end{aligned} \tag{17.76}$$

with both equality sign holding if and only if $\hat{S}_L(e_1) = \hat{\theta}_k(p)$ and $a_1 = K_{1j}^{r^*} = 0$ for $r = 2, \ldots, n$; $j = 2, \ldots, n$. The same inequality holds if the lower index 1 in (17.76) were replaced by any $j \in \{2, \ldots, n\}$. Hence

$$K_{T^{\#}} \geq \frac{n-1}{n}\big(\varepsilon - \hat{\theta}_k(p)\big)I. \tag{17.77}$$

If $K_{T^{\#}} u = \frac{n-1}{n}\big(\varepsilon - \hat{\theta}_k(p)\big)u$ holds for some h-unit vector $u \in T_p^1 N$, then u is an eigenvector of $K_{T^{\#}}$ with eigenvalue $(n-1)\big(\varepsilon - \hat{\theta}_k(p)\big)/n$. Without loss of generality, we may choose $e_1 = u$. In this case we get

$$a_1(a_1 + \cdots + a_n) = (n-1)\big(\varepsilon - \hat{\theta}_k(p)\big). \tag{17.78}$$

On the other hand, (17.76) and (17.78) yield $a_1 = 0$ and $\hat{\theta}_k(p) = \varepsilon$. Also, (17.76) implies that $e_1 \in \mathcal{N}_p^K$. Hence, we have statements (1) and (2) and one part of (3). The remaining part of statement (3) is obvious. □

When $k = 2$, Theorem 17.8(1) implies the following.

Corollary 17.12. *Let $\phi : N \to \mathbf{R}^{n+1}$ be a locally strongly convex centroaffine hypersurface. If $\sup \hat{K} \neq \varepsilon$ at a point $p \in N$, then every eigenvalue of Tchebychev's operator $K_{T^\#}$ at p is greater than $\left(\frac{n-1}{n}\right)(\epsilon - \sup \hat{K}(p))$.*

Similarly, if we denote by $\sup \hat{S}(p)$ the supremum of the Ricci curvature of (N, h) at a point $p \in N$, then Theorem 17.8(1) with $k = n$ implies immediately the following.

Corollary 17.13. *Let $\phi : N \to \mathbf{R}^{n+1}$ be a locally strongly convex centroaffine hypersurface. If $\sup \hat{S} \neq \varepsilon$ at a point $p \in N$, then every eigenvalue of Tchebychev's operator $K_{T^\#}$ at p is greater than $\left(\frac{n-1}{n}\right)(\epsilon - \sup \hat{S}(p))$.*

From Theorem 17.8 we also obtain the following.

Corollary 17.14. *Let $\phi : N \to \mathbf{R}^{n+1}$ be a locally strongly convex centroaffine hypersurface. If $\hat{\theta}_k < \varepsilon$ on N for some integer $k \in [2, n]$, then every eigenvalue of Tchebychev's operator $K_{T^\#}$ is positive.*

Theorem 17.8 also gives rise to the following simple geometric characterization of hyper-ellipsoids and two-sheeted hyperboloids in term of Tchebychev's operator.

Corollary 17.15. *An elliptic centroaffine hypersurface N in $\mathbf{R}^{n+1}$ is centroaffinely equivalent to an open portion of a hyperellipsoid if and only if $nK_{T^\#} = (n-1)(1-\hat{\theta}_k)I$ for some integer $k \in [2, n]$.*

Proof. Let $\phi : N \to \mathbf{R}^{n+1}$ be an elliptic centroaffine hypersurface. If N is an open part of a hyperellipsoid, then K vanishes identically. Thus $K_{T^\#} = 0$. Hence (N, h) is of constant curvature one. So, $\hat{\theta}_2 = \ldots = \hat{\theta}_n = 1$. Consequently, $nK_{T^\#} = (n-1)(1-\hat{\theta}_k)I$ holds identically.

Conversely, assume that $nK_{T^\#} = (n-1)(1-\hat{\theta}_k)I$ holds for some integer $k \in [2, n]$, then Theorem 17.8(3) implies that every tangent vectors of N lie in the relative K-null subbundle. In this case K vanishes identically. Thus, by applying a theorem of Berwald (cf. [Simon et al. (1991)]), N is centroaffinely equivalent to an open portion of a hyperellipsoid centered at the origin. □

Corollary 17.16. *A hyperbolic centroaffine hypersurface N in $\mathbf{R}^{n+1}$ is centroaffinely equivalent to an open part of a two-sheeted hyperboloid if and only if $nK_{T^\#} = (1-n)(1+\hat{\theta}_k)I$ for some integer $k \in [2, n]$.*

Proof. This can be done in the same way as Corollary 17.15. □

The following examples of locally strongly convex centroaffine hypersurfaces show that the eigenvalue estimates given in Theorem 17.8 are best possible.

Example 17.5. Let N be the elliptic locally strongly convex centroaffine hypersurface defined by:

$$e^{bs}\left(e^{(b^{-1}-b)s}, \sin(ax_2), \ldots, \sin(ax_n)\prod_{j=2}^{n-1}\cos(ax_j), \prod_{j=2}^{n}\cos(ax_j)\right), \tag{17.79}$$

with $a = \sqrt{1-b^2}$, $b \in (0,1)$. Then the affine metric h on N is

$$h = ds^2 + dx_2^2 + \cos^2(ax_2)dx_3^2 + \cdots + \prod_{j=2}^{n-1}\cos^2(ax_j)dx_n^2. \tag{17.80}$$

The Levi-Civita connection of h satisfies

$$\begin{aligned}
&\hat{\nabla}_{\partial/\partial s}\frac{\partial}{\partial s} = \hat{\nabla}_{\partial/\partial s}\frac{\partial}{\partial x_k} = \hat{\nabla}_{\partial/\partial x_2}\frac{\partial}{\partial x_2} = 0,\\
&\hat{\nabla}_{\partial/\partial x_i}\frac{\partial}{\partial x_j} = -a\tan(ax_i)\frac{\partial}{\partial x_j}, \quad 2 \le i < j,\\
&\hat{\nabla}_{\partial/\partial x_j}\frac{\partial}{\partial x_j} = a\sum_{k=2}^{j-1}\left(\frac{\sin(2ax_k)}{2}\prod_{l=k+1}^{j-1}\cos^2(ax_l)\right)\frac{\partial}{\partial x_k},\\
&\qquad j = 3, \ldots, n.
\end{aligned} \tag{17.81}$$

It follows from (17.79) and (17.80) that $\hat{K}_{1j} = 0$, $\hat{K}_{jk} = a^2$, $2 \le j \ne k \le n$. Hence we have

$$\hat{\theta}_n = \left(\frac{n-2}{n-1}\right)(1-b^2). \tag{17.82}$$

On the other hand, it follows from (17.79) that

$$\begin{aligned}
&\nabla_{\partial/\partial s}\frac{\partial}{\partial s} = \left(b + \frac{1}{b}\right)\frac{\partial}{\partial s}, \quad \nabla_{\partial/\partial s}\frac{\partial}{\partial x_j} = b\frac{\partial}{\partial x_j},\\
&\nabla_{\partial/\partial x_i}\frac{\partial}{\partial x_j} = -a\tan(ax_i)\frac{\partial}{\partial x_j}, \quad 2 \le i < j \le n,\\
&\nabla_{\partial/\partial x_j}\frac{\partial}{\partial x_j} = b\prod_{i=2}^{j-1}\cos^2(ax_i)\frac{\partial}{\partial s}\\
&\qquad + a\sum_{k=2}^{j-1}\left(\frac{\sin(2ax_k)}{2}\prod_{l=k+1}^{j-1}\cos^2(ax_l)\right)\frac{\partial}{\partial x_k}
\end{aligned} \tag{17.83}$$

for $j = 2, \ldots, n$. By applying (17.5), (17.81) and (17.83) we find

$$\begin{aligned} &K\left(\frac{\partial}{\partial s}, \frac{\partial}{\partial s}\right) = \left(b + \frac{1}{b}\right)\frac{\partial}{\partial s}, \; K\left(\frac{\partial}{\partial s}, \frac{\partial}{\partial x_j}\right) = b\,\frac{\partial}{\partial x_j}, \\ &K\left(\frac{\partial}{\partial x_j}, \frac{\partial}{\partial x_j}\right) = b\prod_{i=2}^{j-1}\cos^2(ax_i)\frac{\partial}{\partial s}, \\ &K\left(\frac{\partial}{\partial x_i}, \frac{\partial}{\partial x_j}\right) = 0 \end{aligned} \tag{17.84}$$

for $2 \leq i \neq j \leq n$. Thus, we obtain from (2.13), (17.80) and(17.84) that

$$T^{\#} = \left(b + \frac{1}{nb}\right)\frac{\partial}{\partial s}, \;\; K_{T^{\#}}\left(\frac{\partial}{\partial x_j}\right) = \lambda_j \frac{\partial}{\partial x_j}, \;\; \lambda_j = b^2 + \frac{1}{n} \tag{17.85}$$

for $j = 2, \ldots, n$. Consequently, the eigenvalue λ_j of Tchebychev's operator $K_{T^{\#}}$ associated with eigenvector $\partial/\partial x_j$ satisfies

$$\lambda_j - \frac{n-1}{n}(1 - \hat{\theta}_n) = \frac{2b^2}{n} \longrightarrow 0 \quad \text{as } b \to 0.$$

Example 17.6. Consider the hyperbolic locally strongly convex centroaffine hypersurface defined by:

$$e^{bs}\left(e^{-(b^{-1}+b)s}, \sinh(ax_2), \ldots, \sinh(ax_n)\prod_{j=2}^{n-1}\cosh(ax_j), \prod_{j=2}^{n}\cosh(ax_j)\right) \tag{17.86}$$

with $a = \sqrt{1+b^2}$, $b \in (0, \infty)$. The induced affine metric h is given by

$$h = ds^2 + dx_2^2 + \cosh^2(ax_2)dx_3^2 + \cdots + \prod_{j=2}^{n-1}\cosh^2(ax_j)dx_n^2, \tag{17.87}$$

which implies that $\hat{K}_{1j} = 0, \hat{K}_{jk} = -a^2$ for $2 \leq j \neq k \leq n$. Hence

$$\hat{\theta}_n = \left(\frac{2-n}{n-1}\right)(1 + b^2). \tag{17.88}$$

From (17.5), (17.13), (17.86) and a straightforward computation we find

$$\begin{aligned} &K\left(\frac{\partial}{\partial s}, \frac{\partial}{\partial s}\right) = \left(b - \frac{1}{b}\right)\frac{\partial}{\partial s}, \; K\left(\frac{\partial}{\partial s}, \frac{\partial}{\partial x_j}\right) = b\frac{\partial}{\partial x_j}, \\ &K\left(\frac{\partial}{\partial x_j}, \frac{\partial}{\partial x_j}\right) = b\prod_{i=2}^{j-1}\cosh^2(ax_i)\frac{\partial}{\partial s}, \; K\left(\frac{\partial}{\partial x_i}, \frac{\partial}{\partial x_j}\right) = 0 \end{aligned} \tag{17.89}$$

for $2 \le i \ne j \le n$. Therefore

$$T^{\#} = \left(b - \frac{1}{nb}\right)\frac{\partial}{\partial s},$$
$$K_{T^{\#}}\left(\frac{\partial}{\partial x_j}\right) = \left(b^2 - \frac{1}{n}\right)\frac{\partial}{\partial x_j}, \quad j = 2, \dots, n.$$

Consequently, the eigenvalue λ_j of Tchebychev's operator $K_{T^{\#}}$ associated with eigenvector $\partial/\partial x_j$ satisfies

$$\lambda_j + \frac{n-1}{n}(1 + \hat{\theta}_n) = \frac{2b^2}{n} \longrightarrow 0 \quad \text{as } b \to 0.$$

Examples 17.5 and 17.6 illustrate that the eigenvalue estimate given in Theorem 17.8(1) is optimal for locally strongly convex centroaffine hypersurfaces of both elliptic and hyperbolic types.

Example 17.7. Consider the following elliptic centroaffine locally strongly convex hypersurface:

$$\begin{aligned}\Bigg(&\sin x_1, \sin x_2 \cos x_1, \dots, \sin x_{n-1} \prod_{j=1}^{n-2} \cos x_j, \\ &e^{\frac{1}{2}(b+\sqrt{b^2-4})x_n} \prod_{j=1}^{n-1} \cos x_j, e^{\frac{1}{2}(b+\sqrt{b^2-4})x_n} \prod_{j=1}^{n-1} \cos x_j\Bigg)\end{aligned} \tag{17.90}$$

with $b > 2$. The affine metric of this hypersurface is given by

$$h = dx_1^2 + \cos^2 x_1 dx_2^2 + \cdots + \prod_{j=1}^{n-1} \cos^2 x_j dx_n^2. \tag{17.91}$$

It follows from (17.91) that $\hat{\theta}_k = 1$ for $k = 2, \dots, n$.

On the other hand, from (17.90) and a direct computation, we have

$$K\left(\frac{\partial}{\partial x_i}, \frac{\partial}{\partial x_j}\right) = K\left(\frac{\partial}{\partial x_j}, \frac{\partial}{\partial x_n}\right) = 0,$$
$$K\left(\frac{\partial}{\partial x_n}, \frac{\partial}{\partial x_n}\right) = b\frac{\partial}{\partial x_n}$$

for $1 \le i, j \le n-1$, which ensures that

$$T^{\#} = \left(\frac{b}{n}\prod_{j=1}^{n-1} \sec^2 x_j\right)\frac{\partial}{\partial x_n},$$
$$K_{T^{\#}}\left(\frac{\partial}{\partial x_j}\right) = 0, \quad j = 1, \dots, n-1.$$

Example 17.8. Let N denote the hyperbolic locally strongly convex centroaffine hypersurface defined by

$$\begin{aligned}\Bigg(&\sinh x_1, \sinh x_2 \cosh x_1, \ldots, \sinh x_{n-1} \prod_{j=1}^{n-2} \cos x_j, \\ &e^{\frac{1}{2}(b+\sqrt{b^2+4})x_n} \prod_{j=1}^{n-1} \cosh x_j, e^{\frac{1}{2}(b-\sqrt{b^2+4})x_n} \prod_{j=1}^{n-1} \cosh x_j \Bigg),\end{aligned} \tag{17.92}$$

with nonzero b. Since he induced affine metric is given by

$$h = dx_1^2 + \cosh^2 x_1 dx_2^2 + \cdots + \prod_{j=1}^{n-1} \cosh^2 x_j dx_n^2, \tag{17.93}$$

thus we have $\hat{\theta}_2 = \cdots = \hat{\theta}_n = -1$.

On the other hand, by (17.92) and a straightforward computation, we find

$$\begin{aligned} &K\left(\frac{\partial}{\partial x_i}, \frac{\partial}{\partial x_j}\right) = K\left(\frac{\partial}{\partial x_j}, \frac{\partial}{\partial x_n}\right) = 0, \\ &K\left(\frac{\partial}{\partial x_n}, \frac{\partial}{\partial x_n}\right) = b\,\frac{\partial}{\partial x_n}, \quad 1 \leq i, j \leq n-1. \end{aligned} \tag{17.94}$$

Hence we obtain

$$T^{\#} = \frac{b}{n} \prod_{j=1}^{n-1} \operatorname{sech}^2 x_j \frac{\partial}{\partial x_n}, \tag{17.95}$$

$$K_{T^{\#}}\left(\frac{\partial}{\partial x_j}\right) = 0, \tag{17.96}$$

for $j = 1, \ldots, n-1$.

Clearly, Examples 17.7 and 17.8 illustrate that the estimate given in Theorem 17.8(2) is optimal for locally strongly convex centroaffine hypersurfaces of both elliptic and hyperbolic types.

17.7.2 *Graph hypersurfaces*

For graph hypersurfaces we have the following [Chen (2006)].

Theorem 17.9. *Let $\phi : N \to \mathbf{R}^{n+1}$ be a graph hypersurface in $\mathbf{R}^{n+1}$ with positive definite Calabi metric. Then for any integer $k \in [2, n]$ we have:*

(1) *If $\hat{\theta}_k \neq 0$ at a point $p \in M$, then each eigenvalue of $K_{T^{\#}}$ at p is greater than $\left(\frac{1-n}{n}\right) \hat{\theta}_k(p)$;*

(2) *If* $\hat{\theta}_k = 0$ *at* p, *then every eigenvalue of* $K_{T^\#}$ *at* p *is* ≥ 0;
(3) *A nonzero vector* $X \in T_pN$ *is an eigenvector of Tchebychev's operator* $K_{T^\#}$ *with eigenvalue* $\left(\frac{1-n}{n}\right)\hat{\theta}_k(p)$ *if and only if we have* $\hat{\theta}_k(p) = 0$ *and* $X \in \mathcal{N}_p^K$.

Proof. For graph hypersurfaces in $\mathbf{R}^{n+1}$ we have

$$\hat{R}(X,Y)Z = K_Y K_X Z - K_X K_Y Z.$$

Thus, this theorem can be proved in the same way as Theorem 17.8. □

Theorem 17.9 implies the following.

Corollary 17.17. *Let* $\phi : N \to \mathbf{R}^{n+1}$ *be a graph hypersurface with positive definite Calabi metric. If* $\sup \hat{K} \neq 0$ *or* $\sup \hat{S} \neq 0$ *at a point* $p \in N$, *then every eigenvalue of Tchebychev's operator* $K_{T^\#}$ *is greater than* $\left(\frac{1-n}{n}\right)\sup \hat{K}$ *at* p.

Corollary 17.18. *Let* $\phi : N \to \mathbf{R}^{n+1}$ *be a graph hypersurface with positive definite Calabi metric. If there is an integer* $k \in [2, n]$ *such that* $\hat{\theta}_k < 0$, *then every eigenvalue of* $K_{T^\#}$ *is positive.*

From Corollaries 17.14 and 17.18 we obtain the following.

Corollary 17.19. *Let* N *be a Riemannian* n*-manifold. If there exists an integer* $k \in [2, n]$ *such that* $\hat{\theta}_k(p) < 1$ *at some point* $p \in N$, *then* N *cannot be realized as an elliptic proper affine hypersphere in* $\mathbf{R}^{n+1}$.

Corollary 17.20. *Let* N *be a Riemannian* n*-manifold. If there exists an integer* $k \in [2, n]$ *such that* $\hat{\theta}_k(p) < -1$ *at some point* $p \in N$, *then* N *cannot be realized as a hyperbolic proper affine hypersphere in* $\mathbf{R}^{n+1}$.

Corollary 17.21. *Let* N *be a Riemannian* n*-manifold. If there exists an integer* $k \in [2, n]$ *such that* $\hat{\theta}_k(p) < 0$ *at some point* p, *then* N *cannot be realized as an improper affine hypersphere in* $\mathbf{R}^{n+1}$.

The next examples show that Theorem 17.9 is the best possible as well.

Example 17.9. Consider the graph hypersurface N in $\mathbf{R}^{n+1}$:

$$\left(u_2, \ldots, u_n, \frac{s^4}{4} + \sum_{j=2}^{n} u_j^2, \frac{s^2}{4}\right) \tag{17.97}$$

with constant affine normal $\xi = (0, \ldots, 0, -1)$ and Calabi metric given by

$$h = ds^2 + \frac{1}{s^2}\left(du_2^2 + \cdots + du_n^2\right).$$

A direct computation shows that

$$\hat{K}_{1j} = -\frac{1}{s^2}, \;\; \hat{K}_{ij} = -1, \; 2 \leq i \neq j \leq n.$$

Thus we get

$$\hat{\theta}_2 = \cdots = \hat{\theta}_n = \begin{cases} -\dfrac{1}{s^2} & \text{if } s^2 \geq 1; \\ -1 & \text{if } s^2 < 1. \end{cases} \tag{17.98}$$

From (17.97) and a straightforward computation, we find

$$\begin{aligned} &K\left(\frac{\partial}{\partial s}, \frac{\partial}{\partial s}\right) = \frac{3}{s}\frac{\partial}{\partial s}, \;\; K\left(\frac{\partial}{\partial s}, \frac{\partial}{\partial u_j}\right) = \frac{1}{s}\frac{\partial}{\partial u_j}, \\ &K\left(\frac{\partial}{\partial u_i}, \frac{\partial}{\partial u_j}\right) = 0, \;\; K\left(\frac{\partial}{\partial u_j}, \frac{\partial}{\partial u_j}\right) = \frac{1}{s^3}\frac{\partial}{\partial s}, \end{aligned} \tag{17.99}$$

for $2 \leq i \neq j \leq n$, which yields

$$\begin{aligned} &T^{\#} = \frac{(n+2)}{ns}\frac{\partial}{\partial s}, \\ &K_{T^{\#}}\left(\frac{\partial}{\partial u_j}\right) = \lambda_j \frac{\partial}{\partial u_j}, \;\; \lambda_j = \frac{(n+2)}{ns^2} \end{aligned}$$

for $j = 2, \ldots, n$. Hence we obtain

$$\lambda_j - \left(\frac{1-n}{n}\right)\hat{\theta}_k = \frac{3}{ns^2} \longrightarrow 0 \quad \text{as } s \to \infty.$$

This example shows that the estimate given in Theorem 17.9(1) is optimal.

Example 17.10. Consider the graph hypersurface N in $\mathbf{R}^{n+1}$

$$\left(u_2, \ldots, u_n, e^{u_1}, u_1 - \frac{1}{2}\sum_{j=2}^{n} u_j^2\right) \tag{17.100}$$

with affine normal $\xi = (0, \ldots, 0, -1)$ and Calabi metric $h = du_1^2 + \cdots + du_n^2$. Obviously, we have $\hat{\theta}_2 = \cdots = \hat{\theta}_n = 0$. It follows from (17.100) that

$$\begin{aligned} &K\left(\frac{\partial}{\partial u_1}, \frac{\partial}{\partial u_1}\right) = \frac{\partial}{\partial u_1}, \\ &K\left(\frac{\partial}{\partial u_1}, \frac{\partial}{\partial u_j}\right) = K\left(\frac{\partial}{\partial u_i}, \frac{\partial}{\partial u_j}\right) = 0, \end{aligned}$$

for $i, j = 2, \ldots, n$. Thus we find

$$T^{\#} = \frac{1}{n}\frac{\partial}{\partial u_1}, \;\; K_{T^{\#}}\left(\frac{\partial}{\partial u_j}\right) = 0, \;\; j = 2, \ldots, n.$$

This example shows that the estimate given in Theorem 17.9(2) is optimal as well.

Chapter 18

Applications to Riemannian Submersions

18.1 A submersion δ-invariant

The simplest type of pseudo-Riemannian submersions is the orthogonal projection of a direct product of two pseudo-Riemannian manifolds onto one of its two factors, for which both horizontal and vertical distributions are *totally geodesic distributions.*

A pseudo-Riemannian manifold M is said to admit a *non-trivial pseudo-Riemannian submersion* if there exists a pseudo-Riemannian submersion $\pi : M \to B$ of M onto another pseudo-Riemannian manifold B such that the horizontal and vertical distributions are not both totally geodesic.

Obviously, if a pseudo-Riemannian submersion has totally geodesic fibers, then the submersion is non-trivial if and only if the horizontal distribution $\mathcal{H}$ is a non-totally geodesic distribution.

Let $\pi : M \to B$ be a pseudo-Riemannian submersion. We choose an orthonormal basis $e_1, \ldots, e_b, e_{b+1}, \ldots, e_n$ of T_pM, $p \in M$, such that $e_1, \ldots, e_b$ are horizontal and $e_{b+1}, \ldots, e_n$ are vertical. Define the invariant $\breve{A}_\pi$ of π as in (11.14) by

$$\breve{A}_\pi(p) = \sum_{i=1}^{b} \sum_{r=b+1}^{n} \langle A_{e_i} e_r, A_{e_i} e_r \rangle , \tag{18.1}$$

where A is O'Neill's integrability tensor. It is direct to verify that $\breve{A}_\pi(p)$ is well-defined, called the *submersion δ-invariant* of $\pi : M \to B$.

Example 18.1. Let $\pi : S^{2n+1}(1) \to CP^n(4)$ be the Hopf fibration. For a submanifold N of $CP^n(4)$, the restriction $\pi_N : \pi^{-1}(N) \to N$ is a Riemannian submersion with totally geodesic fibers. The invariant $\breve{A}_\pi$ of π_N is $\breve{A}_\pi = ||P||^2$, where $P : \mathcal{H} \to \mathcal{H}$ is the projection of ϕ onto the horizontal distribution $\mathcal{H}$ and ϕ is the $(1,1)$-tensor of the Sasakian $(S^{2n+1}(1), \phi, \xi, \eta, g)$.

Example 18.2. Let G be a Lie group with a bi-invariant Riemannian metric and H is a closed subgroup of G. Then the usual Riemannian structure on the homogeneous space G/H is characterized by the fact that the natural mapping $\pi : G \to G/H$ is a Riemannian submersion.

The fibers of such a Riemannian submersion are the left cosets of G (mod H) which are totally geodesic. The invariant $\breve{A}_\pi$ is given by

$$\breve{A}_\pi = \frac{1}{4} \sum_{i,j=1}^{b} \sum_{b+1}^{n} \langle [e_i, e_j], e_s \rangle^2, \quad b = \dim H,$$

where $e_1, \ldots, e_b$ are orthonormal left-invariant horizontal vector fields and $e_{b+1}, \ldots, e_n$ an orthonormal basis of the vertical distribution $\mathcal{V}$.

18.2 An optimal inequality for Riemannian submersions

The following is an optimal inequality involving $\breve{A}_\pi$ [Chen (2005b)].

Theorem 18.1. *Let $\pi : M \to B$ be a Riemannian submersion with totally geodesic fibers. Then, for an isometric immersion ψ of M into a Riemannian manifold $\tilde{M}$, we have*

$$\breve{A}_\pi \leq \frac{n^2}{4} ||H||^2 + b(n-b) \max \tilde{K}, \tag{18.2}$$

where $||H||^2$ is the squared mean curvature of M in $\tilde{M}$ and $\max \tilde{K}(p)$ is the maximum value of the sectional curvature of $\tilde{M}$ restricted to plane sections in $T_pM \subset T_{\psi(p)}\tilde{M}$, $p \in M$.

Proof. Let M be a Riemannian n-manifold. Denote by R, K and τ the Riemann curvature tensor, the sectional curvature and the scalar curvature of M, respectively. Assume that M admits a Riemannian immersion: $\pi : M \to B$ with $\dim M > \dim B > 0$. Then for any horizontal vector field X and any vertical vector field V, we have

$$A_X V = \mathcal{H} \nabla_X V, \tag{18.3}$$

where A is O'Neill's integrability tensor.

Now, assume that the Riemannian submersion $\pi : M \to B$ has totally geodesic fibers. Then, by Lemma 11.3(2), the sectional curvature $K(X \wedge V)$ on M of the plane section spanned by a unit horizontal vector X and a unit vertical vector V is given by

$$K(X \wedge V) = ||A_X V||^2. \tag{18.4}$$

Next, assume that $\phi : M \to \tilde{M}$ is an isometric immersion of M into a Riemannian m-manifold $\tilde{M}$. Denote by $\tilde{R}$ and $\tilde{K}$ the Riemann curvature tensor and the sectional curvature of $\tilde{M}$, respectively.

From Gauss's equation, we find

$$2\tau(p) = n^2||H||^2 - ||h||^2 + 2\tilde{\tau}(T_pM), \tag{18.5}$$

where $||h||^2$ and $||H||^2$ are the squared norm of the second fundamental form and the squared mean curvature of M in $\tilde{M}$, respectively, and $\tilde{\tau}(T_pM)$ is the scalar curvature of $\tilde{M}$ restricted to T_pM.

Let us put

$$\eta = 2\tau - \frac{n^2}{2}||H||^2 - 2\tilde{\tau}(T_pM). \tag{18.6}$$

Then (18.5) becomes

$$n^2||H||^2 = 2\eta + 2||h||^2. \tag{18.7}$$

If we choose a local orthonormal frame $e_1, \ldots, e_b, e_{b+1}, \ldots, e_n, e_{n+1}, \ldots, e_m$ such that $e_1, \ldots, e_b$ are horizontal, $e_{b+1}, \ldots, e_n$ are vertical, and that e_{n+1} is a unit normal vector field parallel to the mean curvature vector field of M, then (18.7) becomes

$$\begin{aligned}\left(\sum_{i=1}^{n} h_{ii}^{n+1}\right)^2 = 2\Bigg\{\delta + \sum_{i=1}^{n}(h_{ii}^{n+1})^2 + \sum_{i\neq j}(h_{ij}^{n+1})^2 \\ + \sum_{r=n+2}^{m}\sum_{i,j=1}^{n}(h_{ij}^{r})^2\Bigg\},\end{aligned} \tag{18.8}$$

where $h_{ij}^r = \langle h(e_i, e_j), e_r\rangle, n+1 \leq r \leq m; 1 \leq i, j \leq n$. Equation (18.8) is equivalent to

$$\begin{aligned}(\bar{a}_1 + \bar{a}_2 + \bar{a}_3)^2 = 2\Bigg[\delta + \bar{a}_1^2 + \bar{a}_2^2 + \bar{a}_3^2 + 2\sum_{1\leq i<j\leq n}(h_{ij}^{n+1})^2 \\ + \sum_{r=n+2}^{m}\sum_{i,j=1}^{n}(h_{ij}^{r})^2 - 2\sum_{2\leq j<k\leq b} h_{jj}^{n+1}h_{kk}^{n+1} - 2\sum_{b+1\leq s<t\leq n} h_{ss}^{n+1}h_{tt}^{n+1}\Bigg],\end{aligned} \tag{18.9}$$

where

$$\bar{a}_1 = h_{11}^{n+1},\ \bar{a}_2 = h_{22}^{n+1} + \cdots + h_{bb}^{n+1},\ \bar{a}_3 = h_{b+1b+1}^{n+1} + \cdots + h_{nn}^{n+1}.$$

By applying Corollary 13.1 to (18.9), we obtain

$$\begin{aligned}\sum_{1\leq j<k\leq b} h_{jj}^{n+1}h_{kk}^{n+1} + \sum_{b+1\leq s<t\leq n} h_{ss}^{n+1}h_{tt}^{n+1} \\ \geq \frac{\delta}{2} + \sum_{1\leq s<t\leq n}(h_{st}^{n+1})^2 + \frac{1}{2}\sum_{r=n+2}^{m}\sum_{s,t=1}^{n}(h_{st}^{r})^2,\end{aligned} \tag{18.10}$$

with equality holding if and only if

$$\sum_{i=1}^{b} h_{ii}^{n+1} = \sum_{s=b+1}^{n} h_{ss}^{n+1}. \tag{18.11}$$

From the equation of Gauss and (18.4), we obtain

$$\begin{aligned}
\breve{A}_\pi &= \tau - \sum_{1\le j<k\le b} K(e_j \wedge e_k) - \sum_{b+1\le s<t\le n} K(e_s \wedge e_t) \\
&= \tau - \sum_{r=n+1}^{m} \sum_{1\le j<k\le b} (h_{jj}^r h_{kk}^r - (h_{jk}^r)^2) \\
&\quad - \sum_{1\le i<j\le b} \tilde{K}(e_i \wedge e_j) - \sum_{b+1\le s<t\le n} \tilde{K}(e_s \wedge e_t) \\
&\quad - \sum_{r=n+1}^{m} \sum_{b+1\le s<t<n} \left(h_{ss}^r h_{tt}^r - (h_{st}^r)^2\right).
\end{aligned} \tag{18.12}$$

Thus, after applying (18.6), (18.10) and (18.12), we find

$$\begin{aligned}
\breve{A}_\pi \le\ & \tau - \tilde{\tau}(T_pM) + \sum_{i=1}^{b} \sum_{s=b+1}^{n} \tilde{K}(e_i \wedge e_s) - \sum_{j=1}^{n} \sum_{t=b+1}^{n} (h_{jt}^{n+1})^2 \\
& - \frac{\delta}{2} + \sum_{r=n+2}^{m} \Bigg\{ \sum_{1\le j<k\le b} ((h_{jk}^r)^2 - h_{jj}^r h_{kk}^r) \\
& + \sum_{b+1\le s<t<n} \left((h_{st}^r)^2 - h_{ss}^r h_{tt}^r\right) - \frac{1}{2} \sum_{\alpha,\beta=1}^{n} (h_{\alpha\beta}^r)^2 \Bigg\} \\
=\ & \tau - \tilde{\tau}(T_pM) + \sum_{i=1}^{b} \sum_{s=b+1}^{n} \tilde{K}(e_i \wedge e_s) \\
& - \sum_{r=n+1}^{m} \sum_{j=1}^{b} \sum_{t=b+1}^{n} (h_{jt}^r)^2 - \frac{\delta}{2} \\
& - \frac{1}{2} \sum_{r=n+2}^{m} \left\{ \left(\sum_{j=1}^{b} h_{jj}^r \right)^2 + \left(\sum_{t=b+1}^{n} h_{tt}^r \right)^2 \right\} \\
\le\ & \tau - \tilde{\tau}(T_pM) + \sum_{i=1}^{b} \sum_{s=b+1}^{n} \tilde{K}(e_i \wedge e_s) - \frac{\delta}{2} \\
=\ & \frac{n^2}{4} ||H||^2 + \sum_{i=1}^{b} \sum_{s=b+1}^{n} \tilde{K}(e_i \wedge e_s).
\end{aligned} \tag{18.13}$$

This implies (18.2). □

18.3 Some applications

An important application of Theorem 18.1 is the following [Chen (2005b)].

Theorem 18.2. *If a Riemannian manifold M admits a non-trivial Riemannian submersion with totally geodesic fibers, then M cannot be isometrically immersed into any Riemannian manifold of non-positive curvature as a minimal submanifold.*

Proof. Let $\pi : M \to B$ be a Riemannian submersion with totally geodesic fibers. If M admits a minimal immersion into a Riemannian manifold $\tilde{M}$ of non-positive curvature, then (18.2) and (18.4) imply that $A_X V = 0$ for all horizontal vector field X and all vertical vector field V in M. Hence, (18.3) implies that $\nabla_X V$ is always vertical. Therefore, for any horizontal vector fields X, Y, $\nabla_X Y$ is horizontal. Consequently, the horizontal distribution of the submersion is a totally geodesic foliation. As a consequence, $\pi : M \to B$ is a trivial submersion. $\square$

Remark 18.1. The condition "*non-positive curvature*" imposed on $\tilde{M}$ in Theorem 18.2 is necessary. For instance, the Hopf fibration $\pi : S^{2n+1} \to CP^n(4)$ is a non-trivial Riemannian submersion with totally geodesic fibers; and S^{2n+1} can be embedded in S^{2n+2} as a totally geodesic hypersurface.

Remark 18.2. Also, the condition "*Riemannian submersion has totally geodesic fibers*" in Theorem 18.2 cannot be omitted, since there exist minimal submanifolds in Euclidean spaces which admit non-trivial Riemannian submersions. The simplest examples are catenoid $\mathcal{C}$ and helicoid $\mathcal{H}_a$ in $\mathbb{E}^3$.

A catenoid $\mathcal{C}$ in $\mathbb{E}^3$ is defined as

$$(\cosh u \cos v, \cosh u \sin v, u),$$

which is a minimal surface admitting a nontrivial submersion $\pi : \mathcal{C} \to B$ such that the base B is the profile curve and the projection $\pi : \mathcal{C} \to B$ is the mapping which carries $(\cosh u \cos v, \cosh u \sin v, u) \in \mathcal{C}$ to $(\cosh u, u) \in B$. Fibers of this submersion are the circles of latitude.

For a real number $a > 0$, the helicoid $\mathcal{H}_a$ is defined by

$$(t \cos s, t \sin s, as), \quad t > 0.$$

$\mathcal{H}_a$ is a minimal surface which admits a non-trivial submersion $\pi : \mathcal{H}_a \to \ell_+$, where ℓ_+ is the half line $\{t : t > 0\}$ and the projection $\pi : \mathcal{H}_a \to \ell_+$ carries $(t \cos s, t \sin s, as) \in \mathcal{H}_a$ to $t \in \ell_+$. Fibers of this submersion are helices.

If $\phi_F : F \to \mathbb{E}^{m_1}$ and $\phi_B : B \to \mathbb{E}^{m_2}$ are minimal immersions of F and B into Euclidean spaces, then the direct product immersion of ϕ_F and ϕ_B is

$$(\phi_F, \phi_B) : F \times B \to \mathbb{E}^{m_1} \oplus \mathbb{E}^{m_2}, \tag{18.14}$$

which carries $(q, p) \in F \times B$ to $(\phi_F(q), \phi_B(b))$. It is easy to see that (ϕ_F, ϕ_B) is also a minimal immersion.

Another application of Theorem 18.1 is the following [Chen (2005b)].

Theorem 18.3. *Let $\pi : M \to B$ be a Riemannian submersion with totally geodesic fibers. If M admits a minimal isometric immersion into a Euclidean space, then locally M is the direct product of a fiber F and the base B and the immersion is the direct product immersion (ϕ_F, ϕ_B) of some minimal isometric immersions $\phi_F : F \to \mathbb{E}^{m_1}$ and $\phi_B : B \to \mathbb{E}^{m_2}$ into some Euclidean spaces.*

Proof. Let $\pi : M \to B$ be a Riemannian submersion with totally geodesic fibers. If $\phi : M \to \mathbb{E}^m$ is an isometric minimal immersion, then Theorem 18.1 implies that $\breve{A}$ vanishes identically and inequality (18.2) is actually an equality. Thus $A_X V = 0$ for any horizontal vector field X and vertical vector field V. As we already seen in the proof of Theorem 18.2, this implies that the horizontal distribution $\mathcal{H}$ is a totally geodesic foliation. Therefore, M is locally the direct product $F \times B$ of a (totally geodesic) fiber F and the base B.

On the other hand, since (18.2) is an equality, all of the inequalities in (18.10) and (18.13) become equalities. Hence, the second fundamental form h of M in $\mathbb{E}^m$ satisfies the following two conditions:

$$\sum_{i=1}^{b} h(e_i, e_i) = \sum_{s=b+1}^{n} h(e_s, e_s), \tag{18.15}$$

$$h(X, V) = 0, \ \ \forall X \in \mathcal{H}, \ \forall V \in \mathcal{V}, \tag{18.16}$$

where $e_1 \ldots, e_b$ and $e_{b+1}, \ldots, e_n$ are orthonormal horizontal and vertical frames, respectively.

Now, it follows from (18.16) and Moore's lemma that $\phi : F \times B \to \mathbb{E}^m$ is locally the direct product immersion of $\phi_F : F \to \mathbb{E}^{m_1}$ and $\phi_B : B \to \mathbb{E}^{m_2}$. Since ϕ is minimal, (18.2) implies that both ϕ_F and ϕ_B are minimal immersions. □

18.4 Submersions satisfying the basic equality

When the ambient space $\tilde{M}$ is a unit sphere, Theorem 18.1 reduces to the following.

Theorem 18.4. *Let $\pi : M \to B$ be a Riemannian submersion with totally geodesic fibers. Then, for any isometric immersion of M into the unit m-sphere $S^m(1)$, we have*

$$\breve{A}_\pi \leq \frac{n^2}{4}||H||^2 + b(n-b) \tag{18.17}$$

with $b = \dim B$ and $n = \dim M$.

The following lemma can be found in [Lawson (1970b)].

Lemma 18.1. *Let $\pi : S^{2m+1}(1) \to CP^m(4)$ denote the Hopf fibration and let N be a submanifold of $CP^m(4)$. Then $\pi^{-1}(N)$ is minimal in $S^{2m+1}(1)$ if and only if N is minimal in $CP^m(4)$.*

Proof. For a given fixed point $p \in \pi^{-1}(N)$, we choose a set $v_1, \dots, v_r$ of local orthonormal vertical vector fields about p. Let $e_1, \dots, e_n$ be local orthonormal vector fields on N about $\pi(p)$, and let $\bar{e}_1, \dots, \bar{e}_n$ be their horizontal lifts. Then $\pi^{-1}(N)$ is minimal if and only if

$$\begin{aligned} 0 &= \sum_{i=1}^{r} nor(\nabla_{v_i} v_i) + \sum_{j=1}^{n} nor(\nabla_{\bar{e}_j} \bar{e}_j) \\ &= \sum_{j=1}^{n} nor(\nabla_{\bar{e}_j} \bar{e}_j) = \sum_{j=1}^{n} nor(\overline{\nabla'_{e_j} e_j}), \end{aligned} \tag{18.18}$$

where $nor(\cdot)$ denotes the normal component of $(\cdot)$ and ∇ and ∇' are the Levi-Civita connection of $S^{2m+1}(1)$ and $CP^m(4)$, respectively. Clearly, (18.18) is equivalent to the minimality of N. □

Next, we provide classification results on Riemannian submersions which verify the equality case of (18.17).

Theorem 18.5. *Let B be a Kähler submanifold of $CP^m(4)$ and let $\pi_B : \pi^{-1}(B) \to B$ be the restriction of Hopf's fibration to $\pi^{-1}(B) \subset S^{2m+1}(1)$. Then the inclusion $\iota : \pi^{-1}(B) \subset S^{2m+1}(1)$ satisfies the equality case of* (18.17) *identically (with $n = b + 1$; $b = \dim_{\mathbb{R}} B$).*

Proof. Under the hypothesis, let us consider $S^{2m+1}(1)$ as the ordinary unit hypersphere of $\mathbf{C}^{m+1}$. Let z be the unit outward normal of $S^{2m+1}(1)$ and J the almost complex structure of $\mathbf{C}^{m+1}$. Then fibers of π_B are great circles in $S^{2m+1}(1)$ and the characteristic vector field ξ on the Sasakian $S^{2m+1}(1)$ is Jz. Clearly, the restriction of ξ to $\pi^{-1}(B)$ is a vertical vector field of the Riemannian submersion $\pi^B : \pi^{-1}(B) \to B$ and the fibers of π are fibers of the Hopf fibration $\pi : S^{2m+1}(1) \to CP^m(4)$.

Let $\tilde{\nabla}^{\mathbf{C}}$ and ∇ denote the Levi-Civita connections on $\mathbf{C}^{\mathbf{m+1}}$ and $\pi^{-1}(B)$, respectively. Then

$$\tilde{\nabla}^{\mathbf{C}}_X(Jz) = JX \tag{18.19}$$

for any horizontal vector X of $\pi_B : \pi^{-1}(B) \to B$. For any horizontal vector E, we have $0 = \langle E, z\rangle = \langle E, Jz\rangle = 0$. Thus

$$0 = \langle JE, z\rangle = \langle JE, Jz\rangle = 0$$

which implies that the tangential component, PE, of JE is also a horizontal vector on $\pi^{-1}(B)$. Hence, we find from (18.3) and (18.19) that

$$A_X\xi = PX. \tag{18.20}$$

Consequently, we find from (18.1) that

$$\breve{A}_\pi = ||P_{\mathcal{H}}||^2, \tag{18.21}$$

where $P_{\mathcal{H}}$ is the endomorphism of the horizontal distribution $\mathcal{H}$ defined to be the restriction of P on $\mathcal{H}$.

Now, if B is a Kähler submanifold of $CP^m(4)$, then B is minimal in $CP^m(4)$. Thus, by Lemma 18.1, $\pi^{-1}(B)$ is minimal in $S^{2n+1}(1)$. Because the horizontal distribution $\mathcal{H}$ is J-invariant, (18.21) implies that $\breve{A}_\pi = \dim_{\mathbb{R}} B = b$. From these we conclude that $\iota : \pi^{-1}(B) \to S^{2n+1}(1)$ satisfies the equality case of (18.17). □

Definition 18.1. Let $\pi : M \to B$ and $\pi' : M' \to B'$ be two Riemannian submersions with totally geodesic fibers. Then π and π' are said to be *equivalent* provided that there exists an isometry $\Phi : M \to M'$ which induces an isometry $\Phi_B : B \to B'$ such that the diagram

$$\begin{array}{ccc} M & \xrightarrow{\Phi} & M' \\ \pi\downarrow & & \downarrow\pi' \\ B & \xrightarrow{\Phi_B} & B' \end{array}$$

commutes.

As a converse to Theorem 18.5, we have the following.

Theorem 18.6. *Let $\pi : M \to B$ be a Riemannian submersion with totally geodesic fibers. If M admits an isometric embedding $\phi : M \to S^{2n+1}(1)$ which carries fibers of π to fibers of $\pi : S^{2n+1}(1) \to CP^n(4)$ and if ϕ satisfies the equality sign of* (18.17)*, then there exists a Kähler submanifold $B_1 \subset CP^n(4)$ such that $\pi : M \to B$ is equivalent to $\pi_{B_1} : \pi^{-1}(B_1) \to B_1$ and ϕ is congruent to the inclusion map: $\iota : \pi^{-1}(B_1) \subset S^{2n+1}(1)$.*

Proof. Let $\pi : M \to B$ be a Riemannian submersion with totally geodesic fibers. Suppose that M admits an isometric embedding $\phi : M \to S^{2n+1}(1)$ which carries the fibers of π to fibers of the Hopf fibration.

Let z be the position vector field of M in $\mathbf{C}^{2m+1}$. Then Jz is a unit vertical vector field on M. For any horizontal vector X of M, we have $\tilde{\nabla}^{\mathbf{C}}_X Jz = JX$ as before. Thus, we find from (18.3) that

$$||A_X(Jz)|| \leq 1 \tag{18.22}$$

for each unit horizontal vector X on M. Moreover, we know that the equality sign of (18.22) holds if and only if JX is a horizontal vector.

Assume that the embedding $\phi : M \to S^{2n+1}(1)$ satisfies the equality case of (18.17) identically. Then we find from (18.1), (18.17) and (18.22) that

(α) M is embedded as a minimal submanifold of $S^{2n+1}(1)$ and

(β) the horizontal distribution $\mathcal{H}$ on M is J-invariant.

Since $\phi : M \to S^{2n+1}(1)$ carries fibers of $\pi : M \to B$ to fibers of Hopf's fibration, the embedding ϕ gives rise to an isometric embedding of B into $CP^n(4)$. Further, since the horizontal distribution $\mathcal{H}$ is J-invariant, the embedding of B into $CP^n(4)$ is a Kählerian. Thus, if we put

$$M_1 = \phi(M), \quad B_1 = \pi(M_1), \quad \pi_1 = \pi|_{M_1},$$

then $\pi : M \to B$ and $\pi_1 : M_1 \to B_1$ are equivalent and also that ϕ is congruent to the inclusion map: $\iota : \pi_1^{-1}(B_1) \to S^{n+1}(1)$. □

Let $\pi : M \to B$ be a Riemannian submersion. An isometric immersion $\phi : M \to S^N$ is called *mixed-totally geodesic* if its second fundamental form h satisfies

$$h(X, V) = 0 \tag{18.23}$$

for any horizontal vector X and vertical vector V on M.

Lemma 18.2. *Let $\pi : M \to B$ be a Riemannian submersion with totally geodesic fibers. If an isometric immersion $\phi : M \to S^{\bar{m}}(1)$ carries fibers of π to totally geodesic submanifolds of the unit $\bar{m}$-sphere $S^{\bar{m}}(1)$ and if it satisfies the equality case of* (18.17), *then*

(1) $\dim M \geq 3$;
(2) *M is a minimal mixed-totally geodesic submanifold of $S^{\bar{m}}(1)$.*

Proof. Let $\pi : M \to B$ be a Riemannian submersion with totally geodesic fibers. Denote by ∇ and R the Levi-Civita connection and the Riemann curvature tensor of M. Assume that M admits an isometric immersion $\phi : M \to S^{\bar{m}}(1)$ which carries fibers of π to totally geodesic submanifolds of $S^{\bar{m}}(1)$. Then the second fundamental form h of M in $S^{\bar{m}}(1)$ satisfies

$$h(V, W) = 0 \tag{18.24}$$

for vertical vectors V, W on M. From Gauss's equation we know that the sectional curvature of M satisfies

$$K(X \wedge V) = 1 + \langle h(X, X), h(V, V)\rangle - ||h(X, V)||^2 \tag{18.25}$$

for unit horizontal vector X and unit vertical vector V on M. Hence, we find from (18.24) and (18.25) that

$$K(X \wedge V) \leq 1, \tag{18.26}$$

with the equality sign of (18.26) holding if and only if $h(X, V) = 0$.

On the other hand, since the submersion π has totally geodesic fibers,

$$||A_X V||^2 = K(X \wedge V) \leq 1 \tag{18.27}$$

for every unit horizontal vector X and unit vertical vector V. Hence, by applying (18.1), (18.26) and (18.27), we obtain

$$\breve{A}_\pi \leq b(n - b), \quad b = \dim B, \quad n = \dim M, \tag{18.28}$$

with equality holding if and only if M is mixed-totally geodesic in $S^{\bar{m}}(1)$, i.e., $h(X, V) = 0$ holds for any unit horizontal vector X and unit vertical vector V on M. Thus, the equality sign of (18.17) holds implies that

$$\breve{A}_\pi = b(n - b), \quad H = 0. \tag{18.29}$$

Hence, M is immersed as a minimal mixed-totally geodesic submanifold of $S^{\bar{m}}(1)$. The converse is easy to verify. This gives statement (2).

Next, assume that $\dim M = 2$. Then (18.29) implies that the Gaussian curvature G of M satisfying $G = 1$ and M is immersed as a minimal surface in $S^{\bar{m}}$ via ϕ. From these we conclude that M is totally geodesic in $S^{\bar{m}}(1)$. Hence, M is an open portion of a unit 2-sphere $S^2(1)$.

Let e_1 be a unit vertical vector field and e_2 be a unit horizontal vector field on M. Then, it follows from the totally geodesy of fibers of π that $\nabla_{e_1} e_1 = 0$. On the other hand, it follows from (11.1) and (18.29) that $\nabla_{e_2} e_1 = \pm e_2$. By applying these we conclude that

$$G = \langle R(e_2, e_1)e_1, e_2 \rangle = -1,$$

which is a contradiction. Thus we have statement (1). □

18.5 A characterization of Cartan hypersurface

The *Cartan hypersurface* in $S^4(1) \subset \mathbb{E}^5$ is defined by the equation:

$$\begin{aligned} 2x_5^3 + 3(x_1^2 + x_2^2)x_5 - 6(x_3^2 + x_4^2)x_5 + 3\sqrt{3}(x_1^2 - x_2^2)x_4 \\ + 6\sqrt{3}x_1x_2x_3 = 0. \end{aligned} \tag{18.30}$$

É. Cartan proved that this hypersurface is the homogeneous Riemannian manifold $SU(3)/SO(3)$ (equipped with a suitable metric) and its principal curvatures in $S^4(1)$ are given by $0, \sqrt{3}, -\sqrt{3}$.

Theorem 18.7. *Let $\pi : M \to B$ be a Riemannian submersion with totally geodesic fibers. If $\dim M = 3$ and $\phi : M \to S^4(1)$ is a non-totally geodesic immersion carrying fibers to totally geodesic submanifolds of $S^4(1)$, then ϕ satisfies the equality case of* (18.17) *if and only if ϕ is congruent to the Cartan hypersurface.*

Proof. Clearly, $S^3(1)$ admits the Hopf fibration $\pi : S^3(1) \to CP^1(4)$ with totally geodesic fibers and the inclusion map $S^3(1) \subset S^4(1)$ satisfies the equality case of (18.17) identically.

Conversely, assume that M is a Riemannian 3-manifold and $\pi : M \to B$ is a Riemannian submersion with totally geodesic fibers. Suppose that M admits a non-totally geodesic isometric immersion $\phi : M \to S^4(1)$ which carries fibers into totally geodesic submanifolds of $S^4(1)$ and which satisfies the equality case of (18.17). Then, by Lemma 18.2, M is a minimal mixed-totally geodesic hypersurface of $S^4(1)$.

Case (i): $\dim B = 1$. The fibers of π are 2-dimensional and the fibers are immersed as totally geodesic surfaces in $S^4(1)$. Thus, by applying the mixed-totally geodesy and minimality of ϕ, we know that ϕ is totally geodesic, which is a contradiction.

Case (ii): $\dim B = 2$. The fibers of π are carried to great circles by ϕ. Thus the second fundamental form h of ϕ satisfies $h(V,V) = 0$ for any vertical vector V. So, by applying the mixed-totally geodesy and minimality of ϕ, we may choose e_1 to be a unit vertical vector field and e_2, e_3 to be orthonormal horizontal vector fields so that e_1, e_2, e_3 diagonalize the shape operator. Thus

$$\begin{aligned} &h(e_1,e_1) = h(e_1,e_2) = h(e_1,e_2) = h(e_2,e_3) = 0, \\ &h(e_2,e_2) = -h(e_3,e_3) = \lambda e_4, \end{aligned} \tag{18.31}$$

for some function $\lambda \neq 0$, where e_4 is a unit normal vector field of M in $S^4(1)$. Let ∇' be the Levi-Civita connection of M and put

$$\nabla'_X e_i = \sum_{j=1}^{3} \omega_i^j(X) e_j. \tag{18.32}$$

Since the submersion has totally geodesic fibers, we have

$$\nabla'_{e_1} e_1 = 0. \tag{18.33}$$

By applying (18.31)-(18.33) and Codazzi's equation, we find

$$\begin{aligned} &e_1\lambda = 0, \quad e_2\lambda = 2\lambda\omega_3^2(e_3), \quad e_3\lambda = 2\lambda\omega_2^3(e_3), \\ &\omega_1^3(e_2) = \omega_2^1(e_3) = 2\omega_2^3(e_1). \end{aligned} \tag{18.34}$$

Because the O'neill integrability tensor A satisfies the *alternation property* $A_X Y = -A_Y X$ for horizontal vector fields X, Y, we obtain

$$\left\langle \nabla'_{e_i} e_1, e_i \right\rangle = -\left\langle \nabla'_{e_i} e_i, e_1 \right\rangle = -\left\langle A_{e_i} e_i, e_1 \right\rangle = 0, \quad i = 2,3. \tag{18.35}$$

Since the equality sign of (18.17) holds, we find from (18.27) and (18.29) that $||A_{e_2} e_1|| = ||A_{e_3} e_1|| = 1$. Thus, by applying (18.34) and (18.35), we find $\nabla'_{e_2} e_1 = \pm e_3, \nabla'_{e_3} e_1 = \mp e_2$. Without loss of generality, we may put

$$\nabla_{e_2} e_1 = e_3, \quad \nabla_{e_3} e_1 = -e_2. \tag{18.36}$$

Now, by combining (18.33), (18.34) and (18.36), we obtain

$$\begin{aligned} &\nabla'_{e_1} e_1 = 0, \quad \nabla'_{e_2} e_1 = e_3, \quad \nabla'_{e_3} e_1 = -e_2, \\ &\nabla'_{e_1} e_2 = \frac{e_3}{2}, \quad \nabla'_{e_2} e_2 = \frac{1}{2} e_3(\ln\lambda) e_3, \\ &\nabla'_{e_3} e_2 = e_1 - \frac{1}{2} e_2(\ln\lambda) e_3, \quad \nabla'_{e_1} e_3 = -\frac{e_2}{2}, \\ &\nabla'_{e_2} e_3 = -e_1 - \frac{1}{2} e_3(\ln\lambda) e_2, \\ &\nabla'_{e_3} e_3 = \frac{1}{2} e_2(\ln\lambda) e_2. \end{aligned} \tag{18.37}$$

It follows from $e_1\lambda = 0$, (18.31), (18.37), and Gauss's equation that

$$0 = 4\,\langle R(e_1, e_2)e_3, e_2\rangle = e_2(\ln \lambda). \tag{18.38}$$

Similarly, we find

$$0 = 4\,\langle R(e_1, e_3)e_3, e_2\rangle = e_3(\ln \lambda). \tag{18.39}$$

From $e_1\lambda = 0$, (18.38) and (18.39), we conclude that λ is constant. Thus, M is immersed as a minimal isoparametric hypersurface in $S^4(1)$, i.e., M has constant principal curvatures. Now, by applying (18.31), (18.37), Gauss' equation, and by the constancy of λ, we get

$$1 - \lambda^2 = \langle R(e_2, e_3)e_3, e_2\rangle = -2.$$

Hence, $\lambda^2 = 3$. Consequently, $\phi : M \to S^4(1)$ is congruent to the Cartan hypersurface of $S^4(1)$. □

18.6 Links between submersions and affine hypersurfaces

The following two theorems provide links between Riemannian submersions and affine hypersurfaces via the submersion δ-invariant $\breve{A}_\pi$ [Chen (2007b)].

Theorem 18.8. *If a Riemannian manifold M can be realized as an elliptic proper centroaffine hypersphere centered at the origin in some affine space, then every Riemannian submersion $\pi : M \to B$ with minimal fibers has non-totally geodesic horizontal distribution.*

Proof. Let $\pi : M \to B$ be a Riemannian submersion with minimal fibers. Assume that $v_1, \ldots, v_{n-b}$ are orthonormal vertical vector fields and $X_1, \ldots, X_b$ are orthonormal horizontal vector fields. Then

$$\sum_{i,r} K(X_i \wedge v_r) = \breve{A}_\pi - ||h_F||^2, \tag{18.40}$$

where h_F is the second fundamental form of the fibers in M.

Let us assume that M is realized as a definite centroaffine hypersurface in an affine $(n+1)$-space $\mathbf{R}^{n+1}$. Let us put

$$\delta = \frac{n^2}{2} h(T^\#, T^\#) + n(n-1)\hat{\kappa} - 2b(n-b)\varepsilon, \tag{18.41}$$

where $\varepsilon = 1$ or -1 according to the centroaffine hypersurface is positive-definite or negative-definite. Then (17.21) and (18.41) imply that

$$\frac{n^2}{2} h(T^\#, T^\#) = h(K, K) - \delta + \big(n(n-1) - 2b(n-b)\big)\varepsilon. \tag{18.42}$$

For a point $p \in M$, let us put

$$e_i = X_i, \quad e_{b+r} = v_r, \quad i = 1, \ldots, b;\ r = 1, \ldots, n-b.$$

Let us choose an orthonormal basis $e_1^*, \ldots, e_n^*$ of T_pM such that e_1^* is in the direction of the Tchebychev vector $T^\#$ at p (When $T^\# = 0$ at p, we may choose $e_1^*, \ldots, e_n^*$ to be any arbitrary orthonormal basis).

If we put $K_{\alpha\beta}^{\gamma^*} = h(K(e_\alpha, e_\beta), e_\gamma^*)$, then (18.42) becomes

$$\begin{aligned}\frac{n^2}{2} h(T^\#, T^\#) - \sum_{\alpha=1}^{n} \left(K_{\alpha\alpha}^{1^*}\right)^2 &= \sum_{\alpha \neq \beta} \left(K_{\alpha\beta}^{1^*}\right)^2 \\ &\quad + \sum_{r=2}^{n} \sum_{\alpha,\beta} (K_{\alpha\beta}^{r^*})^2 + \big(n(n-1) - 2b(n-b)\big)\varepsilon - \delta.\end{aligned} \tag{18.43}$$

If we apply Corollary 13.1 to the left-hand-side of (18.43), we find

$$\begin{aligned}\sum_{i<j} K_{ii}^{1^*} K_{jj}^{1^*} + \sum_{s<t} K_{ss}^{1^*} K_{tt}^{1^*} &\geq \sum_{\alpha<\beta} \left(K_{\alpha\beta}^{1^*}\right)^2 + \frac{1}{2} \sum_{r=2}^{n} \sum_{\alpha,\beta}^{n} (K_{\alpha\beta}^{r^*})^2 \\ &\quad + \frac{n(n-1) - 2b(n-b)}{2}\varepsilon - \frac{\delta}{2},\end{aligned} \tag{18.44}$$

with equality holding if and only if $K_{11}^{1^*} + \cdots + K_{bb}^{1^*} = K_{b+1b+1}^{1^*} + \cdots + K_{nn}^{1^*}$.

On the other hand, (17.19) implies that

$$\begin{aligned}\hat{K}(e_\alpha \wedge e_\beta) &= \varepsilon - h(K(e_\alpha, e_\alpha), K(e_\beta, e_\beta)) \\ &\quad + h(K(e_\alpha, e_\beta), K(e_\alpha, e_\beta)).\end{aligned} \tag{18.45}$$

Hence, by combining (18.40) and (18.45), we find

$$\begin{aligned}\breve{A}_\pi - ||h_F||^2 &= \frac{n(n-1)}{2}\hat{\kappa} - \sum_{j<k} \hat{K}(e_j \wedge e_k) - \sum_{s<t} \hat{K}(e_s \wedge e_t) \\ &= \frac{n(n-1)}{2}\hat{\kappa} - \frac{b(b-1)}{2}\varepsilon + \sum_{\alpha} \sum_{j<k} \left(K_{jj}^{\alpha^*} K_{kk}^{\alpha^*} - (K_{jk}^{\alpha^*})^2\right) \\ &\quad - \frac{(n-b)(n-b-1)}{2}\varepsilon + \sum_{\alpha} \sum_{s<t} \left(K_{ss}^{\alpha^*} K_{tt}^{\alpha^*} - (K_{st}^{\alpha^*})^2\right) \\ &= \frac{n(n-1)}{2}(\hat{\kappa} - \varepsilon) + b(n-b)\varepsilon + \sum_{r=2}^{n} \sum_{j<k} \left(K_{jj}^{r^*} K_{kk}^{r^*} - (K_{jk}^{r^*})^2\right) \\ &\quad + \sum_{r=2}^{n} \sum_{s<t} \left(K_{ss}^{r^*} K_{tt}^{r^*} - (K_{st}^{r^*})^2\right) + \sum_{j<k} K_{jj}^{1^*} K_{kk}^{1^*} \\ &\quad + \sum_{s<t} K_{ss}^{1^*} K_{tt}^{1^*} - \sum_{j<k} (K_{jk}^{1^*})^2 - \sum_{s<t} (K_{st}^{1^*})^2.\end{aligned} \tag{18.46}$$

Therefore, by (18.44) and (18.46), we obtain

$$
\begin{aligned}
\breve{A}_\pi - ||h_F||^2 \geq\ & \frac{n(n-1)}{2}\hat{\kappa} - \frac{\delta}{2} \\
& + \sum_{r=2}^{n} \sum_{j<k} \left(K^{r^*}_{jj} K^{r^*}_{kk} - (K^{r^*}_{jk})^2\right) + \frac{1}{2}\sum_{r=2}^{n}\sum_{\alpha,\beta}(K^{r^*}_{\alpha\beta})^2 \\
& + \sum_{r=2}^{n} \sum_{s<t} \left(K^{r^*}_{ss} K^{r^*}_{tt} - (K^{r^*}_{st})^2\right) + \sum_{j,t}(K^{1^*}_{jt})^2 \\
=\ & \frac{n(n-1)}{2}\hat{\kappa} - \frac{\delta}{2} + \sum_{r=1}^{n}\sum_{j,t}(K^{r^*}_{jt})^2 + \frac{1}{2}\sum_{r=2}^{n}\Big(\sum_{j} K^{r^*}_{jj}\Big)^2 \\
& + \frac{1}{2}\sum_{r=2}^{n}\Big(\sum_{t} K^{r^*}_{tt}\Big)^2 \\
\geq\ & \frac{n(n-1)}{2}\hat{\kappa} - \frac{\delta}{2}.
\end{aligned}
\tag{18.47}
$$

Thus, by combining (18.41) and (18.47), we obtain

$$\breve{A}_\pi + \frac{n^2}{4}h(T^\#, T^\#) \geq ||h_F||^2 + b(n-b)\varepsilon. \tag{18.48}$$

In particular, if M can be realized as a positive definite centroaffine hypersurface, then

$$\breve{A}_\pi + \frac{n^2}{4}h(T^\#, T^\#) > ||h_F||^2 \geq 0, \tag{18.49}$$

which implies that at each point on M we have either $\breve{A}_\pi > 0$ or $T^\# \neq 0$. Thus, either the horizontal distribution $\mathcal{H}$ is a non-totally geodesic distribution or the centroaffine hypersurface is not a proper affine hypersphere centered at the origin. □

Theorem 18.9. *If a Riemannian manifold M can be realized as an improper hypersphere in some affine space, then every Riemannian submersion $\pi : M \to B$ with non-totally geodesic minimal fibers has non-totally geodesic horizontal distribution.*

Proof. If M can be realized as graph hypersurfaces in some affine space, then

$$\breve{A}_\pi + \frac{n^2}{4}h(T^\#, T^\#) \geq ||h_F||^2 \tag{18.50}$$

instead of (18.48). Thus, by applying (18.50), we obtain the theorem. □

The next two corollaries follow easily from (18.48) and (18.50).

Corollary 18.1. *Let $\pi : M \to B$ be a Riemannian submersion with minimal fibers. If $\breve{A}_\pi \le ||h_F||^2$, then M cannot be realized as an elliptic proper affine hypersphere.*

Corollary 18.2. *Let $\pi : M \to B$ be a Riemannian submersion with minimal fibers. If $\breve{A}_\pi < ||h_F||^2$, then M cannot be realized as an improper affine hypersphere.*

Theorems 18.8 and 18.9 can be applied to many Riemannian manifolds, because there are many elliptic proper affine hyperspheres and improper hyperspheres in affine spaces. Here we give merely one such application.

Corollary 18.3. *Let I be an open subinterval of $(0,\infty)$ and let N be an open portion of $\mathbb{E}^{n-1}$. Denote by $I \times_{s^{1/(n+1)}} N$ the warped product of I and N with the warping function $s^{1/(n+1)}$. Then, for every Riemannian submersion $\pi : I \times_{s^{1/(n+1)}} N \to B$ with non-totally geodesic minimal fibers, the horizontal distribution of π is a non-totally geodesic distribution.*

Proof. Let $I \times_{s^{1/(n+1)}} N$ be the warped product of the open interval I and an open portion N of $\mathbb{E}^{n-1}$ endowed with the warped product metric

$$g = ds^2 + s^{\frac{2}{n+1}}\left(du_2^2 + \cdots + du_n^2\right). \tag{18.51}$$

Consider the immersion of $I \times_{s^{1/(n+1)}} N$ into the affine $(n+1)$-space $\mathbf{R}^{n+1}$ defined by

$$\left(u_2, \ldots, u_n, \frac{(n+1)^2 s^{\frac{2n}{n+1}}}{2n} - \sum_{j=2}^{n} u_j^2, \frac{(n+1)s^2}{2}\right). \tag{18.52}$$

Then $I \times_{s^{1/(n+1)}} N$ is a graph hypersurface with affine normal vector field $\xi = (0, \ldots, 0, 2)$.

It is straightforward to verify that the Tchebychev vector field $T^{\#}$ vanishes identically on $I \times_{s^{1/(n+1)}} N$ and its induced affine metric is exactly the warped product metric (18.51). Hence, after applying Theorem 18.9, we conclude that, for every Riemannian submersion

$$\pi : I \times_{s^{1/(n+1)}} N \to B$$

from $I \times_{s^{1/(n+1)}} N$ onto a Riemannian manifold B with non-totally geodesic minimal fibers, the horizontal distribution is always a non-totally geodesic distribution. □

Chapter 19

Nearly Kähler Manifolds and Nearly Kähler $S^6(1)$

An almost Hermitian manifold M with metric tensor g and almost complex structure J is called *nearly Kähler* provided that [Gray (1970)]

$$(\nabla_X J)X = 0, \quad \forall X \in TM. \tag{19.1}$$

Nearly Kähler manifolds are exactly *almost Tachibana manifolds* initially studied in [Tachibana (1959)].

The best known example of nearly Kähler manifolds, but not Kählerian, is $S^6(1)$ with the nearly Kählerian structure induced from the vector cross product on the space of purely imaginary Cayley numbers $\mathcal{O}$. More general examples of nearly Kähler manifolds are the homogeneous spaces G/K, where G is a compact semisimple Lie group and K is the fixed point set of an automorphism of G of order 3 [Wolf and Gray (1968)].

Nearly Kähler manifolds are characterized by the following geometric property: Let γ be a piecewise differentiable loop at $p \in M$ and let π be the almost complex section of the tangent space T_pM which contains $\gamma'(0)$. Denote by τ_γ the parallel translation along γ. Then there exists $h \in U(n)$ such that $\tau_\gamma|_\pi = h|_\pi$, where $U(n)$ is regarded as the structure group of the tangent bundle of M. Conversely, any almost Hermitian manifold with this property is a nearly Kähler manifold [Gray (1970)].

A. Gray showed in [Gray (1970)] that a compact nearly Kähler manifold of positive holomorphic sectional curvature is simply-connected; generalizing a well known result of [Tsukamoto (1957)]. He also proved the following:

(a) A nearly Kähler manifold M is Kählerian (resp. Einstein) if $\dim_{\mathbf{R}} M = 4$ (resp. if $\dim_{\mathbf{R}} = 6$);
(b) a complete simply-connected nearly Kähler manifold with $\dim_{\mathbf{R}} M = 8$ is the direct product $M_1 \times M_2$ of 2-dimensional Kähler manifold M_1 and a 6-dimensional nearly Kähler manifold M_2;

(c) $\{X \in T_pM : (\nabla_X J)Y = 0,\ \forall Y \in T_pM,\ p \in M\}$ defines an integrable distribution on a nearly Kähler manifold M, whose leaves are Kähler manifolds;
(d) if a non-Kählerian, nearly Kähler manifold M has sufficiently large sectional curvature (or holomorphic pinching), then $H^2(M;\mathbf{R}) = 0$.

19.1 Real hypersurfaces of nearly Kähler manifolds

Let N be a real hypersurface of a nearly Kähler manifold $(M^{2n}, g, J), n \geq 2$, and ξ a local unit normal vector field of N. Then $J\xi$ is a unit tangent vector field of N. Denote by $\mathcal{D}^\perp$ the distribution spanned by $J\xi$. Then $\mathcal{D}^\perp$ is a totally real distribution on N. Let $\mathcal{D}$ be the complementary orthogonal distribution of $\mathcal{D}^\perp$ in TN. Then $\mathcal{D}$ is J-invariant. Thus $TN = \mathcal{D} \oplus \mathcal{D}^\perp$, which shows that every real hypersurface of (M^{2n}, g, J) is a CR-submanifold such that $\text{rank}(\mathcal{D}) = 2n-2$ and $\text{rank}(\mathcal{D}^\perp) = 1$.

For real hypersurfaces we have the following general result.

Lemma 19.1. *The invariant distribution $\mathcal{D}$ of a real hypersurface N in a nearly Kähler manifold $(M^{2n}, g, J), n \geq 2$, is a minimal distribution.*

Proof. Since M^{2n} is nearly Kählerian, we have

$$0 = (\nabla_X J)X = \nabla'_X JX + h(X, JX) - J(\nabla'_X X) - Jh(X,X) \qquad (19.2)$$

for any vector field X in $\mathcal{D}$, where ∇' is the Levi-Civita connection of N. Let ξ be a unit normal vector field of N. Then $J\xi$ spans $\mathcal{D}^\perp$. Thus, after taking the inner product of (19.2) with ξ, we find

$$\langle \nabla'_X X, J\xi \rangle = -\langle h(X, JX), \xi \rangle . \qquad (19.3)$$

If we replace X in (19.3) by JX, then (19.3) becomes

$$\langle \nabla'_{JX} JX, J\xi \rangle = \langle h(JX, X), \xi \rangle .$$

Combining these two equations yields $\langle \nabla'_X X + \nabla'_{JX} JX, J\xi \rangle = 0$, which implies that $\mathcal{D}$ is a minimal distribution. □

For a real hypersurface N of a nearly Kähler manifold M^{2n}, let us choose a local orthonormal frame $\{e_1, \ldots, e_{2n-1}\}$ of N with $e_2 = Je_1, \ldots, e_{2n-2} = Je_{2n-3}$ so that $e_1, \ldots, e_{2n-2} \in \mathcal{D}$ and $e_{2n-1} \in \mathcal{D}^\perp$.

Let $\omega^1, \ldots, \omega^{2n-1}$ be the dual 1-forms of $e_1, \ldots, e_{2n-1}$ such that

$$\omega^r(e_s) = \delta^r_s, \quad r, s = 1, \ldots, 2n-1. \qquad (19.4)$$

Put

$$\omega = \omega^1 \wedge \cdots \wedge \omega^{2n-2}. \tag{19.5}$$

Theorem 19.1. *Let N be an orientable closed real hypersurface of a nearly Kähler manifold (M^{2n}, g, J), $n \geq 2$. Then ω is a closed form; thus ω defines a canonical de Rham cohomology class $[\omega] \in H^{2n-2}(N, \mathbf{R})$. Moreover, this cohomology class is nontrivial if $\mathcal{D}$ is integrable and $\mathcal{D}^\perp$ is a totally geodesic distribution.*

Proof. It follows from (6.1) and (19.5) that

$$\begin{aligned} &d\omega(e_1, \ldots, e_{2n-1}) \\ &= \sum_{r=1}^{2n-1} (-1)^{r-1} e_r(\omega^1 \wedge \cdots \wedge \omega^{2n-2}(e_1, \cdots, \hat{e}_r, \cdots, e_{2n-1})) \\ &\quad + \sum_{r<s} (-1)^{r+s-1} \omega^1 \wedge \cdots \wedge \omega^{2n-2}([e_r, e_s], e_1, \cdots, \hat{e}_r, \cdots, \hat{e}_s, \cdots, e_{2n-1}), \end{aligned}$$

where $\hat{}$ denotes the omission of that element. Thus, by applying (19.4), we find

$$\begin{aligned} &d\omega(e_1, \ldots, e_{2n-1}) \\ &= \sum_{i=1}^{2n-2} (-1)^i \omega^1 \wedge \cdots \wedge \omega^{2n-2}([e_i, e_{2n-1}], e_1, \cdots, \hat{e}_i, \cdots, e_{2n-2}) \\ &= -\sum_{i=1}^{2n-2} \omega_i^{2n-1}(e_1)(\omega^1 \wedge \cdots \wedge \omega^{2n-2}(e_1, \cdots, e_{2n-2})) \\ &= \sum_{i=1}^{2n-2} \left\langle \nabla'_{e_i} e_i, e_{2n-1} \right\rangle. \end{aligned}$$

After applying Lemma 19.1, we obtain $d\omega = 0$. Thus ω is a closed form which defines a canonical de Rham cohomology class $[\omega] \in H^{2n-2}(N, \mathbf{R})$.

Next, let us consider the 1-form $\omega^\perp = \omega^{2n-1}$. Then $d\omega^\perp = 0$ if and only if the following two conditions hold:

$$d\omega^\perp(e_i, e_j) = 0, \quad 1 \leq i < j \leq 2n-2, \tag{19.6}$$

$$d\omega^\perp(e_i, e_{2n-1}) = 0, \quad i = 1, \ldots, 2n-2. \tag{19.7}$$

On the other hand, it follows from (6.1) that

$$2d\omega^\perp(e_i, e_j) = \omega^{2n-1}([e_j, e_i]), \tag{19.8}$$

$$2d\omega^\perp(e_i, e_{2n-1}) = \omega^{2n-1}(\nabla'_{e_{2n-1}} e_i). \tag{19.9}$$

From (19.8) we know that (19.6) holds if and only if the invariant distribution $\mathcal{D}$ is integrable. Also, it follows from (19.7) that (19.9) holds if and only if $\langle \nabla'_{e_{2n-1}} e_{2n-1}, e_i \rangle = 0, \quad i = 1, \ldots, 2n-2$. Therefore, $d\omega^{\perp} = 0$ (or equivalently $\delta\omega = 0$) if $\mathcal{D}$ is an integrable distribution and $\mathcal{D}^{\perp}$ is a totally geodesic distribution. Hence, if $\mathcal{D}$ is integrable and $\mathcal{D}^{\perp}$ is a totally geodesic distribution, then ω is an exact form. Consequently, the theorem follows easily from the Hodge-de Rham decomposition theorem. □

Proposition 19.1. *Let N be an orientable closed hypersurface of a nearly Kähler manifold M. If $H^{2n-2}(N;\mathbf{R}) = 0$ with $n = \frac{1}{2}\dim M \geq 2$, then either the invariant distribution $\mathcal{D}$ is non-integrable or the totally real distribution $\mathcal{D}^{\perp}$ is non-totally geodesic.*

Proof. Follows immediately from Theorem 19.1. □

Recall that a real hypersurface N of almost Hermitian manifold (M^{2n}, g, J) is called a *Hopf hypersurface* if $U = -J\xi$ is a principal curvature vector, where ξ is a unit normal vector of N in M^{2n}. The one-dimensional integrable distribution spanned by U is called the *Hopf foliation.*

The following lemma can be found in [Berndt et al. (1995)].

Lemma 19.2. *An orientable hypersurface N of a nearly Kähler manifold (M^{2n}, g, J) is a Hopf hypersurface if and only if the Hopf fibration is totally geodesic.*

Proof. Since M^{2n} is nearly Kählerian,

$$(\nabla_X J)(JX) = -\nabla_X X - J(\nabla_X JX) = 0 \tag{19.10}$$

for any $X \in \mathcal{X}(M^{2n})$. Thus, by Weingarten's formula,

$$\nabla_U U = -\nabla_U(J\xi) = -(\nabla_U J)\xi - J(\nabla_U \xi) = (\nabla_U J)JU + JA_\xi U.$$

Thus, by (19.10) and Gauss's formulas, we find $\nabla'_U U + h(U,U) = JA_\xi U$, which shows that $\nabla'_U U = 0$ if and only if N is a Hopf hypersurface. □

Combining Proposition 19.1 and Lemma 19.2 gives the following.

Corollary 19.1. *Let N be an orientable closed Hopf hypersurface of a nearly Kähler manifold M. If $H^{2n-2}(N;\mathbf{R}) = 0$ with $n = \frac{1}{2}\dim M \geq 2$, then the invariant distribution $\mathcal{D}$ of N is never integrable.*

19.2 Nearly Kähler structure on $S^6(1)$

The unit 6-sphere $S^6(1)$ can be considered as a homogeneous space $G_2/SU(3)$, where G_2 is the group of automorphisms of the Cayley numbers $\mathcal{O}$. From this representation, one can define a nearly Kählerian structure on $S^6(1)$ by making use of the vector cross product of $\mathcal{O}$.

Now, we give a brief explanation of how this standard nearly Kähler structure on $S^6(1)$ arises in a natural manner from Cayley multiplication. The multiplication of the Cayley algebra $\mathcal{O}$ may be used to define a vector cross-product on the purely imaginary Cayley numbers $\mathbb{R}^7$ using the formula

$$u \times v = \tfrac{1}{2}(uv - vu), \tag{19.11}$$

while the standard inner product on $\mathbb{R}^7$ is given by

$$(u, v) = -\tfrac{1}{2}(uv + vu). \tag{19.12}$$

It is now elementary to show that

$$u \times (v \times w) + (u \times v) \times w = 2(u, w)v - (u, v)w - (w, v)u, \tag{19.13}$$

and that the triple scalar product $\langle u \times v, w\rangle$ is skew symmetric in u, v, w.

Conversely, Cayley multiplication on $\mathcal{O}$ is given in terms of the vector cross product and the inner product by

$$(r + u)(s + v) = rs + rv + su - (u, v) + (u \times v) \tag{19.14}$$

for $r, s \in \operatorname{Re}\mathcal{O}$, $u, v \in \operatorname{Im}\mathcal{O}$. In view of (19.11), (19.12) and (19.14), it is clear that the group G_2 of automorphisms of $\mathcal{O}$ is precisely the group of isometries of $\mathbb{R}^7$ which preserve the vector cross-product.

An ordered orthonormal basis $\{e_1, \ldots, e_7\}$ of $\mathbb{R}^7$ is called *canonical* if

$$e_3 = e_1 \times e_2, \quad e_5 = e_1 \times e_4, \quad e_6 = e_2 \times e_4, \quad e_7 = e_3 \times e_4. \tag{19.15}$$

For example, the standard basis of $\mathbb{R}^7$ is canonical. Moreover, $(\mathbb{R}^7, \times)$ is generated by e_1, e_2, e_4 subject to the relations

$$e_i \times (e_j \times e_k) + (e_i \times e_j) \times e_k = 2\delta_{ik}e_j - \delta_{ij}e_k - \delta_{jk}e_i. \tag{19.16}$$

Given two canonical bases $\{e_1, \ldots, e_7\}$, $\{f_1, \ldots, f_7\}$ of $\mathbb{R}^7$, there is a unique $g \in G_2$ such that $ge_i = f_i$; thus g is uniquely determined by ge_1, ge_2, ge_4.

For any G_2-frame, we have the following multiplication table.

$\times$	e_1	e_2	e_3	e_4	e_5	e_6	e_7
e_1	0	e_3	$-e_2$	e_5	$-e_4$	$-e_7$	e_6
e_2	$-e_3$	0	e_1	e_6	e_7	$-e_4$	$-e_5$
e_3	e_2	$-e_1$	0	e_7	$-e_6$	e_5	$-e_4$
e_4	$-e_5$	$-e_6$	$-e_7$	0	e_1	e_2	e_3
e_5	e_4	$-e_7$	e_6	$-e_1$	0	$-e_3$	e_2
e_6	e_7	e_4	$-e_5$	$-e_2$	e_3	0	$-e_1$
e_7	$-e_6$	e_5	e_4	$-e_3$	$-e_2$	e_1	0

It follows from the above that G_2 acts transitively on $S^6(1)$ and that the stabilizer of the point $(1, 0, \ldots, 0)$ is $SU(3)$. Thus, $S^6(1) = G_2/SU(3)$. Moreover, it follows that G_2, a connected subgroup of $SO(7)$ of dimension 14, is the group of automorphisms of the nearly Kähler structure J.

Let J be the automorphism of $TS^6(1)$ defined by

$$Ju = x \times u, \qquad u \in T_x S^6(1), \quad x \in S^6(1). \tag{19.17}$$

Then J is an almost complex structure on $S^6(1)$, which is not integrable; but in fact J is a nearly Kähler structure on $S^6(1)$. The fundamental 2-form Ω of $(S^6(1), g, J)$ is clearly not closed, but it satisfy $\Omega \wedge d\Omega = 0$. Moreover, it was shown in [Calabi and Gluck (1993)] that J is the best almost complex structure on $S^6(1)$ from variational point of view.

If we define a skew-symmetric $(2, 1)$-tensor field G by

$$G(X, Y) = (\nabla_X J)(Y), \tag{19.18}$$

then it is direct to show that

$$G(X, Y) = X \times Y - g(x \times X, Y)x. \tag{19.19}$$

It is also known that G satisfies the following properties [Sekigawa (1983)]:

$$G(X, JY) = -JG(X, Y), \tag{19.20}$$

$$(\nabla G)(X, Y, Z) = g(Y, JZ)X + g(X, Z)JY - g(X, Y)JZ, \tag{19.21}$$

$$g(G(X, Y), Z) + g(G(X, Z), Y) = 0, \tag{19.22}$$

$$\begin{aligned} g(G(X, Y), G(Z, W)) = {} & g(X, Z)g(Y, W) - g(X, W)g(Z, Y) \\ & + g(JX, Z)g(Y, JW) - g(JX, W)g(Y, JZ). \end{aligned} \tag{19.23}$$

19.3 Almost complex submanifolds of nearly Kähler manifolds

A submanifold N of the nearly Kähler $(S^6(1), g, J)$ is called an *almost complex* submanifold if the tangent bundle TN is invariant under the action of J, i.e., $J(T_pN) = T_pN$, $p \in N$.

The following lemmas and Theorem are due to [Gray (1969)].

Lemma 19.3. *Let M be a nearly Kähler manifold with almost complex structure J. If N is an almost complex submanifold of M, then*

(i) *N is nearly Kählerian;*
(ii) *$h(X, JX) = Jh(X, X)$ for all $X \in TN$;*

(iii) *N is an austere, and hence minimal, submanifold of M.*

Proof. For $X \in \mathcal{X}(N)$ we have

$$0 = (\nabla_X J)X = (\nabla'_X J)X + h(X, JX) - Jh(X, X).$$

Hence (i) and (ii) follow. (iii) is an easy consequence of (ii). □

Lemma 19.4. *Every 4-dimensional nearly Kähler manifold is Kählerian.*

Proof. Let M be a 4-dimensional nearly Kähler manifold. Let us choose an orthonormal frame field on an open subset of M to be of the form $\{X, JX, Y, JY\}$. Clearly, $(\nabla'_X J)Y$ is perpendicular to X and Y. It is also perpendicular to JX and JY because $(\nabla'_X J)Y = J(\nabla'_X J)(JY)$. Hence $(\nabla'_X J)Y = 0$. Thus, M is Kählerian. □

Lemma 19.5. *Let M be a nearly Kähler manifold. Then, for X, Y, Z, W tangent to M, we have*

(i) $||(\nabla_X J)Y||^2 = g(R(X,Y)Y, X) - g(R(X,Y)JY, JX)$;
(ii) $g(R(W,X)Y,Z) = g(R(JW,JX)JY,JZ)$.

Proof. Let F be the 2-form defined by

$$F(X,Y) = g(JX,Y), \quad X, Y \in TM.$$

Also define $\nabla F, \nabla^2 F$, and $R(W,X)F$ by

$$\begin{aligned}(\nabla F)(X,Y,Z) &= XF(Y,Z) - F(\nabla_X Y, Z) - F(Y, \nabla_X Z),\\ (\nabla^2 F)(W,X,Y,Z) &= W(\nabla F)(X,Y,Z) - (\nabla F)(\nabla_W X, Y, Z)\\ &\quad - (\nabla F)(X, \nabla_W Y, Z) - (\nabla F)(X, Y, \nabla_W Z),\\ (R(W,X)F)(Y,Z) &= F(R(W,X)Y,Z) - F(Y, R(W,X)Z),\end{aligned}$$

for W, X, Y, Z tangent to M. Calculations show first that

$$(R(W,X)F)(Y,Z) = -(\nabla^2 F)(W,X,Y,Z) + (\nabla^2 F)(X,W,Y,Z),$$

and secondly $||(\nabla_X J)Y||^2 = -(\nabla^2 F)(X,X,Y,JY)$. These computations do not depend on the fact that M is nearly Kählerian. However, the fact that M is nearly Kählerian is used to prove that

$$(\nabla^2 F)(W,X,Y,Z) = -(\nabla^2 F)(W,Y,X,Z).$$

We now observe that

$$\begin{aligned}||(\nabla_X J)Y||^2 &= (\nabla^2 F)(X,Y,X,JY) = -(R(X,Y)FX, JY)\\ &= g(R(X,Y)X,Y) - g(R(X,Y)JX,JY),\end{aligned}$$

which proves statement (i). Statement (ii) follows from statement (i) and the symmetry properties of the curvature tensor. □

Theorem 19.2. *With respect to the usual nearly Kähler structure, $S^6(1)$ has no 4-dimensional almost complex submanifolds.*

Proof. Suppose that N is a 4-dimensional almost complex submanifold of $S^6(1)$. Assume that there exists $p \in N$ and a unit vector $X \in T_pN$ such that $h(X,X) \neq 0$. Then $h(X,JX) = Jh(X,X) \neq 0$. We may assume that X is chosen so that $||h(X,X)||^2$ is maximum. If $g(X,Y) = g(JX,Y) = 0$, then $g(h(X,X),h(X,Y)) = 0$. Also, $g(Jh(X,X),h(X,Y)) = 0$ because $Jh(X,X) = h(U,U)$ with $U = (X + JY)/\sqrt{2}$.

Similarly, $h(X,JY)$ is perpendicular to both $h(X,X)$and $Jh(X,X)$. Since $h(X,X)$ and $Jh(X,X)$ span $T_p^\perp N$, we have $h(X,Y) = h(X,JY) = 0$. Let $\tilde{K}$ denote the (constant) sectional curvature of the curvature operator of $S^6(1)$. Then if $||Y|| = 1$, we have

$$\begin{aligned}\bar{K} &= g(\tilde{R}(X,Y)Y,X) - g(\tilde{R}(X,Y)JY,JX) = ||(\nabla_X J)Y||^2 \\ &= ||(\nabla_X J)Y + h(X,JY) - Jh(X,Y)||^2 = 0.\end{aligned}$$

This is a contradiction, thus $h(X,X) = 0$ for all $X \in T_pN$. Hence N is totally geodesic in $S^6(1)$, and so it must be an open submanifold of a 4-sphere with constant curvature one, and also must be Kählerian. We then have $0 = ||(\nabla_X J)Y||^2 = 1$, which is impossible. □

Let $\psi : N \to S^6(1)$ be an isometric immersion of a surface N into $S^6(1) \subset \mathbb{E}^7$. The *ellipse of curvature* at a point $p \in N$ is defined as the image of the unit circle of T_pN under the second fundamental h, i.e.,

$$E_p = \{h(u,u) \,|\, u \in T_pM, \|u\| = 1\},$$

where h is the second fundamental form. The *third fundamental form III* of ψ is defined by

$$I\!I\!I(X,Y,Z) = (\tilde{\nabla}_X h(Y,Z))^\perp, \quad X,Y,Z \in \mathcal{X}(N),$$

where $\tilde{\nabla}$ is the flat connection on $\mathbb{E}^7$ and $(\tilde{\nabla}_X h(Y,Z))^\perp$ is the normal component of $\tilde{\nabla}_X h(Y,Z)$ orthogonal to the first normal subspace of ψ (see, for instance, [Bolton et al. (2007)]). The image of the unit circle of T_pN under *III* is called the *ellipse of second curvature.*

Definition 19.1. A minimal surface N in $S^6(1)$ is called *superminimal* if, at each point $p \in N$, the ellipse of curvature and the ellipse of second curvature are either circles or points.

Almost complex curves in $S^6(1)$ have been studied among others in [Bolton et al. (2007); Byrant (1982); Sekigawa (1983)]. Bryant's results states that superminimal almost complex curves can be completely described by means of a Weierstrass's formula and that curves of every genus can occur. It was proved in [Sekigawa (1983)] that if the Gauss curvature G of an almost complex curve in $S^6(1)$ is constant, then $G = 0, \frac{1}{6}$ or 1.

Almost complex curves in $S^6(1)$ were classified into four classes in [Bolton et al. (2007)]; namely,

(a) linearly full in $S^6(1)$ and superminimal;
(b) linearly full in $S^6(1)$ but not superminimal;
(c) linearly full in some total geodesic $S^5(1) \subset S^6(1)$ necessarily not superminimal;
(d) totally geodesic.

Remark 19.1. An oriented almost complex curves are also known as *pseudo-holomorphic curves.*

19.4 Ejiri's theorem for Lagrangian submanifolds of $S^6(1)$

A submanifold N in the nearly Kähler manifold is called *Lagrangian* if $\Omega|_N = 0$, or equivalently, $JT_p(N) = T_p^{\perp}N$, $p \in N$.

Lemma 19.6. *Let N be a Lagrangian submanifold of $S^6(1)$. Then $G(X,Y)$ is normal to N for $X, Y \in TN$.*

Proof. Gauss's and Weingarten's formulas imply that

$$\begin{aligned} g((\nabla_X J)Y, Z) &= g(Jh(X,Z), Y) - g(Jh(X,Y), Z), \\ g((\nabla_Z J)X, Y) &= g(Jh(Z,Y), X) - g(Jh(Z,X), Y), \\ g((\nabla_Y J)Z, X) &= g(Jh(Y,X), Z) - g(Jh(Y,Z), X) \end{aligned}$$

for $X, Y, Z \in TN$. Since g is Hermitian, $\nabla_X J$ is skew-symmetric with respect to g. This, together with (19.20) and (19.21), implies that the left-hand sides of the above three equations are equal to each other. Therefore, $g((\nabla_X J)Y, Z) = 0$, which means $G(X,Y)$ is orthogonal to N. □

The following result is fundamental [Ejiri (1981)].

Theorem 19.3. *Every Lagrangian submanifold N of $S^6(1)$ is orientable and minimal.*

Proof. By (19.20) we obtain

$$\begin{aligned}(\nabla_X G)(JY,JZ) &= \nabla_X G(JY,JZ) - G(\nabla_X JY,JZ) - G(JY,\nabla_X JZ)\\ &= -\nabla_X G(Y,Z) - G((\nabla_X J)Y,JZ) - G(J\nabla_X Y,JZ)\\ &\quad - G(JY,(\nabla_X J)Z) - G(JY,J\nabla_X Z) - \nabla_X G(Y,Z)\\ &\quad + JG(G(X,Y),Z) + G(\nabla_X Y,Z)\\ &\quad + JG(Y,G(X,Z)) + G(Y,\nabla_X Z)\\ &= -(\nabla_X G)(Y,Z) + JG(G(X,Y),Z) + JG(Y,G(X,Z))\end{aligned}$$

for X,Y,Z tangent to N. Combining this with (19.21) gives

$$G(Y,G(Z,X)) + G(Z,G(X,Y)) = g(X,Y)Z - g(X,Z)Y$$

and hence

$$JG(X,JG(Y,Z)) = g(X,Z)Y - g(X,Y)Z. \tag{19.24}$$

Since $JG(X,Y)$ is tangent to N by Lemma 19.3, (19.24) implies that

$$\begin{aligned}&g(JG(X,Y),Y)X - g(JG(X,Y),X)Y\\ &\qquad = JG(JG(X,Y),JG(X,Y)) = 0.\end{aligned}$$

Thus $JG(X,Y)$ is orthogonal to X,Y whenever X,Y are linearly independent. This together with (19.24) implies that N is orientable, because the orientation can be defined by regarding $JG(X,Y)$ as the vector product of X and Y at each $p \in N$.

Next, we prove that N is minimal. It follows immediately from the formulas of Gauss and Weingarten and Lemma 19.6 that

$$D_X JY = G(X,Y) + J\nabla'_X Y, \tag{19.25}$$

$$A_{JX} = -Jh(X,Y). \tag{19.26}$$

By (19.21), (19.25) and (19.26) we find

$$\begin{aligned}(\nabla_X G)(Y,Z) &= \nabla_X G(Y,Z) - G(\nabla_X Y,Z) - G(Y,\nabla_X Z)\\ &= -A_{G(Y,Z)}X + D_X G(Y,Z) - G(\nabla_X Y,Z) - G(Y,\nabla_X Z)\\ &= Jh(JG(Y,Z),X) + JG(X,G(Y,Z)) - J(\nabla'_X JG)(Y,Z)\\ &\quad - G(h(X,Y),Z) - G(Y,h(X,Z)).\end{aligned}$$

This, combined with (19.21), implies that

$$\begin{aligned}(\nabla'_X JG)(Y,Z) &= g(X,Y)Z - g(X,Z)Y + G(X,G(Y,Z))\\ &\quad + h(X,JG(Y,Z)) + JG(h(X,Y),Z) + JG(Y,h(Z,X)).\end{aligned}$$

By taking the normal component we find

$$h(X, JG(Y, Z)) + JG(h(X, Y), Z) + JG(Y, h(Z, X)) = 0. \tag{19.27}$$

Let e_1, e_2, e_3 be a local field of orthonormal frames on N. Then, without loss of generality, we may assume that

$$JG(e_1, e_2) = e_3, \quad JG(e_2, e_3) = e_1, \quad JG(e_3, e_1) = e_2.$$

Hence we derive from (19.27) that the trace of h is zero, which shows that N is minimal. □

19.5 Dillen–Vrancken's theorem for Lagrangian submanifolds

Let N be a Riemannian n-manifold isometrically immersed in a real space form $R^m(c)$. Then Theorem 13.5 implies that the δ-invariant $\delta(2)$ satisfies

$$\delta(2) \leq \frac{n^2(n-2)}{2(n-1)}||H||^2 + \frac{1}{2}(n+1)(n-2)c. \tag{19.28}$$

It is a very natural and interesting problem to classify submanifolds which verify the equality case of this inequality; namely, they satisfy

$$\delta(2) = \frac{n^2(n-2)}{2(n-1)}||H||^2 + \frac{1}{2}(n+1)(n-2)c, \tag{19.29}$$

which is also known as *Chen's basic equality* [Dajczer and Florit (1998)] .

In order to distinguish equality (19.29) from other similar equalities associated with different k-tuples, we simply call (19.29) the $\delta(2)$-*equality*.

Definition 19.2. An n-dimensional submanifold of a real space form $R^m(c)$ is called a $\delta(2)$-*ideal submanifold* if it satisfies equality (19.29).

There exists a canonical distribution $\mathcal{E}$ for n-dimensional $\delta(2)$-ideal submanifolds in $R^m(c)$ defined by

$$\mathcal{E}(p) = \{X \in T_pN : (n-1)h(X, Y) = n\langle X, Y\rangle H, \ \forall Y \in T_pN\}. \tag{19.30}$$

Since Lagrangian submanifolds in $S^6(1)$ are always minimal, inequality (19.28) reduces to $\delta(2) \leq 2$ for Lagrangian submanifolds of $S^6(1)$. Thus, after applying Theorem 13.5, we obtain the following.

Theorem 19.4. *Let N be a Lagrangian submanifold of $S^6(1)$. Then*

$$\delta(2) \leq 2, \tag{19.31}$$

with the equality holding if and only if

$$h(e_1,e_1) = -h(e_2,e_2) = \lambda Je_1,\ h(e_1,e_2) = -\lambda Je_2,$$
$$h(e_1,e_3) = h(e_2,e_3) = h(e_3,e_3) = 0,$$

with respect to a suitable tangent basis $\{e_1,e_2,e_3\}$, *where* λ *is a function determined by the scalar curvature* τ *according to* $2\lambda^2 = 3-\tau$

G_2-equivariant Lagrangian submanifolds of the nearly Kähler $S^6(1)$ have been classified in [Mashimo (1985)]. It turns out that there are five models, so that every equivariant Lagrangian submanifold in the nearly Kähler $S^6(1)$ is G_2-congruent to one of the five models. These five models can be distinguished by the following curvature properties:

(1) N is totally geodesic ($\delta(2)=2$);
(2) N has constant curvature $\frac{1}{16}$ ($\delta(2)=\frac{1}{8}$);
(3) the curvature of N satisfies $\frac{1}{16} \le K \le \frac{21}{16}$ ($\delta(2)=\frac{11}{8}$);
(4) the curvature of N satisfies $-\frac{7}{3} \le K \le 1$ ($\delta(2)=2$);
(5) the curvature of N satisfies $-1 \le K \le 1$ ($\delta(2)=2$).

Dillen, Verstraelen and Vrancken characterized models (1), (2) and (3) as the only compact Lagrangian submanifolds in $S^6(1)$ whose sectional curvatures satisfy $K \ge \frac{1}{16}$ [Dillen et al. (1990)]. They also obtained an explicit expression for the Lagrangian submanifold of constant curvature $\frac{1}{16}$ in terms of harmonic homogeneous polynomials of degree 6. Using their expression, it follows that the immersion has degree 24. Further, they also obtained an explicit expression for model (3).

It follows from Mashimo's list that models (1), (4) and (5) satisfy $\delta(2)=2$ identically. On the other hand, it was proved in [Chen et al. (1995a)] that models (1), (2) and (3) are the only Lagrangian submanifolds of the nearly Kähler $S^6(1)$ with constant scalar curvature that satisfy $\delta(2)=2$.

Definition 19.3. A Riemannian n-manifold N whose Ricci tensor admits an eigenvalue of multiplicity at least $n-1$ is called *quasi-Einstein*.

It was proved in [Deszcz et al. (1999)] that Lagrangian submanifolds of $S^6(1)$ satisfying $\delta(2)=2$ are always quasi-Einstein.

It is known that the homogeneous space $G_2/U(2)$ coincides with the Grassmann manifold of all oriented 2-planes of a Euclidean 7-space $\mathbb{E}^7$. By using this representation, one can define an almost contact metric structure on any 5-dimensional oriented submanifold immersed in $S^6(1) \subset \mathbb{E}^7$. In particular, let us consider $S^5(1)$ as the unit hypersphere in $S^6(1) \subset \mathbb{E}^7$ given by $x_4 = 0$ and define $j : S^5(1) \to \mathbf{C}^3$ by

$$j : (x_1, x_2, x_3, 0, x_5, x_6, x_7) \mapsto (x_1 + ix_5, x_2 + ix_6, x_3 + ix_7).$$

Then at a point $p = (x_1, \ldots, x_7)$ the structure vector field ξ is given by

$$\xi(p) = e_4 \times p = (x_5, x_6, x_7, 0, -x_1, -x_2, -x_3).$$

And for a tangent vector $v = (v_1, v_2, v_3, 0, v_5, v_6, v_7)$, orthogonal to ξ, we have that the endomorphism ϕ of the Sasakian structure satisfies

$$\phi(v) = (-v_5, -v_6, -v_7, 0, v_1, v_2, v_3) = v \times e_4.$$

Since

$$\phi\xi(p) = 0 = (e_4 \times p) \times e_4 - \langle (e_4 \times p) \times e_4, p\rangle\, p,$$

we deduce that, for any tangent vector w to $S^5(1)$,

$$\phi(w) = w \times e_4 - \langle w \times e_4, p\rangle\, p. \tag{19.32}$$

We call a submanifold N^n of S^5 an *invariant submanifold* if $\phi(T_pN^n) \subset T_pN^n$ for each $p \in N^n$. If n is odd, then ξ is automatically tangent to N^n.

Assume $n = 3$. The Hopf fibration $\pi : S^5(1) \to \mathbb{C}P^2(4)$ annihilates ξ, i.e. $\pi_*(\xi) = 0$. So, if N^3 is invariant, $\pi(N^3)$ is an almost complex curve.

Conversely, let $\phi : N_1 \to CP^2(4)$ be an almost complex curve and let PN_1 be the circle bundle over N_1 induced by the Hopf fibration and let ψ be the immersion such that the following diagram commutes:

$$\begin{array}{ccc} PN_1 & \xrightarrow{\ \psi\ } & S^5(1) \\ \big\downarrow \pi & & \big\downarrow \pi \\ N_1 & \xrightarrow{\ \phi\ } & \mathbb{C}P^2(4). \end{array} \tag{19.33}$$

Then ψ is an invariant immersion in the Sasakian space form $S^5(1)$ with structure vector field ξ tangent along ψ.

The complete classification of Lagrangian submanifolds in the nearly Kähler $S^6(1)$ satisfying $\delta(2) = 2$ was established in [Dillen and Vrancken (1996)]. More precisely, they proved the following:

Theorem 19.5. *We have:*

(1) *let $\phi : N_1 \to CP^2(4)$ be an almost complex curve in $CP^2(4)$, PN_1 the circle bundle over N_1 induced by $\pi : S^5(1) \to CP^2(4)$, and $\psi : PN_1 \to S^5(1)$ be the isometric immersion such that diagram* (19.33) *commutes. Then there exists a totally geodesic embedding $i : S^5(1) \to S^6(1)$ such that the composition $i \circ \psi : PN_1 \to S^6(1)$ is a Lagrangian immersion satisfying the equality $\delta(2) = 2$;*

(2) *let* $\bar{\phi} : N_2 \to S^6(1)$ *be an almost complex curve without totally geodesic points. Denote by* T^1N_2 *the unit tangent bundle of* N_2 *and by* h *the second fundamental form of* $\bar{\phi}$, *and define a map*

$$\bar{\psi} : T^1N_2 \to S^6(1) : v \mapsto \bar{\phi}_\star(v) \times \frac{h(v,v)}{||h(v,v)||}.$$

Then $\bar{\psi}$ *is a (possibly branched) Lagrangian immersion satisfying* $\delta(2) = 2$. *Moreover, the immersion is linearly full in* $S^6(1)$;

(3) *let* $\bar{\phi} : N_2 \to S^6(1)$ *be a (branched) almost complex immersion and let* SN_2 *be the tube of radius* $\frac{\pi}{2}$ *over* N_2 *in the direction* $(L_0 \oplus L_1)^\perp$, *where* L_0 *is* $\bar{\phi}_*(TN_2)$ *and* L_1 *is the first normal space, except at the (isolated) branch points. Then* SN_2 *is a (possibly branched) Lagrangian submanifold of* $S^6(1)$ *satisfying* $\delta(2) = 2$;

(4) *let* $\eta : N \to S^6(1)$ *be a Lagrangian immersion which is not linearly full in* $S^6(1)$. *Then* N *satisfies* $\delta(2) = 2$ *and there exist a totally geodesic* S^5 *and an almost complex immersion* $\phi : N_1 \to CP^2(4)$ *such that* η *is congruent to* ψ, *which is obtained from* ϕ *as in* (1);

(5) *let* $\eta : N \to S^6(1)$ *be a linearly full Lagrangian immersion satisfying* $\delta(2) = 2$. *Let* p *be a non-totally geodesic point. Then there is a (possibly branched) almost complex curve* $\bar{\phi} : N_2 \to S^6(1)$ *such that* η *is locally around* p *congruent to* $\bar{\psi}$, *which is obtained from* $\bar{\phi}$ *as in* (3).

The following result was proved in [Vrancken (1998)].

Theorem 19.6. *We have:*

(i) *let* $\phi : S \to S^6$ *be an almost complex curve without totally geodesic points. Define*

$$F : T^1S \to S^6(1) : v \mapsto \frac{h(v,v)}{||h(v,v)||}, \tag{19.34}$$

where T^1S *denotes the unit tangent bundle of* S. *Then* F *defines a Lagrangian immersion if and only if* ϕ *is superminimal;*

(ii) *let* $\psi : N \to S^6(1)$ *be a Lagrangian immersion which admits a unit length Killing vector field whose integral curves are great circles. Then there exists an open dense subset* U *of* N *such that each point* p *of* U *has a neighborhood* V *such that* $\psi : V \to S^6(1)$ *satisfies* $\delta(2) = 2$, *or* $\psi = F : V \to S^6(1)$ *is obtained as in* (i).

19.6 $\delta(2)$ and CR-submanifolds of $S^6(1)$

Every CR-submanifold of the nearly Kähler $S^6(1)$ is of dimension 3, 4 or 5. On the other hand, every hypersurface of $S^6(1)$ is automatically a CR-submanifold. For 3-dimensional submanifolds, the $\delta(2)$-equality reduces to

$$\delta(2) = 2 + \frac{9}{4}||H||^2. \tag{19.35}$$

The following result is due to [Sasahara (2000)].

Theorem 19.7. *There exist no 3-dimensional proper CR-submanifolds in the nearly Kähler $S^6(1)$ satisfying* (19.35) *such that $\mathcal{E}$ is totally real, where $\mathcal{E}$ is defined by* (19.30).

T. Sasahara also proved the following [Sasahara (2001b)].

Theorem 19.8. *Let N be a 3-dimensional proper CR-submanifold in $S^6(1)$. If N satisfies* (19.35) *identically, then N is minimal and quasi-Einsteinian.*

3-dimensional CR-submanifolds satisfying (19.35) were determined in [Djorić and Vrancken (2006, 2010)].

Theorem 19.9. *Let N be a 3-dimensional CR-submanifold of the nearly Kähler $S^6(1)$ satisfying* (19.35). *Then N is either a Lagrangian submanifold or locally congruent to the immersion*

$$\Big(\cos t\cos u\cos v, \sin t, \cos t\sin u\cos v, \cos t\cos u\sin v,$$
$$0, -\cos t\sin u\sin v, 0\Big).$$

Notice that this immersion can be described algebraically by the equations

$$x_1^2 + x_2^2 + x_3^2 + x_4^2 + x_6^2 = 1,\ x_3x_4 + x_1x_6 = 0,\ x_5 = x_7 = 0,$$

thus it can be seen as a hypersurface lying in a totally geodesic $S^4(1)$.

For 4-dimensional submanifolds of $S^6(1)$, $\delta(2)$-equality becomes

$$\delta(2) = \frac{16}{3}||H||^2 + 5. \tag{19.36}$$

According to Theorem 19.2, 4-dimensional CR-minimal submanifolds in $S^6(1)$ are proper. For 4-dimensional CR-minimal submanifolds in $S^6(1)$ the following result was proved in [Antić et al. (2007)].

Theorem 19.10. *Let N be a 4-dimensional minimal CR-submanifold in the nearly Kähler $S^6(1)$ satisfying (19.36) identically. Then N is locally congruent to the following immersion:*

$$\begin{aligned}&\big(\cos x_4 \cos x_1 \cos x_2 \cos x_3, \sin x_4 \sin x_1 \cos x_2 \cos x_3,\\ &\sin 2x_4 \sin x_3 \cos x_2 + \cos 2x_4 \sin x_2, 0, \sin x_4 \cos x_1 \cos x_2 \cos x_3,\\ &\cos x_4 \sin x_1 \cos x_2 \cos x_3, \cos 2x_4 \sin x_3 \cos x_2 - \sin 2x_4 \sin x_2\big).\end{aligned} \tag{19.37}$$

There do not exist non-minimal CR-submanifolds of $S^6(1)$ satisfying the equality (19.36) [Antić (2010b)]. 4-dimensional minimal CR-submanifolds in a totally geodesic $S^5(1) \subset S^6(1)$ were classified in [Antić (2010a)].

Theorem 19.11. *If N is a 4-dimensional minimal CR-submanifold lying in a totally geodesic sphere $S^5(1) \subset S^6(1)$, then N is locally congruent to the one of the following three immersions:*

(1) *the $\delta(2)$-ideal immersion given by (19.37);*
(2) *the immersion defined by*

$$\begin{aligned}&\frac{\sqrt{6}}{8}\bigg(\frac{1}{\sqrt{3}}(3(\cos x \cos y + \cos u \sin x) + \sqrt{2}\cos(\sqrt{3}v)),\\ &\sqrt{2}\cos(\sqrt{3}v) - \cos x \cos y - \cos u \sin x, 2\sin x \sin u - 2\cos x \sin y,\\ &\qquad 0, \frac{1}{\sqrt{3}}\big(\sqrt{2}\sin(\sqrt{3}v) - 3\cos x \sin y - 3\sin x \sin u\big),\\ &\cos x \sin y + \sin x \sin u + \sqrt{2}\sin(\sqrt{3}v), 2\cos u \sin x - 2\cos x \cos y\bigg);\end{aligned}$$

(3) *the immersion defined by*

$$\begin{aligned}&\frac{1}{4}\bigg(\big((1+\sqrt{2})\cos y + \cos u\big)\cos(x-v)\\ &\qquad + \big(\cos y + \cos u - \sqrt{2}\cos u\big)\cos(x+v)\\ &\qquad\quad + \sin y \sin u\big((\sqrt{2}-1)\sin(x+v) - \sin(x-v)\big),\\ &\big((1+\sqrt{2})\cos u - \cos y\big)\cos(x-v) - \big((\sqrt{2}-1)\cos y + \cos u\big)\cos(x+v)\\ &\qquad + \sin y \sin u\big(\sin(x+v) - (1+\sqrt{2})\sin(x-v)\big),\\ &\sqrt{8\sqrt{2}-8}\big\{(1+\sqrt{2})(\cos x \cos u \sin y - \sin x \sin u)\sin v\\ &\qquad - (\cos u \sin x \sin y + \cos x \sin u)\cos v\big\}, 0,\\ &\big(\cos y + (1-\sqrt{2})\cos u\big)\sin(x+v)\\ &- \big(\cos(x-v) + (\sqrt{2}-1)\cos(x+v)\big)\sin y \sin u\\ &- \big((1+\sqrt{2})\cos y + \cos u\big)\sin(x-v), -\big((1+\sqrt{2})\cos(x-v)\end{aligned}$$

$$+\cos(x+v)\big)\sin y\sin u+\big(\cos y-(1+\sqrt{2}\,)\cos y\big)\sin(x-v)$$
$$-\big((\sqrt{2}-1)\cos y+\cos u\big)\sin(x+v),$$
$$-4\sqrt{2}\sqrt{\sqrt{2}+1}\big\{\cos v(\cos x\cos u\sin y-\sin x\sin u)$$
$$+(\sqrt{2}-1)(\cos u\sin x\sin y+\cos x\sin u)\sin v\big\}\Big).$$

19.7 Hopf hypersurfaces of $S^6(1)$

Lemma 19.7. *Let X be a parallel vector field along a geodesic γ of $S^6(1)$ which is orthogonal to $\dot\gamma$ and to $J\dot\gamma$, and let*

$$Y(t)=(\cos t)JX-(\sin t)\dot\gamma\times X.$$

If γ is parametrized by arc length, then $Y(t)$ is the parallel vector fields along γ with initial value $JX(0)$.

Proof. The hypotheses on X implies that $\dot X=0$. Further, $Y(0)=JX(0)$ which is orthogonal to $\dot\gamma(0)$ and, using (19.17),

$$\dot Y(t)=(\cos t)\dot\gamma\times X-(\sin t)\gamma\times X-(\cos t)\dot\gamma\times X-(\sin t)\ddot\gamma\times X.$$

However, γ is a great circle in $S^6(1)$, so that $\ddot\gamma=-\gamma$ and hence $\dot Y=0$ and the lemma is proved. □

Let $\gamma(t)$ be a geodesic in a Riemannian manifold. A vector field $J(t)$ along $\gamma(t)$ is called a *Jacobi vector field* if it satisfies the *Jacobi equation*

$$\nabla^2_{\gamma'(t)}J(t)+R(J(t),\gamma'(t))\gamma'(t)=0.$$

Geometrically, a Jacobi vector field describes the difference between the geodesic and an "infinitesimally close" geodesic.

The remaining results of this section on Hopf hypersurfaces of $S^6(1)$ are due to [Berndt et al. (1995)].

Proposition 19.2. *Every geodesic hypersphere and every tube $T^r(H^2)$ with radius r around an almost complex curve H^2 in $S^6(1)$ is an orientable Hopf hypersurface.*

Proof. Since geodesic hyperspheres of $S^6(1)$ are totally umbilical, the Hopf vector field on a geodesic hypersphere is a principal curvature vector field. Thus, all geodesic hyperspheres are Hopf hypersurfaces.

Let H^2 be an almost complex curve in $S^6(1)$. At least locally and for small radii $r \in \mathbf{R}^+$ we may define the tube $T^r(H^2)$ of radius r around H^2. Let η be a unit normal vector of H^2 at some point $p \in H^2$. Then

$$\gamma_\eta : \mathbf{R} \to S^6(1); \quad t \mapsto (\cos t)p + (\sin t)\eta$$

is the maximal geodesic in $S^6(1)$ with $\gamma_\eta(0) = 0, \dot{\gamma_\eta}(0) = \eta$. If $q = \gamma_\eta(r) \in T^r(H^2)$, then the outward unit normal vector ξ of $T^r(H^2)$ at q is given by $\xi := \dot{\gamma_\eta}(r)$. In particular, $T^r(H^2)$ is orientable.

Now, we apply the standard procedure to compute the second fundamental form of tubes $T^r(H^2)$. Let $X \in T_pS^6(1)$ be orthogonal to η and let Y_X be the Jacobi vector field along γ_η with initial values $Y_X(0) = X^T$ and $\nabla_{\dot{\gamma}} Y(0) = X^\perp - A_\eta X^T$, where A_η is the shape operator of H^2 at p with respect to η, and ^{T} and $^\perp$ are the orthogonal projections onto the tangent and normal spaces of H^2 at p, respectively. Then the shape operator A_ξ of $T^r(H^2)$ at q with respect to ξ satisfies

$$A_\xi Y_X(r) = -\nabla_{\dot{\gamma}} Y_X(r).$$

Since $S^6(1)$ is of constant curvature one,

$$Y_X(t) = (\cos t)B_{X^T}(t) + (\sin t)B_{X^\perp - A_\eta X^T}(t),$$

where B_V is the parallel vector field along γ_η with $B_V(0) = V$, for $V \in T_pS^6(1)$. Now, let X be a principal curvature vector of H^2 at p with respect to η, say $A_\eta X = \lambda X$. Then

$$Y_X(t) = (\cos t - \lambda \sin t)B_X(t),$$

and hence

$$A_\xi B_X(r) = \frac{\sin r + \lambda \cos r}{\cos r - \lambda \sin r} B_X(r). \tag{19.38}$$

Further, if X is orthogonal to H^2 at p, we have $Y_X(t) = \sin t B_X(t)$, which implies

$$A_\xi B_X(r) = -\cot r B_X(r). \tag{19.39}$$

Thus, the principal curvature vectors of $T^r(H^2)$ at q are obtained by parallel translation along γ_η from p to q of

(i) the principal curvature vectors of H^2 with respect to η, and

(ii) $N_pH^2 \cap (\mathbf{R}\eta)^\perp$, where N_pH^2 is the normal space to H^2 at p.

Since H^2 is almost complex, N_pH^2 is J-invariant, so $J\eta \in N_pH^2 \cap (\mathbf{R}\eta)^\perp$. However, the nearly Kähler condition implies that $J\dot{\gamma_\eta}$ is the parallel translate of $J\eta$ along γ_η so that $J\xi = J\dot{\gamma_\eta}(r)$ is a principal curvature vector of $T^r(H^2)$ at q. Thus, $T^r(H^2)$ is a Hopf hypersurface of $S^6(1)$. □

Lemma 19.8. *For a Hopf hypersurface N of $S^6(1)$, the principal curvature function α associated with the Hopf vector field U is constant.*

Proof. By applying Codazzi's equation and the symmetry of the shape operator A, we find

$$\begin{aligned} 0 &= \langle (\nabla_X A)Y - (\nabla_Y A)X, U\rangle \\ &= \langle (\nabla_X A)U, Y\rangle - \langle (\nabla_Y A)U, X\rangle \\ &= (X\alpha)\langle U, Y\rangle + \alpha\langle \nabla_X U, Y\rangle - \langle A\nabla_X U, Y\rangle \\ &\quad - (Y\alpha)\langle U, X\rangle - \alpha\langle \nabla_Y U, X\rangle + \langle A\nabla_Y U, X\rangle. \end{aligned} \tag{19.40}$$

Putting $X = U$ and using $AU = \alpha U$, we find $Y\alpha = (U\alpha)\langle U, Y\rangle$, and thus

$$\operatorname{grad}\alpha = (U\alpha)U. \tag{19.41}$$

Hence, for vector fields X, Y in $\mathcal{D}$, we have

$$0 = X(Y\alpha) - Y(X\alpha) = [X,Y]\alpha = (U\alpha)\langle [X,Y], U\rangle. \tag{19.42}$$

However, $\mathcal{D}$ is J-invariant so, if $\mathcal{D}$ were integrable on some open subset of N, then the integral submanifolds would be 4-dimensional almost complex submanifolds of $S^6(1)$, which contradicts to Theorem 19.2 and so, using (19.42), we may deduce that $U\alpha = 0$. Combining this with (19.41) shows that α is constant. □

Lemma 19.9. *For a Hopf hypersurface N of $S^6(1)$, we put $\alpha = -\cot r$, $T_\alpha = \{X \in TN : AX = \alpha X\}$ and $LX = (\cos r)JX - (\sin r)\,\xi \times X$. Then the orthogonal complement $T_\alpha^\perp$ of T_α in TN is invariant under the action of L.*

Proof. For $X, Y \in T_pN$, it follows from $U = -J\xi$ and the formula of Weingarten that

$$\langle \nabla_X U, Y\rangle = -\langle \nabla_X (p\times\xi), Y\rangle = -\langle X\times\xi, Y\rangle + \langle JAX, Y\rangle. \tag{19.43}$$

Now, let $X \in T_\alpha(p)$ and, assuming that N is not totally umbilical at p, and let $Y \in T_pN$ be such that $AY = \lambda Y$ with $\lambda \neq \alpha$. Then JY and $\xi\times Y$ are orthogonal to both p and ξ, so that LY is also orthogonal to p and ξ. Also, (19.40) and Lemma 19.8 may be used to show that $\langle \nabla_X U, Y\rangle = 0$. Hence, it follows from (19.43) that

$$\langle JAX + \xi\times X, Y\rangle = 0. \tag{19.44}$$

Thus

$$\begin{aligned} \langle LY, X\rangle &= \langle (\cos r)JY - (\sin r)\xi\times Y, X\rangle \\ &= (\sin r)\langle (\cot r)JY - \xi\times Y, X\rangle \\ &= (\sin r)\langle Y, -(\cot r)JX + \xi\times X\rangle \\ &= (\sin r)\langle Y, JAX + \xi\times X\rangle. \end{aligned}$$

Hence from (19.44) we obtain $\langle LY, X\rangle = 0$. □

Theorem 19.12. *If N is a Hopf hypersurface in $S^6(1)$, then N is an open part of either a geodesic hypersphere or a tube around an almost complex curve H^2 in $S^6(1)$.*

Proof. We choose a local unit normal field ξ on an open subset W of N such that the principal curvature α is non-positive, and put $\alpha = -\cot r$ with some $r \in (0, \frac{\pi}{2}]$. Let $F : W \to S^6(1)$, $p \mapsto \exp(-r\xi_p)$ be the map assigning to each $p \in W$ the point $\exp(-r\xi_p) = \gamma_p(-\gamma)$, where $\gamma_p := \gamma_{\xi_p}$. If $X \in T_pW$ is a principal curvature vector with corresponding principal curvature κ, then

$$F_*X = (\cos r + \kappa \sin r)B_r(X), \tag{19.45}$$

where B_r denotes parallel translation along γ_p from point p to point $F(p)$. From (19.45) we see that F is singular at all points of W and $\ker(F_*) = T_\alpha$, while $\mathrm{Im}(F_*) = B_r(T_\alpha^\perp(p))$. It follows from Lemma 19.7 (applied to the geodesic $\gamma(t) = \gamma_p(t-r)$) that $JB_r = B_rL$ and hence, by Lemma 19.9, $\mathrm{Im}(F_*)$ is J-invariant. Hence, $\mathrm{Im}(F_*)(p)$ has dimension 0, 2 or 4 at each point $p \in E$.

First, we assume that $\mathrm{Im}(F_*)(p_o)$ has dimension 4 at some point $p_o \in W$, i.e, $\alpha(p_o)$ has multiplicity 1. The continuity of the principal curvature functions implies that there is an open neighborhood V of p_o in W such that $\dim T_\alpha = 1$ on V so that F has constant rank 4 on V. Thus there is a 4-dimensional submanifold B of $S^6(1)$ and an open neighborhood $\tilde{V}$ of $p_o \in V$ such that $F|_{\tilde{V}} : \tilde{V} \to B$ is a submersion onto B. But then B of $S^6(1)$ is almost complex, which contradicts Theorem 19.2.

Next, we assume that $\mathrm{Im}(F_*)(p_o)$ has dimension 2 at some point $p_o \in W$. Reasoning as above, we deduce the existence of an open neighborhood V of $p_o \in W$ which is an open part of a tube of radius r around an almost complex curve H^2 (whence we may apply the results discussed following Proposition 19.2 to V, as will be done below).

Finally, assume that $\mathrm{Im}(F_*)(p_o)$ has dimension 0 at some point $p_o \in W$. Let us first suppose that in every open neighborhood of p_o there is a point p with $\mathrm{Im}(F_*)(p)$ having dimension 2. Then we may choose a sequence of points $p_n \in W$ converging towards p_o and with $\mathrm{Im}(F_*)(p_n)$ having dimension 2. According to the preceding case and the discussion following Proposition 19.2, the principal curvatures of N at p_n are $-\cot r, \tan(r+\theta_n)$ and $\tan(r-\theta_n)$ with some $\theta_n \in [0, \frac{\pi}{2})$.

As the principal curvature functions are continuous the two sequences $(\tan(r+\theta_n))_{n\in\mathbb{N}}$ and $(\tan(r-\theta_n))_{n\in\mathbb{N}}$ must converge towards $-\cot r$, which is impossible. Thus there exists a connected open neighborhood V of $p_o \in$

W such that $\dim T_\alpha = 5$ on V. Hence $F(V)$ is a single point $q \in S^6(1)$ and V is an open part of a geodesic hypersphere of radius r and with center q. As we have seen above, the dimension of $\mathrm{Im}(F_*)$ is locally constant with value either 0 or 2. As N is connected by definition, $\mathrm{Im}(F*)$ has constant dimension on the entire hypersurface.

Now, by applying a continuity argument we can now conclude that the whole of N is an open part of either a geodesic hypersphere of $S^6(1)$ or of a tube around an almost complex curve in $S^6(1)$. □

Since geodesic hyperspheres and tubes over an almost complex curves in the nearly Kähler $S^6(1)$ are both orientable, Theorem 19.12 implies immediately the following.

Corollary 19.2. *Every Hopf hypersurface of $S^6(1)$ is orientable.*

19.8 Ideal real hypersurfaces of $S^6(1)$

Let $R^{n+1}(c)$ denote $S^{n+1}(c)$ if $c > 0$ (resp. be $H^{n+1}(c)$ if $c < 0$). Let us regard $R^{n+1}(c)$ as a hypersurface of $\mathbb{E}^{2n+2}$ via (1.16) if $c > 0$ (resp. $R^{n+1}(c)$ as a hypersurface in $\mathbb{E}^{2n+2}_1$ via (1.17) if $c < 0$). A rotational hypersurface of $R^{n+1}(c)$ $(c \neq 0)$ is defined as follows:

Let P^3 be a linear 3-subspace of $\mathbb{E}^{n+2}$ or of $\mathbb{E}^{n+2}_1$ that intersects $R^{n+1}(c)$ and denote the intersection by $R^2(c)$ (if $c < 0$ we take only the upper part). Let P^2 be a linear subspace in P^3 and let $O(P^2)$ be the group of orthogonal transformations with positive determinant that leave P^2 pointwise fixed. We take a curve γ in $R^2(c)$ which does not intersect P^2. The orbit of γ under $O(P^2)$ is called the *rotational hypersurface* with profile curve γ and with axis P^2 [do Carmo and Dajczer (1983)].

As we already seen, every real hypersurface of the nearly Kähler $S^6(1)$ is a proper CR-submanifold. For real hypersurfaces of $S^6(1)$, the $\delta(2)$-equality becomes

$$\delta(2) = \frac{75}{8}||H||^2 + 9. \tag{19.46}$$

For hypersurfaces of the nearly Kähler $S^6(1)$, we have [Chen (1993a)].

Theorem 19.13. *Let N be a real hypersurface of $S^6(1)$ without totally geodesic points. If N satisfies* (19.46) *identically and* $\dim \mathcal{E}(p)$ *is independent of $p \in N$, then N is either a rotational hypersurface with a geodesic of $S^6(1)$ as its profile curve or N is foliated by 3-dimensional totally umbilical submanifolds of $S^6(1)$.*

Proof. Under the hypothesis, Theorem 13.5 implies that there exists an orthonormal local frames $e_1, \dots, e_5, e_6$ such that, with respect to this frame field, the shape operator $A = A_{e_6}$ of N in $S^6(1)$ satisfies

$$A = \begin{pmatrix} a & 0 & 0 & 0 & 0 \\ 0 & b & 0 & 0 & 0 \\ 0 & 0 & \mu & 0 & 0 \\ 0 & 0 & 0 & \mu & 0 \\ 0 & 0 & 0 & 0 & \mu \end{pmatrix}, \quad a + b = \mu, \tag{19.47}$$

for some functions a, b. Since N contains no totally geodesic points, it follows from (19.30) and (19.47) that either

(i) $\dim \mathcal{E}(p) = 4$, $p \in N$ or
(ii) $\mathcal{E} = \operatorname{Span}\{e_3, e_4, e_5\}$.

If case (i) occurs, A has one eigenvalue μ of multiplicity $n-1$ and the other simple eigenvalue is zero. Hence it follows from a result of [do Carmo and Dajczer (1983)] that N is a rotational hypersurface whose profile curve is a geodesic of $S^6(1)$.

Next, assume that $\mathcal{E} = \operatorname{Span}\{e_3, e_4, e_5\}$. Denote by $\mathcal{E}^\perp$ the orthogonal complementary distribution of $\mathcal{E}$ in TN. For any vector fields X, Y in $\mathcal{D}$ and Z in $\mathcal{E}^\perp$, we have

$$(\bar{\nabla}_X h)(Y, Z) = -h(\nabla'_X Y, Z) - h(Y, \nabla'_X Z) \tag{19.48}$$

by (19.47). Thus, by Codazzi's equation and (19.48), we find

$$\begin{aligned} h([X, Y], Z) &= h(X, \nabla'_Y Z) - h(Y, \nabla'_X Z) \\ &= \mu\{\langle X, \nabla'_Y Z\rangle - \langle Y, \nabla'_X Z\rangle\} e_6 \\ &= \mu \langle [X, Y], Z\rangle\, e_6. \end{aligned} \tag{19.49}$$

Hence, the distribution $\mathcal{E}$ is integrable by applying (19.47).

Next, we prove that each leaf of $\mathcal{E}$ is totally umbilical in $S^6(1)$. Let a, b, μ be defined by (19.45). Then, for any vector fields X, Y in $\mathcal{E}$, we find from Codazzi's equation and (19.47) that

$$\begin{aligned} a \langle \nabla'_X Y, e_1\rangle &= \langle \nabla'_X Y, A_{n+1} e_1\rangle \\ &= - \langle Y, \nabla'_X (A_{n+1} e_1)\rangle \\ &= - \langle Y, (\nabla_X A_{n+1}) e_1\rangle - \langle Y, A_{n+1}(\nabla_X e_1)\rangle + \langle Y, A_{D_X e_{n+1}} e_1\rangle \\ &= - \langle Y, (\nabla'_{e_1} A_{n+1}) X\rangle - \langle Y, A_{n+1}(\nabla'_X e_1)\rangle + \langle Y, A_{D_{e_1} e_{n+1}} X\rangle \\ &= - \langle Y, \nabla_{e_1}(A_{n+1} X)\rangle + \langle Y, A_{n+1}(\nabla_{e_1} X)\rangle - \langle Y, A_{n+1}(\nabla_X e_1)\rangle \\ &= -(e_1 \mu) \langle X, Y\rangle + \mu \langle \nabla_X Y, e_1\rangle . \end{aligned}$$

Since $a+b=\mu$, this yields $b\,\langle \nabla'_X Y, e_1\rangle = (e_1\mu)\,\langle X,Y\rangle$.

Similarly, we have $a\,\langle \nabla'_X Y, e_2\rangle = (e_2\mu)\,\langle X,Y\rangle$. The last two equations, together with (19.47), imply that each leaf of $\mathcal{E}$ is totally umbilical in $S^6(1)$. Consequently, N is foliated by 3-dimensional totally umbilical submanifolds of $S^6(1)$. □

For ideal Hopf hypersurfaces we have the following.

Theorem 19.14. *A Hopf hypersurface of the nearly Kähler $S^6(1)$ is an ideal hypersurface if and only if it is either*

(1) *a totally geodesic hypersurface, or*
(2) *an open part of a tube $T^{\frac{\pi}{2}}(H^2)$ with radius $\frac{\pi}{2}$ around an almost complex curve H^2 of $S^6(1)$.*

Proof. Let N be a Hopf hypersurface of the nearly Kähler $S^6(1)$. Then Theorem 19.12 implies that N is an open part of a geodesic hypersphere or a tube around an almost complex curve in $S^6(1)$. Assume that N is an ideal hypersurface of $S^6(1)$. Then there are five possibilities:

(a) N is $\delta(2)$-ideal;
(b) N is $\delta(2,2)$-ideal;
(c) N is $\delta(3)$-ideal;
(d) N is $\delta(2,3)$-ideal;
(e) N is $\delta(4)$-ideal.

Thus, if N is totally umbilical, N cannot be an ideal hypersurface unless it is totally geodesic. In this case, we have case (1).

Next, assume that N is an open part of a tube $T^r(H^2)$ with radius r around an almost complex curve H^2 in $S^6(1)$. For each unit normal vector $e\in T_p^{\perp}H^2$ at $p\in H^2$, we denote by $\kappa_1(e),\kappa_2(e)$ the eigenvalues of the shape operator A_e of H^2 in $S^6(1)$. Put

$$\kappa_i(e)=\tan c_i(e),\ \ c_i(e)\in(-\tfrac{\pi}{2},\tfrac{\pi}{2}),\ \ i=1,2. \tag{19.50}$$

Since N^2 is an almost complex curve, it follows from Lemma 19.3(iii) that $\kappa_2(e)=-\kappa_2(e)$. Thus $c_2(e)=-c_1(e)$. Hence, it follows from (19.38) and (19.39) that the principal curvatures of the tube $T^r(H^2)$ in $S^6(1)$ at $\gamma_e(r)$ are given by

$$\tan(r+c(e)),\ \tan(r-c(e)),\ -\cot r,\ -\cot r,\ -\cot r, \tag{19.51}$$

with $c(e)=\tan^{-1}\kappa_1(e)$ and r is a positive constant.

Case (a): *N is $\delta(2)$-ideal.* From (13.30) and (19.51) we get

$$\tan(r+c)+\tan(r-c)=-\cot r.$$

Thus $(\cot r)(\cot^2 r + 2 + k_1^2) = 0$, which implies $\cot r = 0$. Hence, $r = \frac{\pi}{2}$, which gives case (2).

Case (b): *N is* $\delta(2,2)$*-ideal.* It follows from (13.30) and (19.51) that there are only two possibilities:

$$\tan(r + c) = -2\cot r \text{ and } \tan(r - c) = -\cot r;$$
$$\tan(r - c) = -2\cot r \text{ and } \tan(r + c) = -\cot r.$$

But both cases are impossible.

Case (c): *N is* $\delta(3)$*-ideal.* It follows from (13.30) and (19.51) that there are four possibilities:

$$\tan(r + c) + \tan(r - c) = 0; \quad (19.52)$$
$$\tan(r + c) = \tan(r - c) = -3\cot r; \quad (19.53)$$
$$\tan(r + c) = 0 \text{ and } \tan(r - c) = -\cot r; \quad (19.54)$$
$$\tan(r - c) = 0 \text{ and } \tan(r + c) = -\cot r. \quad (19.55)$$

If (19.52) holds, then $\cot r = 0$. Thus, $r = \frac{\pi}{2}$, so we have case (2). It is easy to verify that other cases are impossible.

Case (d): *N is* $\delta(2,3)$*-ideal.* It follows from (13.30) and (19.51) that there are three possibilities:

$$\tan(r + c(e)) + \tan(r - c(e)) = -3\cot r; \quad (19.56)$$
$$\tan(r + c(e)) - \cot r = \tan(r - c(e)); \quad (19.57)$$
$$\tan(r - c(e)) - \cot r = \tan(r + c(e)). \quad (19.58)$$

If (19.56) holds, then either $r = \frac{\pi}{2}$ or $\kappa_1^2(e) = 2 + 3\cot^2 r$. If $r = \frac{\pi}{2}$, we obtain case (2). Otherwise, κ_1 and κ_2 are constant.

On the other hand, since the codimension of H^2 is 4, $\kappa_i(-e) = -\kappa_i(e)$. Thus κ_1, κ_2 cannot be constant, unless H^2 is totally geodesic. But this is also impossible since $\kappa_1^2(e) = 2 + 3\cot^2 r$. Consequently, the later case cannot occur. Similar argument applies to the remaining cases.

Case (e): *N is* $\delta(4)$*-ideal.* It follows from (13.30) and (19.51) that one of the following three cases must occurs:

$$\tan(r + c) + \tan(r - c) - \cot r = 0;$$
$$\tan(r + c) - 3\cot r = \tan(r - c);$$
$$\tan(r - c) - 3\cot r = \tan(r + c).$$

By applying a similar argument as case (d), we obtain case (2). □

Chapter 20

$\delta(2)$-ideal Immersions

The simplest non-trivial δ-curvature invariant is $\delta(2)$. In this chapter we summarize known results on $\delta(2)$-ideal submanifolds of space forms.

20.1 $\delta(2)$-ideal submanifolds of real space forms

Let N be an n-dimensional submanifold of a real space form $R^m(c)$. Then Theorem 13.5 implies that

$$\delta(2) \leq \frac{n^2(n-2)}{2(n-1)}||H||^2 + \frac{1}{2}(n+1)(n-2)c. \tag{20.1}$$

For a $\delta(2)$-ideal submanifold N in $R^m(c)$, let $\mathcal{E}(p)$ be the subspace of T_pN defined by (19.30).

A submanifold N of a Riemannian manifold M is called *k-ruled* if N is foliated by k-dimensional totally geodesic submanifolds of M.

For $\delta(2)$-ideal submanifolds in $R^m(c)$, we have [Chen (1993a)].

Theorem 20.1. *Let N be an n-dimensional submanifold of a real space form $R^m(c)$ with $n \geq 3$. If N is $\delta(2)$-ideal and $r = \dim \mathcal{E}(p)$ is independent of $p \in N$, then N is foliated by r-dimensional totally umbilical submanifold of $R^m(c)$. In particular, if N is minimal and $\delta(2)$-ideal in $R^m(c)$, then either N is totally geodesic, or $(n-2)$-ruled.*

Theorem 20.2. *Let N be a $\delta(2)$-ideal submanifold of a complete simply-connected real space form $R^m(c)$ with $\dim N \geq 3$. If $\dim \mathcal{E} \equiv n-1$, then N is a rotational hypersurface of an $(n+1)$-dimensional totally geodesic submanifold of $R^m(c)$ with a geodesic of $R^m(c)$ as its profile curve.*

A Riemannian manifold is called *reducible* if it is locally the direct product of two Riemannian manifolds.

Reducible $\delta(2)$-submanifolds in Euclidean spaces were classified in [Chen (1993b)].

Theorem 20.3. *Let $\phi : N \to \mathbb{E}^m$ be an isometric immersion of a reducible Riemannian n-manifold into $\mathbb{E}^m$. Then N is $\delta(2)$-ideal if and only if one of the following statement holds:*

(1) *ϕ is totally geodesic;*
(2) *locally, N is a spherical cylinder $\mathbf{R} \times S^{n-1}(c)$, $c > 0$;*
(3) *locally, N is the direct product of a Euclidean k-space $\mathbb{E}^k$ and an irreducible Riemannian $(n-k)$-manifold P with vanishing $\delta(2)$-curvature, i.e., $\delta_P(2) \equiv 0$, and $\phi : \mathbb{E}^k \times P \to \mathbb{E}^m$ is a minimal direct product immersion.*

If N is an n-dimensional k-ruled submanifold of $\mathbb{E}^m$, then there is an $(n-k)$-dimensional submanifold B such that N is generated by a moving k-dimensional totally geodesic submanifold along B. In particular, if the moving k-dimensional totally geodesic submanifold is normal to B at each point of B, the ruled submanifold is called a *normal k-ruled submanifold.*

For $\delta(2)$-ideal minimal submanifolds of $\mathbb{E}^m$ we have [Chen (1993a)].

Theorem 20.4. *Let N be an n-dimensional minimal submanifold of $\mathbb{E}^m$. If N is $\delta(2)$-ideal, then*

(a) *at each point $p \in N$, either $\dim \mathcal{E}(p) = n$ or $\dim \mathcal{E}(p) = n-2$;*
(b) *if $\dim \mathcal{E} \equiv n$, N is totally geodesic in $\mathbb{E}^m$;*
(c) *if $\dim \mathcal{E} \equiv n-2$, then N is an $(n-2)$-ruled submanifold. In particular, if N is normal $(n-2)$-ruled, then N is an open portion of one the following submanifolds:*
  (c.1) *the product $N \times \mathbb{E}^{n-2}$ of a minimal surface N of $\mathbb{E}^{m-n+2}$ and a linear $(n-2)$-subspace $\mathbb{E}^{n-2}$ of $\mathbb{E}^m$;*
  (c.2) *the product $CN \times \mathbb{E}^{n-3}$ of a 3-dimensional minimal cone CN in $\mathbb{E}^{m-n+3}$ with a linear $(n-3)$-subspace $\mathbb{E}^{n-3}$ of $\mathbb{E}^m$.*

(d) *Conversely, if N is a minimal submanifold of $\mathbb{E}^m$ given either by (c.1) or by (c.2), then N is $\delta(2)$-ideal.*

For Einstein submanifolds we have [Özgür and Tripathi (2008)].

Theorem 20.5. *Let N be a $\delta(2)$-ideal submanifold of a real space form $R^m(c)$ with $\dim N \geq 3$. Then N is Einsteinian if and only if N is totally geodesic.*

This theorem is due to [Dillen et al. (1997)] when $R^m(c)$ is $\mathbb{E}^m$. Also, from [Dillen et al. (1997)] we have the following.

Theorem 20.6. *Let N be a $\delta(2)$-ideal hypersurface of $\mathbb{E}^{n+1}$ with $n \geq 3$. Then N is a conformally flat manifold with null scalar curvature if and only if N is totally geodesic.*

Theorem 20.7. *A $\delta(2)$-ideal submanifold of dimension $n \geq 4$ in $\mathbb{E}^m$ is conformally flat if and only if $K_{\inf} \equiv 0$.*

Remark 20.1. Theorem 20.5 was extended in [Li (2002)] to $\delta(n_1, \ldots, n_k)$-ideal submanifolds N in real space form with $\sum_{j=1}^k n_j < n = \dim N$; and Theorem 20.7 to $(n_1, \ldots, n_k)$-ideal submanifolds with $\sum_{j=1}^k n_j < n - 1$.

20.2 δ(2)-ideal tubes in real space forms

$\delta(2)$-ideal tubes in real space forms were classified in [Chen (1994b)].

Theorem 20.8. *A tube $T^r(B^\ell)$ in $\mathbb{E}^{n+1}$, $n \geq 3$, is $\delta(2)$-ideal if and only if $\ell = 1$ and B^ℓ is an open portion of a line, i.e., $T^r(B^\ell)$ is an open portion of a spherical hypercylinder $\mathbb{R} \times S^{n-1}(r)$.*

Theorem 20.9. *Let B^ℓ be an embedded ℓ-dimensional submanifold in $S^{n+1}(1)\,(n \geq 3)$. Then a tube $T^r(B^\ell)$ is $\delta(2)$-ideal if and only if either*

(1) *$\ell = 0$ and $r = \frac{\pi}{2}$ or*
(2) *$\ell = 2$, $r = \frac{\pi}{2}$, and B^2 is a minimal surface in $S^{n+1}(1)$.*

Theorem 20.10. *Let $T^r(B^\ell)$ be a tube over an embedded submanifold B^ℓ in $H^{n+1}(-1)\,(n \geq 3)$. Then $T^r(B^\ell)$ is $\delta(2)$-ideal if and only if*

(1) *$\ell = 2$,*
(2) *B^ℓ is totally geodesic in $H^{n+1}(-1)$, and*
(3) $\coth r = \sqrt{2}$.

Remark 20.2. Theorems 20.8-20.10 were extended to $\bar{\delta}_2^k$-ideal tubes in real space forms in [Fastenakels (2006)].

Definition 20.1. When B^2 is a minimal surface with constant curvature in $S^{n+1}(1)$, the tube $T^{\frac{\pi}{2}}(B^2)$ with radius $\frac{\pi}{2}$ is called a *generalized Cartan hypersurface* of $S^{n+1}(1)$.

The following result was obtained in [Chen (1993a)].

Theorem 20.11. *Let B^ℓ be a topological embedded ℓ-dimensional submanifold of $S^{n+1}(1)$ with $0 < \ell < n$. Then the tube $T^r(B^\ell)$ is a generalized Cartan hypersurface if and only if $T^r(B^\ell)$ is $\delta(2)$-ideal with constant scalar curvature.*

20.3 $\delta(2)$-ideal isoparametric hypersurfaces in real space forms

For $\delta(2)$-ideal isoparametric hypersurfaces in real space forms, we have the following [Chen (1994b)].

Theorem 20.12. *Let N be an isoparametric hypersurface of a complete simply-connected real space form $R^{n+1}(c)$ $(c \in \{-1, 0, 1\}, n \geq 3)$. Then N is $\delta(2)$-ideal if and only if either N is totally geodesic or N is a tube $T^r(B^\ell)$ satisfying one of the following three cases:*

(1) $c = 0$, $\ell = 1$, *B^1 is a geodesic, i.e., N is locally a spherical hypercylinder: $\mathbb{R} \times S^{n-1}(a) \subset \mathbb{E}^{n+1}$;*
(2) $c = 1$, $\ell = 2$, $n = 3$, $r = \frac{\pi}{2}$, *and B^2 is the real projective plane of constant curvature $\frac{1}{3}$ embedded in $S^4(1)$ as a Veronese surface;*
(3) $c = -1$, $\ell = 2$, $\coth r = \sqrt{2}$, *and B^2 is totally geodesic.*

Remark 20.3. An isoparametric hypersurface of $\mathbb{E}^{n+1}$ has q distinct principal curvatures with $q \leq 2$. If $q = 2$, one of principal curvatures must be 0. Isoparametric hypersurfaces of $\mathbb{E}^{n+1}$ are locally hyperspheres, hyperplanes or a standard product embedding of $S^k \times \mathbb{E}^{n-k}$. This result was proved in [Levi-Civita (1937)] for $n = 2$; and in [Segre (1938)] for arbitrary n.

Remark 20.4. É. Cartan proved that the number of the distinct principal curvatures of an isoparametric hypersurface in $H^{n+1}(-1)$ is 1 or 2. Such a hypersurface in $H^{n+1}(-1)$ is either totally umbilical or locally a standard product embedding of $S^k \times H^{n-k} \subset H^{n+1}(-1) \subset \mathbb{E}^{n+2}_1$.

Remark 20.5. An isoparametric hypersurface in S^{n+1} with $q = 2$ is locally a standard embedding of the product of two spheres with appropriate radii. It was proved in [Cartan (1939)] that if $q = 3$, then $n = 3$, 6, 12 or 24; and the multiplicities of principal curvatures are the same in each case. Moreover, the isoparametric hypersurface must be a tube of constant radius over a standard Veronese embedding of a projective plane FP^2 into S^{3m+1}, where F is the division algebra of reals, complex numbers, quaternions, or Cayley numbers for the (common) multiplicity $m = 1, 2, 4, 8$, respectively. Thus, up to congruence, there is only one such family for each value of m. These isoparametric hypersurfaces with 3 principal curvatures are known as Cartan hypersurfaces. If $q = 4$ and if the multiplicities are equal, then $n = 4$ or 8.

For each of the above cases, Cartan gave an example of a hypersurface with constant principal curvatures. All the compact isoparametric hypersurfaces in S^{n+1} given by Cartan are homogeneous; indeed, each of them is the orbit of a certain point by an appropriate closed subgroup of the isometry group of S^{n+1}.

20.4 2-type $\delta(2)$-ideal hypersurfaces of real space forms

For $\delta(2)$-ideal, 2-type hypersurfaces in non-flat real space forms, we have the following [Chen (1993b)]

Proposition 20.1. *Every 2-type $\delta(2)$-ideal hypersurface of a real space form $R^{n+1}(c)$ $(c \neq 0;\ n \geq 3)$ is isoparametric.*

Proof. Follows from Theorems 7.17, 7.22 and 13.5. □

Proposition 20.2. *Every 2-type $\delta(2)$-ideal hypersurface in the hypersphere $S^{n+1}(1) \subset \mathbb{E}^{n+2}$ is an open portion of the tube $T^{\frac{\pi}{2}}(V)$ in $S^4(1)$, where V is the Veronese surface in $S^4(1)$.*

Proposition 20.3. *Every 2-type $\delta(2)$-ideal hypersurface in the hypersphere $H^{n+1}(-1) \subset \mathbb{E}^{n+2}_1$ is an open portion of the tube $T^r(B^2)$ over a totally geodesic surface B^2 with $r = \coth^{-1}\sqrt{2}$.*

Propositions 20.2 and 20.3 follow from Proposition 20.1 and Theorem 20.12.

For 2-type, $\delta(2)$-ideal hypersurface of the Euclidean $(n+1)$-space $\mathbb{E}^{n+1}$ we have the following.

Proposition 20.4. *Let N be a 2-type $\delta(2)$-ideal hypersurface of $\mathbb{E}^{n+1}$. Then N has constant mean curvature if and only if it is an open portion of a spherical hypercylinder $S^{n-1} \times \mathbf{R} \subset \mathbb{E}^{n+1}$.*

Proof. Let N be hypersurface of $\mathbb{E}^{n+1}$ with ξ as a unit normal vector field. If we put $H = \alpha\xi$ and $A = A_\xi$, then

$$\Delta^D H = (\Delta\alpha)\xi, \quad \text{trace}\,(A_{DH}) = A(\nabla\alpha),$$
$$\sum_{i=1}^n h(A_H e_i, e_i) = \alpha||A||^2\xi,$$

where $\{e_1, \dots, e_n\}$ is an orthonormal local frame of N. Thus, formula (2.45) of Proposition 2.7 reduces to

$$\Delta H = \{\Delta\alpha + \alpha||A||^2\}\xi + \frac{n}{2}\nabla\alpha^2 + 2\,A(\nabla\alpha). \tag{20.2}$$

Now, let us assume that N is a 2-type hypersurface with constant mean curvature. Then (20.2) reduces to

$$\Delta H = \alpha||A||^2\xi. \tag{20.3}$$

Since $DH = 0$, it follows from Corollary 7.5 that N is of null 2-type. Hence, the position vector field of N in $\mathbb{E}^{n+1}$ satisfies the following spectral decomposition:

$$\mathbf{x} = \mathbf{x}_0 + \mathbf{x}_p, \quad \Delta\mathbf{x}_0 = 0, \quad \Delta\mathbf{x}_p = \lambda_p\mathbf{x}_p, \tag{20.4}$$

where λ_p is a nonzero real number. Therefore, after applying Betrami's formula, we find $\Delta H = \lambda_p H$. Now, by comparing this with (20.3), we obtain $\alpha||H||^2 = \lambda_p$. Thus, $||A||^2$ is constant.

On the other hand, since N is assumed to be $\delta(2)$-ideal, Theorem 13.5 implies that the principal curvature of N are given by

$$\beta, \gamma, \beta+\gamma, \dots, \beta+\gamma, \quad (\beta+\gamma \text{ repeats } n-2 \text{ times}), \tag{20.5}$$

for some functions β, γ. Since N has constant mean curvature and constant squared norm $||A||^2$, β and γ are both constant. Therefore, we conclude from Theorem 20.12 that N is an open portion of a spherical hypercylinder $S^{n-1} \times \mathbf{R} \subset \mathbb{E}^{n+1}$.

The converse is clear. □

20.5 $\delta(2)$ and CMC hypersurfaces of real space forms

A hypersurface of a real space form with constant mean curvature is called a *CMC hypersurface*.

$\delta(2)$-ideal CMC hypersurfaces in real space forms were classified in [Chen and Yang (1999)].

Theorem 20.13. *A CMC hypersurface N of $\mathbb{E}^{n+1}$ $(n > 2)$ is $\delta(2)$-ideal if and only if either N is minimal or N is an open portion of a spherical hypercylinder $\mathbb{R} \times S^{n-1}(r)$.*

Theorem 20.14. *Let N be a CMC hypersurface in $S^{n+1}(1)$ $(n > 2)$. Then N is $\delta(2)$-ideal if and only if one of the following two cases occurs.*

(1) *N is a totally geodesic hypersurface;*
(2) *there is an open dense subset U of N and a non-totally geodesic isometric minimal immersion $\phi : B^2 \to S^{n+1}(1)$ of a surface B^2 such that U is an open subset of the unit normal bundle NB^2 defined by*

$$N_pB^2 = \big\{\xi \in T_{\phi(p)}S^{n+1}(1) : \langle \xi, \xi \rangle = 1 \text{ and } \big\langle \xi, \phi_\star(T_pB^2)\big\rangle = 0\big\}.$$

Theorem 20.15. *Let N be a CMC hypersurface in $H^{n+1}(-1)$ $(n > 2)$. Then N is $\delta(2)$-ideal if and only if one of the following three cases occurs.*

(1) *N is a totally geodesic hypersurface;*
(2) *N is a tube $T^r(B^2)$ with radius $r = \coth^{-1}(\sqrt{2})$ over a totally geodesic surface B^2 of $H^{n+1}(-1)$;*
(3) *there is an open dense subset U of N and a non-totally geodesic minimal immersion $\phi : B^2 \to S_1^{n+1}$ of a surface B^2 into $S_1^{n+1}(1)$ such that U is an open subset of the unit normal bundle NB^2 of B^2 defined by*

$$N_pB^2 = \big\{\xi \in T_{\phi(p)}S_1^{n+1}(1) : \langle \xi, \xi \rangle = -1 \text{ and } \big\langle \xi, \phi_\star(T_pB^2)\big\rangle = 0\big\}.$$

Remark 20.6. Minimal $\delta(2)$-ideal hypersurfaces of $S^{n+1}(1)$ are exactly those given in Theorem 20.14(2); and minimal $\delta(2)$-ideal hypersurfaces of $H^{n+1}(-1)$ are those given in Theorem 20.15(3).

Remark 20.7. For minimal $\delta(2)$-ideal hypersurfaces of higher codimension, see Theorem 20.4 and [Dajczer and Florit (1998)].

Theorem 20.14 was extended in [Dajczer and Florit (2001)] to submanifolds of dimension $n \geq 4$ with arbitrary codimension as follows.

Theorem 20.16. *Let $\psi : N \to \mathbb{E}^m$ be an n-dimensional $\delta(2)$-ideal submanifold with constant mean curvature. If $n \geq 4$, then ψ is either minimal or an open subset of a Riemannian product $\mathbb{E} \times S^{n-1}(r) \subset \mathbb{E}^{n+1} \subset \mathbb{E}^m$.*

The next two results were proved in [Chen and Garay (2003)].

Theorem 20.17. *A CMC hypersurface in $\mathbb{E}^4$ has constant $\delta(2)$-invariant if and only if it is one of the following three hypersurfaces:*

(1) *an isoparametric hypersurface;*
(2) *a minimal hypersurface with relative nullity* ≥ 1*;*
(3) *an open portion of a hypercylinder* $B^2 \times \mathbf{R}$ *over a CMC surface* B^2 *in* $\mathbb{E}^3$ *with nonpositive Gauss curvature.*

Theorem 20.18. *A CMC hypersurface N in* $S^4(1)$ *has constant* $\delta(2)$*-invariant if and only if one of the following two statements holds:*

(1) *N is an isoparametric hypersurface;*
(2) *There is an open dense subset U of N and a non-totally geodesic minimal immersion* $\phi : B^2 \to S^4$ *of a surface* B^2 *into* $S^4(1)$ *such that U is an open subset of the unit normal bundle* NB^2 *of* B^2 *defined by*

$$N_pB^2 = \{\xi \in T_{\phi(p)}S^4(1) : \langle \xi, \xi \rangle = -1 \text{ and } \langle \xi, \phi_\star(T_pB^2) \rangle = 0\}.$$

20.6 $\delta(2)$-ideal conformally flat hypersurfaces

Conformally flat $\delta(2)$-ideal hypersurfaces in a real space forms were classified in [Chen and Yang (1999)].

Theorem 20.19. *A conformally flat hypersurface N of* $\mathbb{E}^{n+1}, n > 2$, *is* $\delta(2)$*-ideal if and only if one of the following four cases occurs.*

(1) *N is an open portion of a totally geodesic hyperplane;*
(2) *N is an open portion of a spherical hypercylinder* $S^{n-1} \times \mathbb{R}$*;*
(3) *N is an open portion of a round hypercone;*
(4) $n = 3$ *and up to dilation it is congruent to*

$$\left(\frac{1}{2}\int_0^x \mathrm{sd}^2\left(\tfrac{\sqrt{2}}{a}x\right)dx, ay_1\,\mathrm{sd}\left(\tfrac{\sqrt{2}}{a}x\right), ay_2\,\mathrm{sd}\left(\tfrac{\sqrt{2}}{a}x\right), ay_3\,\mathrm{sd}\left(\tfrac{\sqrt{2}}{a}x\right)\right),$$

where $y_1^2 + y_2^2 + y_3^2 = \frac{1}{2}$ *and* $k = 1/\sqrt{2}$ *is the modulus of the Jacobi's elliptic functions.*

Theorem 20.20. *A conformally flat hypersurface* $\phi : N \to S^{n+1}(1) \subset \mathbb{E}^{n+2}$ $(n > 2)$ *is* $\delta(2)$*-ideal if and only if it is one of the following three cases occurs:*

(1) *N is a totally geodesic hypersurface.*
(2) ϕ *is congruent to*

$$\left(\sin x, a\cos x, \sqrt{1-a^2}\,y_1\cos x, \dots, \sqrt{1-a^2}\,y_n\cos x\right)$$

with $y_1^2 + y_2^2 + \cdots + y_n^2 = 1$*;*

(3) *$n = 3$ and ϕ is congruent to*

$$\frac{1}{ak'}\Bigg(y_1 \operatorname{cn}(ax), y_2 \operatorname{cn}(ax), y_3 \operatorname{cn}(ax),$$

$$\sqrt{a^2k'^2 - \operatorname{cn}^2(ax)} \cos\left(\frac{k'}{k}x + \frac{\mathrm{i}}{2}\left(\ln\frac{\Theta(ax-\gamma)}{\Theta(ax+\gamma)} + 2aZ(\gamma)x\right)\right)$$

$$\sqrt{a^2k'^2 - \operatorname{cn}^2(ax)} \sin\left(\frac{k'}{k}x + \frac{\mathrm{i}}{2}\left(\ln\frac{\Theta(ax-\gamma)}{\Theta(ax+\gamma)} + 2aZ(\gamma)x\right)\right)\Bigg),$$

$$y_1^2 + y_2^2 + y_3^2 = 1, \;\; \gamma = \operatorname{sn}^{-1}\left(\frac{\mathrm{i}}{ak^2}\right), \;\; \mathrm{i} = \sqrt{-1},$$

where $k = \sqrt{a^2-1}/\sqrt{2}a$, $k' = \sqrt{a^2+1}/\sqrt{2}a$ are the modulus and the complementary modulus of elliptic functions and Θ, Z are the Theta and Zeta functions.

Theorem 20.21. *A conformally flat hypersurface $\phi : N \to H^{n+1}(-1) \subset \mathbb{E}_1^{n+2}$ $(n > 2)$ is $\delta(2)$-ideal if and only if it is one of the following ten cases occurs:*

(1) *N is a totally geodesic hypersurface;*
(2) *up to rigid motions ϕ is given by*

$$\left(a\cosh x, \sinh x, \sqrt{a^2-1}\, y_1 \cosh x, \ldots, \sqrt{a^2-1}\, y_n \cosh x\right)$$

with $y_1^2 + y_2^2 + \cdots + y_n^2 = 1$;
(3) *up to rigid motions ϕ is given by*

$$\left(\left(1 + \frac{1}{2}(u_2^2 + \cdots + u_n^2)\right)\cosh x, \frac{1}{2}(u_2^2 + \cdots + u_n^2)\cosh x,\right.$$

$$\left.\sinh x, u_2\cosh x, \ldots, u_n \cosh x\right);$$

(4) *up to rigid motions ϕ is given by*

$$\left(\sqrt{a^2+1}\, y_1 \cosh x, \ldots, \sqrt{a^2+1}\, y_n \cosh x, a\cosh x, \sinh x\right)$$

with $y_1^2 - y_2^2 + \cdots - y_n^2 = 1$;
(5) *up to rigid motions ϕ is given by*

$$\left(\cosh x, a\sinh x, \sqrt{1-a^2}\, y_1 \sinh x, \ldots, \sqrt{1-a^2}\, y_n \sinh x\right)$$

with $y_1^2 + y_2^2 + \cdots + y_n^2 = 1$;
(6) *up to rigid motions ϕ is given by*

$$\left(e^x + \frac{1}{2}e^{-x}, \frac{1}{2}e^{-x}, e^x y_1, \ldots, e^x y_n\right)$$

with $y_1^2 + y_2^2 + \cdots + y_n^2 = 1$;

(7) *$n = 3$ and up to rigid motions ϕ is given by*

$$\left(\sqrt{2}\cosh u \cosh v, \sqrt{2}\cosh u \sinh v, \sqrt{2}\sinh u, \cos x, \sin x\right);$$

(8) *$n = 3$ and up to rigid motions ϕ is given by*

$$\operatorname{sech} x\left(x^2+u^2+v^2+\cosh^2 x+\tfrac{1}{4}, x^2+u^2+v^2+\cosh^2 x-\tfrac{1}{4}, x, u, v\right);$$

(9) *$n = 3$ and up to rigid motions ϕ is given by*

$$\frac{1}{ak'}\left(\sqrt{a^2k'^2+\operatorname{cn}^2(ax)}\cosh\left(\frac{k'}{k}x-\frac{1}{2}\ln\frac{\Theta(ax-\gamma)}{\Theta(ax+\gamma)}-aZ(\gamma)x\right),\right.$$
$$\sqrt{a^2k'^2+\operatorname{cn}^2(ax)}\sinh\left(\frac{k'}{k}x-\frac{1}{2}\ln\frac{\Theta(ax-\gamma)}{\Theta(ax+\gamma)}-aZ(\gamma)x\right),$$
$$\left. y_1\operatorname{cn}(ax), y_2\operatorname{cn}(ax), y_3\operatorname{cn}(ax)\right),\quad y_1^2+y_2^2+y_3^2=1,$$

where $\gamma = \operatorname{sn}^{-1}\left(1/(ak^2)\right)$, $k = \sqrt{a^2+1}/\sqrt{2}a$ and $k' = \sqrt{a^2-1}/\sqrt{2}a$;

(10) *$n = 3$ and, up to rigid motions, ϕ is given by*

$$\frac{1}{ak'}\left(k\operatorname{dn}(\tfrac{a}{k}x)\cosh u\cosh v, k\operatorname{dn}(\tfrac{a}{k}x)\cosh u\sinh v, k\operatorname{dn}(\tfrac{a}{k}x)\sinh u,\right.$$
$$\sqrt{k^2\operatorname{dn}^2(\tfrac{a}{k}x)-a^2k'^2}\cos\left(k'x-\frac{\mathrm{i}}{2}\ln\frac{\Theta(\frac{a}{k}x-\gamma)}{\Theta(\frac{a}{k}x+\gamma)}-\mathrm{i}\frac{a}{k}Z(\gamma)x\right)$$
$$\left.\sqrt{k^2\operatorname{dn}^2(\tfrac{a}{k}x)-a^2k'^2}\sin\left(k'x-\frac{\mathrm{i}}{2}\ln\frac{\Theta(\frac{a}{k}x-\gamma)}{\Theta(\frac{a}{k}x+\gamma)}-\mathrm{i}\frac{a}{k}Z(\gamma)x\right)\right)$$

where $\gamma = \operatorname{sn}^{-1}(k/a)$, $k = \sqrt{2}a/\sqrt{1+a^2}$ and $k' = \sqrt{1-a^2}/\sqrt{1+a^2}$.

Remark 20.8. Let N be a conformally flat n-manifolds with $n \geq 4$ isometrically immersed in a real space form $R^{n+1}(c)$. Then according to a well known result of Cartan and Schouten, N has a principal curvature μ with multiplicity at least $n-1$ (cf. [Chen (1973b)]). Denote by λ the remaining principal curvature (where we put $\lambda(p) = \mu(p)$ when the multiplicity of μ is n at a point $p \in N$).

Assume that N is an ideal hypersurface in $R^{n+1}(c)$ and N is non-totally umbilical. Then, by Theorem 13.5, we may prove that there exist three integers $t \geq 0, r, k \geq 1$ such that

$$\lambda = (r-t)\mu, \quad n = t+1+(k-1)r. \tag{20.6}$$

Thus, by applying a result of [do Carmo and Dajczer (1983)], we conclude that N is a rotational hypersurface of $R^{n+1}(c)$. Hence, every ideal immersion of a conformally flat n-manifold in a complete simply-connected real space form $R^{n+1}(c)$ is a rotational hypersurface when $n \geq 4$.

Moreover, up to rigid motions, the profile curve of the rotational hypersurface is completely determined by (20.6). In particular, if $R^{n+1}(c) = \mathbb{E}^{n+1}$, the profile curve is congruent to the graph of a function f of one variable which satisfies the following differential equation:

$$ff'' + (r-t)(1+f'^2) = 0, \tag{20.7}$$

where $r-t$ is the difference of two integers satisfying $n = t+1+(k-1)r$ for some $t \geq 0, r \geq 1$, and $k \geq 2$.

Based on these facts, conformally flat ideal hypersurfaces in real space forms were completely classified in [Chen et al. (2004a)].

20.7 Symmetries on δ(2)-ideal submanifolds

A pseudo-Riemannian manifold is called *semi-symmetric* if $R \cdot R = 0$, i.e.,

$$\begin{aligned}
&(R(X,Y)\cdot R)(U,V)W := \\
&\quad R(X,Y)(R(U,V)W) - R(R(X,Y)U,V)W \\
&\qquad - R(U,R(X,Y)V)W - R(U,V)(R(X,Y)W) = 0.
\end{aligned} \tag{20.8}$$

The next two results were proved in [Dillen et al. (1997)].

Theorem 20.22. *Let N be a $\delta(2)$-ideal submanifold of $\mathbb{E}^m$ with* $\dim N \geq 3$. *Then N is semi-symmetric if and only if either N is a minimal submanifold or N is a round hypercone in a totally geodesic subspace $\mathbb{E}^{n+1}$ of $\mathbb{E}^m$.*

Remark 20.9. Ideal semi-symmetric hypersurfaces in real space forms were classified in [Li (2002)].

A pseudo-Riemannian submanifold N of a pseudo-Riemannian manifold is called *semi-parallel* if $\bar{R}(X,Y) \cdot h = 0, X, Y \in TN$, where h is the second fundamental form and $\bar{R}$ is the curvature tensor of the van der Waerden-Bortolotti connection $\bar{\nabla} = \nabla \oplus D$.

Theorem 20.23. *Let N be a $\delta(2)$-ideal submanifold of $\mathbb{E}^m$ with* $\dim N \geq 3$. *Then N is semi-parallel if and only if N either is totally geodesic or a round hypercone in a totally geodesic subspace $\mathbb{E}^{n+1}$ of $\mathbb{E}^m$.*

Given any symmetric $(0,2)$-tensor A and any pair of vector field X, Y on a pseudo-Riemannian manifold (N, g), one may consider the associated endomorphism

$$(X \wedge_A Y)Z = A(Y,Z)X - A(X,Z)Y.$$

The action of the natural metrical endomorphism $X \wedge_g Y$ as a derivation on the curvature tensor R results the $(0,6)$-tensor $Q(g,R) = -\wedge_g \cdot R$, known as *Tachibana tensor.*

Definition 20.2. A pseudo-Riemannian manifold N of dimension ≥ 3 is called *pseudo-symmetric in the sense of Deszcz* if $R \cdot R = L \cdot Q(g,R)$ for some function $L \in \mathcal{F}(N)$, where the dot is the derivation on the tensor algebra. A pseudo-Riemannian manifold is called *Weyl semi-symmetric* if $R \cdot C = 0$, where C is the Weyl conformal curvature tensor.

Similarly, a pseudo-Riemannian manifold N is said to have *pseudo-symmetric Weyl tensor* if $C \cdot C = L_C Q(g,C)$ for a function $L_C \in \mathcal{F}(N)$.

The next result was proved in [Deszcz et al. (1998)].

Theorem 20.24. *Every $\delta(2)$-ideal hypersurface of a real space form $R^{n+1}(c)$, $n \geq 4$, is a hypersurface with pseudo-symmetric Weyl tensor.*

For pseudo-symmetric hypersurfaces we have [Özgür and Arslan (2002)].

Theorem 20.25. *Let N be a $\delta(2)$-ideal hypersurface of $\mathbb{E}^{n+1}$ with $n \geq 3$. If N is pseudo-symmetric, then N is one of the following hypersurfaces:*

(1) *a totally geodesic hypersurface;*
(2) *a round hypercone;*
(3) *a minimal hypersurface in which case N is $(n-2)$-ruled.*

It was proved in [Hinić (2008)] that

Theorem 20.26. *A $\delta(2)$-ideal submanifold N of $\mathbb{E}^m$ with $\dim N \geq 4$ is Weyl semi-symmetric if and only if either N is conformally flat or N is minimal in $\mathbb{E}^m$.*

The following two results were obtained in [Deszcz et al. (2008)].

Theorem 20.27. *A $\delta(2)$-ideal submanifold N of the Euclidean m-space $\mathbb{E}^m$ with $\dim N \geq 4$ and $m \geq 5$ has pseudo-symmetric Weyl conformal curvature tensor C and*

$$L_C = \frac{3-n}{(n-1)(n-2)} K_{\inf}.$$

Theorem 20.28. *A $\delta(2)$-ideal submanifold N of $\mathbb{E}^m$ with $\dim N \geq 3$ is pseudo-symmetric if and only if either it is semi-symmetric or at each point $p \in N$ where $R \cdot R \neq 0$, the 2-dimensional normal section $\sigma_{\tilde{\pi}} \subset \mathbb{E}^{2+m}$ of N at p in the direction of the tangent plane $\tilde{\pi} \subset T_pN$ for which the sectional curvature $K(\pi)$ at p attains its minimal value $K_{\inf}$ is pseudo-umbilical.*

20.8 G_2-structure on $S^7(1)$

A G_2-structure on a 7-manifold is a geometric structure identifying each tangent space with the pure imaginary Cayley numbers $\mathcal{O}$. This automatically induces a Riemannian metric and a nowhere vanishing 3-form.

Let $(x_1, \ldots, x_7)$ and $(x_0, x_1, \ldots, x_7)$ be natural coordinate systems of $\mathbf{R}^7$ and of $\mathbf{R}^8$. We simply denote the form $dx_i \wedge dx_j \wedge \cdots \wedge dx_k$ by $dx_{ij\cdots k}$. Define a 3-form φ_0 on $\mathbf{R}^7$ and a 4-form Φ_0 on $\mathbf{R}^8$ by

$$\begin{aligned} \varphi_0 &= dx_{123} + dx_{145} + dx_{167} + dx_{246} - dx_{257} \\ &\quad - dx_{347} - dx_{356}, \\ \Phi_0 &= dx_{0123} + dx_{0145} + dx_{0167} + dx_{0246} - dx_{0257} \\ &\quad - dx_{0347} - dx_{0356} + dx_{4567} + dx_{2367} + dx_{2345} \\ &\quad + dx_{1357} - dx_{1346} - dx_{1256} - dx_{1247}. \end{aligned} \tag{20.9}$$

If we decompose $\mathbf{R}^8 = \mathbf{R} \oplus \mathbf{R}^7$, then $\Phi_0 = dx_0 \wedge \varphi_0 + *\varphi_0$, where $*\varphi_0$ is the Hodge dual of φ_0 in $\mathbf{R}^7$. Geometrically, G_2 is the 14-dimensional simple Lie group acting on $\mathbf{R}^7$ preserving the 3-form φ_0 and $Spin(7)$ is the simple 21-dimensional Lie group preserving the 4-form Φ_0.

Definition 20.3. Let M be an oriented 7-manifold. We say that a 3-form φ on M is *positive* if, for each $p \in M$, $\varphi|_p = \iota_p^*(\varphi_0)$ for some orientation-preserving isomorphism $\iota_p : T_pN \to \mathbf{R}^7$.

Let us denote the bundle of positive 3-forms on M by $\wedge_+^3 T^*M$. A positive 3-form φ on M determines a unique metric g_φ on M such that, if $*\varphi$ is the Hodge dual to φ with respect to g_φ and g_7 is the Euclidean metric on $\mathbf{R}^7$, the triple $(\varphi, *\varphi, g_\varphi)$ is identified with $(\varphi_0, *\varphi_0, g_7)$ at each point. A *G_2-structure* on M is a choice of $\varphi \in \Gamma(\wedge_+^3 T^*M)$, where $\Gamma(\wedge_+^3 T^*M)$ is the space of cross sections of $\wedge_+^3 T^*M$.

Definition 20.4. Let Q be an oriented 8-manifold. We say that a 4-form Φ on Q is *positive* if, for each $q \in Q$, $\Phi|_p = \iota_q^*(\Phi_0)$ for some orientation-preserving isomorphism $\iota_q : T_qN \to \mathbf{R}^8$.

Denote the bundle of positive 4-forms on Q by $\wedge^4_+ T^*Q$. A positive 4-form Φ on Q determines a unique metric g_Φ on Q such that, if g_8 is the Euclidean metric on $\mathbf{R}^8$, the pair (Φ, g_Φ) is identified with (Φ_0, g_8) at each point. A $Spin(7)$*-structure* on Q is a choice of $\Phi \in \Gamma(\wedge^4_+ T^*Q)$.

Since G_2 is the automorphism group of the Cayley numbers $\mathcal{O}$, there is a natural link between the G_2 and $Spin(7)$ forms and the cross product structure on $\mathcal{O}$. The imaginary Cayley numbers are endowed with a cross product structure as described in section 19.2, which may be extended to a skew-symmetric product on $\mathcal{O}$, and one may further define triple and 4-fold cross products on $\mathcal{O}$. We may identify $\operatorname{Im}\mathcal{O}$ with $\mathbf{R}^7$ such that

$$\varphi_0(u, v, w) = g_7(u \times v, w), \quad \forall u, v, w \in \mathbf{R}^7. \tag{20.10}$$

Moreover, we may identify $\mathcal{O}$ with $\mathbf{R}^8$ so that

$$\Phi_0(u, v, w, z) = g_8(u \times v \times w, z), \quad \forall u, v, w, z \in \mathbf{R}^8. \tag{20.11}$$

20.9 $\delta(2)$-ideal associative submanifolds of $S^7(1)$

Associative submanifolds of $S^7(1)$ are 3-dimensional minimal submanifolds which are related to the G_2-structure that $S^7(1)$ naturally inherits from the standard $Spin(7)$-structure on $\mathbb{E}^8$. They were introduced in [Harvey and Lawson (1982)]. The key to defining associative submanifolds of $S^7(1)$ is the G_2-structure on $S^7(1)$.

Examples of associative submanifolds of $S^7(1)$ include Lagrangian submanifolds of the nearly Kähler $S^6(1)$, Hopf lifts of holomorphic curves in $CP^3(4)$ and minimal Legendre submanifolds of the Sasakian $S^7(1)$.

Definition 20.5. Let M be an oriented 7-manifold with a G_2-structure φ. Define a vector-valued 3-form $\mathcal{K}$ on M by

$$*\varphi(u, v, w, z) = g_\varphi(\mathcal{K}(u, v, w), z), \quad \forall u, v, w, z \in \mathcal{X}(M),$$

where g_φ is defined in Definition 20.3.

Definition 20.6. A 3-dimensional oriented submanifold N of M is called an *associative submanifold* if $\mathcal{K}|_N \equiv 0$ and $\varphi_N > 0$.

Definition 20.7. A submanifold N of a Riemannian manifold is called *isotropic* if, for each given point $p \in N$, $||h(u, u)||$ is independent of the choice of $u \in T^1_pN$. If $||h(u, u)||$ is also independent of $p \in N$, then N is said to be *constant isotropic*.

The following results of [Lotay (2010)] provide interesting links between $\delta(2)$-ideal and associative submanifolds of $S^7(1)$.

Proposition 20.5. *Let S be an oriented real analytic surface in $S^7(1)$. Then locally there exists a unique associative submanifold in $S^7(1)$ which contains S.*

Theorem 20.29. *$\delta(2)$-ideal associative submanifolds in $S^7(1)$ are in one-to-one correspondence with "linear" pseudo-holomorphic curves in $\mathcal{C}$, where $\mathcal{C}$ is the space of oriented geodesic circles in $S^7(1)$.*

Theorem 20.30. *Let N be an associative submanifold in $S^7(1)$ which is $\delta(2)$-ideal. Then either*

(a) *N is the Hopf lift of a holomorphic curve in $CP^3(4)$ or*
(b) *N is constructed from an isotropic minimal surface Σ in $S^6(1)$ and a holomorphic curve Γ in a CP^1-bundle over Σ such that the pseudo-holomorphic lift of Σ in $\mathcal{C}$ is "linear".*

Theorem 20.31. *Given a minimal 2-sphere S^2 of nonconstant curvature in $S^6(1)$, there is a Riemannian 3-manifold (N, g_N) which is an S^1-bundle over S^2 and an S^1-family of non-congruent isometric associative embeddings of (N, g_N) in $S^7(1)$ which are $\delta(2)$-ideal.*

20.10 $\delta(2)$-ideal Lagrangian submanifolds of complex space forms

Let N be a Lagrangian submanifold of a complex space form $M^n(4c)$ of constant holomorphic sectional curvature $4c$. It follows from Theorem 13.6 that inequality (20.1) holds as well. On the other hand, it follows from Theorem 15.3 that N is minimal. Therefore, we have the following.

Lemma 20.1. *Let N be a Lagrangian submanifold of a complex space form $M^n(4c)$. Then*

$$\delta(2) \leq \frac{n^2(n-2)}{2(n-1)}||H||^2 + \frac{1}{2}(n+1)(n-2)c. \tag{20.12}$$

If the equality sign of (20.12) holds identically, then N is minimal in $M^n(4c)$. Thus N satisfies

$$\delta(2) = \frac{1}{2}(n+1)(n-2)c. \tag{20.13}$$

To state the next result, we provide the following example of minimal Lagrangian submanifold with constant scalar curvature in $CP^3(4)$ which verifies (20.13) identically.

Example 20.1. Define two complex structures on $\mathbf{C}^4$ by

$$\begin{aligned} I(v_1, v_2, v_3, v_4) &= (iv_1, iv_2, iv_3, iv_4), \\ J(\bar{v}_1, \bar{v}_2, \bar{v}_3, \bar{v}_4) &= (-\bar{v}_4, \bar{v}_3, -\bar{v}_2, \bar{v}_1). \end{aligned} \tag{20.14}$$

Clearly, I is the standard complex structure. The corresponding Sasakian structures on $S^7(1)$ have characteristic vector fields $\xi_1 = -I(x)$ and $\xi_2 = -J(x)$. Since we consider two different complex structures on $\mathbf{C}^4$, we can consider the two corresponding Hopf fibrations $\pi_j : S^7(1) \to CP^3(4)$. The vector field ξ_j is vertical for π_j for $j = 1, 2$.

Now, consider the *Calabi curve* $\mathcal{C}_3$ of constant curvature $\frac{4}{3}$ in $CP^3(4)$ defined by

$$\mathcal{C}_3(z) = \left[1, \sqrt{3}z, \sqrt{3}z^2, z^3\right].$$

Since $\mathcal{C}_3$ is a holomorphic curve with respect to I, there exists a circle bundle $\pi : N^3 \to CP^1$ and an isometric minimal immersion $\mathcal{I} : N^3 \to S^7(1)$ such that $\pi_1(\mathcal{I}) = \mathcal{C}_3(\pi)$. It is direct to show that N^3 has constant scalar curvature and $\mathcal{I}$ is horizontal with respect to π_2 such that the immersion $\mathcal{J}$: $N^3 \to CP^3(4)$ defined by $\mathcal{J} = \pi_2(\mathcal{I})$ is a minimal Lagrangian immersion which satisfies the equality (20.13) identically with $n = 3$. The Hopf's lift of this minimal Lagrangian submanifold is a Legendre submanifolds of $S^7(1)$ studied in [Baikoussis and Blair (1995)].

The following result was proved in [Chen et al. (1996)].

Theorem 20.32. *Let $\phi : N \to M^n(4c)$ be a Lagrangian minimal immersion with constant scalar curvature into a complex space form $M^n(4c)$ ($n \geq 3$, $c \in \{-1, 0, 1\}$). Then N satisfies $\delta(2) = \frac{1}{2}(n+1)(n-2)c$ if and only if either*

(1) *N is a totally geodesic immersion or*
(2) *$n = 3$, $c = 1$, and ϕ is locally congruent to the immersion $\mathcal{J} : N \to CP^3(4)$ defined above.*

The next example shows that there exists a Lagrangian submanifold in $CP^3(4)$ *with non-constant scalar curvature* which satisfies the equality (20.13). From the example, it follows that the condition of constant scalar curvature in Theorem 20.32 cannot be omitted.

Example 20.2. Consider a totally geodesic $S^5(1) \subset S^7(1)$. Let ξ be a unit vector orthogonal to the linear subspace containing $S^5(1)$. Let $\psi : B^2 \to S^5(1)$ be a minimal surface in $S^5(1)$. We define an isometric immersion of warped product $N = (-\frac{\pi}{2}, \frac{\pi}{2}) \times_{\cos t} B^2$ into $S^7(1)$ by

$$x : N \to S^7(1) : (t, p) \mapsto (\sin t)\xi + (\cos t)\psi(p).$$

This immersion is minimal and it satisfies (20.13). In order for x to be Legendrian, we have to assume that $S^5(1)$ lies in a complex hyperplane $\mathbf{C}^3$ of $\mathbf{C}^4 \supset S^7(1)$ and that ψ is Legendrian. It is direct to verify that N does not have constant scalar curvature whenever B^2 is non-totally geodesic.

The family of Lagrangian submanifolds of $CP^n(4)$ with integrable $\mathcal{E}^\perp$ satisfying (20.13) were determined in [Chen et al. (1994)].

Theorem 20.33. *Let $\phi : N \to CP^n(4)$ be a Lagrangian immersion satisfying equality (20.13) and that the dimension of $\mathcal{E}$ is constant and $\mathcal{E}^\perp$ is an integrable distribution. Then either*

(a) *ϕ is totally geodesic or*
(b) *ϕ has no totally geodesic points and, up to holomorphic transformations, $\phi(N)$ lies in the image under Hopf's fibration $\pi : S^{2n+1}(1) \to CP^n(4)$ of the image of one of the immersions described in the next proposition.*

Proposition 20.6. *Let $S^{2n+1}(1)$ be the unit hypersphere of $\mathbf{C}^{n+1}$. Consider the orthogonal decomposition $\mathbf{C}^{n+1} = \mathbf{C}^3 \oplus J(\mathbb{E}^{n-2}) \oplus \mathbb{E}^{n-2}$. Let $\psi : M^2 \to S^5(1) \subset \mathbf{C}^3$ be a minimal Legendre immersion and consider the hypersphere $S^{n-3}(1)$ in $\mathbb{E}^{n-2}$. Then*

$$F : (0, \tfrac{\pi}{2}) \times_{\cos t} M^2 \times_{\sin t} S^{n-3}(1) \to S^{2n+1}(1);$$
$$(t, p, q) \mapsto (\cos t)\psi(p) + (\sin t)q,$$

is a minimal Legendre immersion satisfying equality (20.13). Moreover, if ψ has no totally geodesic points, then the dimension of $\mathcal{E}$ is exactly $n-2$. Moreover, if we extend F to a map

$$\tilde{F} : (-\tfrac{\pi}{2}, \tfrac{\pi}{2}) \times M^2 \times S^{n-3}(1) \to S^{2n+1}(1);$$
$$(t, p, q) \mapsto (\cos t)\psi(p) + (\sin t)q.$$

Then $\tilde{F}$ fails to be immersive at $t = 0$, but the image of $\tilde{F}$ is an immersed minimal Legendre submanifold. Furthermore, if ψ is not totally geodesic, then this image can not be extended further.

Lagrangian submanifolds of $CH^n(-4)$ with integrable $\mathcal{E}^{\perp}$ satisfying the equality case of (20.12) were classified in [Chen and Vrancken (2002b)].

Theorem 20.34. *Let $F : N \to CH^n(-4)$ be a Lagrangian immersion without geodesic points which satisfies*

$$\delta(2) = \frac{n^2(n-2)}{2(n-1)}||H||^2 - \frac{1}{2}(n+1)(n-2). \tag{20.15}$$

Assume that $\mathcal{E}^{\perp}$ is integrable. Then, up to rigid motions, every point p of an open dense subset of N has a neighborhood U_p such that F is given by one of the following:

(a) $F(t,u,v) = \pi(\cosh t(\psi(u), 0, \cdots, 0) + \sinh t(0,0,0,\phi(v)))$, *where*

$$\phi : (v_1, \ldots, v_{n-3}) \mapsto \phi(v)$$

describes the standard totally real $(n-3)$-sphere S^{n-3} in $\mathbb{E}^{n-2} \subset \mathbf{C}^{n-2}$ and $\psi : (u_1, u_2) \mapsto \psi(u)$ describes a minimal horizontal immersion in $H_1^5(-1)$, or

(b) $F = \pi(\cosh t(\phi(v), 0,0,0)) - \sinh t(0, \cdots, 0, \psi(u)))$, *where*

$$\phi : (v_1, \ldots, v_{n-3}) \mapsto \phi(v)$$

describes the standard totally real hyperbolic space $H_1^{n-3}(-1)$ in $\mathbb{E}_1^{n-2} \subset \mathbf{C}_1^{n-2}$ and $\psi : (u_1, u_2) \mapsto \psi(u)$ describes a minimal horizontal immersion in $S^5(1)$, or

(c) $F(t,u,v) = \pi((\cosh t, -\sinh t, 0, \cdots, 0) + \frac{1}{2}e^{-t}z(u,v)(1,-1,0,\cdots,0) + \frac{1}{2}e^{-t}(0,0,w_1(u), w_2(u), v_1, \cdots, v_{n-3}))$, *where*

$$w : D \subset \mathbf{R}^2 \to \mathbf{C}^2 : (u_1,u_2) \mapsto (w_1(u_1,u_2), w_2(u_1,u_2))$$

is a minimal Lagrangian immersion and z is a complex-valued function determined by the condition that

$$2(z+\bar{z}) = w_1\bar{w}_1 + w_2\bar{w}_2 + \sum_{i=1}^{n-3} v_i^2,$$

and by the condition that its imaginary part depends only on u and satisfies the following system of differential equations :

$$2(z-\bar{z})_{u_1} = w_1(\bar{w}_1)_{u_1} + w_2(\bar{w}_2)_{u_1} - \bar{w}_1(w_1)_{u_1} - \bar{w}_2(w_2)_{u_1},$$
$$2(z-\bar{z})_{u_2} = w_1(\bar{w}_1)_{u_2} + w_2(\bar{w}_2)_{u_2} - \bar{w}_1(w_1)_{u_2} - \bar{w}_2(w_2)_{u_2},$$

where $\pi : H_1^{2m+1}(-1) \to CH^m(-4)$ is a generalized Hopf fibration defined in Example 12.6.

Lagrangian submanifolds in $CH^3(-4)$ satisfying (20.15) with non-integrable $\mathcal{E}^{\perp}$ were also determined in [Chen and Vrancken (2002b)].

20.11 $\delta(2)$-ideal CR-submanifolds of complex space forms

For CR-submanifolds in complex space forms, there is a sharp relationship between $\delta(2)$ and the squared mean curvature $||H||^2$ (cf. [Chen (1996b, 2008a)]).

Theorem 20.35. *Let N be an n-dimensional CR-submanifold in a complex space form $M^m(4c)$. Then*

$$\delta(2) \leq \begin{cases} \dfrac{n^2(n-2)}{2(n-1)}||H||^2 + \left\{\dfrac{1}{2}(n+1)(n-2) + 3\mathfrak{h}\right\}c, & \text{if } c > 0; \\ \dfrac{n^2(n-2)}{2(n-1)}||H||^2, & \text{if } c = 0; \\ \dfrac{n^2(n-2)}{2(n-1)}||H||^2 + \dfrac{1}{2}(n+1)(n-2)c, & \text{if } c < 0, \end{cases} \tag{20.16}$$

where $\mathfrak{h}$ is the complex dimension of the holomorphic distribution.

Real hypersurfaces in $CP^m(4)$ and $CH^m(-4)$ which satisfy the equality case of (20.16) were already determined in Theorems 15.11 and 15.12.

For proper CR-submanifolds of $CH^m(-4)$, we have the following result from [Chen and Vrancken (1999)].

Theorem 20.36. *Let U be a domain of $\mathbf{C}$ and let $\Psi : U \to \mathbf{C}^{m-1}$ be a non-constant holomorphic curve in $\mathbf{C}^{m-1}$. Define $z : \mathbb{E}^2 \times U \to \mathbf{C}_1^{m+1}$ by*

$$z(u,t,w) = \mathrm{e}^{\mathrm{i}t}\left(\mathrm{i}u - 1 - \frac{1}{2}\Psi(w)\bar{\Psi}(w), \mathrm{i}u - \frac{1}{2}\Psi(w)\bar{\Psi}(w), \Psi(w)\right).$$

Then $\langle z, z\rangle = -1$ and $z(\mathbb{E}^2 \times U)$ in $H_1^{2m+1}(-1)$ is invariant under the action of $H_1^1 = \{\lambda \in \mathbf{C} : \lambda\bar{\lambda} = 1\}$. Away from points where $\Psi'(w) = 0$, the image $\pi(\mathbb{E}^2 \times U)$ via the projection $\pi : H_1^{2m+1}(-1) \to CH^m(-4)$ is a proper CR-submanifold of $CH^m(-4)$ which satisfies (20.15).

Conversely, up to rigid motions, every proper CR-submanifold of $CH^m(-4)$ satisfying (20.15) *is obtained in such way.*

A submanifold is said to be *linearly full* in $CH^m(-4)$ if it does not lie in any totally geodesic complex submanifold of $CH^m(-4)$.

The next two results are due to [Sasahara (2001b)].

Proposition 20.7. *Let N be an n-dimensional submanifold of a complex space form $M^m(4c)$. Then*

$$\delta(n-1) \leq \frac{n^2}{4}||H||^2 + (n+2)c, \quad \text{if } c \geq 0; \tag{20.17}$$

$$\delta(n-1) \leq \frac{n^2}{4}||H||^2 + (n-1)c, \quad \text{if } c < 0. \tag{20.18}$$

The equality case of (20.17) *holds at a point* p *if and only if there exists an orthonormal basis* $e_1, \ldots, e_{2m}$ *at* p *such that* $e_1, \ldots, e_n \in T_pN$ *and*

(a) $\overline{Ric} = S(e_n, e_n)$;
(b) $c\sum_{i=1}^{n-1} \langle Je_i, e_n\rangle^2 = c$;
(c) $h^s_{in} = 0$, $h^s_{nn} = \sum_i^{n-1} h^s_{ii}$; $1 \le i \le n-1$, $n+1 \le s \le 2m$,

where $\overline{Ric}$ *denotes the maximum Ricci curvature function of* N.

The equality case of (20.18) *holds at a point* p *if and only if there exists an orthonormal basis* $e_1, \ldots, e_{2m}$ *at* p *such that* $e_1, \ldots, e_n \in T_pN$ *and*

(a) $\overline{Ric} = S(e_n, e_n)$;
(b) $\sum_{i=1}^{n-1} \langle Je_i, e_n\rangle^2 = 0$;
(c) $h^s_{in} = 0$, $h^s_{nn} = \sum_i^{n-1} h^s_{ii}$; $1 \le i \le n-1$, $n+1 \le s \le 2m$.

Proof. Put

$$\delta = \tau - n(n-1)c - \frac{n^2}{2}||H||^2 - 3c\sum_{i,j=1}^{n} \langle e_i, Je_j\rangle^2 .$$

Then Gauss's equation gives

$$n^2H^2 = 2(\delta + ||h||^2).$$

In a similar way as the proof of Theorem 13.3, we have

$$S(e_n, e_n) \le (n-1)c + \frac{n^2}{4}H^2 + 3c\sum_{i=1}^{n-1} \langle Je_i, e_n\rangle^2 ,$$

which implies the proposition. □

Theorem 20.37. *Let* U *be a domain of a Cartesian* $2n$*-space* $\mathbf{R}^{2n}$. *Define* $z : \mathbf{R}^2 \times U \to \mathbf{C}^{m+1}_1$ *by*

$$z(s, t, x_1, x_2, \ldots, y_1, y_2) = \left(\psi(x_1, \ldots, y_2)e^{is}, \frac{e^{it}}{\sqrt{2n-2}}\right) \tag{20.19}$$

where $|\psi|^2 = \frac{1-2n}{2n-2}$ *and* $\psi(x_1, \ldots, y_2)e^{is}$ *is a* CR*-submanifold of* $\mathbf{C}^m_1$ *such that the unit vector field* η *in the totally real distribution* $\mathcal{D}^\perp$ *is* $\sqrt{\frac{2n-2}{2n-1}}\frac{\partial}{\partial s}$. *Then* $\langle z, z\rangle = -1$ *and* $z(\mathbf{R}^2 \times U)$ *is* H^1_1*-invariant. Moreover,* $z(\mathbf{R}^2 \times U)/_\sim$ *is a* $(2n+1)$*-dimensional* CR*-submanifold with* $\dim \mathcal{D}^\perp = 1$ *satisfying the equality case of* (20.18) *under the condition that the shape operator* A_η *has constant principal curvatures.*

Conversely, in case $n > 1$ *and* $m > n+1$, *up to rigid motions of* $CH^m(-4)$, *every linearly full* $(2n+1)$*-dimensional* CR*-submanifold with* $\dim \mathcal{D}^\perp = 1$ *satisfying the equality case of* (20.18) *under the condition that* A_η *has constant principal curvatures is obtained in such way.*

20.12 $\delta(2)$-ideal Kähler hypersurfaces in complex space forms

Since every Kähler submanifold of a Kähler manifold is a minimal submanifold, the equation of Gauss implies the following.

Proposition 20.8. *The scalar curvature τ of a complex n-dimensional Kaehler submanifold N of a complex space form $M^m(4c)$ of constant holomorphic sectional curvature $4c$ satisfies*

$$\tau \leq 2n(n+1)c,$$

with the equality holding identically if and only if N is totally geodesic.

Recall from (15.68) that $\delta^c(2)$ on a Kähler manifold is defined by

$$\delta^c(2) = \tau - \max_{u \in T^1 N} H(u),$$

where $H(u)$ is the holomorphic sectional curvature of u.

Proposition 20.9. *Let N be a Kähler hypersurface of a complex space form $M^{n+1}(4c)$ of constant holomorphic sectional $4c$. Then*

$$\delta^c(2) \leq 2(n-1)(n+2)c, \tag{20.20}$$

with the equality holding at a point $p \in N$ if and only if $\operatorname{rank}(A) \leq 2$ *at p.*

Proof. This proposition is a special case of Theorem 15.10. □

For Kähler hypersurfaces of $CP^n(4)$, we have the following proposition by combining Theorem 4.2.1 of [Abe (1971)] and Proposition 20.9.

Proposition 20.10. *If N is a complete Kähler hypersurface of $CP^{n+1}(4)$, then we have*

$$\delta^c(2) \leq 2(n-1)(n+2), \tag{20.21}$$

with the equality holding if and only if N is totally geodesic, unless $n \leq 2$.

Proof. Inequality (20.21) is a special case of inequality (20.20).

If the equality sign of (20.21) holds identically, Theorem 15.10 implies that the complex dimension of the relative nullity subspace is at least $n-1$. Thus, after applying Theorem 4.2.1 of [Abe (1971)], we conclude that N is totally geodesic unless $n = 2$.

The converse is clear. □

Proposition 20.9 can be rephrased as follows.

Proposition 20.11. *Let N be a Kähler hypersurface of a complex space form $M^{n+1}(4c)$. Then*

$$\delta(2) \leq 2(n-1)(n+2)c, \tag{20.22}$$

with the equality holding at a point $p \in N$ if and only if $\operatorname{rank}(A) \leq 2$ *at p.*

Proof. Follows from Proposition 20.9 and the fact that, for a Kähler hypersurface N of $M^{n+1}(4c)$, the maximum sectional curvature of N at a point $p \in N$ is equal to the maximum holomorphic sectional curvature of N at p. □

Remark 20.10. Proposition 20.11 with $c = 0$ can be found in [Sentürk and Verstraelen (2008)].

By a *complex hypercylinder* $\mathcal{C}^n$ in $\mathbf{C}^{n+1}$ we mean the direct product $\mathcal{C} \times (\mathbf{C}^2)^{\perp}$ of a complex curve $\mathcal{C}$ in $\mathbf{C}^2 \subset \mathbf{C}^{n+1}$ and the orthogonal complementary complex subspace $(\mathbf{C}^2)^{\perp}$ of $\mathbf{C}^2 \subset \mathbf{C}^{n+1}$.

By applying Proposition 20.11 with $c = 0$ and a result of [Abe (1972)], we have the following [Sentürk and Verstraelen (2008)].

Proposition 20.12. *If N is a complete Kähler hypersurface of $\mathbf{C}^{n+1}$ which satisfies $\delta(2) = 0$, then N is a complex hypercylinder.*

Consider a Kähler hypersurface N of $CP^{n+1}(4c)$. Let $\{e_1, \ldots, e_{2n}\}$ be an orthonormal tangent frame. Choose an orthonormal normal frame $\xi, J\xi$ of N. Put

$$K^{\perp}_{\alpha\beta} = R^D(e_\alpha, e_\beta, \xi, J\xi), \tag{20.23}$$

where R^D is the curvature tensor of the normal connection. Define the scalar normal curvature $\tau^{\perp}$ by

$$\tau^{\perp} = \sum_{\alpha<\beta} |K_{\alpha\beta}|^{\perp}. \tag{20.24}$$

The *scalar normal δ-curvature* $\hat{\delta}^{\perp}(2)$ is defined by

$$\hat{\delta}^{\perp}(2) = \tau^{\perp} - \max |K^{\perp}|. \tag{20.25}$$

The following result was proved in [Sentürk and Verstraelen (2008)].

Theorem 20.38. *Let N be a Kähler hypersurface of the complex projective space $CP^{n+1}(4)$. Then*

$$\hat{\delta}^{\perp}(2) + \delta^c(2) \leq 2(n-1)(4n+3), \tag{20.26}$$

with the equality holding identically if and only if either

(a) *N is a totally geodesic Kähler hypersurface or*
(b) *N is an open portion of a complex quadrics Q^n in $CP^{n+1}(4)$.*

Bibliography

Abe, K. (1971). A characterization of totally geodesic submanifolds in S^N and CP^N by an inequality, *Tohoku Math. J.* **23**, 219–244.

Abe, K. (1972). The complex version of Hartman-Nirenberg cylinder theorem, *J. Differential Geom.* **7**, 453–460.

Ahn, S. S., Kim D. S, and Kim, Y. H. (1996). Totally umbilical Lorentzian submanifolds, *J. Korean Math. Soc.* **33**, 507–512.

Aktan, N., Sarıkaya, M. Z. and Özüsağlam, E. (2008). B.-Y. Chen's inequality for semi-slant submanifolds in T-space forms, *Balkan J. Geom. Appl.* **13**, 1–10.

Alegre, P., Carriazo, A., Kim, Y. H. and Yoon, D. W. (2007). B.-Y. Chen's inequality for submanifolds of generalized space forms, *Indian J. Pure Appl. Math.* **38**, 185–201.

Amari, T. (1985). *Differential Geometrical Methods in Statistics*, (Springer-Verlag, Berlin).

Antić, M. (2007). CR submanifolds of the sphere S^6 and Chen's equality, *Symposium on the Differential Geometry of Submanifolds*, 17–23.

Antić, M. (2010a). 4-dimensional minimal CR submanifolds of the sphere $\mathbf{S}^6$ contained in a totally geodesic sphere $\mathbf{S}^5$, *J. Geom. Phys.* **60**, 96–110.

Antić, M. (2010b). A note on four-dimensional CR submanifolds of the sphere S^6 and Chen's equality, preprint.

Antić, M., Djorić, M. and Vrancken, L. (2007). 4-dimensional minimal CR submanifolds of the sphere S^6 satisfying Chen's equality, *Differential Geom. Appl.* **25**, 290–298.

Arroyo, J., Garay, O. J. and Barros, M. (1999). Willmore-Chen tubes on homogeneous spaces in warped product spaces, *Pacific J. Math.* **188**, 201–207.

Arroyo, J., Garay, O. J. and Barros, M. (2003). Closed free hyperelastic curves in the hyperbolic plane and Chen-Willmore rotational hypersurfaces, *Israel J. Math.* **138**, 171–187.

Arroyo, J.; Garay, O. J. and Mencia, J. J. (1998). On a family of surfaces of revolution of finite Chen-type, *Kodai Math. J.* **21**, 73–80.

Arslan, K., Carriazo, A., Chen, B. Y. and Murathan, C. (2010). On slant submanifolds of neutral Kaehler manifolds, *Taiwanese J. Math.* **14**, 561–584.

Arslan, K., Ezentas, R., Mihai, I., Murathan, C. and Özgür, C. (2001). B. Y. Chen inequalities for submanifolds in locally conformal almost cosymplectic manifolds, *Bull. Inst. Math. Acad. Sinica*, **29**, 231–242.

Arslan, K., Ezentas, R., Murathan, C. and Sasahara, T. (2007). Biharmonic anti-invariant submanifolds in Sasakian space forms, *Beitrge Algebra Geom.* **48**, 191–207.

Arvanitoyeorgos, A., Kaimakamis, G. and Magid, M. (2009). Lorentz hypersurfaces in E_1^4 satisfying $\Delta H = \alpha H$, *Illinois J. Math.* **53** (2009), 581–590.

Arvanitoyeorgos, A., Kaimakamis, G. and Palamourdas, D. (2007). Chen's conjecture and generalizations, *Symposium on the Differential Geometry of Submanifolds*, 109–112.

Bădiţoiu, G. (2004). Semi-Riemannian submersions with totally geodesic fibres, *Tohoku Math. J.* **56**, 179–204.

Bădiţoiu, G. and Ianuş, S. (2002). Semi-Riemannian submersions from real and complex pseudo-hyperbolic spaces, *Differential Geom. Appl.* **16**, 79–94.

Baikoussis, C. and Blair, D. E. (1995). On the geometry of the 7-sphere, *Results Math.* **27**, 5–16.

Balmuş, A. (2004). Biharmonic properties and conformal changes, *An. Stiint. Univ. Al.I. Cuza Iasi Mat. (N.S.)*, **50**, 361–372.

Balmuş, A. and Oniciuc, C. (2005). Some remarks on the biharmonic submanifolds of S^3 and their stability, *An. Stiint. Univ. Al.I. Cuza Iasi, Mat. (N.S)*, **51**, 171–190.

Balmuş, A., Montaldo, S. and Oniciuc, C. (2007). Biharmonic maps between warped product manifolds, *J. Geom. Phys.* **57**, 449–466.

Balmuş, A., Montaldo, S. and Oniciuc, C. (2008). Classification results for biharmonic submanifolds in spheres, *Israel J. Math.* **168**, 201–220.

Barros, M. (1999). Superelasticity and conformal total tension in unified theories. Implications for the Willmore-Chen variational problem in particle physics, *Rev. Acad. Canaria Cienc.* **11** 193–221.

Barros, M. (2000). Willmore-Chen branes and Hopf T-duality, *Classical Quantum Gravity*, **17** (2000), 1979–1988.

Barros, M. (2001). Critical points of Willmore-Chen tension functionals, *Differential Geometry*, Valencia, 72–83.

Barros, M., Cabrerizo, J. L. and Fernández, M. (1999). Reduction of variables for Willmore-Chen submanifolds in seven spheres, *Israel J. Math.* **113**, 29–43.

Barros, M., Ferrández, A., Lucas, P. and Merono, M. A. (2000). A criterion for reduction of variables in the Willmore-Chen variational problem and its applications, *Trans. Amer. Math. Soc.* **352**, 3015–3027.

Barros, M. and Garay, O. J. (1987). 2-type surfaces of S^3, *Geometriae Dedicata*, **24**, 329–336.

Barros, M. and Garay, O. J. (1998). Hopf submanifolds in S^7 which are Willmore-Chen submanfolds, *Math. Z.* **228**, 121–129.

Barros, M. and Romero, A. (1982). Indefinite Kähler manifolds, *Math. Ann.* **261**, 55–62.

Bejan, C.-L. (1988). A classification of the almost para-Hermitian manifolds, *Differential Geometry and its Applications* (Dubrovnik), 23–27.

Bejancu, A. (1978). CR-submanifolds of a Kaehler manifold I, *Proc. Amer. Math. Soc.* **69**, 134–142.

Bejancu, A. (1979). On the geometry of leaves on a CR-submanifold, *Ann. St. Univ. Al. I. Cuza Iasi*, **25**, 393–398.

Bejancu, A. (1986). *Geometry of CR-Submanifolds*, (Reidel Publ, Dordrecht).

Bejancu, A. and Duggal, K. L. (1993). Real hypersurfaces of indefinite Kaehler manifolds, *Int. J. Math. Math. Sci.* **16**, 545–556.

Berger, M. (2003). *A panoramic view of Riemannian geometry*, (Springer-Verlag, Berlin).

Berger, M., Gauduchon, P. and Mazet, E. (1971). *Le Spectre d'une Variété Riemannienne*, Lecture Notes in Mathematics, Vol. 194.

Berndt, J. (1989). Real hypersurfaces with constant principal curvatures in complex hyperbolic space, *J. Reine Angew. Math.* **395**, 132–141.

Berndt, J., Bolton, J. and Woodward, L. M. (1995). Almost complex curves and Hopf hypersurfaces in the nearly Kähler 6-sphere, *Geom. Dedicata*, **56**, 237–247.

Besse, A. L. (1987). *Einstein manifolds*, (Springer-Verlag, Berlin).

Bishop, R. L. and Crittenden, R. J. (1964). *Geometry of Manifolds*, (Academic Press, New York).

Bishop, R. L. and O'Neill, B. (1969). Manifolds of negative curvature, *Trans. Amer. Math. Soc.* **145**, 1–49.

Blair, D. E. (2010). *Riemannian Geometry of Contact and Symplectic Manifolds*, 2nd edition (Birkhäuser Boston, Inc., MA).

Blair, D. E. and Carriazo, A. (2001). The contact Whitney sphere, *Note di Mat.* **20**, 125–133.

Blair, D. E. and Chen, B. Y. (1979). On CR-submanifolds of Hermitian manifold, *Israel J. Math.* **34**, 353–363.

Blair, D. E., Dillen, F., Verstraelen, L. and Vrancken, L. (1996). Calabi curves as holomorphic Legendre curves and Chen's inequality, *Kyungpook Math. J.* **35**, 407–416.

Blair, D. E. and Dragomir, S. (2002). CR products in locally conformal Kähler manifolds, *Kyushu J. Math.* **56**, 337–362.

Blaschke, W. (1929). *Vorlesungen über Differentialgeometrie, vol. III*, (Springer Verlag, Berlin).

Bolton, J., Dillen, F., Fastenakels, J. and Vrancken, L. (2009). A best possible inequality for curvature-like tensor fields, *Math. Inequal. Appl.* **12**, 663–681.

Bolton, J., Rodriguez Montealegre, C. and Vrancken, L. (2009a). Characterizing warped-product Lagrangian immersions in complex projective space, *Proc. Edinb. Math. Soc.* **52**, 273–286.

Bolton, J., Scharlach, C. and Vrancken L. (2002). From surfaces in the 5-sphere to 3-manifolds in complex projective 3-space, *Bull. Austral. Math. Soc.* **66**, 465–475.

Bolton, J. and Vrancken, L. (2005). Ruled minimal Lagrangian submanifolds of complex projective 3-space, *Asian J. Math.* **9**, 45–56.

Bolton, J. and Vrancken, L. (2007). Lagrangian submanifolds attaining equality in the improved Chen's inequality, *Bull. Belg. Math. Soc. Simon Stevin*, **14**, 311–315.

Bolton, J. and Vrancken, L. (2009). Transforms for minimal surfaces in the 5-sphere, *Differential Geom. Appl.* **27**, 34–46.

Bolton, J., Vrancken, L. and Woodward, L. M. (2007). On almost complex curves in the nearly Kähler 6-sphere, *Quart. J. Math.* **45**, 407–427.

Bonnet, O. (1867). Mémoire sur la theéorie des surfaces applicables sur une surface donnée, *École Polytech.* **42**, 31–151.

Borrelli, V., Chen, B. Y. and Morvan, J. M. (1995). Une caractérisation géométrique de la sphère de Whitney, *C.R. Acad. Sci. Paris Sér. I Math.* **321**, 1485–1490.

Bryant, R. L. (1982). Submanifolds and special structure on the octonians, *J. Diff. Geom.*, **17**, 185–232.

Burstin, C. (1931). Ein Betrag zum Problem der Einbettung der Riemannschen Räume in eukildische Räume, *Math Sb.* **38**, 74–85.

Cabrerizo, J. L. (2001). A bridge between Willmore-Chen submanifolds and elastic curves, *Nonlinear Anal.* **47**, 5145–5156.

Cabrerizo, J. L., Carriazo, A., Fernández, L. M. and Fernández, M. (2002). Riemannian submersions and slant submanifolds, *Publ. Math. Debrecen*, **61**, 523–532.

Caddeo, R., Montaldo, S. and Oniciuc, C. (2001). Biharmonic submanifolds of S^3, *Int. J. Math.* **12**, 867–876.

Caddeo, R., Montaldo, S. and Oniciuc, C. (2002). Biharmonic submanifolds in spheres, *Israel J. Math.* **130**, 109–123.

Caddeo, R., Montaldo, S. and Piu, P. (2001). On Biharmonic Maps, *Contemp. Math.* **288**, 286–290.

Calabi, E. (1953). Isometric imbedding of complex manifolds, *Ann. Math.* **58**, 1–23.

Calabi, E. (1958). Construction and properties of some 6-dimensional almost complex manifolds, *Trans. Amer. Math. Soc.* **87**, 407–438.

Calabi, E. and Gluck, H. (1993). What are the best almost complex structures on the 6-sphere in differential Geometry, *Geometry in Mathematical Physics and Relatd Topics*, Amer. Math. Soc. 99–106.

Calvaruso, G. and and Perrone, D. (2010). Contact pseudo-metric manifolds, *Differential Geom. Appl.* **28**, 615–634.

Campbell, J., (1926). *A Course of Differential Geometry*, (Claredon, Oxford).

Carriazo, A. (1999). A contact version of B.-Y. Chen's inequality and its applications to slant immersions, *Kyungpook Math. J.* **39**, 465–476.

Carriazo, A. (2007). Chen's inequalities in generalized space forms, *Pure and Applied Differential Geometry - PADGE*, 32–38.

Carriazo, A., Fernandez, L. M. and Hans-Uber, M. B. (2003). B. Y. Chen's inequality for S-space-forms: applications to slant immersions, *Indian J. Pure Appl. Math.* **34**, 1287–1298.

Cartan, É. (1927). Sur la possibilité de plonger un espace riemannien donné dans un espace euclidien, *Ann. Soc. Math. Polon.* **6**, 1–17.

Cartan, É. (1939). Sur les familles remarquables d'hypersurfaces isoparamétriques dans les espaces sphériques, *Math. Z.* **45**, 335–367.

Casorati, F. (1890). Mesure de la courbure des surfaces suivant l'ind'ee commune, *Acta Math.* **14**, 95–110.

Castro, I. and Urbano, F. (1993). Lagrangian surfaces in the complex Euclidean plane with conformal Maslov form, *Tôhoku Math. J.* **45**, 656–582.
Castro, I. and Urbano, F. (1995). Twistor holomorphic Lagrangian surface in the complex projective and hyperbolic planes, *Ann. Global Anal. Geom.* **13**, 59–67.
Chang, S. (1993). A closed hypersurface with constant scalar and mean curvatures in S4 is isoparametric, *Comm. Anal. Geom.* **1**, 71–100.
Chen, B. Y. (1971). On total curvature of immersed manifolds, I, *Amer. J. Math.* **93**, 148–162.
Chen, B. Y. (1972). On total curvature of immersed manifolds, II: Mean curvature and length of second fundamental form, *Amer. J. Math.* **94**, 899–907.
Chen, B. Y. (1973a). On total curvature of immersed manifolds, III: Surfaces in euclidean 4-space, *Amer. J. Math.* **95**, 636–642.
Chen, B. Y. (1973b). *Geometry of Submanifolds*, (M. Dekker, New York).
Chen, B. Y. (1973c). An invariant of conformal mappings, *Proc. Amer. Math. Soc.* **40**, 563–564.
Chen, B. Y. (1974). Some conformal invariants of submanifolds and their applications, *Boll. Un. Mat. Ital.* **10**, 380–385.
Chen, B. Y. (1975). Mean curvature vector of a submanifold, *Proc. Symp. Pure Math.* Part I, **27**, 119–123.
Chen, B. Y. (1979a). Conformal mappings and first eigenvalue of Laplacian on surfaces, *Bull. Inst. Math. Acad. Sinica*, **7**, 395–400.
Chen, B. Y. (1979b). On the total curvature of immersed manifolds, IV: Spectrum and total mean curvature, *Bull. Inst. Math. Acad. Sinica*, **7**, 301–311.
Chen, B. Y. (1979c). On submanifolds of finite type, *Soochow J. Math.* **9**, 65–81.
Chen, B. Y. (1981a). *Geometry of Submanifolds and Its Applications*, (Science University of Tokyo).
Chen, B. Y. (1981b). *CR*-submanifolds of a Kähler manifold, *J. Differential Geom.* **16**, 305–322; —, II, *ibid*, **16**, 493–509.
Chen, B. Y. (1981c). Cohomology of *CR*-submanifolds, *Ann. Fac. Sc. Toulouse Math.* **3**, 167–172.
Chen, B. Y. (1981d). Differential geometry of real submanifolds in a Kähler manifold, *Monatsh. Math.* **91**, 257–274.
Chen, B. Y. (1981e). On the total curvature of immersed manifolds, V: C-surfaces in Euclidean m-space, *Bull. Inst. Math. Acad. Sinica*, **9**, 509–516.
Chen, B. Y. (1983). On the total curvature of immersed manifolds, VI. Submanifolds of finite type and their applications, *Bull. Inst. Math. Acad. Sinica*, **11**, 309–328.
Chen, B. Y. (1984). *Total Mean Curvature and Submanifolds of Finite Type*, (World Scientific Publ., New Jersey).
Chen, B. Y. (1986). Finite type pseudo-Riemannian submanifolds, *Tamkang J. Math.*, **17**, 137–151.
Chen, B. Y. (1987a). Some estimates of total tension and their applications, *Kodai Math. J.*, **10**, 93–101.
Chen, B. Y. (1987b). Surfaces of finite type in Euclidean 3-space, *Bull. Soc. Math. Belg. Sér. B*, **39**, 243–254.

Chen, B. Y. (1987c). *A new approach to compact symmetric spaces and applications. A report on joint work with Professor T. Nagano*, (Katholieke Universiteit Leuven).
Chen, B. Y. (1988a). Null 2-type surfaces in Euclidean space, *Algebra, Analysis and Geometry*, 1–18 (Proc. Symp. in Honor of C. J. Hsu and K. S. Shih).
Chen, B. Y. (1988b). Null 2-type surfaces in E^3 are circular cylinders, *Kodai Math. J.* **11**, 295–299.
Chen, B. Y. (1990). *Geometry of Slant Submanifolds*, (Katholieke Universiteit Leuven, Belgium).
Chen, B. Y. (1991a). Some open problems and conjectures on submanifolds of finite type, *Soochow J. Math.* **17**, 169–188.
Chen, B. Y. (1991b). Linearly independent, orthogonal, and equivariant immersions, *Kodai Math. J.* **14**, 341–349.
Chen, B. Y. (1991c). Local rigidity theorems of 2-type hypersurfaces in a hypersphere, *Nagoya Math. J.* **122**, 139–148.
Chen, B. Y. (1992). Submanifolds of finite type in hyperbolic spaces, *Chinese J. Math.* **20**, 5–21.
Chen, B. Y. (1993a). Some pinching and classification theorems for minimal submanifolds, *Arch. Math.* **60**, 568–578.
Chen, B. Y. (1993b). A Riemannian invariant for submanfiolds in space forms and its applications, *Gemetry and Topology of Submanifolds*, **VI**, 1–25.
Chen, B. Y. (1994a). Some classification theorems for submanifolds in Minkowski space-time, *Arch. Math.* **62**, 177–182.
Chen, B. Y. (1994b). Tubular hypersurfaces satisfying a basic equality, *Soochow J. Math.* **20**, 569–586.
Chen, B. Y. (1995a). Submanfiolds in de Sitter space-time satisfying $\Delta H = \lambda H$, *Israel J. Math.* **91**, 373–391.
Chen, B. Y. (1995b). A Riemannian invariant and its applications to submanifold theory, *Results Math.* **27**, 17–28.
Chen, B. Y. (1996a). Mean curvature and shape operator of isometric immersions in real-space-form, *Glasgow Math. J.* **38**, 87–97.
Chen, B. Y. (1996b). A general inequality for submanifolds in complex space forms and its applications, *Arch. Math.* **67**, 519–528.
Chen, B. Y. (1996c). A report of submanifolds of finite type, *Soochow J. Math.* **22**, 117–337.
Chen, B. Y. (1996d). Jacobi's elliptic functions and Lagrangian immersions, *Proc. Roy. Soc. Edinburgh Sect. A*, **126**, 687–704.
Chen, B. Y. (1997a). Complex extensors and Lagrangian submanifolds in complex Euclidean spaces, *Tohoku Math. J.* **49**, 277–297.
Chen, B. Y. (1997b). Interaction of Legendre curves and Lagrangian submanifolds, *Israel J. Math.* **99**, 69–108.
Chen, B. Y. (1998a). A vanishing theorem for Lagrangian immersions into Einstein-Kaehler manifolds and its applications, *Soochow J. Math.* **24**, 155–162.
Chen, B. Y. (1998b). Special slant surfaces and a basic inequality, *Results Math.* **33**, 65–78.

Chen, B. Y. (1998c). Strings of Riemannian invariants, inequalities, ideal immersions and their applications, in: *Third Pacific Rim Geom. Conf.* 7–60.

Chen, B. Y. (1999a). Relations between Ricci curvature and shape operator for submanifolds with arbitrary codimensions, *Glasgow Math. J.* **41**, 33–41.

Chen, B. Y. (1999b). New types of Riemannian curvature invariants and their applications, *Geometry and Topology of Submanifolds*, IX, 80–92, World Sci. Publ., River Edge, NJ.

Chen, B. Y. (1999c). On slant surfaces, *Taiwanese J. Math.* **3**, 163–179.

Chen, B. Y. (2000a). Some new obstructions to minimal and Lagrangian isometric immersions, *Japan. J. Math.* **26**, 105–127.

Chen, B. Y. (2000b). On Ricci curvature of isotropic and Lagrangian submanifolds in complex space forms, *Arch. Math.* **74**, 154–160.

Chen, B. Y. (2000c). Riemannian Submanifolds, in: *Handbook of Differential Geometry* (eds. F. Dillen and L. Verstraelen), Vol. I, 187–418.

Chen, B. Y. (2000d). Ideal Lagrangian immersions in complex space forms, *Math. Proc. Cambridge Philos. Soc.* **128**, 511–533.

Chen, B. Y. (2001a). A general inequality for Kählerian slant submanifolds and related results, *Geom. Dedicata*, **85**, 253–271.

Chen, B. Y. (2001b). Geometry of warped product CR-submanifolds in Kähler manifolds, *Monatsh. Math.* **133**, 177–195.

Chen, B. Y. (2001c). Geometry of warped product CR-submanifolds in Kähler manifolds. II, *Monatsh. Math.* **134**, 103–119.

Chen, B. Y. (2001d). A series of Kählerian invariants and their applications to Kählerian geometry, *Beiträge Algebra Geom.* **42**, 165–178.

Chen, B. Y. (2001e). Riemannian geometry of Lagrangian submanifolds, *Taiwanese J. Math.* **5**, 681–723.

Chen, B. Y. (2002a). On isometric minimal immersions from warped products into real space forms, *Proc. Edinburgh Math. Soc.* **45**, 579–587.

Chen, B. Y. (2002b). Ricci curvature of real hypersurfaces in complex hyperbolic space, *Arch. Math. (Brno)*, **38**, 73–80.

Chen, B. Y. (2002c). Geometry of warped products as Riemannian submanifolds and related problems, *Soochow J. Math.* **28**, 125–156.

Chen, B. Y. (2002d). Non-immersion theorems for warped products in complex hyperbolic spaces, *Proc. Japan Acad. Ser. A Math. Sci.* **79**, 96–100.

Chen, B. Y. (2003a). Another general inequality for CR-warped products in complex space forms, *Hokkaido Math. J.* **32**, 415–444.

Chen, B. Y. (2003b). A general optimal inequality for warped products in complex projective spaces and its applications, *Proc. Japan Acad. Ser. A Math.* **79**, 89–94.

Chen, B. Y. (2003c). What can we do with Nash's embedding theorem?, *Soochow J. Math.* **30**, 303–338.

Chen, B. Y. (2004a). Warped products in real space forms, *Rocky Mountain J. Math.* **34**, 551–563.

Chen, B. Y. (2004b). A Riemannian invariant and its application to Einstein manifolds, *Bull. Austral. Math. Soc.* **70**, 55–65.

Chen, B. Y. (2004c). CR-warped products in complex projective spaces with compact holomorphic factor, *Monatsh. Math.* **141**, 177–186.
Chen, B. Y. (2005a). A general inequality for conformally flat submanifolds and its applications, *Acta Math. Hungar.* **106**, 239–252.
Chen, B. Y. (2005b). Riemannian submersions, minimal immersions and cohomology class, *Proc. Japan Acad. Ser. A Math. Sci.* **81**, 162–167.
Chen, B. Y. (2005c). Geometry of affine warped product hypersurfaces, *Results Math.* **48**, 9–28.
Chen, B. Y. (2005d). A general optimal inequality for arbitrary Riemannian submanifolds, *J. Inequal. Pure Appl. Math.* **6**, no. 3, Article 77, 10 pages.
Chen, B. Y. (2005e). Examples and classification of Riemannian submersions satisfying a basic equality, *Bull. Austral. Math. Soc.* **72**, 391–402.
Chen, B. Y. (2005f). First normal bundle of ideal Lagrangian immersions in complex space forms, *Math. Proc. Camb. Phil. Soc.* **138**, 461–464.
Chen, B. Y. (2006). Eigenvalues of a natural operator of centro-affine and graph hypersurfaces, *Beiträge Algebra Geom.* **47**, 15–27.
Chen, B. Y. (2007a). Realization of Robertson-Walker spacetimes as affine hypersurfaces, *J. Phys. A*, **40**, 4241–4250.
Chen, B. Y. (2007b). Riemannian submersions with minimal fibers and affine hypersurfaces, *Monatsh. Math.* **151**, 143–152.
Chen, B. Y. (2008a). δ-invariants, inequalities of submanifolds and their applications, in *Topics in differential geometry*, 29–155 (Ed. Acad. Române, Bucharest).
Chen, B. Y. (2008b). Classification of marginally trapped Lorentzian flat surfaces in $\mathbb{E}^4_2$ and its application to biharmonic surfaces, *J. Math. Anal. Appl.* **340**, 861–875.
Chen, B. Y. (2009a). Dependence of the Gauss-Codazzi equations and the Ricci equation of Lorentz surfaces, *Publ. Math. Debrecen*, **74**, 341–349.
Chen, B. Y. (2009b). Classification of spatial surfaces with parallel mean curvature vector in pseudo-Euclidean spaces with arbitrary codimension, *J. Math. Phys.* **50**, 043503, 14 pages.
Chen, B. Y. (2009c). Complete classification of spatial surfaces with parallel mean curvature vector in arbitrary non-flat pseudo-Riemannian space forms, *Cent. Eur. J. Math.* **7**, 400–428.
Chen, B. Y. (2009d). Black holes, marginally trapped surfaces and quasi-minimal surfaces, *Tamkang J. Math.* **40**, 313–341.
Chen, B. Y. (2009e). Classification of spatial surfaces with parallel mean curvature vector in pseudo-Euclidean spaces with arbitrary codimension, *J. Math. Phys.* **50** (2009), 043503, 14 pages.
Chen, B. Y. (2009f). Classification of marginally trapped surfaces of constant curvature in Lorentzian complex plane, *Hokkaido Math. J.* **38**, 361–408.
Chen, B. Y. (2009g). Marginally trapped surfaces and Kaluza-Klein theory, *Int. Electron. J. Geom.* **2**, 1–16.
Chen, B. Y. (2010a). Complete classification of parallel spatial surfaces in pseudo-Riemannian space forms with arbitrary index and dimension, *J. Geom. Phys.* **60**, 260–280.

Chen, B. Y. (2010b). Complete explicit classification of parallel Lorentz surfaces in arbitrary pseudo-Euclidean spaces, *J. Geom. Phys.* **60**, 1333–1351.

Chen, B. Y. (2010c). Complete classification of Lorentz surfaces with parallel mean curvature vector in arbitrary pseudo-Euclidean space, *Kyushu J. Math.* **64**, 261–279.

Chen, B. Y. (2010d). Complete classification of parallel Lorentz surfaces in neutral pseudo hyperbolic 4-space, *Cent. Eur. J. Math.* **8**, 706–734.

Chen, B. Y. (2010e). Complete classification of parallel Lorentz surfaces in 4D neutral pseudo-sphere, *J. Math. Phys.* **51**, no. 8, 083518, 22 pages.

Chen, B. Y. (2010f). On purely real surfaces in Kaehler surfaces, *Turk. J. Math.* **34**, 275–292.

Chen, B. Y. (2010g). Explicit classification of parallel Lorentz surfaces in 4D indefinite space forms with index 3, *Bull. Inst. Math. Acad. Sinica (N.S.)*, **5**, 311–345.

Chen, B. Y. (2010h). Lagrangian submanifolds in para-Kähler manifolds, *Nonlinear Analysis*, **73**, 3561–3571.

Chen, B. Y. (2010i). Lagrangian H-umbiilical submanifolds of para-Kähler manifolds, *Taiwanese J. Math.* (in press).

Chen, B. Y. (2010j). Submanifolds with parallel mean curvature vector in Riemannian and indefinite space forms, *Arab J. Math. Sci.* **16**, 1–45.

Chen, B. Y. (2010k). Classification of Wintgen ideal surfaces in Euclidean 4-space with equal Gauss and normal curvatures, *Ann. Global Anal. Geom.* **38**, 145–160.

Chen, B. Y. (2011a). Classification of minimal Lorentz surfaces in indefinite space forms with arbitrary codimension and arbitrary index, *Publ. Math. Debrecen*, **78**, 485–503.

Chen, B. Y. (2011b). Classification of flat Lagrangian H-umbilical submanifolds in para-Kähler n-plane, *Int. Electron. J. Geom.* **4**, (in press).

Chen, B. Y., Deprez, J. and Verheyen, P. (1987). Immersions, dans un espace euclidien, d'un espace symétrique compact de rang un à géodésiques simples, *C. R. Acad. Sci. Paris Sér. I Math.* **304**, 567–570.

Chen, B. Y. and Dillen, F. (2007). Classification of marginally trapped Lagrangian surfaces in Lorentzian complex space forms, *J. Math. Phys.* **48**, no. 1, 013509, 23 pp; Erratum, *J. Math. Phys.* **49** (2008), no. 5, 059901.

Chen, B. Y. and Dillen, F. (2008). Optimal inequalities for multiply warped product, *Int. Electron. J. Geom.* **1**, 1–11. Erratum, *ibid,* **4** (2011).

Chen, B. Y. and Dillen, F. (2010). Optimal general inequalities for Lagrangian submanifolds in complex space forms, *J. Math. Anal. Appl.* (in press).

Chen, B. Y., Dillen, D., Fastenakels, J. and Verstraelen, L. (2004). Conformally flat hypersurfaces in real space forms with least tension, *Taiwanese J. Math.* **8**, 285–325.

Chen, B. Y., Dillen, F. and Van der Veken, J. (2010). Complete classification of parallel Lorentzian surfaces in Lorentzian complex space forms, *Intern. J. Math.* **21**, 665–686.

Chen, B. Y., Dillen, D. and Verstraelen, L. (2005). δ-invariants and their applications to centroaffine geometry, *Differential Geom. Appl.* **22**, 341–354.

Chen, B. Y., Dillen, D., Verstraelen, L. and Vrancken, L. (1993). A variational minimal principle characterizes submanifolds of finite type, *C.R. Acad. Sc. Paris*, **317**, 961–965
Chen, B. Y., Dillen, D., Verstraelen, L. and Vrancken, L. (1994). Totally real submanifolds of CP^n satisfying a basic equality, *Arch. Math.* **63**, 553–564.
Chen, B. Y., Dillen, D., Verstraelen, L. and Vrancken, L. (1995a). Characterizing a class of totally real submanifolds of S^6 by their sectional curvatures, *Tohoku Math. J.* **47**, 185–198.
Chen, B. Y., Dillen, D., Verstraelen, L. and Vrancken, L. (1995b). Two equivariant totally real immersions into the nearly Kähler 6-sphere and their characterization, *Japan. J. Math. (N.S.)*, **21**, 207–222.
Chen, B. Y., Dillen, D., Verstraelen, L. and Vrancken, L. (1996). An exotic totally real minimal immersion of S^3 in CP^3 and its characterization, *Proc. Roy. Soc. Edinburgh Sec. A,* **126**, 153–165.
Chen, B. Y., Dillen, D., Verstraelen, L. and Vrancken, L. (2000). Characterizations of Riemannian space forms, Einstein spaces and conformally flat spaces, *Proc. Amer. Math. Soc.* **128**, 589–598.
Chen, B. Y. and Garay, O. J. (2003). Constant mean curvature hypersurfaces with constant δ-invariant, *Int. J. Math. Math. Sci.* 2003, no. 67, 4205–4216.
Chen, B. Y. and Garay, O. J. (2006). An extremal class of conformally flat submanifolds in Euclidean spaces, *Acta Math. Hungar.* **111**, 263–303.
Chen, B. Y. and Garay, O. J. (2009). Complete classification of quasi-minimal surfaces with parallel mean curvature vector in neutral pseudo-Euclidean 4-space $\mathbb{E}^4_2$, *Results. Math.* **55**, 23–38.
Chen, B. Y. and Ishikawa, S. (1991). Biharmonic surfaces in pseudo-Euclidean spaces, *Memoirs Fac. Sci. Kyushu Univ. Ser. A, Math.* **45**, 323–347.
Chen, B. Y. and Ishikawa, S. (1998). Biharmonic pseduo-Riemannian submanifolds in pseudo-Euclidean spaces, *Kyushu J. Math.* **52**, 1–18.
Chen, B. Y. and Jiang, S. (1995). Inequalities between volume, center of mass, circumscribed radius, order and mean curvature, *Bull. Belg. Math. Soc.* **2**, 75–85.
Chen, B. Y. and Kuan, W. E. (1981). The Segre imbedding and its converse, *Ann. Fac. Sc. Toulouse Math.* **7**, 1–28.
Chen, B. Y., Ludden D. and Montiel, S. (1984). Real submanifolds of a Kähler manifold, *Algebra, Groups Geom.* **1**, 176–212.
Chen, B. Y. and Lue, H.. S. (1988). Some 2-type submanifolds and applications, *Ann. Fac. Sc. Toulouse*, Sér. V, **9**, 121–131.
Chen, B. Y. and Martín-Molina, V. (2010). Optimal inequalities, contact δ-invariants and their applications, preprint.
Chen, B. Y. and Mihai, I. (2005). Isometric immersions of contact Riemannian manifolds in real space forms, *Houston J. Math.* **31**, 743–764
Chen, B. Y. and Mihai, I. (2009). Classification of quasi-minimal slant surfaces in Lorentzian complex space forms, *Acta Math. Hungar.* **122**, 307–328.
Chen, B. Y. and Nagano, T. (1977). Totally geodesic submanifolds of symmetric spaces. I, *Duke Math. J.* **44**, 745–755.

Chen, B. Y. and Nagano, T. (1978). Totally geodesic submanifolds of symmetric spaces. II, *Duke Math. J.* **45**, 405–425.

Chen, B. Y. and Nagano, T. (1988). A Riemannian geometric invariant and its applications to a problem of Borel and Serre, *Trans. Amer. Math. Soc.* **308**, 273–297.

Chen, B. Y. and Ogiue, K. (1974a). On totally real submanifolds, *Trans. Amer. Math. Soc.* **193**, 257–266.

Chen, B. Y. and Ogiue, K. (1974b). Two theorems on Kähler manifolds, *Michigan Math. J.* **21**, 257–266.

Chen, B. Y. and Petrovic, M. (1991). On spectral decomposition of immersions of finite type, *Bull. Austral. Math. Soc.* **44**, 117–129.

Chen, B. Y. and Piccinni, P. (1987). Submanifolds with finite type Gauss map, *Bull. Austral. Math. Soc.* **44**, 161–186.

Chen, B. Y. and Song, H.-Z. (1989a). Null 2–type surfaces in Minkowski space-time,*Algebras, Groups and Geometries*, **6**, 333–352.

Chen, B. Y. and Song, H.-Z. (1989b). Null 2-type surfaces in Minkowski space-time. II, *Atti Accad. Peloritana Pericolanti Cl. Sci. Fis. Mat. Natur.* **67**, 421–432.

Chen, B. Y. and Tazawa, Y. (1991). Slant submanifolds in complex Euclidean spaces, *Tokyo J. Math.* **14**, 101–120.

Chen, B. Y. and Tazawa, Y. (2000). Slant submanifolds of complex projective and complex hyperbolic spaces, *Glasgow Math. J.* **42**, 439–454.

Chen, B. Y. and Van der Veken, J. (2007a). Marginally trapped surfaces in Lorentzian space forms with positive relative nullity, *Classical Quantum Gravity*, **24** (2007), 551–563.

Chen, B. Y. and Van der Veken, J. (2007b). Spatial and Lorentzian surfaces in Robertson-Walker space times, *J. Math. Phys.* **48**, 073509, 12 pages.

Chen, B. Y. and Van der Veken, J. (2009). Complete classification of parallel surfaces in 4-dimensional Lorentzian space forms, *Tohoku Math. J.* **61**, 1–40.

Chen, B. Y. and Van der Veken, J. (2010). Classification of marginally trapped surfaces with parallel mean curvature vector in Lorentzian space forms, *Houston J. Math.* **36**, 421–449.

Chen, B. Y. and Vrancken, L. (1996). Lagrangian submanifolds satisfying a basic equality, *Math. Proc. Cambridge Philos. Soc.* **120**, 291–307.

Chen, B. Y. and Vrancken, L. (1999). CR-submanifolds of complex hyperbolic spaces satisfying a basic equality, *Israel J. Math.* **110**, 341–358.

Chen, B. Y. and Vrancken, L. (2002). Lagrangian minimal isometric immersions of a Lorentzian real space form into a Lorentzian complex space form, *Tohoku Math. J.* **54**, 121–143..

Chen, B. Y. and Vrancken, L. (2002). Lagrangian submanifolds of the complex hyperbolic space, *Tsukuba J. Math.* **26**, 95–118.

Chen, B. Y. and Wei, S. W. (2008). Differential geometry of submanifolds of warped product manifolds $I \times_f S^{m-1}(k)$, *J. Geom.* **91**, 21–42.

Chen, B. Y. and Wei, S. W. (2009). Growth estimates for warping functions and their geometric applications, *Glasgow Math. J.* **51**, 579–592.

Chen, B. Y. and Yang, D. (2010). Addendum to "Classification of marginally trapped Lorentzian flat surfaces in $\mathbb{E}^4_2$ and its application to biharmonic surfaces", *J. Math. Anal. Appl.* **361**, 280–282.

Chen, B. Y. and Yang, J. (1999). Elliptic functions, theta function and hypersurfaces satisfying a basic equality, *Math. Proc. Cambridge Philos. Soc.* **125**, 463–509; Corrigendum, *ibid,* **130** (2001), 189–191.

Chern, S. S. (1968). *Minimal Submanifolds in a Riemannian Manifold,* (University of Kansas).

Chern, S. S. and Lashof, R. K. (1958). On the total curvature of immersed manifold, II, *Michigan Math. J.* **5**, 5–12.

Chervon, S., Dahia, F. and Romero, C. (2004). Harmonic maps and isometric embeddings of the spacetime, *Phys. Lett. A*, **326**, 171–177.

Choi, T. and Lu, Z. (2008). On the DDVV Conjecture and the Comass in Calibrated Geometry (I), *Math. Z.* **260**, 409–429.

Cioroboiu, D. (2003a). B. Y. Chen inequalities for bi-slant submanifolds in Sasakian space forms, *Demonstratio Math.* **36**, 179–187.

Cioroboiu, D. (2003b). B.-Y. Chen inequalities for semislant submanifolds in Sasakian space forms, *Int. J. Math. Math. Sci.* **2003** no. 27, 1731–1738.

Cioroboiu, D. and Oiagă, A. (2003). B. Y. Chen inequalities for slant submanifolds in Sasakian space forms, *Rend. Circ. Mat. Palermo*, **52**, 367–381.

Clarke, C. J. S. (1970). On the global isometric embedding of pseudo-Riemannian manifolds, *Proc. Roy. Soc. London, A*, **314**, 417–428.

Codazzi, D. (1868). Sulle coordinate curvilinee, *Ann. di Mat. Pura Appl.* **2**, 1101–1119.

Costache, S. (2008). B.-Y. Chen inequalities for slant submanifolds in Kenmosu space forms, *Bull. Transilv. Univ. Brasov* Ser. III, **1(50)**, 87–92.

Costache, S. (2010). B.-Y. Chen inequalities for slant submanifolds in Kenmosu space forms, II, *Sarajevo J. Math.* **6(18)**, 125–135.

Cruceanu, V., Fortuny, P. and Gadea, P. M. (1966). A survey on paracomplex geometry, *Rocky Mountain J. Math.* **26**, 83–115.

Dajczer, M. and Florit, L. A. (1998). On Chen's basic equality, *Illinois J. Math.* **42**, 97–106.

Dajczer, M. and Florit, L. A. (2001). A class of austere submanifolds, *Illinois J. Math.* **45**, 735–755.

Decu, S., Jahanara, B., Petrović-Torgašev, M. and Verstraelen, L. (2009). On the Chen character of $\delta(2)$-ideal submanifolds, *Kragujevac J. Math.* **32**, 37–46.

Decu, S., Petrović-Torgašev, M., Šebeković, A. and Verstraelen, L. (2010). On the intrinsic Deszcz symmetries and the extrinsic Chen character of Wintgen ideal submanifolds, *Tamkang J. Math.* **41**, 109–116.

Defever, F. (1998), Hypersurfaces of $\mathbb{E}^4$ with harmonic mean curvature vector, *Math. Nachr.* **196**, 61–69.

Defever, F., Mihai, I. and Verstraelen, L. (1997). B.-Y. Chen's inequality for C-totally real submanifolds in Sasakian space forms, *Boll. Un. Mat. Ital. Ser. B,* **11**, 365–374.

Defever, F., Mihai, I. and Verstraelen, L. (2001). B.-Y. Chen's inequalities for submanifolds of Sasakian space forms, *Boll. Unione Mat. Ital. Sez. B Artic. Ric. Mat.* (8) **4**, 521–529.

Deng, S. (2009). An improved Chen-Ricci inequality, *Int. Electron. J. Geom.* **2**, no. 2, 39–45.

Deprez, J. (1988). *Immersions of finite type of compact homogeneous Riemannian manifolds*, (Ph.D. Thesis, Katholieke Universiteit Leuven).

De Smet, P. J., Dillen, F., Verstraelen, L. and Vrancken, L. (1998). The normal curvature of totally real submanifolds of $S^6(1)$, *Glasgow Math. J.* **40**, 199–204.

De Smet, P. J., Dillen, F., Verstraelen, L. and Vrancken, L. (1999). A pointwise inequality in submanifold theory, *Arch. Math. (Brno)*, **35**, 115–128.

Deszcz, R., Dillen, F., Verstraelen L. and Vrancken, L. (1999). Quasi-Einstein totally real submanifolds of $S^6(1)$, *Tohoku Math. J.* **51**, 461–478.

Deszcz, R., Glogowska, M. Petrović-Torgasev, M., Verstraelen L. (2010). On the Roter type of Chen ideal submanifolds, *preprint.*

Deszcz, R., Petrović-Torgašev, M., Verstraelen L. and Zafindratafa, G. (2008). On the intrinsic symmetries of Chen ideal submanifolds, *Bull. Transilv. Univ. Brasov* Ser. III, **1(50)**, 99–108.

Deszcz, R., Verstraelen, L. and Yaprak, S. (1998). On 2-quasi-umbilical hypersurfaces in conformally flat spaces, *Acta Math. Hungar.* **78**, 45–57.

Dillen, F. and Fastenakels, J. (2009). On an inequality of Oprea for Lagrangian submanifolds, *Cent. Eur. J. Math.* **7**, 140–144.

Dillen, F., Fastenakels, J. and Van der Veken, J. (2007a). A pinching theorem for the normal scalar curvature of invariant submanifolds, *J. Geom. Phys.* **57**, 833–840.

Dillen, F., Haesen, S., Petrović-Torgašev, M. and Verstraelen, L. (2004). An inequality between intrinsic and extrinsic scalar curvature invariants for codimension 2 embeddings, *J. Geom. Phys.* **52**, 101–112.

Dillen, F., Nomizu, K. and Verstraelen, L. (1990). Conjugate connections and Radons theorem in affine differential geometry, *Monatsh. Math.* **109**, 221–235.

Dillen, F., Petrović, M. and Verstraelen, L. (1997). Einstein, conformally flat and semi-symmetric submanifolds satisfying Chen's equality, *Israel J. Math.* **100**, 163–169.

Dillen, F., Verstraelen, L. and Vrancken, L. (1990). Classification of totally real 3-dimensional submanifolds of $S^6(1)$ with $K \geq 1/16$, *J. Math. Soc. Japan*, **42**, 565–584.

Dillen, F. and Vrancken, L. (1994). Calabi-type composition of affine spheres, *Differential Geom. Appl.* **4**, 303–328.

Dillen, F. and Vrancken, L. (1996). Totally real submanifolds in $S^6(1)$ satisfying Chen's equality, *Trans. Amer. Math. Soc.* **348**, 1633–1646.

Dillen, F. and Vrancken, L. (2005). Improper affine spheres and δ-invariants, *PDEs, Submanifolds and Affine Differential Geometry*, 157–162, Banach Center Publ., 69, Polish Acad. Sci., Warsaw.

Dimitric, I. (1989). *Quadric representation and submanifold of finite type*, (Doctoral thesis, Michigan State University).

Dimitric, I. (1992). Submanifolds of $\mathbb{E}^m$ with harmonic mean curvature vector, *Bull. Inst. Math. Acad. Sinica*, **20**, 53–65.

Dimitric, I. (1997). 1-type submanifolds of non-Euclidean complex space forms, *Bull. Belg. Math. Soc. Simon Stevin*, **4**, 673–684.

Dimitric, I. (2000). CR-submanifolds of HP^m and hypersurfaces of the Cayley plane whose Chen-type is 1, *Kyungpook Math. J.* **40**, 407–429.

Dimitric, I. (2009). Low-type submanifolds of real space forms via the immersions by projectors, *Differential Geom. Appl.* **27**, 507–526.

Djorić, M. and Vrancken, L. (2006). Three-dimensional minimal CR submanifolds in S^6 satisfying Chen's equality, *J. Geom. Phys.* **56**, 2279–2288.

Djorić, M. and Vrancken, L. (2010). Geometric conditions in three dimensional CR submanifolds in S^6, *Adv. Geom.* **10**, 185–196.

Dobarro, F. and Ünal, B. (2005). Curvature of multiply warped products, *J. Geom. Phys.* **55**, 75–106.

do Carmo, M. and Dajczer, M (1983). Rotaion hypersurfaces in spaces of constant curvature, *Trans. Amer. Math. Soc.* **277**, 685–709.

Dragomir, S. and Ornea, L. (1998). *Locally Conformal Kähler Geometry*, (Birkhäuser, Boston-Basel-Berlin).

Eells, J. and Sampson, J. H. (1964). Harmonic mappings of Riemannian manifolds, *Amer. J. Math.* **86**, 109–160.

Einstein, A. (1905). Zur Elektrodynamik bewegter Körper, *Ann. Phys.* **17**, 891–921.

Einstein, A. (1915). Zur allgemeinen Relativittstheorie, *Preuß. Akad. Wissen. Sitz.* Part **2**, 778–786; 799–801.

Einstein, A. (1916). Grundlage der allgemeinen Relativittstheorie, *Ann, Phys.* **49**, 769–822.

Einstein, A. (1939). On a stationary system with spherical symmetry consisting of many gravitating masses, *Ann. of Math.* **40**, 922–936.

Einstein, A. and Laub, J. (1908). Elektromagnetische Grundgleichungen für bewegte Körper, *Ann. Phys.* **26**, 532–540.

Ejiri, N. (1979). Minimal immersions of Riemannian products into real space forms, *Tokyo J. Math.* **2**, 63–70.

Ejiri, N. (1981). Totally real submanifolds in a 6-sphere, *Proc. Amer. Math. Soc.* **83**, 759–763.

Ejiri, N. (1982). Totally real minimal immersions of n-dimensional totally real space forms into n-dimensional complex space forms, *Proc. Amer. Math. Soc.* **84**, 243–246.

Erbacher, J. (1971). Reduction of the codimension of an isometric immersion, *J. Differential Geom.* **5**, 333–340.

Eschenburg, J.-H. and Tribuzy, R. (1993). Existence and uniqueness of maps into affine homogeneous spaces, *Rend. Sem. Mat. Univ. Padova*, **89**, 11–18.

Escobales, R. (1975). Riemannian submersions with totally geodesic fibers, *J. Differential Geom.* **10**, 253–276.

Escobales, R. (1978). Riemannian submersions from complex projective spaces, *J. Differential Geom.* **13**, 93–107.

Etayo, F., Santamaría, R. and Trías, U. (2006). The geometry of a bi-Lagrangian manifold, *Differential Geom. Appl.* **24**, 33–59.

Falcitelli, M., Ianus, S. and Pastore, A. M. (2004). *Riemannian Submersions and Related Topics*, (World Scientific, New Jersey).

Fastenakels, J. (2006). Ideal tubular hypersurfaces in real space forms, *Arch. Math. (Brno)*, **42**, 295–305.

Ferus, D. (1974). Immersions with parallel second fundamental form, *Math. Z.* **140**, 87–93.

Fetcu, D. and Oniciuc, C. (2009a). On the geometry of biharmonic submanifolds in Sasakian space forms, *J. Geom. Symmetry Phys.* **14**, 21–34.

Fetcu, D. and Oniciuc, C. (2009b). Explicit formulas for biharmonic submanifolds in Sasakian space forms, *Pacific J. Math*, **240**, 85–107.

Freudenthal, H. (1954). Zur Berechnung der Charaktere der halbeinfachen Lieschen Gruppen. I, II. *Indag. Math.* **16**, 369–376 & 487–491.

Fu, Y. and Hou, Z. H. (2010). Classification of Lorentzian surfaces with parallel mean curvature vector in pseudo-Euclidean spaces, *J. Math. Anal. Appl.* **371**, 25–40.

Funabashi, S., Kim, Y. M. and Pak, J. S. (2003). On submanifolds of $L \times_f F$ satisfying Chen's basic equality, *Acta Math. Hungr.* **99**, 189–201.

Gadea, P. M. and Montesinos Amilibia, A. (1989). Spaces of constant paraholomorphic sectional curvature, *Pacific J. Math.* **136**, 85–101.

Garay, O. J. (1988). Spherical Chen surfaces which are mass-symmetric and of 2-type. *J. Geom.* **33**, 39–52.

Garay, O. J. (1990). An extension of Takahashi's theorem, *Geom. Dedicata*, **34**, 105–112.

Garay, O. J. (1994). A classification of certain 3-dimensional conformally flat Euclidean hypersurfaces, *Pacific J. Math.* **162**, 13–25.

Gauss, C. F. (1827). Disquisitiones generales circa superficies curvas, *Comment. Soc. Sci. Gotting. Recent. Classis Math.* **6**.

Gheysens, L., Verheyen, P. and Verstraelen, L. (1983). Characterization and examples of Chen submanifolds, *J. Geom.* **20**, 47–62.

Goenner, H. (2004). On the History of Unified Field Theories, Living Rev. Relativity **7** no. 2. (http://www.livingreviews.org/llr-2004-2).

Graves, J. T. (1845). On a connection between the general theory of normal couples and the theory of complete quadratic functions of two variables, *Phil. Magaz.* **26**, 315–320.

Graves, L. K. (1979a). On codimension one isometric immersions between indefinite space forms, *Tsukuba J. Math.* **3**, 17–29.

Graves, L. K. (1979b). Codimension one isometric immersions between Lorentz spaces, *Trans. Amer. Math. Soc.* **252**, 367–392.

Gray, A. (1967). Pseudo-Riemannian almost product manifolds and submersions, *J. Math. Mech.* **16**, 715–737.

Gray, A. (1969). Almost complex submanifolds of the six sphere, *Proc. Amer. Math. Soc.* **20**, 277–279.

Gray, A. (1970). Nearly Kähler manifolds, *J. Differential Geometry*, **4**, 283–309.

Greene, R. (1970). Isometric embeddings of Riemannian and pseudo-Riemannian manifolds, *Mem. Amer. Math. Soc.*, No. **97**, 1–63.

Greenfield, S. (1968). Cauchy-Riemann equations in several variable, *Ann. Sc. Norm. Sup. Pisa*, **22**, 275–314.

Gromoll, D. and Grove, K. (1988). The low-dimensional metric foliations of Euclidean spheres, *J. Differential Geom.* **28**, 143–156.

Gromov, M. L. (1971). A topological technique for the construction of solutions of differential equations and inequalities, *Intern. Congr. Math. (Nice 1970)*, **2**, 221–225.

Gromov, M. L. (1985). Isometric immersions of Riemannian manifolds, in: *Elie Cartan et les Mathématiques d'Aujourd'hui*, Astérisque, 129–133.

Gromov, M. L. and Rokhlin, V. A. (1970). Embeddings and immersions in Riemannian geometry, *Russian Math. Surveys*, **25**, 1–57.

Gupta, R. S., Ahmad, I. and Khursheed Haider, S. M. (2004). B. Y. Chen's inequality and its application to slant immersions into Kenmostu manifolds, *Kyungpook Math. J.* **44**, 101–110.

Haesen, S. (2005). Optimal inequalities for embedded space-times, *Kragujevac J. Math.* **28**, 69–85.

Haesen, S., Kowalczyk, D. and Verstraelen, L. (2008). On the extrinsic principal direction of Riemannian submanifolds, *Note di Mat.* **29**, 41–53.

Haesen, S. and Ortega, M. (2007). Boost invariant marginally trapped surfaces in Minkowski 4-space, *Classical Quantum Gravity*, **24**, 5441–5452.

Haesen, S. and Verstraelen, L. (2004). Ideally embedded space-times, *J. Math. Phys.* **45**, 1497–1510.

Harvey, R. and Lawson, H. B. (1982). Calibrated foliation, *Amer. J. Math.* **104**, 607–633.

Hasanis, T. and Vlachos, T. (1991). Spherical 2-type hypersurfaces, *J. Geom.* **40**, 82–94.

Hasanis, T. and Vlachos, T. (1995). Hypersurfaces in E^4 with harmonic mean curvature vector field, *Math. Nachr.* **172**, 145–169

Hasegawa, I. and Mihai, I. (2003). Contact CR-warped product submanifolds in Sasakian manifolds, *Geom. Dedicata*, **102**, 143–150.

Hawking, S. W. and Ellis, G. F. R. (1973). *The Large Scale Structure of Space-time*, (Cambridge University Press).

Hermann, R. (1960). A sufficient condition that a mapping of Riemannian manifolds be fibre bundle, *Proc.Amer. Math. Soc.* **11**, 236–242.

Hertrich-Jeromin, U. and Pinkall, U. (1992). Ein Beweis der Willmoreschen Vermutung fur Kanaltori, *J. Reine Angew. Math.* **430**, 21–34

Hiepko, S. (1979). Eine innere Kennzeichnung der verzerrten Produkte, *Math. Ann.* **241**, 209-215.

Hinić, A. (2007). Some curvature conditions of the type 2×4 on the submanifolds satisfying Chen's equality, *Mat. Vesnik*, **59**, 189–196.

Hinić, A. (2008). Symmetries on Chen ideal submanifolds, *Bull. Transilv. Univ. Brasov*, Ser. III **1(50)**, 181–189.

Hong, Y. and Houh, C. S. (1998). Lagrangian submanifolds of quaternion Kählerian manifolds satisfying Chen's equality, *Beiträge Algebra Geom.* **39**, 413–421.

Hou, Z., Deng, S. and Kaneyuki, S. (1997). Dipolarizations in compact Lie algebras and homogeneous para-Kähler manifold, *Tokyo J. Math.* **20**, 381–388.

Ikawa, T., Nakagawa, H. and Romero, A. (1988). Product complex submanifold of indefinite complex space forms, *Rocky Mountain J. Math.* **18**, 601–615.

Inoguchi, J. (2004). Submanifolds with harmonic mean curvature in contact 3-manifolds, *Colloq. Math.* **100**, 163–179.

Ishikawa, S. (1992). *Classification Problems of Finite Type Submanifolds and Biharmonic Submanifolds*, (Doctoral Thesis, Kyushu Uiversity, Japan).

Janet, M. (1926). Sur la possibilité de plonger un espace riemannien donné dans un espace euclidien, *Ann. Soc. Math. Polon.* **5**, 38–43.

Jiang, G. Y. (1986). 2-harmonic maps and their first and second variational formulas, *Chinese Ann. Math. Ser. A*, **7**, 389–402.

Jordan, C. (1874). Généralisation du théoréma d'Euler sur la courbure des surfaces, *C.R. Acad. Sc.* **79**, 909–912.

Kaluza, T. (1921). Zum Unitatsproblem der Physik, *Sitz. Preuss. Akad. der Wiss. Phys. Math. Berlin*, **96**, 966–972.

Kath, I. (2008). Indefinite extrinsic symmetric spaces I, arXiv:0809.4713.

Kath, I. (2010). Indefinite extrinsic symmetric spaces II, arXiv:1004.1786.

Kim, J. K., Song, Y. M. and Tripathi, M. M. (2003a). B.-Y. Chen inequalities for submanifolds in generalized complex space forms, *Bull. Korean Math. Soc.* **40**, 411–423.

Kim, Y. H. and Kim, D. S. (1999). A basic inequality for submanifolds in Sasakian space forms, *Houston J. Math.* **25**, 247–257.

Kim, Y. M. (2003). Chen's basic equalities for submanifolds of Sasakian space form, *Kyungpook Math. J.* **43**, 63–71.

Kim, Y. M. and Pak, J. S. (2002). On the Ricci curvature of submanifolds in the warped product $L \times_f F$, *J. Korean Math. Soc.* **39**, 693–708.

Kimura, M. (1986). Real hypersurfaces and complex submanifolds in complex projective space, *Trans. Amer. Math. Soc.* **296**, 137–149.

Klein, O. (1926). Quantentheorie und fnfdimensionale Relativittstheorie, *Zeits. Phys.* **37**, 895–906.

Kobayashi, S. and Nomizu, K. (1969). *Fondations of Differential Geometry*, I, II, (J. Wiley, New York).

Kriele, M., Scharlach, C. and Vrancken, L. (2001). An extremal class of 3-dimensional elliptic affine spheres, *Hokkaido Math. J.* **30**, 1–23.

Kriele, M. and Vrancken, L. (1999). An extremal class of three-dimensional hyperbolic affine spheres, *Geom. Dedicata*, **77**, 239–252.

Kruchkovich, G. I. (1957). On semi-reducible Riemannian spaces (in Russian), *Dokl. Akad. Nauk SSSR*, **115**, 862–865.

Kulkarni, R. S. (1969). Curvature structures and conformal transformations, *Bull. Amer. Math. Soc.* **75**, 91–94.

Kulkarni, R. S. (1979). The values of sectional curvature in indefinite metrics, *Comm. Math. Helv.* **54**, 173–176.

Kusner, R. (1989). Comparison surfaces for the Willmore problem, *Pacific J. Math.* **138**, 317–345.

Kwon, J. H. and Suh, Y. H. (1997). On sectional and Ricci curvatures of semi-Riemannian submersions, *Kodai Math. J.* **20**, 53–66.

Larsen, J. C. (1996). Complex analysis, maximal immersions and metric singularities, *Monatsh. Math.* **122**, 105–156.
Lawson, H. B. (1970a). Complete minimal surfaces in S^3, *Ann. Math.* **92**, 335–374.
Lawson, H. B. (1970b). Rigidity theorems in rank-1 symmetric spaces, *J. Differential Geom.* **4**, 349–357.
Levi-Civita, T. (1937). Famiglie di superficie isoparametriche nell'orinario spazio euclideo, *Atti Accad. Naz. Lincei Rend. Ser. VI,* **26**, 355–362.
Li, A. M., Simon, U. and Zhao, G. S. (1993). *Global affine differential geometry of hypersurfaces*, (Walter de Gruyter & Co., Berlin).
Li, G. (2001). Semi-parallel, semi-symmetric immersions and Chen's equality, *Results Math.* **40**, 257–264.
Li, G. (2002). Ideal Einstein, conformally flat and semi-symmetric immersions, *Israel J. Math.* **132**, 207–220.
Li, G. and Wu, C. (2005). Slant immersions of complex space forms and Chen's inequality, *Acta Math. Sci. Ser. B Engl. Ed.* **25**, 223–232.
Li, S. J. (1995). Null 2-type Chen surfaces, *Glasgow Math. J.* **37**, 233–242.
Libermann, P. (1954). Sur le probleme d'equivalence de certains structures infinitésimales, *Ann. Mat. Pura Appl.* **36**, 27–120.
Logunov, A. A., Mestvirishvili, M. A. and Petrov, V. A. (2004). How were the Hilbert-Einstein equations discovered?, *arXive:physics/0405075.*
Lotay, J. D. (2010). Associate submanifolds of the 7-sphere, arXiv:1006.0361.
Lupu, C. (2001). Chen-like inequalities for C-totally real submanifolds of Sasakian space forms, *An. Univ. Bucuresti Mat. Inform.* **50**, 83–86.
Magid, M. A. (1984). Isometric immersions of Lorentz space with parallel second fundamental forms, *Tsukuba J. Math.* **8**, 31–54.
Mainardi, G. (1856). Sulla teria generale delle superficie, *Giornale dell' Istituteo Lombardo di Sci. Lett. Art.* **9**, 385–398.
Mars, M. and Senovilla, J. M. M. (2003). Trapped surfaces and symmetries, *Classical Quantum Gravity*, **20**, L293–L300
Mashimo, K. (1985). Homogeneous totally real submanifolds of S^6, *Tsukuba J. Math.* **9**, 185–202.
Matsumoto, K., Mihai, I. and Oiagă, A. (2000). Shape operator for slant submanifolds in complex space forms, *Bull. Yamagata Univ. Natur. Sci.* **14**, 169–177.
Matsumoto, K., Mihai, I. and Oiagă, A. (2001). Ricci curvature of submanifolds in complex space forms, *Rev. Roumaine Math. Pures Appl.* **46**, 775–782.
McKean, H. P. and Singer, I. M. (1967). Curvature and the eigenvalues of the Laplacian, *J. Differential Geom.* **1**, 43–69.
McNertney, L. V. (1980). *One-parameter families of surfaces with constant curvature in Lorentz 3-space*, (Doctoral Thesis, Brown University).
Mihai, A. (2002). Certain Chen-like inequalities for slant submanifolds in generalized complex space forms, *Tensor (N.S.)*, **63**, 101–112.
Mihai, A. (2004a). Warped product submanifolds in complex space forms, *Acta Sci. Math. (Szeged)*, **70**, 419–427.
Mihai, A. (2004b). B. Y. Chen inequalities for slant submanifolds in generalized complex space forms, *Rad. Mat.* **12**, 215–231.

Mihai, A. (2006). *Modern topics in submanifold theory*, (Editura Universităţii din Bucureşti, Bucharest).
Mihai, A. (2008). Chen-like inequalities for purely real submanifolds, *Bull. Transilv. Univ. Brasov*, Ser. III, **1(50)**, 233–240.
Mihai, A. and Ozgur, C. (2010a). Chen inequalities for submanifolds of real space forms with a semi-symmetric metric connection, *Taiwanese J. Math.* **14**, 1465–1477.
Mihai, A. and Ozgur, C. (2010). Chen inequalities for submanifolds of complex space forms and Sasakian space forms endowed with a semi-symmetric non-metric connection, *Rocky Mountain J. Math.* (to appear).
Mihai, I. (2003b). Ideal C-totally real submanifolds in Sasakian space forms, *Ann. Mat. Pura Appl.* (4), **182**, 345–355.
Mihai, I. (2004b). Contact CR-warped product submanifolds in Sasakian space forms, *Geom. Dedicata*, **109**, 165–173.
Mihai, I. (2005). Ideal Kählerian slant submanifolds in complex space forms, *Rocky Mountain J. Math.* **35**, 941–951.
Mihai, I. (2008). A Chen invariant on contact manifolds *Bull. Transilv. Univ. Brasov* Ser. III, **1(50)**, 241–246.
Mihai, I. and Tazawa, Y. (2003). On 3-dimensional contact slant submanifolds in Sasakian space forms, *Bull. Austral. Math. Soc.* **68**, 275–283.
Milnor, J. W. (1964). Eigenvalues of the Laplace operator on certain manifolds, *Proc. Nat. Acad. Sci. U.S.A.* **51**, p. 542.
Minakshisundaram, S. and Pleijel, A. (1949). Some properties of the eigenfunctions of the Laplace-operator on Riemannian manifolds, *Canad. J. Math.* **1**, 242–256.
Montaldo, S. and Oniciuc, C. (2006), A short survey on biharmonic maps between Riemannian manifolds, *Rev. Un. Mat. Argentina,* **47**, 1-22.
Moore, J. D. (1971). Isometric immersions of riemannian products, *J. Differential Geom.* **5**, 159–168.
Morimoto, A. (1964). On normal almost contact structure with regularity, *Tohoku Math. J.* **16**, 90–104.
Munteanu, M. I. (2005). Warped product contact CR-submanifolds of Sasakian space forms, *Publ. Math. Debrecen*, **66**, 75–120.
Munteanu, M. I. (2007). Doubly warped product CR-submanifolds in locally conformal Kähler manifolds, *Monatsh. Math.* **150**, 333–342.
Myers, S. B. (1941). Riemannian manifolds with positive mean curvature, *Duke Math. J.* **8**, 401–404.
Nagano, T. (1960). On fibred Riemann manifolds, *Sci. Papers Coll. Gen. Ed. Univ. Tokyo*, **10**, 17–27.
Nagano, T. (1961). On the miniumum eigenvalues of the Laplacian in Riemannian manifolds, *Sci. Papers College Gen. Edu. Univ. Tokyo*, **11**, 177–182.
Nash, J. F. (1956). The imbedding problem for Riemannian manifolds, *Ann. of Math.* **63**, 20–63.
Nölker, S. (1996). Isometric immersions of warped products, *Differential Geom. Appl.* **6**, 1–30.
Nomizu, K. and Pinkall, U. (1987). On the geometry of affine immersions, *Math. Z.* **195**, 165–178.

Nomizu, K. and Sasaki, T. (1994). *Affine Differential Geometry: Geometry of Affine Immersions*, Cambridge Tracts in Math. No. 111, (Cambridge: Cambridge University Press).

Obata, M. (1966). Riemannian manifolds admitting a solution of a certain system of differential equations, *Proc. U.S.-Japan Seminar in Differential Geometry*, (Kyoto, 1965), 101–114.

Ogiue, K. (1964). On almost contact manifolds admitting axiom of planes or axiom of free mobility, *Kodai Math. Sem. Rep.* **16**, 223–232.

Ogiue, K. (1974). Differential geometry of Kähler submanifolds, *Adv. Math.* **13**, 73–114.

Oiagă, A. and Mihai, I. (1999). Chen inequalities for slant submanifolds in complex space forms, *Demonstratio Math.* **32**, 835–846.

Olteanu, A. (2008). CR-doubly warped product submanifolds in Sasakian space forms, *Bull. Transilv. Univ. Brasov* Ser. III, **1(50)**, 269–277.

Olteanu, A. (2010). A general inequality for doubly warped product submanifolds, *Math. J. Okayama Univ.* **52**, 133–142.

O'Neill, B. (1966). The fundamental equations of a submersion, *Michigan Math. J.* **13**, 459–469.

O'Neill, B. (1983). *Semi-Riemannian Geometry with Applications to Relativity*, (Academic Press, New York).

Oniciuc, C. (2002). Biharmonic maps between Riemannian manifolds, *An. Stiint. Univ. Al.I. Cuza Iasi Mat. (N.S.)*, **48**, 237–248.

Oniciuc, C. (2002b). On the second variation formula for biharmonic maps to a sphere, *Publ. Math. Debrecen*, **61**, 613–622.

Oprea, T. (2005). On a geometric inequality, *arXiv:math.DG/0511088*.

Oprea, T. (2006a). Generalization of Chen's inequality, *Algebras Groups Geom.* **23**, 93–103.

Oprea, T. (2006b). A new obstruction to minimal isometric immersions into a real space form, *J. Inequal. Pure Appl. Math.* **7**, no. 5, Article 174, 7 pages.

Oprea, T. (2007). Chen's inequality in the Lagrangian case, *Colloq. Math.* **108**, 163–169.

Oprea, T. (2008). On a Riemannian invariant of Chen type, *Rocky Mountain J. Math.* **38**, 568–587.

Osserman, R. (1990). Curvature in the eighties, *Amer. Math. Monthly*, **97**, 731–756.

Ou, Y.-L. (2009a). Some constructions of biharmonic maps and Chen's conjecture on biharmonic hypersurfaces, *arXiv:0912.1141*.

Ou, Y.-L. (2009b). On conformal biharmonic immersions, *Ann. Global Anal. Geom.* **36**, 133–142.

Ou, Y.-L. (2010). Biharmonic hypersurfaces in Riemannian manifolds, *Pacific J. Math.* **248**, 217–232.

Ou, Y.-L. and Tang, L. (2010). The generalized Chen's conjecture on biharmonic submanifolds is false, *arXiv:1006.1838*.

Overduina, J. M. and Wesson, P. (1997). Kaluza-Klein gravity, *Phys. Rep.* **283**, 303–378.

Özgür, C. (2010). B. Y. Chen inequalities for submanifolds of a Riemannian manifold of quasi-constant curvature, *Turkish J. Math.* (in press).

Özgür, C. and Arslan, K. (2002). On some class of hypersurfaces in $\mathbb{E}^{n+1}$ satisfying Chen's equality, *Turkish J. Math.* **26**, 283–293.

Özgür, C. and Murathan, C. (2010). Chen inequality for submanifolds of a locally conformal almost cosymplectic manifold with semi-symmetric metric connection, *An. St. Univ. Ovidius Constanta,* **18**, 239–254.

Özgür, C. and Tripathi, M. M. (2008). On submanifolds satisfying Chen's equality in a real space form, *Arab. J. Sci. Eng. Sect. A Sci.* **33**, 321–330.

Pavsić, M. (1986). String model for General Relativity and canonical formalism for minimal surfaces, *Nuovo Cimento A*, **95**, 297–309.

Penrose, R. (1965). Gravitational collapse and space-time singularities, *Phys. Rev. Lett.* **14**, 57–59.

Petrov, A. Z. (1969). *Einstein Spaces*, (Pergamon Press, Hungary).

Petrović-Torgašev, M. and Verstraelen, L. (2008). On Deszcz symmetries of Wintgen ideal submanifolds, *Arch. Math. (Brno)*, **44**, 57–67.

Petrović-Torgašev, M. and Hinić, A. (2007). Some curvature conditions of the type 4×2 on the submanifolds satisfying Chen's equality, *Mat. Vesnik*, **59**, 143–150.

Pinkall, U. (1985). Hopf tori in S^3, *Invent. Math.* **81**, 379–386.

Ponge, R. and Reckziegel, H. (1993). Twisted products in pseudo-Riemannian geometry, *Geom. Dedicata*, **48**, 15–15.

Ranjan, A. (1985). Riemannian submersions of spheres with totally geodesics fibers, *Osaka J. Math.* **22**, 243–260.

Rashevskij, P. K. (1948). The scalar field in a stratified space, *Trudy Sem. Vektor. Tenzor. Anal.* **6**, 225–248.

Reckziegel, H. (1985). Horizontal lifts of isometric immersions into the bundle space of a pseudo-Riemannian submersion, *Lecture Notes in Mathematics,* **1156**, 264–279.

Reilly, R. C. (1977). On the first eigenvalue of the Laplacian for compact submanifolds of Euclidean space, *Comm. Math. Helv.* **52**, 525–533.

Ricci, G. (1888). Sulla classificazione delle forme differenziali quadratiche, *Rend. di Lincei*, **41**, 203–207.

Riemann, G. (1854). Über die Hypothesen welche der Geometrie zu Grunde liegen, *Habili. Abhand. Königlichen Ges. Wissen.* Göttingen, **13**.

Rodriguez Montealegre, C. and Vrancken, L. (2009). Warped product minimal Lagrangian immersions in complex projective space, *Results Math.* **56**, 405–420.

Ros, A. (1983). Spectral geometry of CR-minimal submanifolds in the complex projective space, *Kodai Math. J.* **6**, 88–99.

Ros, A. (1984). On spectral geometry of Kaehler submanifolds, *J. Math. Soc. Japan*, **36**, 433–447.

Ros, A. and Urbano, F. (1998). Lagrangian submanifolds of $\mathbf{C}^n$ with conformal Maslov form and the Whitney sphere, *J. Math. Soc. Japan,* **50**, 203–228.

Rosca, R. (1972). On null hypersurfaces of a Lorentzian manifold, *Tensor (N.S.)*, **23**, 66–74.

Rouxel, B. (1993). Chen submanifolds, *Geometry and Topology of Submanifolds*, **VI**, 185–198.
Rozenfeld, B. A. (1949). On unitary and stratified spaces, *Trudy Sem. Vektor. Tenzor. Anal.* **7**, 260–275.
Ruse, H. S. (1949). On parallel fields of planes in a Riemannian manifold, *Quart. J. Math.* **20**, 218–234.
Sahin, B. (2009). Warped product submanifolds of Kaehler manifolds with a slant factor, *Ann. Polon. Math.* **95**, 207–226.
Sakai, T. (1971). On eigenvalues of Laplacian and curvature of Riemannian manifold, *Tohoku Math. J.* **23**, 589–603.
Sasahara, T. (1999). CR-submanifolds in a complex hyperbolic space satisfying an equality of Chen, *Tsukuba J. Math.* **23**, 565–583.
Sasahara, T. (2000). Three-dimensional CR-submanifolds in the nearly Kähler six-sphere satisfying B. Y. Chen's basic equality, *Tamkang J. Math.* **31**, 289–296.
Sasahara, T. (2001). On Ricci curvature of CR-submanifolds with rank one totally real distribution, *Nihonkai Math. J.* **12**, 47–58.
Sasahara, T. (2002). On Chen invariant of CR-submanifolds in a complex hyperbolic space, *Tsukuba J. Math.* **26**, 119–132.
Sasahara, T. (2008). Stability of biharmonic Legendrian submanifolds in Sasakian space forms, *Canad. Math. Bull.* **51**, 448–459.
Sasaki, S. (1960). On differentiate manifolds with certain structures which are closely related to almost contact structure I, *Tohoku Math. J.* **12**, 459–476.
Scharlach, C., Simon, U., Verstraelen, L. and Vrancken, L. (1997). A new intrinsic curvature invariant for centroaffine hypersurfaces, *Beiträge Algebra Geom.* **38**, 437–458.
Schlafli, L. (1873). Nota alla Memoria del sig. Beltrami, *Ann. Mat. Pura Appl.* **5**, 178–193.
Schouten, J. A. (1954). *Ricci-calculus*, (Springer-Verlag, Berlin).
Schwarzschild, K. (1916). Über das Gravitationsfeld eines Massenpunktes nach der Einstein'schen Theorie. *Preuß. Akad. Wissen. Sitz.*, Part **1**, 189–196.
Segre, B. (1938). Famiglie di ipersuperficie isoparametrische negli spazi euclidei ad un qualunque numero di dimensoni, *Atti Accad. Naz. Lincei Rend. Cl. Sc. Fis Mat. Natur.* **27**, 203–207.
Segre, C. (1891). Sulle varietà che rappresentano le coppie di punti di due piani o spazi, *Rend. Cir. Mat. Palermo*, **5**, 192–204.
Senovilla, J. M. M. (2002). Trapped surfaces, horizons and exact solutions in higher dimensions, *Classical Quantum Gravity*, **19**, L113–L119.
Sekigawa, K. (1983). Almost complex submanifolds of a 6-dimensional sphere, *Kodai Math. J.*, **6**, 174–185.
Sentürk, Z. and Verstraelen, L. (2008). Chen Kähler hypersurface, *Taiwanese J. Math.* **12**, 1597–1608.
Shiohama, K. and Tagaki, R. (1970). A characterization of a standard torus in E^3, *J. Differential Geom.* **4**, 477–485.
Shukla, S. S. and Chaubey, P. K. (2010). B. Y Chen's inequality and its applications to slant immersions into locally conformal almost cosymplectic manifolds, *Yokohama Math. J.* **55**, 113–127.

Sikorav, J. C. (1986). Non-existence de sous-variété lagrangienne exacte dans $\mathbf{C}^n$ (d'après Gromov), *Aspects dynamiques et topologiques des groupes infinis de transformation de la mcanique* (Lyon, 1986), 95–110.

Simon, U. (2000). Affine differential geometry, *Handbook of Differential Geometry*, vol. I, 905–961.

Simon, U., Schwenk-Schellschmidt, A. and Viesel, H. (1991). *Introduction to the Affine Differential Geometry of Hypersurfaces*, (Science Univ. Tokyo).

Singer, I.M. and Thorpe, J.A. (1969). The curvature of 4-dimensional Einstein spaces, *Global Analysis*, Princeton University Press, 355–365.

Strubing, W. (1979). Symmetric submanifolds of Riemannian manifolds, *Math. Ann.* **245**, 37–44.

Suceavă, B. (1999). Some remarks on B. Y. Chen's inequality involving classical invariants, *An. Ştiinţ. Univ. Al. I. Cuza Iaşi. Mat. (N.S.)*, **45**, 405–412.

Suceavă, B. (2001). The Chen invariants of warped products of hyperbolic planes and their applications to immersibility problems, *Tsukuba J. Math.* **25**, 311–320.

Suceavă, B. (2003). Fundamental inequalities and strongly minimal submanifolds, *Contemp. Math.* **337**, 155–170.

Suceavă, B. and Vajiac, M. B. (2008). Remarks on Chen's fundamental inequality with classical curvature invariants in Riemannian spaces, *An. Stiint. Univ. Al. I. Cuza Iasi. Mat. (N.S.)*, **54**, 27–37.

Tachibana, S. (1959). On almost-analytic vectors in certain almost-Hermitian manifolds, *Tohoku Math. J.* **11**, 351–363.

Takagi, R. (1975). Real hypersurfaces in a complex projective space with constant principal curvatures I; II, *J. Math. Soc. Japan* **27**, 43–53; 507–516.

Takahashi, T. (1966). Minimal immersions of Riemannian manifolds, *J. Math. Soc. Japan*, **18**, 380–385

Takahashi, T. (1969). Sasakian manifold with pseudo-Riemannian metric, *Tohoku Math. J.* **21**, 271–290.

Takahashi, T. and Tanno, S. (1971). K-contact Riemannian manifolds isometrically immersed in a space of constant curvature, *Tôhoku Math. J.* **23**, 535–539.

Takeuchi, M. (1981). Parallel submanifolds of space forms, in: *Manifolds and Lie Groups*, 429–447 (Birkhäuser, Boston).

Takeuchi, M. and Kobayashi, S. (1968). Minimal imbeddings of R-spaces, *J. Differential Geom.* **2**, 203–215.

Thomsen, G. (1923). Über konforme Geometrie I: Grundlagen der konformen Fächentheorie, *Abh. Math. Sem. Univ. Hamburg*, **3**, 31–56.

Tripathi, M. M. (2008). Chen-Ricci inequality for submanifolds of contact metric manifolds, *J. Adv. Math. Stud.* **1**, no. 1-2, 111–134.

Tripathi, M. M., Kim, J. S. and Kim, S. B. (2003). A note on Chen's basic equality for submanifolds in a Sasakian space form, *Int. J. Math. Math. Sci.* **2003**, no. 11, 711–716.

Tsukamoto, Y. (1957). On Kählerian manifolds with positive holomorphic curvature, *Proc. Japan Acad.* **33**, 333–335.

Tzitzéica, G. (1908). Sur une nouvelle classe de surfaces, *Rend. Circ. Mat. Palermo*, **25**, 180–187.

Ucci, J. (1983). On the nonexistence of Riemannian submersions from CP^7 and QP^3 *Proc. Amer. Math. Soc.* **88**, 698–700.

Verstraelen, L. (1990). Curves and surfaces of finite Chen type, *Geometry and Topology of Submanifolds*, **III**, 304–311.

Vilcu, G. E. (2010). B. Y. Chen inequalities for slant submanifolds in quaternionic space forms, *Turk. J. Math.* **34**, 115–128.

Voss, A. (1956). Einige differentialgeometrische Kongruenzsätze für geschlossene Flächen und Hyperflächen, *Math. Ann.* **131**, 180–218.

Vrancken, L. (1998). Killing vector fields and Lagrangian submanifolds of the nearly Kähler S^6, *J. Math. Pure Appl.* **77**, 631–645.

Vrancken, L. (2001). Minimal Lagrangian submanfiolds with constant sectional curvature in indefinite complex space forms, *Proc. Amer. Math. Soc.* **130**, 1459–1466.

Weil, A. (1947). Sur la théorie des formes différentielles attachété analytique complexe, *Comm. Math. Helv.*, **20**, 110-116.

Weiner, J. (1978). On a problem of Chen, Willmore, et al., *Indiana Univ. Math. J.* **27**, 19–35.

Wettstein, B. (1978). *Congruence and existence of differentiable map,* (Doctoral Thesis, ETH Zürich, no. 6252).

Weyl, H. (1918). Reine Infinitesimalgeometrie, *Math. Z.* **26**, 384–411.

Wilking, B. (2001). Index parity of closed geodesics and rigidity of Hopf fibrations, *Invent. Math.* **144**, 281–295.

Willmore, T. L. (1968). Mean curvature of immersed surfaces, *An. Sti. Univ. "Al. I. Cuza" Iasi, Sec. I. a Mat. (N.S.)*, **14**, 99–103

Willmore, T. L. (1971). Mean curvature of Riemannian immersions, *J. London Math. Soc.* **3**, 307–310.

Wintgen, P. (1979). Sur l'inégalité de Chen-Willmore, *C. R. Acad. Sci. Paris Sér.* A-B, **288**, A993–A995.

Witt, E. (1941). Eine Identität zwischen Modulformen zweiten Grades, *Abh. Math. Sem. Univ. Hamburg*, **14**, 323–337.

Wolf, J. and Gray, A. (1968). Homogeneous spaces defined by Lie group automorphisms. II, *J. Differential Geom.* **2**, 115–159.

Yang, C. N. and Mills, R. (1954). Conservation of isotopic spin and isotopic gauge invariance, *Phys. Rev.* **96**, 191–195.

Yang, J. (1999). On hypersurfaces satisfying a basic equality, *Proc. Sympos. Pure Math.* **64**, 335–341.

Yano, K. and Ishihara, S. (1973). *Tangent and Cotangent Bundles: Differerntial Geometry* (M. Dekker, New York)

Yano, K. and Kon, M. (1983). *CR-submanifolds of Kählerian and Sasakian manifolds* (Birkäuser, Boston-Basel).

Yoon, D. W. (2003). B.-Y. Chen's inequality for CR-submanifolds of locally conformal Kähler space forms, *Demonstratio Math.* **36**, 189–198.

General Index

Author Index

www.ingramcontent.com/pod-product-compliance
Ingram Content Group UK Ltd.
Pitfield, Milton Keynes, MK11 3LW, UK
UKHW020404150926
8624IPUK00004B/339

9 789814 329637